Biotechnologies of Crop Improvement, Volume 2

Satbir Singh Gosal • Shabir Hussain Wani

Editors

Biotechnologies of Crop Improvement, Volume 2

Transgenic Approaches

 Springer

Editors
Satbir Singh Gosal
School of Agricultural Biotechnology
Punjab Agricultural University
Ludhiana, Punjab, India

Shabir Hussain Wani
Mountain Research Centre
for Field Crops, Khudwani
Sher-e-Kashmir University of Agricultural
Sciences and Technology of Kashmir
Srinagar, Jammu and Kashmir, India

ISBN 978-3-030-08065-5   ISBN 978-3-319-90650-8   (eBook)
https://doi.org/10.1007/978-3-319-90650-8

This Springer imprint is published by the registered company Springer International Publishing AG part of Springer Nature.
The registered company address is: Gewerbestrasse 11, 6330 Cham, Switzerland

Professor Paul Christou

Paul Christou obtained his Ph.D. in Organic Chemistry in 1980 at the University of London. He subsequently undertook postdoctoral research (1980–1982) at University College London in plant biochemistry with emphasis on the elucidation of the biosynthetic pathway of a number of irregular monoterpenoids useful in the perfumery and flavoring industries. In 1982, he was recruited by one of the first plant biotechnology companies in the USA, Agracetus Inc. (Madison, Wisconsin), as senior scientist with responsibilities for molecular and cellular plant biology.

At Agracetus, he led a team that developed the first transgenic staple crop, soybean, currently being sold by Monsanto. His team also developed a facile genotype-independent method for rice transformation. In 1994, he moved to the John Innes Centre (JIC), Norwich, UK, as Head of Molecular Biotechnology Unit and Director of Tropical Maize and Rice Biotechnology Training Laboratory funded by the Rockefeller Foundation, USA. During his tenure at the JIC, he trained many Ph.D. students and hosted a number of postdoctoral fellows from 27 different countries in Asia, Africa, and Central and South America, funded by the Rockefeller Foundation, to pursue research in diverse aspects of rice and maize biotechnology. While at the JIC his team unraveled underpinning mechanisms controlling transgenic locus organization in cereal crops and pioneered a multigene transformation system useful in plant metabolic engineering. In 2001, he joined the Fraunhofer Institute of Molecular Biotechnology and Applied Ecology Schmallenberg-Aachen, Germany, as full professor. While at Fraunhofer he expanded his research to molecular pharming with emphasis on the production of recombinant proteins active against infectious diseases such as HIV. In 2004, he joined the University of Lleida, Spain, as an ICREA professor. He was the founding director of Agrotecnio Center, Lleida, Spain (2012), a position he held till 2015. He is currently the Editor in Chief of Molecular Breeding and Transgenic Research. He has mentored over 130 graduate students, and he is the senior author of over 250 scientific publications in the general area of plant biotechnology. He was awarded a European Research Council Advanced Grant (2009–2014) and a subsequent Proof of Concept Grant, also by the ERC. He has been a PI in 12 different EU-funded projects and has participated in 2 projects funded by the Bill & Melinda Gates Foundation, 1 currently in progress. He has given over 300 invited, plenary, or keynote lectures at international meetings. His current research focuses on metabolic engineering, genome editing and synthetic biology in cereals (rice and maize), production of pharmaceutical macromolecules in plants, and engineering novel agronomic traits in crops. He is active in training and technology transfer for developing country biotechnology, intellectual property issues, and regulatory and biosafety issues of transgenic crops, focusing on developing countries. He is also interested in science policy issues and strategic planning covering the interphase between fundamental and applied research. This book is dedicated to Prof. Paul Christou – A pioneer in particle gun technology for plant transformation.

Foreword

Agricultural production continues to face many challenges. Major constraints imposed by biotic and abiotic stresses on crop quality, yield, and productivity in conjunction with the urgent need to provide healthier and more nutritious food and animal feed in a durable and sustainable manner constitute one of the grand global challenges of our times. The challenge is even more severe in regions of the world which are less able to cope with increasing populations, changes in environmental and climatic conditions, pollution, and political and social instability. Agriculture has always been driven by technology. Its efficiency reflects game-changing discoveries in biology and new developments in chemistry and engineering. We seek to increase food, feed, and fiber production while minimizing agriculture's environmental footprint, and also to develop new products that branch into the chemical, pharmaceutical, and other industries. A key driver in this transition will be our ability to control plant metabolism, development, biochemistry, and physiology holistically, with more precision and predictability. Recent genome editing technologies are already beginning to make an impact on agriculture and are transforming the way improved and more resilient crop varieties are, and will continue to be created in the near future. However, it is important to recognize that older, more mature technologies still have the potential to play an enormous role in addressing the challenges agriculture faces. *Biotechnologies of Crop Improvement – Transgenic Approaches*, edited by Drs. Satbir Singh Gosal and Shabir Hussain Wani, provides an invaluable resource especially for students and younger practitioners who may be new to the field. The book comprises 16 chapters that address a number of important technologies all relying on transgenesis. The introductory chapter is a general overview of plant transformation methods. The chapter also provides examples of target traits under development. The subsequent four chapters have the common theme of gene silencing as a means to improve crops in a number of different ways. RNA interference and virus-induced gene silencing are being widely used to create plants which are more tolerant to biotic and abiotic stresses. These technologies have also been utilized to improve nutritional quality and overall physiology, development, and metabolism at the whole plant level. The next two chapters discuss antifungal plant defensins, small molecules with critical roles in plant immunity and recent examples of plants with improved tolerance to salinity and drought. Specific chapters are devoted to the engineering of disease resistance in rice, an overview of the state of the art of sugarcane transformation and individual chapters on crops important for small-scale subsistence farmers such as pulses and millets. Tomato biotechnology is discussed in a dedicated chapter with emphasis on tolerance to environmental stresses and improvement of fruit quality traits. Eucalyptus genetic transformation is discussed in a separate chapter and target traits such as productivity and quality for fiber production are highlighted. The importance of the enzyme glutamine synthetase in a number of different aspects of crop improvement is then discussed. This chapter highlights the importance of this enzyme in overall nitrogen metabolism in plants and by extension its role in improving nitrogen use efficiency a paramount challenge in agriculture. Its usefulness in creating new modes of herbicide tolerance in crops is discussed. The authors conclude the chapter by discussing inconsistent results reported in the literature concerning expression of glutamine synthase encoding genes in different plants. The penultimate chapter discusses the role of phytohormones in developing crops tolerant to environmental stresses. It focuses on underpinning mechanisms responsible for the role of phytohormones in plant metabolism. The last chapter provides a general historical overview of the development of transgenic crops, their adoption, and remaining challenges that still prevent such crops to reach those who need them the most, the small-scale subsistence farmers in the developing world. The last chapter also discusses next-generation precision engineering technologies using site-specific nucleases.

Paul Christou
University of Lleida
Lleida, Spain

Preface

The combined use of recombinant DNA technology, gene-transfer methods, and tissue-culture techniques has led to the efficient transformation and production of transgenic plants in a wide variety of crop plants. In fact, transgenesis has emerged as an additional tool to carry out single-gene breeding or transgenic breeding of crops. The transgenic approach provides access to a larger gene pool, as the gene(s) may come from viruses, bacteria, fungi, insects, animals, human beings, unrelated plants, and even from chemical synthesis in the laboratory. Unlike conventional breeding, only the cloned gene(s) of agronomic importance is/are being introduced without co-transfer of undesirable genes/alleles from the donor parent. The recipient genotype is least disturbed, which eliminates the need for repeated backcrosses. Various gene-transfer methods such as *Agrobacterium*, physicochemical uptake of DNA, liposome encapsulation, electroporation of protoplast, microinjection, DNA injection into intact plants, incubation of seeds with DNA, pollen tube pathway, laser microbeam, electroporation into tissues/embryos, silicon carbide fiber/whiskers method, particle bombardment, and "in planta" transformation have been developed. Using different gene-transfer methods and strategies, transgenics carrying useful agronomic traits have been developed and released for commercial cultivation. Attempts are being made to develop transgenic crop varieties resistant to abiotic stresses, such as drought, low and high temperature, salts, and heavy metals, and also to develop transgenic varieties possessing better nutrient-use efficiency and better keeping, nutritional and processing qualities. Genetically modified foods, such as tomato containing high lycopene, tomato with high flavonols as antioxidants, edible vaccines, are leading examples of genetically engineered crops. Several genes of agronomic importance have been isolated from various organisms; cloned and suitable constructs have been developed for plant transformation. *Agrobacterium* and "particle gun" methods have been refined and now being used for genetic transformation of a wide variety of field, fruit, vegetable, forest crops, and ornamental plant species. Transgenic crops such as maize, cotton, soybean, potato, tomato, papaya, and rice carrying mainly insect resistance, herbicide resistance, or both are now being commercially grown in several countries.

This book includes 16 chapters prepared by specialists dealing with genetic transformation in relation to crop improvement. First chapter is a general overview that introduces various methods for plant genetic transformation and their applications to crop improvement. Further, a separate chapter deals with virus-based transformation method, i.e., VIGS for functional characterization of plant genes involving gene transcript suppression. Three chapters exclusively deal with RNA interference (RNAi) and its relevance for developing resistance to biotic and abiotic stresses. A special chapter dealing with antifungal plant defensins has been included. A separate chapter has been included regarding the application of transgenic approach for developing resistance to abiotic stresses. Six chapters deal with important crop plants such as rice, sugarcane, millets, pulses, tomato, and eucalyptus. A particular chapter deals with the potential of transgenic overexpression of glutamine synthetase in relation to crop improvement. Understanding the phytohormones biosynthetic pathways for developing engineered environmental stress-tolerant crops has been dealt separately. Furthermore, the status, potential, and challenges of transgenic crops have been covered in the last chapter.

The book provides state-of-the-art information on genetic transformation in relation to crop improvement. We earnestly feel that this book will be highly useful for students, research scholars, and scientists working in the area of crop improvement and biotechnology at universities, research institutes, R&Ds of agricultural MNCs for conducting research, and various funding agencies for planning future strategies.

We are highly grateful to all learned contributors, each of who has attempted to update scientific information of their respective area and expertise and has kindly spared valuable time and knowledge.

We apologize wholeheartedly for any mistakes, omissions, or failure to acknowledge fully.

We thank our families {Dr. Satwant Kaur Gosal (wife of SSG); Sana Ullah Wani, Taja Begum, Yasir Wani, and Shazia (father, mother, brother, and wife of SHW)} for their continuous support and encouragement throughout the completion of this book.

We highly appreciate the all-round cooperation and support of Springer International Publishing AG, Cham, for their careful and speedy publication of this book.

Ludhiana, Punjab, India   Satbir Singh Gosal
Srinagar, Jammu and Kashmir, India   Shabir Hussain Wani

Contents

Contributors

V. M. M. Achary International Centre for Genetic Engineering and Biotechnology, New Delhi, India

Pawan Kumar Agrawal ICAR-NASF, KAB-I, IARI Campus, New Delhi, India

Lalitkumar Aher ICAR-National Institute of Abiotic Stress Management, Malegaon, Maharashtra, India

Raghavendra Aminedi ICAR-NRC on Plant Biotechnology, IARI Campus, New Delhi, India

L. Arul Department of Plant Biotechnology, Centre for Plant Molecular Biology & Biotechnology, Tamil Nadu Agricultural University, Coimbatore, India

Harinath Babu Kassa Molecular Biology and Genetic Engineering Laboratory, Vasantdada Sugar Institute, Pune, Maharashtra, India

Joydeep Banerjee Department of Genetics and Plant Breeding, Faculty of Agriculture, Bidhan Chandra Krishi Viswavidyalaya, Mohanpur, Nadia, West Bengal, India

Ramcharan Bhattacharya ICAR-NRC on Plant Biotechnology, IARI Campus, New Delhi, India

Prabir Kumar Bhattacharyya Department of Genetics and Plant Breeding, Faculty of Agriculture, Bidhan Chandra Krishi Viswavidyalaya, Mohanpur, Nadia, West Bengal, India

Bhabesh Borphukan International Centre for Genetic Engineering and Biotechnology, New Delhi, India

H. S. Chawla Genetics & Plant Breeding Department, G.B. Pant University of Agriculture & Technology, Pantnagar, India

Deepika Choudhary ICAR-National Institute of Abiotic Stress Management, Malegaon, Baramati, Maharashtra, India

Cory Christensen Dow AgroSciences LLC, West Lafayette, IN, USA

Alok Das Division of Plant Biotechnology, ICAR-Indian Institute of Pulses Research, Kanpur, India

Rachayya Devarumath Molecular Biology and Genetic Engineering Laboratory, Vasantdada Sugar Institute, Pune, Maharashtra, India

Sweta Dosad Genetics & Plant Breeding Department, G.B. Pant University of Agriculture & Technology, Pantnagar, India

Ali El-Keblawy Department of Applied Biology, College of Sciences, University of Sharjah, Sharjah, UAE

Dhirendra Fartyal International Centre for Genetic Engineering and Biotechnology, New Delhi, India

Saikat Gantait Department of Genetics and Plant Breeding, Faculty of Agriculture, Bidhan Chandra Krishi Viswavidyalaya, Mohanpur, Nadia, West Bengal, India

Satbir Singh Gosal Punjab Agricultural University, Ludhiana, India

Lei Han Dow AgroSciences LLC, Indianapolis, IN, USA

Mohamed Helmy The Donnelly Centre for Cellular and Biomedical Research, University of Toronto, Toronto, ON, Canada

Pradeep Kumar Jain ICAR-NRC on Plant Biotechnology, IARI Campus, New Delhi, India

Donald James International Centre for Genetic Engineering and Biotechnology, New Delhi, India

Jagdeep Kaur Donald Danforth Plant Science Center, St. Louis, MO, USA
Monsanto Company, Chesterfield, MO, USA

Shuchishweta Vinay Kendurkar Plant Tissue Culture Division, National Chemical Laboratory, Pune, India

Deshika Kohli ICAR-NRC on Plant Biotechnology, IARI Campus, New Delhi, India

E. Kokiladevi Department of Plant Biotechnology, Centre for Plant Molecular Biology & Biotechnology, Tamil Nadu Agricultural University, Coimbatore, India

K. K. Kumar Department of Plant Biotechnology, Centre for Plant Molecular Biology & Biotechnology, Tamil Nadu Agricultural University, Coimbatore, India

Mahesh Kumar ICAR-National Institute of Abiotic Stress Management, Malegaon, Baramati, Maharashtra, India

Neetu S. Kushwah Division of Plant Biotechnology, ICAR-Indian Institute of Pulses Research, Kanpur, India

Tejinder Mall Dow AgroSciences LLC, West Lafayette, IN, USA

B. Mamta Department of Genetics, University of Delhi South Campus, New Delhi, India

Kareem A. Mosa Department of Applied Biology, College of Sciences, University of Sharjah, Sharjah, UAE

Department of Biotechnology, Faculty of Agriculture, Al-Azhar University, Cairo, Egypt

Gauri Nerkar Molecular Biology and Genetic Engineering Laboratory, Vasantdada Sugar Institute, Pune, Maharashtra, India

Saikat Paul Department of Biotechnology, St. Xavier's College (Autonomous), Kolkata, West Bengal, India

Manchikatla V. Rajam Department of Genetics, University of Delhi South Campus, New Delhi, India

Jagadish Rane ICAR-National Institute of Abiotic Stress Management, Malegaon, Baramati, Maharashtra, India

Mamatha Rangaswamy Plant Tissue Culture Division, National Chemical Laboratory, Pune, India

Meenal Rathore Division of Plant Biotechnology, ICAR-Indian Institute of Pulses Research, Kanpur, India

M. K. Reddy International Centre for Genetic Engineering and Biotechnology, New Delhi, India

Aryadeep Roychoudhury Department of Biotechnology, St. Xavier's College (Autonomous), Kolkata, West Bengal, India

Sutanu Sarkar Department of Genetics and Plant Breeding, Faculty of Agriculture, Bidhan Chandra Krishi Viswavidyalaya, Mohanpur, Nadia, West Bengal, India

Dilip Shah Donald Danforth Plant Science Center, St. Louis, MO, USA

Suman Sheelavantmath Molecular Biology and Genetic Engineering Laboratory, Vasantdada Sugar Institute, Pune, Maharashtra, India

Department of Biotechnology, Sinhgad College of Science, Ambegaon, Pune, Maharashtra, India

Ajay Kumar Singh ICAR-National Institute of Abiotic Stress Management, Malegaon, Baramati, Maharashtra, India

Narendra Pratap Singh ICAR-National Institute of Abiotic Stress Management, Malegaon, Baramati, Maharashtra, India

Division of Plant Biotechnology, ICAR-Indian Institute of Pulses Research, Kanpur, India

Sameh Soliman Department of Medicinal Chemistry, College of Pharmacy, University of Sharjah, Sharjah, UAE

Department of Pharmacognosy, Faculty of Pharmacy, University of Zagazig, Zagazig, Egypt

D. Sudhakar Department of Plant Biotechnology, Centre for Plant Molecular Biology & Biotechnology, Tamil Nadu Agricultural University, Coimbatore, India

Laura Tagliani Dow AgroSciences LLC, Indianapolis, IN, USA

Avinash Thorat Molecular Biology and Genetic Engineering Laboratory, Vasantdada Sugar Institute, Pune, Maharashtra, India

S. Varanavasiappan Department of Plant Biotechnology, Centre for Plant Molecular Biology & Biotechnology, Tamil Nadu Agricultural University, Coimbatore, India

Siva Ls Velivelli Donald Danforth Plant Science Center, St. Louis, MO, USA

Shabir Hussain Wani MRCFC, Khudwani & Division of Genetics and Plant Breeding, Sher-e-Kashmir University of Agricultural Sciences & Technology of Kashmir, Shalimar, Srinagar, India

About the Editors

Satbir Singh Gosal received his B.Sc. (Med.) from Panjab University, Chandigarh, India, and M.Sc. and Ph.D. (Plant Breeding) from Punjab Agricultural University, Ludhiana, India. He was awarded fellowships by the Royal Society London and the Rockefeller Foundation (USA) for his postdoctoral research at the University of Nottingham, England, and John Innes Centre, Norwich, England. Dr. Gosal has served in Punjab Agricultural University as Professor Biotechnology, Director School of Agricultural Biotechnology, Additional Director Research, and Director of Research. He has also served in FAO/IAEA, Vienna, Austria, and took tissue culture expert mission to Iraq in 1997. Dr. Gosal had rigorous training on "Biosafety of GM crops" from Danforth Centre for Plant Science Research, St. Louis, and APHIS, EPA (USDA), USTDA, Washington, DC, USA. He has been an Honorary Member of the Board of Assessors, Australian Research Council, Canberra; President of the Punjab Academy of Sciences; elected member (Fellow) of the Plant Tissue Culture Association (India); and Fellow of Indian Society of Genetics and Plant Breeding. He is a recipient of Distinction Award from Society for the Promotion of Plant Science Research, Jaipur, India (2009), Fellow of the Punjab Academy of Sciences, and advisory member of several universities/institutes in the area of biotechnology. He served as a member of Review Committee on Genetic Manipulation (RCGM) for 3 years in the Department of Biotechnology (DBT), Government of India, New Delhi, and is a member of panel of experts in the area of Biotechnology for National Fund for Strategic Research of Indian Council of Agricultural Research, New Delhi. He has participated in more than 125 national/international conferences/meetings held in India, England, Scotland, Yugoslavia, Philippines, Indonesia, Thailand, the Netherlands, Malaysia, Singapore, Austria, Iraq, P R China, Australia, Mexico, Germany, and the USA. He has guided more than 75 (M.Sc. and Ph.D.) students for theses research on various aspects of plant tissue culture and plant transformation. He executed more than 20 externally funded research projects funded by Punjab State Government and various national and international organizations such as ICAR, DBT, DAC NATP, FAO/IAEA, and the Rockefeller Foundation, USA. He has more than 200 research papers in refereed journals of high repute, 135 research papers in conference proceedings, several T.V./ radio talks, and 30 book chapters. He has coauthored five laboratory manuals, one textbook, and two edited books.

Shabir Hussain Wani is Assistant Professor cum Scientist, Plant Breeding and Genetics, at the Mountain Research Centre for Field Crops, Khudwani Sher-e-Kashmir University of Agricultural Sciences and Technology of Kashmir, Srinagar, Jammu and Kashmir, India, since May 2013. He received his B.Sc. in Agriculture from Bhim Rao Agricultural University, Agra, India, and M.Sc. and Ph.D. in Genetics and Plant Breeding from Central Agricultural University, Manipur, and Punjab Agricultural University, Ludhiana, India respectively. His Ph.D. research fetched the first prize in north zone at national level competition in India. After obtaining his Ph.D., he worked as research associate in the Biotechnology Laboratory, ICAR-Central Institute of Temperate Horticulture, Rangreth, Srinagar, India, for 2 years, up to October 2011. In November 2011, he joined the Krishi Vigyan Kendra (Farm Science Centre) as program coordinator (i/c) at Senapati, Manipur, India. He teaches courses related to plant breeding, seed science and technology, and stress breeding. He has published more than 100 scientific papers in peer-reviewed journals and chapters in books of international and national repute. He has served as review editor of *Frontiers in Plant Sciences*, Switzerland, from 2015 to 2017. He is an editor of *SKUAST Journal of Research* and *LS: An International Journal of Life Sciences*. He has also edited ten books on current topics in crop improvement, published by reputed publishers including CRC press; Taylor and Francis Group, USA; and Springer. He is a fellow of the Linnean Society of London and Society for Plant Research, India. He received various awards including Young Scientist Award (Agriculture) 2015, Young Scientist Award 2016, and Young Achiever Award 2016 by various prestigious scientific societies. He has also worked as visiting scientist in the Department of Plant Soil and Microbial Sciences, Michigan State University, USA, during 2016–2017 under the Raman Post Doctoral Research Fellowship program sponsored by University Grants Commission, Government of India, New Delhi. He is a member of the Crop Science Society of America.

Chapter 1
Plant Genetic Transformation and Transgenic Crops: Methods and Applications

Satbir Singh Gosal and Shabir Hussain Wani

Abstract The combined use of recombinant DNA technology, gene transfer methods, and tissue culture techniques has led to the efficient transformation and production of transgenics in a wide variety of crop plants. In fact, transgenesis has emerged as an additional tool to carry out single-gene breeding or transgenic breeding of crops. Unlike conventional breeding, only the cloned gene(s) of agronomic importance is/are being introduced without cotransfer of undesirable genes from the donor. The recipient genotype is least disturbed, which eliminates the need for repeated backcrosses. Above all, the transformation methods provide access to a large gene pool, as the gene(s) may come from viruses, bacteria, fungi, insects, animals, human beings, unrelated plants, and even from chemical synthesis in the laboratory. Various gene transfer methods such as *Agrobacterium*, physicochemical uptake of DNA, liposome encapsulation, electroporation of protoplasts, microinjection, DNA injection into intact plants, incubation of seeds with DNA, pollen tube pathway, use of laser microbeam, electroporation into tissues/embryos, silicon carbide fiber method, particle bombardment, and "in planta" transformation have been developed. Among these, *Agrobacterium* and "particle gun" methods are being widely used. Recently RNAi and CRISPR/Cas9 systems have further expanded the scope for genome engineering. Using different gene transfer methods and strategies, transgenics carrying useful agronomic traits have been developed and released. Attempts are being made to develop transgenic varieties resistant to abiotic stresses, such as drought, low and high temperature, salts, and heavy metals, and also to develop transgenic varieties possessing better nutrient-use efficiency and better keeping and nutritional and processing qualities. Genetically modified foods, such as tomato containing high lycopene, tomato with high flavonols as antioxidants, edible vaccines, are leading examples of genetically engineered crops. Several genes of agronomic importance have been isolated from various organisms; cloned and suitable constructs have been developed for plant transformation. *Agrobacterium* and "particle gun" methods have been refined and

S. S. Gosal (✉)
Punjab Agricultural University, Ludhiana, India

S. H. Wani
MRCFC, Khudwani, Sher-e-Kashmir University of Agricultural Sciences & Technology of Kashmir, Shalimar, Srinagar, India

© Springer International Publishing AG, part of Springer Nature 2018
S. S. Gosal, S. H. Wani (eds.), *Biotechnologies of Crop Improvement, Volume 2*,
https://doi.org/10.1007/978-3-319-90650-8_1

now being used for genetic transformation of a wide variety of field, fruit, vegetable, forest crops, and ornamental plant species. Transgenic crops such as cotton, maize, papaya, potato, rice, soybean, and tomato, carrying mainly insect resistance, herbicide resistance, or both, are now being grown over an area of 185 million hectares spread over 28 countries of the world.

Keywords Genetic transformation · GM crops · GMOs · Recombinant DNA technology · Transgenesis · Transgenic breeding · Transgenic crops

1.1 Introduction

Plant genetic transformation leads to the production of transgenic plants (transgenics) which carry additional, stably integrated, and expressed foreign gene(s) usually from trans species. Such plants are commonly called genetically modified organisms (GMOs) or living modified organisms (LMOs). The whole process involving introduction, integration, and expression of foreign gene(s) in the host is called genetic transformation or transgenesis. The combined use of recombinant DNA technology, gene transfer methods, and tissue culture techniques has led to the efficient transformation and production of transgenics in a wide variety of crop plants (Yang and Christou 1994; Mathews et al. 1995; Hilder and Boulter 1999; Gosal and Gosal 2000; Chahal and Gosal 2002; Altman 2003; Grewal et al. 2006; Kerr 2011; Nayak et al. 2011; Bakshi and Dewan 2013; Kamthan et al. 2016; Arora and Narula 2017; Cardi et al. 2017; Tanuja and Kumar 2017). In fact, transgenesis has emerged as an additional tool to carry out single-gene breeding or transgenic breeding of crops. Unlike conventional breeding, only the cloned gene(s) of agronomic importance is/are being introduced without cotransfer of undesirable genes from the donor. The recipient genotype is least disturbed, which eliminates the need for repeated backcrosses. Above all, the transformation method provides access to a large gene pool, as the gene(s) may come from viruses, bacteria, fungi, insects, animals, human beings, unrelated plants, and even from chemical synthesis in the laboratory. Various gene transfer methods such as *Agrobacterium*, physicochemical uptake of DNA, liposome encapsulation, electroporation of protoplasts, microinjection, DNA injection into intact plants, incubation of seeds with DNA, pollen tube pathway, the use of laser microbeam, electroporation into tissues/embryos, silicon carbide fiber method, particle bombardment, and "in planta" transformation have been developed. Among these, *Agrobacterium* and "particle gun" methods are being widely used for plant genetic transformation.

1.2   Making Transgenic Plants

The appropriate gene construct carrying gene of interest, selectable marker/reporter gene, promoter, and terminator sequences are introduced into plants using standard procedures and suitable gene transfer method, and the resulting plants are characterized following phenotypic assays and various molecular techniques (Zhu et al. 2010).

1.2.1   Gene Transfer Methods in Plants

Various gene transfer methods (Ledoux 1965; Fraley et al. 1980; Herrera-Estrella 1983; Paszkowski et al. 1984; Tepfer 1984; Fromm et al. 1985; Lörz et al. 1985; Sanford et al. 1985; Uchimiya et al. 1986; Feldmann and Marks 1987; Grimsley et al. 1987; Klein et al. 1987; Sanford 1988; Sanford 1990; Weber et al. 1989; Kaeppler et al. 1990; Gunther and Spangenberg 1990; Saul and Potrykus 1990; Hooykaas and Schilperoort 1992; Bechtold et al. 1993; Kloti et al. 1993; Frame et al. 1994; Christou 1994; Hiei et al. 1994; Pescitelli and Sukhpinda 1995; Rhodes et al. 1995; Christou 1996; Trick and Finer 1997; Sanford 1988; Leelavati et al. 2004; Junjie et al. 2006; Keshamma et al. 2008; Takahashi et al. 2008; Rasul et al. 2014) are being used for developing transgenic plants (Table 1.1). Recently RNA interference (RNAi) in which RNA molecules inhibit gene expression or translation by neutralizing targeted mRNA molecules (Kim and Rossi 2008; Gupta et al. 2013; Younis et al. 2014) and CRISPR/Cas9 system (Wang et al. 2016; Arora and Narula 2017) have further expanded the scope for genome engineering.

Following the introduction of transgenes, the resulting putative transgenics are screened using various screenable markers (Rakosy Tican et al. 2007; Shimada et al. 2010; Shimada et al. 2011). Among the scorable markers/reporters genes (Table 1.2), GUS expression is the easiest way of assessing transformation. Transformed tissues are kept in X-gluc solution at 37 °C (in dark) for 1–12 h. Appearance of blue spots/sectors indicates their transgenic nature. Transgenic tissues are selected by growing them on medium containing selective agents (antibiotics/herbicides) at appropriate concentrations for at least two cycles of selection of 2 weeks each. Thus, selected tissues are cultured on suitable medium to regenerate the entire plants in the presence of respective selective agent. Regenerated plants are subjected to phenotypic assays, molecular analysis (Deom et al. 1990; Guttikonda et al. 2016), and insect bioassays using the following methods:

Table 1.1 Various methods for genetic transformation of plants

Transformation method	Remarks	Reference
DNA uptake	Uptake of DNA by living cells	Ledoux (1965)
Liposome encapsulation	Introduction of liposome-encapsulated SV40 DNA into cells	Fraley et al. (1980)
Agrobacterium tumefaciens (dicot plant)	First record on transgenic tobacco plant expressing foreign genes	Herrera-Estrella (1983)
Agrobacterium rhizogenes	Transformation of several species of higher plants by *Agrobacterium rhizogenes* and sexual transmission of the transformed genotype and phenotype	Tepfer (1984)
Electroporation	Expression of genes transferred into monocot and dicot plant cells by electroporation	Fromm et al. (1985).
Pollen-mediated transformation	Pollen-mediated plant transformation employing genomic donor DNA	Sanford et al. (1985)
PEG-mediated DNA uptake by protoplasts	Expression of a foreign gene in callus derived from DNA-treated protoplasts of rice (*Oryza sativa*)	Uchimiya et al. (1986)
Electroporation into protoplasts	Electroporation of DNA and RNA into plant protoplasts	Fromm et al. (1987)
Agrobacterium-mediated virus transfer	*Agrobacterium*-mediated delivery of infectious maize streak virus into maize plants	Grimsley et al. (1987)
Microprojectiles	High-velocity microprojectiles for delivering nucleic acids into living cells	Klein et al. (1987)
Microinjection	Transgenic rapeseed plants obtained by the microinjection of DNA into microspore-derived embryoids	Neuhaus et al. (1987)
Laser microbeam	A laser microbeam as a tool to introduce genes into cells and organelles of higher plants	Weber et al. (1989)
Silicon carbide fiber method	Silicon carbide fiber-mediated DNA delivery into plant cells	Kaeppler et al. (1990)
In planta transformation	In planta *Agrobacterium*-mediated gene transfer by infiltration of adult *Arabidopsis thaliana* plants	Bechtold et al. (1993)
Whiskers method	Production of fertile transgenic maize plants by silicon carbide whisker-mediated transformation	Frame et al. (1994)
Agrobacterium tumefaciens (monocot plant)	Efficient transformation of rice (*Oryza sativa*) mediated by *Agrobacterium* and sequence analysis of the boundaries of the T-DNA	Hiei et al. (1994)
Particle bombardment	Plant transformation using particle gun	Christou (1996)
Agrobacterium-based virus-induced gene silencing (VIGS)	Virus-induced gene silencing (VIGS) is a method that takes advantage of the plant RNAi-mediated antiviral defense mechanism	Lu et al. (2003)
SAAT(sonication-assisted *Agrobacterium* transformation)	Sonication-assisted *Agrobacterium*-mediated transformation	Trick and Finer (1997)
Agrobacterium based CRISPR/cas genome editing	Genome editing with CRISPR/Cas9 system	Gaj et al. (2013), and Gao et al. (2015)

Table 1.2 Reporter genes used in plant transformation

Reporter gene	Substrate and assay	Identification
UidA, GUS gene (β-glucuronidase)	X- GLUC	Histochemical assay
Chloramphenicol acetyl transferase (CAT)	^{14}Chloramphenicol + acetyl Co-A, TLC separation	Detection of acetyl chloramphenicol by autoradiography
Octopine synthase	Arginine pyruvate + NADH	Electrophoresis
Nopaline synthase	Arginine + ketoglutaric acid + NADH	Electrophoresis
β-Galactosidase (Lac Z)	β-Galactoside (X-gal)	Color of cells
Luciferase (LUC)	Decanal and FMNH$_2$ ATP + O2 + luciferin	Bioluminescence (exposure of X-ray films)
GFP	Green fluorescent protein	Fluorescent

Table 1.3 Selectable marker genes used in plant transformation

Marker gene	Enzyme	Selectable marker
Antibiotics		
Npt-II	Neomycin phosphotransferase	Kanamycin, neomycin, G418, Geneticin
Aad A	Aminoglycoside-3-adenyl transferase	Streptomycin, spectinomycin
hpt	Hygromycin phosphotransferase	Hygromycin B
ble	Bleomycin resistance	Bleomycin
Herbicides		
Aro A	5-enolpyruvylshikimate-3-phosphate synthase	Glyphosate
Bar	Phosphinothricin acetyltransferase	Phosphinothricin
bxn	Bromoxynil nitrilase	Bromoxynil
DHFR	Dihydrofolate reductase	Methotrexate
als	Acetolactate synthase	Chlorsulfuron
Epsps/aroa	Enolpyruvylshikimate phosphate synthase	Glyphosate

1.2.2 Characterization of Putative Transgenic Plants

1.2.2.1 Phenotypic Assay

A large number of selectable marker genes (Table 1.3) have become available which include antibiotics, herbicide-resistant genes, antimetabolites, hormone biosynthetic genes, and genes conferring resistance to toxic levels of amino acids or their analogs (Perl et al. 1993). The selection agent should fully inhibit the growth of untransformed cells. In general, the lowest concentration of the selection agent that suppresses growth of untransformed cells is used. The sensitivity of plant cells to the selection agent depends on the nature of explants, the plant genotype, the developmental stage, and the tissue culture conditions. Finally, the level of resistance also depends on the transcriptional and translational control signals to which the resistance gene is fused. It thus may be necessary to test several gene constructs. A plant is transformed if it grows in the presence of elevated concentration of selective compounds such as antibiotics and herbicides. Transgenic plants exhibit profuse hairy roots, lack of geotropism, and wrinkled leaves when a wild-type *Agrobacterium rhizogenes* is used for transformation.

1.2.2.2   Enzyme Assays

Enzyme assay of a genetic marker (nos, cat) is done to check the expression of the foreign DNA in the transformed tissue. Different genes express at different levels in different tissues, but enzyme assays are generally done using rapidly expanding tissues.

1.2.2.3   PCR Analysis

The polymerase chain reaction (PCR) amplifies DNA sequences between defined synthetic primers (Waters and Shapter 2014). A set of primers (forward primer and reverse primer) which is specific for the transgene is used to selectively amplify the transgene sequence from the total genomic DNA isolated from putative transgenic tissues/plant. PCR product can indicate the presence or absence of the transgene, but PCR usually amplifies a part of the gene and not the whole cassette. It is good for preliminary screening, but due to DNA contamination, it may lead to false positives. The PCR fails to tell anything about the transgene copy number, the integration sites and intactness of the cassette, and the expression level of the transgene.

1.2.2.4   Southern Blot Analysis

Southern blot hybridization (Southern 1975) is an efficient method for transferring DNA from agarose gels onto membranes prior to hybridization (using either radioactive or nonradioactive probes). It is a very sensitive technique which is used to detect the transgene in the genomic DNA even without any amplification. Southern analysis tells about the (1) stable integration of transgene into the genome, (2) copy number of the transgene, and (3) number of integration sites. However, it does not tell anything about the expression of transgene.

1.2.2.5 Western Blot Analysis

This method involves the detection of proteins produced by transgene in transgenic plant and is a reliable technique for analyzing the expression of transgenes (Burnette 1981). The level of gene expression is estimated by calculating the amount of protein produced by the transgene and its proportion in the total soluble plant protein.

1.2.2.6 Next-Generation Sequencing (NGS) Technologies

The emergence of next-generation sequencing (NGS) technologies has provided highly sensitive and cost- and labor-effective alternative for molecular characterization compared to traditional Southern blot analysis. This technique helps to determine the copy number, integrity, and stability of a transgene; characterize the integration site within a host genome; and confirm the absence of vector DNA. It has become a robust approach to characterize the transgenic crops (Guttikonda et al. 2016).

1.2.2.7 Progeny Analysis

The heritability of introduced gene can be easily determined by selfing or backcrossing of the putative transgenics. Transgene segregation can be studied by analyzing T_1 and subsequent generations.

1.2.2.8 Bioassay

Finally the bioassay is performed using greenhouse-/field-grown transgenic plants. For instance, if insect-resistant gene(s) has been introduced, then the larvae of target insect are allowed to feed on transgenic tissues/plants, and the extent of mortality is recorded. The transgenic lines with better transgene expression and causing higher mortality of larvae are selected, tested, and released for commercial cultivation.

We have produced transgenic sugarcane using *Agrobacterium* method (Fig. 1.1) and "particle gun" method (Figs. 1.2 and 1.3). PCR analysis (Fig. 1.4) has confirmed the presence of *Cry1Ac* gene in some of the regenerated plants which are being maintained for further studies.

Fig. 1.1 (**a–d**) *Agrobacterium*-mediated genetic transformation of sugarcane (**a**) Direct plant regeneration from young leaves after cocultivation (**b**) Shoot proliferation (**c**) Shoot elongation and rooting (**d**) GUS assay of regenerated shoots

1.3   Engineering Crops for Agronomic Traits

The production of first transgenic plant of tobacco (*Nicotiana plumbaginifolia* L.) using *Agrobacterium tumefaciens* strain containing a tumor-inducing plasmid with a chimeric gene for kanamycin resistance (De Block et al. 1984; Horsch et al. 1984) generated lot of interest using this technique for crop improvement (Gasser and Fraley 1989). Rapid and remarkable achievements have been made in the production, characterization, and field evaluation of transgenic plants in several field, fruit, and forest plant species, the world over (Dale et al. 1993; Sharfudeen et al. 2014; ISAAA 2016; Kamthan et al. 2016). However, the major interest has been in the introduction of cloned gene(s) into the commercial cultivars for their incremental improvement. Using different gene transfer methods and strategies, transgenics in several crops carrying useful agronomic traits have been developed (Table 1.4).

Fig. 1.2 (**a–f**) Particle gun-mediated genetic transformation of sugarcane (**a**) Cultured young leaf segments (target tissue) (**b**) Direct shoot regeneration from bombarded leaf segments (**c**) GUS assay of regenerated shoots (**d**) Embryogenic callus (target tissue) (**e**) GUS assay of bombarded embryogenic callus (**f**) Shoot regeneration from bombarded and selected calli

Fig. 1.3 T0 sugarcane plants in the glasshouse

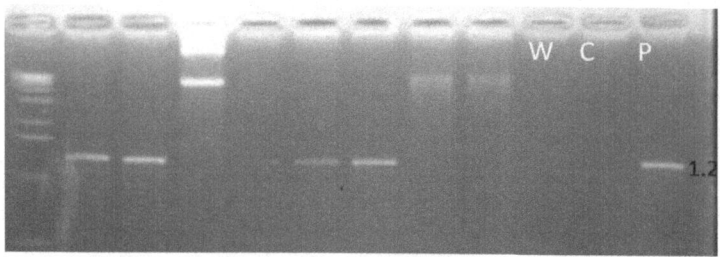

Fig. 1.4 PCR analysis of putative sugarcane transgenic plants showing amplification of Cry1Ac gene in some of the plants

1.3.1 Development of Insect-Resistant Plants

There have been two approaches to develop insect-resistant transgenic plants by transferring insect control protein genes.

1.3.1.1 Introduction of Bacterial Gene(s)

Bacillus thuringiensis synthesizes an insecticidal crystal protein, which resides in the inclusion bodies produced by the *Bacillus* during sporulation. This crystal protein when ingested by insect larvae is solubilized in the alkaline conditions of the midgut of insect and processed by midgut proteases to produce a protease-resistant polypeptide which is toxic to the insect. Lepidopteran-specific Bt gene from *Bacillus thuringiensis subsp. Kurstaki* has been widely and successfully used in tobacco, tomato, potato, maize, cotton, and rice (Gosal et al. 2001; Ahmad et al. 2002; Gómez et al. 2010; Sawardekar et al. 2012; Bakhsh et al. 2015; Abbas et al. 2016) for developing resistance against several lepidopteran insect pests. The use of redesigned synthetic Bt gene has also been used in some of these crops, and in several instances, the synthetic versions have exhibited up to 500-fold increase in the Bt gene expression.

1.3.1.2 Introduction of Plant Gene(s) for Insecticidal Proteins

Several insecticidal proteins of plant origin such as lectins, amylase inhibitors, and protease inhibitors can retard insect growth and development when ingested at high doses. Some genes like CpTi, PIN-1, PIN 11, α A-1, and GNA have been cloned and are being used in the transformation programs aiming at insect resistance (Xu et al. 2005; Gao et al. 2006; Zhang and Pang 2009; Yu et al. 2007; McCafferty et al. 2008; Ismail et al. 2010; Wang et al. 2011; Yue et al. 2011; Mi et al. 2017).

Table 1.4 Engineering crops for agronomic traits

Crop	Remarks	Reference
Maize	*Agrobacterium*-mediated delivery of infectious maize streak virus into maize plants	Grimsley et al. (1987)
Citrus	Production of transgenic citrus plants expressing the citrus tristeza virus coat protein gene	Moore et al. (1993)
Basmati rice	Transgenic basmati rice carrying genes for stem borer and bacterial leaf blight resistance	Gosal et al. (2001)
Basmati rice variety 370	Expression of synthetic *Cry*1AB and *Cry*1AC genes in basmati rice (*Oryza sativa* L.) variety 370 via *Agrobacterium*-mediated transformation for the control of the European corn borer (*Ostrinia nubilalis*)	Ahmad et al. (2002)
Potato	PVY-resistant transgenic plants of cv. Claustar expressing the viral coat protein	Gargouri-Bouzid et al. (2005)
Tomato	Evaluation of agronomic traits and environmental biosafety of a transgenic tomato plant expressing satellite RNA of cucumber mosaic virus	Iwasaki et al. (2005)
Wheat	Molecular test and aphid resistance identification of a new transgenic wheat line with the GNA gene	Xu et al. (2005)
Wheat	Expression of synthesized snowdrop lectin(gna) gene in transgenic wheat and its resistance analysis against aphid	Gao et al. (2006)
Oryza sativa (rice)	Genetic engineering of *Oryza sativa* by particle bombardment	Grewal et al. (2006)
Brassica rapa subsp. *chinensis*	Vacuum infiltration transformation of pakchoi (*B. rapa* subsp. chinensis) with gene *pin* II and the bioassay for *Plutella xylostella*	Zhang and Pang (2009)
Lemon	Enhanced resistance to *Phoma tracheiphila* and *Botrytis cinerea* in transgenic lemon plants expressing a *Trichoderma harzianum* chitinase gene.	Gentile et al. (2007)
Wine grape	*Agrobacterium*-mediated transformation and regeneration of transgenic "chancellor" plants expressing the *tfdA* gene	Mulwa et al. (2007)
Rice	Breeding of transgenic rice lines with GNA and *Bar* genes resistance to both brown planthopper and herbicide	Yu et al. (2007)
Papaya	Papaya transformed with the *Galanthus nivalis* GNA gene produces a biologically active lectin with spider mite control activity	McCafferty et al. (2008)
Rice	Expression of a bacterial flagellin gene triggers plant immune responses and confers disease resistance in transgenic plants	Takakura et al. (2008)
Capsicum annuum L. (pepper)	Transformation of a trivalent antifungal recombinant into pepper (*Capsicum annuum* L.)	Jing et al. (2009)
Maize	Transformation of the salt tolerance gene BIGST into Egyptian maize inbred lines	Assem et al. (2010)
Elaeis guineensis	Molecular and expression analysis of cowpea trypsin inhibitor (CpTI) gene in transgenic *Elaeis guineensis* Jacq leaves	Ismail et al. (2010)

(continued)

Table 1.4 (continued)

Crop	Remarks	Reference
Grapevine (*Vitis vinifera* L.)	Expression of a rice chitinase gene enhances antifungal potential in transgenics	Nirala et al. (2010)
Papaya	Developing transgenic papaya with improved fungal disease resistance	Zhu et al. (2010)
Tomato	*Agrobacterium*-mediated transformation of tomato plants expressing defensin gene	El-Siddig et al. (2011)
Pea (*Pisum sativum* L.)	Enhancing transgenic pea (*Pisum sativum* L.) resistance against fungal diseases through stacking of two antifungal genes (chitinase and glucanase)	Amian et al. (2011)
Chinese cabbage	Inheritance and expression of *pin* II gene in DH transgenic lines and F_1 hybrids	Yue et al. (2011)
Brassica napus (spring rape)	Response of transgenic rape plants bearing the *Osmyb4* gene from rice encoding a trans-factor to low above-zero temperature	Gomaa et al. (2012)
Chinese cabbage	Overexpression of rice leucine-rich repeat protein results in activation of defense response, thereby enhancing resistance to bacterial soft rot in Chinese cabbage	Park et al. (2012)
Pigeonpea	*Agrobacterium*-mediated genetic transformation of pigeonpea [*Cajanus cajan* (L.) Millsp] for pod borer resistance: optimization of protocol	Sawardekar et al. (2012)
Rice	Transgenic rice with inducible ethylene production exhibits broad-spectrum disease resistance to the fungal pathogens *Magnaporthe oryzae* and *Rhizoctonia solani*	Helliwell et al. (2013)
Peanut	Coat protein-mediated transgenic resistance of peanut (*Arachis hypogaea* L.) to peanut stem necrosis disease through *Agrobacterium*-mediated genetic transformation	Mehta et al. (2013)
Tomato	Heterologous expression of the yeast *HAL5* gene in tomato enhances salt tolerance by reducing shoot Na + accumulation in the long term	GarcíaAbellan et al. (2014)
Eggplant	Enhancing salt tolerance in eggplant by introduction of foreign halotolerance gene, *HAL1* isolated from yeast	Kumar et al. (2014)
Brassica juncea	Chitinase gene conferring resistance against fungal infections	Bashir et al. (2015)
Potato	Analysis of drought tolerance and herbicide resistance in transgenic potato plants overexpressing *DREB1A/Bar*	Jia et al. (2015)
Wheat	*Arabidopsis* EFTu receptor enhances bacterial disease resistance in transgenic wheat	Schoonbeek et al. (2015)
Maize	Breeding of transgenic maize with resistance to the Asian corn borer (*Ostrinia furnacalis*) and tolerance to glyphosate	Sun et al. (2015)
Cotton	Transgenic expression of translational fusion of synthetic *Cry1Ac* and *Hvt* genes in tobacco confers resistance to *Helicoverpa armigera* and *Spodoptera littoralis* larvae	Abbas et al. (2016)
Cucumber	Development of broad virus resistance in non-transgenic cucumber using CRISPR/Cas9 technology	Chandrasekaran et al. (2016)

(continued)

Table 1.4 (continued)

Crop	Remarks	Reference
Potato	Transgenic potato plants expressing the cold inducible transcription factor SCOF1 display enhanced tolerance to freezing stress	Kim et al. (2016)
Carrot	Transgenic approaches to enhance disease resistance in carrot plants to fungal pathogens	Punja et al. (2016)
Cotton	Transgenic upland cotton lines of *Gastrodia* antifungal protein gene and their performance of resistance to *Verticillium* wilt	Xiao et al. (2016)
Potato	Expression of the *Galanthus nivalis* agglutinin (GNA) gene in transgenic potato plants confers resistance to aphids	Mi et al. (2017)

1.3.2 Development of Disease-Resistant Plants

1.3.2.1 Virus Resistance

Genetic engineering for developing virus-resistant plants has exploited new genes derived from viruses themselves in a concept referred to as pathogen-derived resistance (PDR).

Coat Protein-Mediated Resistance (CP-MR)

Introduction of viral coat protein gene into the plant makes the plant resistant to virus from which the gene for the CP was derived (Shah et al. 1995). It was first demonstrated for tobacco mosaic virus (TMV) in tobacco. Subsequently, virus-resistant transgenics have been developed in tomato, melon, rice, papaya, potato, sugar beet, and some other plants (Gargouri-Bouzid et al. 2005; Pratap et al. 2012; Mehta et al. 2013). A variety of yellow squash called Freedom II has been released in the USA. Likewise, transgenic papaya resistant to papaya ringspot virus has been released for general cultivation in the USA. Several CP-MR varieties of potato, cucumber, and tomato are under field evaluation.

Satellite RNA-Mediated Resistance

Satellite RNAs are molecules which show little, if any, sequence homologies with the virus to which they are associated, yet are replicated by the virus polymerase and appear to affect 70 of the infections produced by the virus. It has been demonstrated that engineering cucumber, using cucumber mosaic virus (CMV) satellite RNA, leads to transgenics resistant to CMV. This approach has been extended to several other crops (Iwasaki et al. 2005).

Antisense-Mediated Protection

It is now established that gene expression can be controlled by antisense RNA. It has been proposed that antisense RNA technology can also play a role in cross protection. cDNAs representing viral RNA genomes were cloned in an antisense orientation to a promoter and transferred to plants. This approach has been effective against TMV although the protection was not as effective as with coat protein gene (Tang et al. 2005; Araújo et al. 2011).

Development of Resistance Using CRISPR/Cas9 Technology

Genome editing in plants has been boosted tremendously by the development of clustered regularly interspaced short palindromic repeats (CRISPR/Cas9) technology. This powerful tool allows substantial improvement in plant traits in addition to those provided by classical breeding. The development of virus resistance in cucumber (*Cucumis sativus* L.) using Cas9/subgenomic RNA (sgRNA) technology disrupts the function of the recessive eIF4E (eukaryotic translation initiation factor 4E) gene. Cas9/sgRNA constructs were targeted to the N' and C' termini of the eIF4E gene. Small deletions and single nucleotide polymorphisms (SNPs) were observed in the eIF4E gene-targeted sites of transformed T1 generation cucumber plants (Chandrasekaran et al. 2016).

1.3.2.2 Fungal Resistance

Genetic engineering for fungal resistance has been limited. But several new advances in this area now present an optimistic outlook.

Antifungal Protein-Mediated Resistance

Introduction of chitinase gene in tobacco and rice has been shown to enhance fungal resistance in plants. Chitinase enzymes degrade the major constituents of the fungal cell wall (chitin and α-1, 3 glucan). Co-expression of chitinase and glucanase genes in tobacco and tomato plants confers higher level of resistance than either gene alone. Use of genes for ribosome-inactivating proteins (RIP) along with chitinase has also shown synergistic effects. A radish gene encoding antifungal protein 2 (Rs-AFP2) was expressed in transgenic tobacco, and resistance to *Alternaria longipes* was observed. Other pathogenesis-related proteins/peptides include osmotin, thionins, lectins, etc. (Gentile et al. 2007; Jing et al. 2009; Nirala et al. 2010; Amian et al. 2011; El-Siddig et al. 2011; Fang et al. 2012; Bashir et al. 2015; Punja et al. 2016; Xiao et al. 2016).

Antifungal Compound-Mediated Resistance

Low-molecular-weight compounds such as phytoalexins possess antimicrobial properties and have been postulated to play an important role in plant resistance to fungal and bacterial pathogens. Expression of a stilbene synthase gene from grapevine in tobacco resulted in the production of new phytoalexin (resveratrol) and enhanced resistance to infection by *Botrytis cinerea*. Active oxygen species (AOS) including hydrogen peroxide also play an important role in plant defense responses to pathogen infection. Transgenic potato plants expressing an H_2O_2-generating fungal gene for glucose oxidase were found to have elevated levels of H_2O_2 and enhanced levels of resistance both to fungal and bacterial pathogens particularly to *Verticillium* wilt (Zhu et al. 2010; Helliwell et al. 2013).

1.3.2.3 Bacterial Resistance

Genetic engineering for bacterial resistance has relatively met with little success. The expression of a bacteriophage T4 lysozyme in transgenic potato tubers led to increased resistance to *Erwinia carotovora*. Besides, the expression of barley ά-thionin gene significantly enhanced the resistance of transgenic tobacco to bacteria *Pseudomonas syringae*. Advances in the cloning of several new bacterial resistance genes such as the *Arabidopsis* RPS2 gene, tomato *Cf9*, and tomato *Pto* gene may provide better understanding in the area of plant-bacteria interactions (Takakura et al. 2008; Schoonbeek et al. 2015).

1.3.3 Development of Herbicide-Resistant Plants

There have been two approaches to develop herbicide-resistant transgenic plants (Sun et al. 2015).

1.3.3.1 Transfer of Gene Whose Enzyme Product Detoxifies the Herbicide (Detoxification)

Using this approach, the introduced gene produces an enzyme which degrades the herbicide sprayed on the plant. For instance, introduction of *bar* gene cloned from bacteria *Streptomyces hygroscopicus* into plants makes them resistant to herbicides based on phosphinothricin (ppt). *Bar* gene produces an enzyme, phosphinothricin acetyltransferase (PAT), which degrades phosphinothricin into a nontoxic acetylated form. Plants engineered with *bar* gene were found to grow in phosphinothricin (ppt) at levels four to ten times higher than normal field application. Likewise, *bxn* gene of *Klebsiella ozaenae* which produces nitrilase enzyme imparts resistance to plants against herbicide bromoxynil. Other genes including *tfdA* for 2-, 4-D tolerance and

GST gene for atrazine tolerance have also been used. Among these, bar gene has been successfully introduced to develop herbicide-resistant soybean and cotton that have been commercially released in the USA (Hérouet et al. 2005; Mulwa et al. 2007).

1.3.3.2 Transfer of Gene Whose Enzyme Product Becomes Insensitive to Herbicide (Target Modification)

Using this approach, a mutated gene is introduced which produces modified enzyme in the plant which is not recognized by the herbicide; hence, the herbicide cannot kill the plant. For instance, a mutant *aroA* gene from bacteria *Salmonella typhimurium* has been used for developing tolerance to herbicide, glyphosate. The target site of glyphosate is a chloroplast enzyme 5-enolpyruvylshikimate-3-phosphate synthase (EPSPS). Introduction of mutant *aroA* gene produces modified EPSPS, not recognizable to glyphosate. Likewise, sulphonylurea and imidazolinone herbicides inhibit acetolactate synthase (ALS) chloroplast protein. Tolerance to these herbicides has been achieved by engineering the expression of the mutant herbicide ALS gene derived from plant (Sato and Takamizo 2009; Cao et al. 2012).

1.3.4 Development of Plants Resistant to Various Abiotic Stresses

Transfer of cloned genes has resulted in the transgenics which are tolerant to some abiotic stresses. For instance, for frost protection, an antifreeze protein gene from fish has been transferred into tomato and tobacco. Likewise, a gene coding for glycerol-3-phosphate acyltransferase from *Arabidopsis* has been transferred to tobacco for enhancing cold tolerance. *Hal2* gene is being tried for developing salt tolerance in rice (Assem et al. 2010; Gomaa et al. 2012; Duman et al. 2014; García Abellan et al. 2014; Kumar et al. 2014; Jia et al. 2015; Kim et al. 2016).

1.3.5 Development of Male Sterile and Restorer Lines for Hybrid Seed Production

The introduction of bacterial barnase gene results into male sterility, whereas the introduction of the bacterial barstar gene into another plant results into the development of restorer line. The resulting hybrid is fully fertile. This system has been commercially exploited in maize and oilseed rape. Thus, produced hybrids of *Brassica napus* are under field evaluation in India. Likewise, it can be exploited for production of hybrid wheat and rice (Ray et al. 2007).

1.3.6 Improvement in Nutritional Quality and Molecular Farming/Pharming

High protein "phaseolin" and *AmA-1* genes have been introduced to heterologous systems. Introduction of *AmA-1* gene into potato has caused improvement in the yield and protein content. Introduction of provitamin A and carotene genes has resulted into the production of "golden rice." Vitamin-producing transgenic plants have also been developed (Herbers 2003), and more emphasis is given to multigene engineering (Daniell and Dhingra 2002). Besides, transgenic plants producing specialty chemicals and biopharmaceuticals have been produced for molecular farming/pharming (Fischer and Emans 2000). The main objective of these crops is to add value to foods, such as tomato containing high lycopene, flavonols as antioxidants, cavity-fighting apples, rice enriched with carotene and vitamin A (golden rice), iron-pumping rice, canola rich in vitamin E (golden brassica), proteinaceous potatoes, edible vaccines, and decaffeinated coffee, which are some leading examples of genetically modified foods for the future (Doshi et al. 2013).

Thus, several genes of agronomic importance have been isolated from various organisms; cloned and suitable constructs have been developed for plant transformation. *Agrobacterium* and "particle gun" methods have been refined and now being used for genetic transformation of a wide variety of field, fruit, vegetable, forest crops, and ornamental plant species. Transgenic crops such as cotton, maize, papaya, potato, rice, soybean, and tomato, carrying mainly insect resistance, herbicide resistance, or both, are now being grown over an area of 185 million hectares spread over 28 countries of the world.

1.3.7 Biosafety Concerns of Transgenic Plants

The potential risks from the use of transgenics and their products fall under three categories including human health, environmental concerns, and social and ethical grounds. Risk to human health is related mainly to toxicity, allergenicity, and antibiotic resistance, whereas ecological risks include the gene flow to other plants, development of resistance in insects/pathogens, unintended secondary effects on nontarget organisms, and potential effects on biodiversity. In order to address these concerns, there are standard biosafety guidelines, and issues are addressed through deregulation of transgenic varieties for commercial cultivation. BCIL-DBT (2004).

References

Abbas Z, Zafar Y, Khan SA, Mukhtar Z (2016) Transgenic expression of translational fusion of synthetic Cry1Ac and Hvt genes in tobacco confers resistance to *Helicoverpa armigera* and *Spodoptera littoralis* larvae. Pak J Agric Sci 53(4):809–816

Ahmad A, Maqbool SB, Riazudin S, Sticklen B (2002) Expression of synthetic Cry1AB and Cry1AC genes in basmati rice (*Oryza sativa* L.) variety 370 via *Agrobacterium* mediated transformation for the control of the European corn borer (*Ostrinia nubilalis*). In Vitro Cell Dev Biol Plant 38:213–220

Altman A (2003) From plant tissue culture to biotechnology: scientific revolutions, abiotic stress tolerance and forestry. In Vitro Cell Dev Biol Plant 39:75–84

Amian AA, Papenbrock J, Jacobsen HJ, Hassan F (2011) Enhancing transgenic pea (*Pisum sativum* L.) resistance against fungal diseases through stacking of two antifungal genes (Chitinase and Glucanase). GM Crops 2(2):104–109

Araújo WL, Nunes Nesi A, Sonia Osorio BU, Fuentes D, Nagy R, Balbo I, Lehmann M, Studart Witkowski C, Tohge T, Martinoia E, Jordana X, DaMatta FM, Fernie AR (2011) Antisense inhibition of the ironsulphur subunit of succinate dehydrogenase enhances photosynthesis and growth in tomato via an organic acid-mediated effect on stomatal aperture. Plant Cell 23(2):600–627

Arora L, Narula A (2017) Gene editing and crop improvement using CRISPR-Cas9 system. Front Plant Sci 2017(8). https://doi.org/10.3389/fpls.2017.01932

Assem SK, Hussein EHA, Hussein HA, Awaly SB (2010) Transformation of the salt tolerance gene BIGST into Egyptian maize inbred lines. Arab J Biotechnol 13(1):99–114

Bakhsh A, Khabbaz SD, Baloch FS, Demirel U, Caliskan ME, Hatipoglu R, Özcan S, Özkan H, Kandemir N (2015) Insect resistant transgenic crops: retrospect and challenges. Turk J Agric For 39(4):531–548

Bakshi S, Dewan D (2013) Status of transgenic cereal crops: a review. Clon Transgen 3:119. https://doi.org/10.4172/2168-9849.1000119

Bashir A, Khan A, Ali H, Khan I (2015) *Agrobacterium* mediated transformation of *Brassica juncea* (L.) Czern. with chitinase gene conferring resistance against fungal infections. Pak J Bot 47(1):211–216

BCIL-DBT (2004) National consultation on biosafety aspects related to genetically modified organisms. Biotech Consortium India Limited, New Delhi

Bechtold N, Ellis J, Pelletier G (1993) In planta Agrobacterium mediated gene transfer by infiltration of adult Arabidopsis thaliana plants. C R Acad Sci Paris Life Sciences 316:1194–1199.

Burnette WN (1981) "Western blotting": electrophoretic transfer of proteins from sodium dodecyl sulfate polyacrylamide gels to unmodified nitrocellulose and radiographic detection with antibody and radioiodinated protein. Anal Biochem 112:195–203

Cao G, Liu Y, Zhang S, Yang X, Chen R, Zhang Y et al (2012) A novel 5-enolpyruvylshikimate-3-phosphate synthase shows high glyphosate tolerance in *Escherichia coli* and tobacco plants. PLoS One 7(6):e38718. doi.org/10.1371/journal.pone.0038718

Cardi T, D'Agostino N, Tripodi P (2017) Genetic transformation and genomic resources for next-generation precise genome engineering in vegetable crops. Front Plant Sci 8:241

Chahal GS, Gosal SS (2002) Principles and procedures of plant breeding: biotechnological and conventional approaches. Narosa, Publ.House, New Delhi

Chandrasekaran J, Brumin M, Wolf D, Leibman D, Klab C, Pearlsman M, Sherman A, Arazi T, Galon A (2016) Development of broad virus resistance in non-transgenic cucumber using CRISPR/Cas9 technology. Mol Plant Pathol 17(7):1140–1153

Christou P (1994) Applications to plants. In: Yang NS, Christou P (eds) Particle bombardment technology for gene transfer. Oxford Univ. Press, New York, pp 71–99

Christou P (1996) Transformation technology. Trends Plant Sci 1:423–431

Dale P, Irwin J, Scheffler JA (1993) The experimental and commercial release of transgenic crop plants. Plant Breed 111:1–22

Daniell H, Dhingra A (2002) Multigene engineering: dawn of an exciting new era in biotechnology. Curr Opin Biotechnol 13:136–141

De Block M, Herrera-Estrella L, Van Montagu M, Schell J, Zambryski P (1984) Expression of foreign genes in regenerated plants and in their progeny. EMBO J 3(8):1681–1689

Deom CM, Schubert KR, Wolfs S, Holt CA, Lucas WJ, Beachy RN (1990) Molecular characterization and biological function of the movement protein of tobacco mosaic virus in transgenic plants. Proc Nation Acad Sci USA 87:3284–3288

Doshi V, Rawal H, Mukherjee S (2013) Edible vaccines from GM crops: current status and future scope. J Pharm Sci Innov 2(3):1–6

Duman JG, Wisniewski MJ, Wisniewski M, Gusta LV (2014) The use of antifreeze proteins for frost protection in sensitive crop plants. Special issue: the biology of plant cold hardiness: adaptive strategies. Environ Exp Bot 106:60–69

El-Siddig MA, El-Hussein AA, Saker MM (2011) *Agrobacterium*-mediated transformation of tomato plants expressing defensin gene. Int J Agric Res 6(4):323–334

Feldmann KA, Marks MD (1987) *Agrobacterium* mediated transformation of germinating seeds of *Arabidopsis thaliana*: a non-tissue culture approach. Mol Gen Genet 208:1–9

Fischer R, Emans N (2000) Molecular pharming of pharmaceutical proteins. Transgenic Res 9:279–299

Fraley R, Wilschut J, Düzgüneş N, Smith C, Papahadjopoulos D (1980) Studies on the mechanism of membrane fusion: role of phosphate in promoting calcium ion induced fusion of phospholipid vesicles. Biochemistry 19(26):6021–6029

Frame BR, Drayton PR, Bagnall V, Lewnau CJ, Bullock WP, Wilson HM, Dunwell JM, Thompson JA, Wang K (1994) Production of fertile transgenic maize plants by silicon carbide whisker-mediated transformation. Plant J 6(6):941–948

Fromm M, Taylor LP, Walbot V (1985) Expression of genes transferred to monocot and dicot plant cells by electroporation. Proc Natl Acad Sci U S A 82:5824–5828

Fromm M, Callis J, Taylor LP, Walbot V (1987) Electroporation of DNA and RNA into plant protoplasts. Methods in Enzymology 153:351–366

Gaj T, Gersbach CA, Barbas CF III (2013) ZFN, TALEN, and CRISPR/Cas-based methods for genome engineering. Trends Biotechnol 31:397–405. https://doi.org/10.1016/j.tibtech.2013.04.004

Gao ZF, Qing CX, Ping YF, Qi LR, Quan ZL, Dong ZX (2006) Expression of synthesized snowdrop lectin (gna) gene in transgenic wheat and its resistance analysis against aphid. J Agric Biotechnol 14(4):559–564

Gao J, Wang G, Ma S, Xie X, Wu X, Zhang X et al (2015) CRISPR/Cas9-mediated targeted mutagenesis in *Nicotiana tabacum*. Plant Mol Biol 87:99–110

GarcíaAbellan JO, Egea I, Pineda B, SanchezBel P, Belver A, GarciaSogo B, Flores FB, Atares A, Moreno V, Bolarin MC (2014) Heterologous expression of the yeast HAL5 gene in tomato enhances salt tolerance by reducing shoot Na+ accumulation in the long term. Physiol Plant 152(4):700–713

Gargouri-Bouzid R, Jaoua L, Mansour RB, Hathat Y, Ayadi M, Ellouz R (2005) PVY resistant transgenic potato plants (*cv* Claustar) expressing the viral coat protein. J Plant Biotechnol 7(3):1–6

Gasser CS, Fraley RT (1989) Genetically engineering plants for crop improvement. Science New Series 244(4910):1293–1299

Gentile A, Deng Z, Malfa SL, Distefano G, Domina F, Vitale A, Polizzi G, Lorito M, Tribulato E (2007) Enhanced resistance to *Phoma tracheiphila* and *Botrytis cinerea* in transgenic lemon plants expressing a *Trichoderma harzianum chitinase* gene. Plant Breed 126(2):146–151

Gomaa AM, Raldugina GN, Burmistrova NA, Radionov NV, Kuznetso VV (2012) Response of transgenic rape plants bearing the *Osmyb4*gene from rice encoding a trans-factor to low above-zero temperature. Russ J Plant Physiol 59(1):105–114

Gómez I, Arenas I, Pacheco S, Bravo A, Soberón M (2010) New insights into the mode of action of Cry1Ab toxin used in transgenic insect resistant crops. Southwest Entomol 35(3):387–390

Gosal SS, Gosal SK (2000) Genetic transformation and production of transgenic plants. In: Trivedi PC (ed) Plant biotechnology–recent advances. Panima Publishers, New Delhi, pp 29–40

Gosal SS, Gill R, Sindhu AS, Deepinder K, Navraj K, Dhaliwal HS (2001) Transgenic basmati rice carrying genes for stem borer and bacterial leaf blight resistance. In: Peng S, Hardy B (eds) Rice research for food security and poverty alleviation. IRRI, Philippines, pp 353–360

Grewal DK, Gill R, Gosal SS (2006) Genetic engineering of Oryza sativa by particle bombardment. Biol Plant 50(2):311–314

Grimsley N, Hohn T, Daview JW, Hohn B (1987) Agrobacterium mediated delivery of infectious maize streak virus into maize plants. Nature 325:177–179

Gunther N, Spangenberg G (1990) Plant transformation by microinjection techniques. Physiol Plant 79:213–217

Gupta B, Saha J, Sengupta A, Gupta K (2013) Recent advances on virus induced gene silencing (VIGS): plant functional genomics. J Plant Biochem Physiol 1:e116. https://doi.org/10.4172/2329-9029.1000e116

Guttikonda SK, Marri P, Mammadov J, Ye L, Soe K, Richey K et al (2016) Molecular characterization of transgenic events using next generation sequencing approach. PLoS One 11(2):e0149515

Helliwell EE, Wang Q, Yang YN (2013) Transgenic rice with inducible ethylene production exhibits broad-spectrum disease resistance to the fungal pathogens Magnaporthe oryzae and Rhizoctonia solani. Plant Biotechnol J 11(1):33–42

Herbers K (2003) Vitamin production in transgenic plants. Plant Physiol 160:821–829

Hérouet C, Esdaile DJ, Mallyon BA, Debruyne E, Schulz A, Currier T, Hendrickx K, van der Klis RJ, Rouan D (2005) Safety evaluation of the phosphinothricin acetyltransferase proteins encoded by the pat and bar sequences that confer tolerance to glufosinate ammonium herbicide in transgenic plants. Regul Toxicol Pharmacol 41(2):134–149

Herrera-Estrella L (1983) Transfer and expression of foreign genes in plants. PhD thesis. Laboratory of Genetics. Gent University, Belgium

Hiei Y, Ohta S, Komari T, Kumashiro T (1994) Efficient transformation of rice (Oriza sativa) mediated by Agrobacterium and sequence analysis of the boundaries of the T-DNA. Plant J 6:271–282

Hilder VA, Boulter D (1999) Genetic engineering of crop plants for insect resistance – a critical review. Crop Prot 18:177–191

Hooykaas PJJ, Schilperoort RA (1992) Agrobacterium and plant genetic engineering. Plant Mol Biol 19:1538

Horsch RB, Fraley RT, Rogers SG, Sanders PR, Lloyd A, Hoffmann N (1984) Inheritance of functional foreign genes in plants. Science 223(4635):496–498

ISAAA (2016) Global status of commercialized biotech/GM crops: 2016. ISAAA Brief No. 52. ISAAA, Ithaca, NY

Ismail I, Lee FS, Abdullah R, Fee CK, Zainal Z, Sidik NM, Zain CRCM (2010) Molecular and expression analysis of cowpea trypsin inhibitor (CpTI) gene in transgenic Elaeis guineensis Jacq leaves. Aust J Crop Sci 4(1):37–48

Iwasaki M, Ito K, Kawabe K, Sugito T, Nitta T, Takigawa S, Ito K, Nakata T, Ogawa Y, Hayano Y, Fukumoto F (2005) Evaluation of agronomic traits and environmental biosafety of a transgenic tomato plant expressing satellite RNA of Cucumber Mosaic Virus. Research Bulletin of the National Agricultural Research Center for Hokkaido Region (182) Sapporo:51–63

Jia XX, Qi EF, Ma S, Hu XY, Wang YH, Wen GH, Gong CW, Li JW (2015) Analysis of drought tolerance and herbicide resistance in transgenic potato plants overexpressing DREB1A/Bar. Acta Prataculturae Sinica 24(11):58–64

Jing GX, Zeng FH, Li FQ, Chen YS, He YM (2009) Transformation of a trivalent antifungal recombinant into pepper (Capsicum annuum L.). Jiangsu J Agric Sci 25(1):165–168

Junjie Z, Fan L, Hong Z, Chen L (2006) Vacuum infiltration transformation of pakchoi (*B. rapa* subsp. *chinensis*) with gene *pin II* and the bioassay for *Plutella xylostella* resistance. Acta Phytophyacica Sin 33(1):17–21

Kaeppler HF, Gu W, Somers DA, Rines HW, Cockburn AF (1990) Silicon carbide fiber-mediated DNA delivery into plant cells. Plant Cell Rep 9(8):415–418

Kamthan A, Chaudhuri A, Kamthan M, Datta A (2016) Genetically modified (GM) crops: milestones and new advances in crop improvement. Theor Appl Genet 129(9):1639–1655

Kerr A (2011) GM crops – a minireview. Australas Plant Pathol 40(5):449–452

Keshamma E, Rohini S, Rao KS, Madhusudhan B, Kumar MU (2008) Tissue culture independent in planta transformation strategy: an *Agrobacterium tumefaciens* mediated gene transfer method to overcome recalcitrance in cotton (*Gossypium hirsutum* L.). J Cotton Sci 12(3):264–272

Kim DH, Rossi JJ (2008) RNAi mechanisms and applications. Biotechniques 44:613–616

Klein TM, Wolf ED, Wu R, Sanford JC (1987) High velocity microprojectiles for delivering nucleic acids into living cells. Nature 327:70–73

Kloti A, Iglesias VA, Wunn J, Burkdardt PK, Datta SK, Potrykus I (1993) Gene transfer by electroporation into intact scutellum cells of wheat embryos. Plant Cell Rep 12:671–675

Kumar SK, Sivanesan I, Murugesan K et al (2014) Enhancing salt tolerance in eggplant by introduction of foreign halotolerance gene, HAL1 isolated from yeast. Horticulture, Environment and Biotechnology 55:222

Nayak L, Pandey H, Ammayappan L, Ray DP (2011) Genetically modified crops – a review. Agricultural Reviews 32(2):112–119

Ledoux L (1965) Uptake of DNA by living cells. Progr Nucl Acid Res Mol Biol 4:231–267

Leelavathi S, Sunnichan VG, Kumria R, Vijaykanth GP, Bhatnagar RK, Reddy VS (2004) A simple and rapid *Agrobacterium* mediated transformation protocol for cotton (*Gossypium hirsutum* L.): embryogenic calli as a source to generate large numbers of transgenic plants. Plant Cell Rep 22:465–470

Lörz H, Baker B, Schell J (1985) Gene transfer to cereal cells mediated by protoplast transformation. Mol Gen Genet 199:178–182

Lu R, Martin-Hernandez AM, Peart JR, Malcuit I, Baulcombe DC (2003) Virus-induced gene silencing in plants. Methods 30(4):296–303

Mathews H, Wagoner W, Cohen C, Kellogg J, Bestwick R (1995) Efficient genetic transformation of red raspberry *Rubus idaeus* L. Plant Cell Rep 14:471–476

McCafferty HRK, Moore PH, Zhu YJ (2008) Papaya transformed with the *Galanthus nivalis* GNA gene produces a biologically active lectin with spider mite control activity. Plant Sci 175(3):385–393

Mehta R, Thankappan R, Kumar A, Yadav R, Dobaria JR, Thirumalaisamy PP, Jain RK, Chigurupati P (2013) Coat protein mediated transgenic resistance of peanut (*Arachis hypogaea* L.) to peanut stem necrosis disease through *Agrobacterium* mediated genetic transformation. Indian J Virol 24(2):205–213

Mi XX, Liu X, Yan HL, Liang L, Zhou XY, Yang JW, Si HJ, Zhang N (2017) Expression of the *Galanthus nivalis* agglutinin (GNA) gene in transgenic potato plants confers resistance to aphids. C R Biol 340(1):7–12

Moore GA, Gutierrez EA, Jacono A, Jacono C, Caffery MC, Cline K (1993) Production of transgenic citrus plants expressing the citrus tristeza virus coat protein gene. HortScience 28:512

Mulwa RMS, Norton MA, Farrand SK, Skirvin RM (2007) *Agrobacterium* mediated transformation and regeneration of transgenic 'Chancellor' wine grape plants expressing the *tfdA* gene. Vitis 46(3):110–115

Neuhas G, Spangenberg G, Mittelsten Scheid O, Schweiger HG (1987) Transgenic rapeseed plants obtained by the microinjection of DNA into microspore-derived embryoids. Theor Appl Genet 75(1):30–36

Nirala NK, Das DK, Srivastava PS, Sopory SK, Upadhyaya KC (2010) Expression of a rice chitinase gene enhances antifungal potential in transgenic grapevine (*Vitis vinifera* L.). Vitis 49(4):181–187

Park YH, Choi CH, Park EM, Kim HS, Park HJ, Bae SC, Ahn I, Kim MG, ParkSR HDJ (2012) Overexpression of rice leucine-rich repeat protein results in activation of defense response, thereby enhancing resistance to bacterial soft rot in Chinese cabbage. Plant Cell Rep 31(10):1845–1850

Paszkowski J, Shillito RD, Saul M, Mandák V, Hohn T, Hohn B, Potrykus I (1984) Direct gene transfer to plants. EMBO J 3(12):2717–2722

Perl A, Galili S, Shaul O, Ben-Tzvi I, Galili G (1993) Bacterial dihydrodipicolinate synthase and desensitized aspartate kinase: two novel selectable markers for plant transformation. Biotechnology 11:715–718

Pescitelli SM, Sukhapinda K (1995) Stable transformation via electroporation into maize type II callus and regeneration of fertile transgenic plants. Plant Cell Rep 14:712–716

Pratap D, Raj SK, Kumar S, Snehi SK, Gautam KK, Sharma AK (2012) Coat protein mediated resistance. Acta Phytophylacica Sin 33(1):17–21

Punja ZK, Wally O, Jayaraj J, Onus AN (2016) Transgenic approaches to enhance disease resistance in carrot plants to fungal pathogens. Acta Hortic (1145):143–152

Rakosy-Tican E, Aurori CM, Dijkstra C, Thieme R, Aurori A, Davey MR (2007) The usefulness of the *gfp* reporter gene for monitoring *Agrobacterium* mediated transformation of potato dihaploid and tetraploid genotypes. Plant Cell Rep 26(5):661–671

Rasul F, Sohail MN, Mansoor S, Asad S (2014) Enhanced transformation efficiency of *Saccharum officinarum* by vacuum infiltration assisted *Agrobacterium*-mediated transformation. Int J Agric Biol 16(6):1147–1152

Ray K, Bisht NC, Pental D, Burma PK (2007) Development of barnase/barstar transgenics for hybrid seed production in Indian oilseed mustard (*Brassica juncea* L. Czern & Coss) using a mutant acetolactate synthase gene conferring resistance to imidazolinone-based herbicide 'Pursuit'. Curr Sci 93(10):1390–1396

Rhodes CA, Marrs KA, Murry LE (1995) Transformation of maize by electroporation of embryos. Methods in molecular biology 55. In: Plant cell electroporation and electrofusion protocols, vol 55. Springer, Totowa, pp 121–131

Sanford JC (1988) The biolistic process. Trends Biotechnol 6:299–302

Sanford JC (1990) Biolistic plant transformation. Physiol Plant 79:206–209

Sanford JC, Skubik KA, Reisch BI (1985) Attempted pollen-mediated plant transformation employing genomic donor DNA. Theor Appl Genet 69:571–574

Sato H, Takamizo T (2009) Conferred resistance to an acetolactate synthase-inhibiting herbicide in transgenic tall fescue (*Festuca arundinacea* Schreb.). Hortscience 44(5):1254–1257

Saul MW, Potrykus I (1990) Direct gene transfer to protoplasts: fate of the transferred genes. Dev Genet 11:176–181

Sawardekar SV, Mhatre NK, Sawant SS, Bhave SG, Gokhale NB, Narangalkar AL, Katageri IS, Kumar PA (2012) *Agrobacterium* mediated genetic transformation of pigeonpea [*Cajanus cajan* (L.) Millisp] for pod borer resistance: optimization of protocol. Indian J Genet Plant Breed 72(3):380–383

Schoonbeek HJ, Wang HH, Stefanato FL, Craze M, Bowden S, Wallington E, Zipfel C, Ridout C (2015) *Arabidopsis* EFTu receptor enhances bacterial disease resistance in transgenic wheat. New Phytol 206(2):606–613

Shah DM, Rommens CMT, Beachy R (1995) Resistance to disease and insects in transgenic plants: progress and applications to agriculture. Trends Biotechnol 13:362–368

Sharfudeen S, Begum MC, Deepthi CDN, Gullapalli L, Sulthana MR, Akula R, Tejaswini SSN (2014) Transgenic technology: an overview, current status & future perspectives. J Pharm Res 8(4):474–485

Shimada TL, Shimada T, Hara-Nishimura I (2010) A rapid and nondestructive screenable marker, FAST, for identifying transformed seeds of *Arabidopsis thaliana*. Plant J 61(3):519–528

Shimada T, Ogawa Y, Shimada T, Hara-Nishimura I (2011) A non-destructive screenable marker, OsFAST, for identifying transgenic rice seeds. Plant Signal Behav 6(10):1454–1456

Southern EM (1975) Detection of specific sequences among DNA fragments separated by gel electrophoresis. J Mol Biol 98(3):503–517

Sun Y, Liu XX, Li L, Guan Y, Zhang J (2015) Breeding of transgenic maize with resistance to the Asian corn borer (*Ostrinia furnacalis*) and tolerance to glyphosate. J Agric Biotechnol 23(1):52–60

Takahashi W, Tanaka O, Rao GP, Zhao Y, Radchuk VV, Bhatnagar SK (2008) Whisker mediated transformation: the simplest method for direct gene transfer in higher plants. Advances in plant biotechnology Houston: Studium Press LLC 2008:63–80

Takakura Y, Fk C, Ishida Y, Tsutsumi F, Kurotani K, Usami S, Isogai A, Imaseki H (2008) Expression of a bacterial flagellin gene triggers plant immune responses and confers disease resistance in transgenic rice plants. Mol Plant Pathol 9(4):525–529

Tang W, Kinken K, Newton RJ (2005) Inducible antisense mediated posttranscriptional gene silencing in transgenic pine cells using green fluorescent protein as a visual marker. Plant Cell Physiol 46(8):255–1263

Tanuja P, Kumar AL (2017) Transgenic fruit crops – a review. Int J Curr Microbiol App Sci 6(8):2030–2037

Tepfer D (1984) Transformation of several species of higher plants by *Agrobacterium rhizogenes*: sexual transmission of the transformed genotype and phenotype. Cell 37:959–967

Trick HN, Finer JJ (1997) SAAT: sonication-assisted *Agrobacterium*-mediated transformation. Transgenic Res 6(5):329–336

Uchimiya H, Fushimi T, Hashimoto H, Harada H, Syono K, Sugawara Y (1986) Expression of a foreign gene in callus derived from DNA treated protoplasts of rice (Oryza sativa L.). Molecular & General Genetics 204:204–207

Wang XJ, Dong L, Miao MM, Tang QL, Wang ZX (2011) Construction of a standard reference plasmid for detecting *CPTI* gene in transgenic cotton. China Biotechnol 31(8):85–91

Wang H, Russa ML, QiLS (2016) CRISPR/Cas9in genome editing and beyond. Annu Rev Biochem 85:227–264

Waters DL, Shapter FM (2014) The polymerase chain reaction (PCR): general methods. Methods Mol Biol 1099:65–75

Weber G, Monajembashi S, Wolfrum J, Greulich KO (1989) A laser microbeam as a tool to introduce genes into cells and organelles of higher plants. Ber Bunsen Phys Chem 93:252–254

Xiao SH, Zhao J, Liu JG, Wu QJ, Wang YQ, Chu CC, Yu JZ, Yu DY (2016) Transgenic upland cotton lines of *Gastrodia* antifungal protein gene and their performance of resistance to *Verticillium wilt*. Acta Agron Sin 42:212–221

Xu QF, Tian F, Chen X, Li LC, Lin ZS, Mo Y et al (2005) Molecular test and aphid resistance identification of a new transgenic wheat line with the GNA gene. J Triticeae Crops 25(3):7–10

Yang NS, Christou P (eds) (1994) Particle bombardment technology for gene transfer. Oxford University Press, New York, pp 143–165

YongFeng F, YongSheng L, YunLing P, Wang F, Wang W, YanZhao M, Wang HN (2012) *Agrobacterium* mediated transformation of maize shoot apical meristem by introducing fused gene Chilinker*Glu* and *bar*. Acta Prataculturae Sinica 21(5):69–76

Younis A, Siddique MI, Kim CK, Lim KB (2014) RNA interference (RNAi) induced gene silencing: a promising approach of hi-tech plant breeding. Int J Biol Sci 10(10):1150–1158

Yu H, Zhao Z, Wang L, Liu QQ, Gong Z, Gu MH (2007) Breeding of transgenic rice lines with GNA and bar genes resistance to both brown planthopper and herbicide. Acta Phytophylacica Sin 34(5):555–556

Yue Y, Kun L, Guixang W, Fan L (2011) Inheritance and expression of *pin II* gene in DH transgenic lines and F1 hybrids of Chinese cabbage. Mol Plant Breed 9(3):350–356

YunHee K, MyoungDuck K, SungChul P, JaeCheol J, SangSoo K, HaengSoon L (2016) Transgenic potato plants expressing the cold inducible transcription factor SCOF1 display enhanced tolerance to freezing stress. Plant Breed 135(4):513–518

Zhang W, Pang Y (2009) Impact of IPM and Transgenics in the Chinese Agriculture. In: Peshin R, Dhawan AK (eds) Integrated Pest Management: Dissemination and Impact. Springer, Dordrecht

Zhu YJ, Agbayani R, Tang CS, Moore PH, Souza M, Drew R (2010) Developing transgenic papaya with improved fungal disease resistance. Acta Hortic (864):39–44

Chapter 2
Virus Induced Gene Silencing Approach: A Potential Functional Genomics Tool for Rapid Validation of Function of Genes Associated with Abiotic Stress Tolerance in Crop Plants

Ajay Kumar Singh, Mahesh Kumar, Deepika Choudhary, Jagadish Rane, and Narendra Pratap Singh

Abstract Virus-induced gene silencing (VIGS) is a versatile tool for functional characterization of plant genes using gene transcript suppression. With increased identification of differentially expressed genes employing high-throughput transcript profiling under various abiotic stresses, functional elucidation of stress-responsive genes is crucial to understand their role in stress tolerance. In recent past, VIGS has been successfully used as reverse genetic elegant tool for gene function analysis in various model plants and also in crop plants. Viral vector-based silencing of gene of interest and studying the gene knockdown plants under stress can be one of the potential options for assessing functional significance of stress-responsive genes. This review provides an overview of how VIGS is used in different crop plants to characterize genes responsive to various kinds of abiotic stresses, viz., drought stress, salinity stress, heat stress, cold stress, and oxidative and nutrient-deficiency stresses. This review also documents examples from studies where abiotic stress-responsive genes have been functionally characterized using VIGS. In addition, we also summarize improvement in abiotic stress tolerance, seed yield, and seed quality traits in crop plants. This review also describes advantages of VIGS over other functional genomics tools, improvement and limitations of VIGS approach, and future prospects of VIGS as efficient tool for studying adaptation and tolerance in crop plants to various kinds of abiotic stresses. In this review, we have also discussed the mechanism of VIGS and novel ways for application of VIGS to carry out functional elucidation of abiotic stress-responsive genes in a wide range of crops.

A. K. Singh (✉) · M. Kumar · D. Choudhary · J. Rane · N. P. Singh
ICAR-National Institute of Abiotic Stress Management, Malegaon, Baramati, Maharashtra, India

Keywords Abiotic stress · Gene function · Posttranscriptional gene silencing · Virus-induced gene silencing · Limited soil moisture · Drought stress · Salt stress · Water-deficit condition

2.1 Introduction

Plant growth and crop yield are greatly affected by abiotic stresses, viz., drought, salinity, and heat stress and low temperature. These stresses are expected to increase in the future due to drastic change in climate, much of which are driven by global warming. Agriculture will be affected greatly by these changes. Plants can acquire tolerance to these environmental stresses through advanced molecular breeding techniques and genetic engineering; therefore, it is important to understand the molecular mechanisms of these responses. In order to survive, plants respond and adapt to unfavorable environmental conditions. Changes occur in plants at morphological, biochemical, physiological, and molecular level to cope with these abiotic stresses. The molecular mechanism underlying plants' response to abiotic stresses has been analyzed by studying a number of genes associated with drought, salinity, heat, and cold stresses at transcriptional level (Ingram and Bartels 1996; Hasegawa et al. 2000). Transcriptome profiling has been successfully used to identify genes associated with abiotic stress responses. Analysis of stress downregulated as well as stress upregulated genes is crucial for understanding molecular responses of crop plants to abiotic stresses. A large number of genes whose expression altered during various abiotic stresses have been identified through expression profiling, Expressed Sequence Tags (ESTs), and cDNA library generated from various plant species (Seki et al. 2002; Bohnert et al. 2006; Govind et al. 2009a, b; Marques et al. 2009; Becker and Lange 2010; Chen et al. 2015; Ramegowda et al. 2017; Abd El-Daim et al. 2018). However, identifying the functional significance of individual differentially expressed genes during abiotic stresses is challenging task. It is utmost important to elucidate the function of these stress-responsive genes to understand the mechanism of stress tolerance and also for characterizing candidate genes contributing tolerance of susceptible species by genetic engineering. An inventory of genes showing altered expression under various abiotic stresses has been established for many crop species employing EST analysis (Gorantla et al. 2007; Wani et al. 2010; Blair et al. 2011). In contrast to the enormous progress made in generating sequence information, functional analysis of stress-responsive genes is lagging behind.

Although comparative genomic strategies have provided initial clues about function of abiotic stress-responsive genes in many crop species (Gorantla et al. 2007; Tran and Mochida 2010; Soares-Cavalcanti et al. 2012), comprehensive functional characterization tools are necessary for understanding the precise role of these genes in combating abiotic stresses. One such tool is virus-induced gene silencing (VIGS) which has emerged as a potential gene knockdown technique in several crop species because it does not require transformation (Baulcombe 1999;

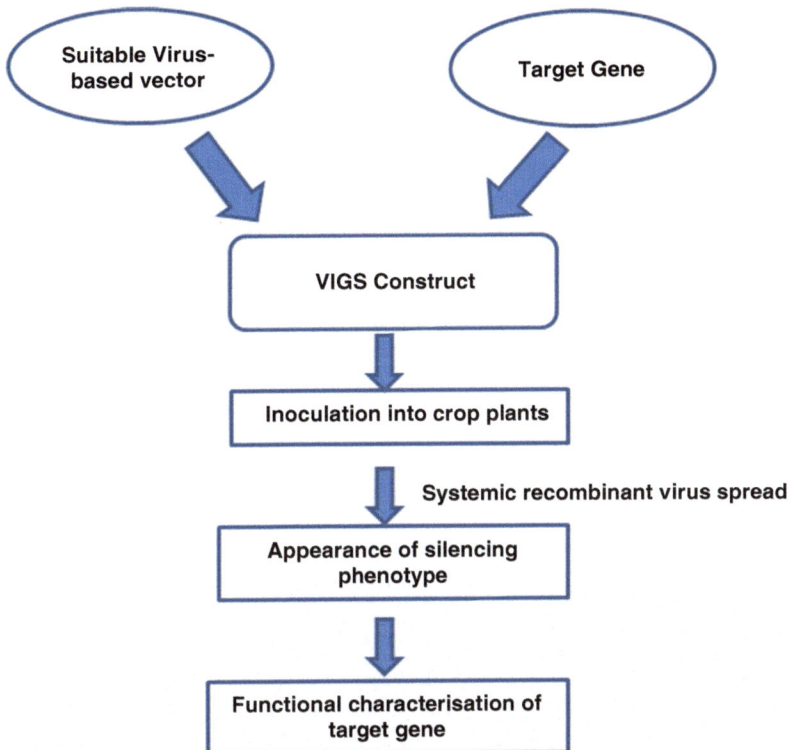

Fig. 2.1 Procedure of virus-induced gene silencing for functional elucidation of abiotic stress-responsive genes

Burch-Smith et al. 2004; Senthil-Kumar and Mysore 2011a). In VIGS system, recombinant virus carrying a partial sequence of a host gene is used to infect the plant (Fig. 2.1). When the virus spreads systemically, the endogenous gene transcripts, which are homologous to the insert in the VIGS vector, are degraded by posttranscriptional gene silencing (PTGS) (Baulcombe 1999).

In recent past, VIGS has been successfully used to unravel the abiotic stress tolerance mechanisms in crop plants (Senthil-Kumar and Udayakumar 2006; Senthil-Kumar et al. 2008; Manmathan et al. 2013). In this review, we document examples from studies where abiotic stress-responsive genes have been functionally characterized using VIGS. In addition, we also summarize improvement in abiotic stress tolerance in crop plants using VIGS technique. This review also describes advantages of VIGS system over other functional genomics tools, improvement and limitations of VIGS approach, and future prospects of VIGS as efficient tool for studying adaptation and tolerance in crop plants to various kinds of abiotic stresses. In this review, we also discuss the mechanism of VIGS and novel ways for application of VIGS to carry out functional elucidation of abiotic stress-responsive genes in a wide range of crop plants.

2.2   Application of VIGS Technology to Study Function of Genes Responsive to Various Abiotic Stresses

VIGS technology has been extensively used to investigate function of genes responsive to various kinds of abiotic stresses (Senthil-Kumar et al. 2007; Cho et al. 2008; Govind et al. 2009a, b; Kuzuoglu-Ozturk et al. 2012; Manmathan et al. 2013; Bao et al. 2015; Wang et al. 2016; Li et al. 2017; Park et al. 2017; Ramegowda et al. 2017; Ullah et al. 2018). Recent development in VIGS vectors has extended the application of VIGS for functional characterization of abiotic stress-responsive genes and also enhancing abiotic stress tolerance in several crops (Table 2.1, Fig. 2.2).

2.2.1   Drought, Salinity, and Osmotic Stress Tolerance

VIGS is a versatile technique for functional characterization of water stress-responsive genes identified from transcriptome profiling of plants exposed to water-deficit conditions. Three potential candidate genes, viz., *Era1* (enhanced response to abscisic acid), *Cyp707a* (ABA 8′-hydroxylase), and *Sal1* (inositol polyphosphate 1-phosphatase) associated with abiotic stress response pathways in *Arabidopsis thaliana* were functionally characterized employing BSMV-based VIGS vector in wheat (Manmathan et al. 2013). The Era1 and Sal1 silenced wheat plants showed increased relative water content, improved water use efficiency, reduced gas exchange, and better vigor compared to water-stressed control plants when subjected to limited soil moisture conditions, whereas the *Cyp707a*-silenced plants showed no improvement in water stress tolerance over BSMV empty vector-inoculated plants under water-deficit condition. These results indicated that *Era1* and *Sal1* genes play important roles in conferring water stress tolerance in wheat. Manmathan et al. (2013) reported delayed seed germination in Era1-silenced plants. From this study it may be concluded that this gene might be useful target for developing resistance to preharvest sprouting. Rao et al. (2014) studied functional relevance of *GmCam4* (*calmodulin*) gene by silencing and overexpression using Bean pod mottle virus (BPMV)-based vector. Silencing of *GmCam4* resulted in susceptible response to salt stress, while overexpression resulted in salinity tolerance in soybean plants at 200 mMNaCl level (Rao et al. 2014). Genes, viz., *glutamate decarboxylases* (*SlGADs*), *GABA transaminases* (*SlGABA-Ts*), and *succinic semialdehyde dehydrogenase* (*SlSSADH*), involved in metabolic pathway of GABA were studied for their function using VIGS approach in tomato (Bao et al. 2015). Silencing of *SlGADs* (GABA biosynthetic genes) and *SlGABA-Ts* (GABA catabolic genes) led to increased accumulation of reactive oxygen species (ROS) as well as salt sensitivity in tomato plants exposed to

Table 2.1 List of genes, associated with abiotic stress tolerance, functionally characterized in crop plants using virus-based vector system

Crop species	Target gene	Abiotic stress	Response of plants to abiotic stresses	References
Wheat	*TaEra1 (enhanced response to abscisic acid), TaSal1 (inositol phosphate 1-phosphatase)*	Drought	Increase in RWC and WUE, reduced stomatal conductance, reduced transpiration rate and higher plant vigor	Manmathan et al. (2013)
	TaBTF3 (basic transcription factor 3)	Drought	Wilting of leaves under drought, higher water loss rate, decrease in RWC and survival rate, lower free proline content, and increase in membrane leakage	Kang et al. (2013)
Barley	*HvHVA1 (H. vulgaris abundant protein)*	Drought	Higher water loss rate in leaves, less survival and retarded growth with reduced height	Liang et al. (2012)
	HvDhn6 (dehydrin)	Drought	Less survival and retarded growth	Liang et al. (2012)
Soybean	GmCam4 *(calmodulin)*	Salinity	Overexpression resulted in salinity stress tolerance and silencing led to susceptible response	Rao et al. (2014)
	GmFAD3	Seed yield increase	Seed size increased in silenced plants	Singh et al. (2011)
Tomato	*SlLea4 (late embryogenesis abundant protein 4)*	Drought or oxidative stress	Leaf wilting, reduced osmotic adjustment and cell viability and higher accumulation of superoxide radicals	Senthil-Kumar and Udayakumar (2006)
	SpMPK1 (mitogen-activated protein kinase 1), SpMPK2 (mitogen-activated protein kinase 2), SpMPK3 (mitogen-activated protein kinase 3)	Drought and oxidative stress	Reduced survival, higher water loss in detached leaves, increased stomatal closure and increased H_2O_2 production	Li et al. (2013)
	SlMPK4 (mitogen-activated protein kinase 4)	Drought	Early leaf wilting	Virk et al. (2013)

(continued)

Table 2.1 (continued)

Crop species	Target gene	Abiotic stress	Response of plants to abiotic stresses	References
	SlGADs (glutamate decarboxylases) SlGABA-Ts (GABA transaminases) SlSSADH (succinic semialdehyde dehydrogenase)	Salt stress	Increased accumulation of ROS as well as salt sensitivity under 200 mm NaCl	Bao et al. (2015)
	SlGRX1 (glutaredoxin 1)	Oxidative or drought or salt stress	Reduced chlorophyll, leaf wilting, and reduced RWC under drought and reduced chlorophyll under salt stress	Guo et al. (2010)
	SlFRO1 (ferric chelate reductase 1)	Nutrient deficiency	Reduced ferric chelate reductase activity in roots	He et al. (2008)
Chili pepper	*CaPO2 (peroxidase 2)*	Salt or osmotic stress	Reduced chlorophyll content and increased lipid peroxidation	Choi and Hwang (2012)
	CaRAV1 (related to ABI3/ VP1), CaOXR1 (oxidoreductase 1)	Salt or osmotic stress	Bleaching of leaf discs, loss of chlorophyll and increased lipid peroxidation	Lee et al. (2010)
	CaDHN1	Salt and osmotic stress	Decreased tolerance to salt and osmotic stresses	Chen et al. (2015)
	CaWDP1	Drought	Low levels of leaf water loss in the drought-treated leaves	Park et al. (2017)
Pyrus belulaefolia	*PbMYB21*	Drought	Higher expression of arginine decarboxylase and accumulated larger amount of polyamine	Li et al. (2017)
Rose	*RhNAC2 (NAC transcription factor 2), RhEXPA4 (A-type expansin 4)*	Dehydration	Reduced fresh weight, petal width and recovery from dehydration	Dai et al. (2012)
	RhNAC3 (NAC transcription factor 3)	Dehydration	Reduced cell expansion	Jiang et al. (2014)
	RhACS1 (ACC synthase 1), RhACS2 (ACC synthase 2)	Dehydration	Reduced ethylene production and cell density decreased	Liu et al. (2013)
	RhETR3 (ethylene receptor)	Dehydration	Inhibition of petal expansion and cell expansion	Liu et al. (2013)

(continued)

Table 2.1 (continued)

Crop species	Target gene	Abiotic stress	Response of plants to abiotic stresses	References
Tomato	*SlGRX1* (*glutaredoxin 1*)	Oxidative or drought or salt stress	Reduced chlorophyll, leaf wilting, and reduced RWC under drought and reduced chlorophyll under salt stress	Guo et al. (2010)
	SlFRO1 (*ferric chelate reductase 1*)	Nutrient deficiency	Reduced ferric chelate reductase activity in roots	He et al. (2008)
Cotton	*GhWRKY27a*	Drought	Enhanced tolerance to drought stress	Yan et al. (2015)
	GhMKK3 (*mitogen-activated protein kinase 3*)	Drought	Enhanced susceptibility to drought stress	Wang et al. (2016)
	GhNAC79	Drought stress	Drought-sensitive phenotype	Guo et al. (2018)
	Gh WRKY6	Drought, osmotic, and salt stress	Higher expression of ROS	Ullah et al. (2018)

200 mm NaCl treatment. Targeted quantitative analysis of metabolites revealed that expression of GABA biosynthetic genes decreased and increased in the SlGADs- and SlGABA-Ts-silenced plants, respectively, whereas succinate which is the final product of GABA metabolism decreased in both silenced plants. In contrast, SlSSADH-silenced plants defective in GABA degradation process showed dwarf phenotype, curled leaves, and higher accumulation of ROS under non-stress conditions. These results indicated that GABA shunt is involved in salt tolerance of tomato and affects homeostasis of metabolites such as succinate and γ-hydroxybutyrate and subsequently ROS accumulation under salt stress (Bao et al. 2015). Chen et al. (2015) isolated *CaDHN1* gene and investigated the response and expression of this gene under various stresses. Loss of function of *CaDHN1* using VIGS technique resulted in decreased tolerance to salt- and osmotic-induced stresses. These results suggest that *CaDHN1* plays an important role in regulating the abiotic stress resistance in pepper plants (Chen et al. 2015). Yan et al. (2015) elucidated the function of *GhWRKY27a* gene in cotton. Silencing of *GhWRKY27a* gene using VIGS approach enhanced drought stress tolerance in cotton. In contrast, *GhWRKY27a* overexpression in *Nicotiana benthamiana* markedly reduced drought stress tolerance. This susceptibility was coupled with reduced stomatal closure in response to abscisic acid and decreased expression of stress-related genes (Yan et al. 2015). Wang et al. (2016) isolated and characterized the cotton group B *MAPKK* gene (*GhMKK3*) using VIGS. Overexpression of *GhMKK3* gene in *Nicotiana benthamiana* enhanced drought stress tolerance. Based on RNA sequencing (RNA-seq) and quantitative Real-time PCR (qRT-PCR) assays, it was observed that GhMKK3 plays an important role in regulating stomatal responses and root hair growth. They further demonstrated that overexpressing

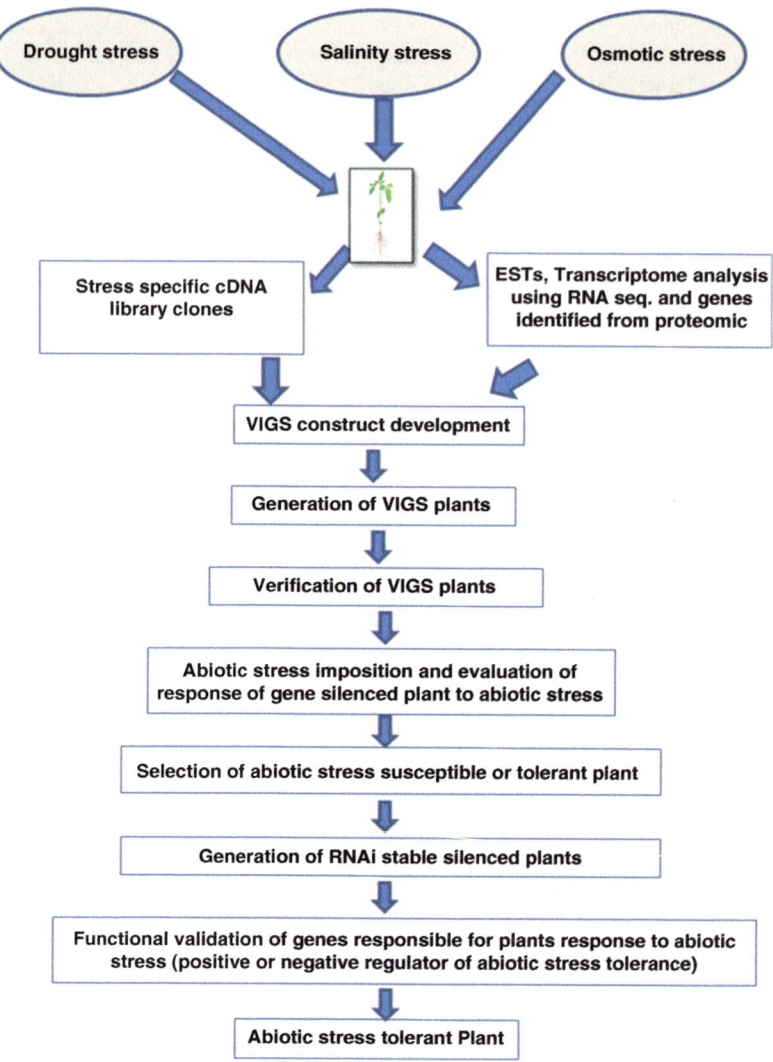

Fig. 2.2 Application of VIGS in understanding the mechanisms of plants response to abiotic stress and improvement of crop plants for abiotic stress tolerance using virus-based vector system

GhMKK3 enhanced root growth and ABA-induced stomatal closure, whereas silencing GhMKK3 in cotton using VIGS resulted in the opposite phenotypes. Li et al. (2017) reported that silencing of PbrMYB21 in *Pyrus betulaefolia* by virus-induced gene silencing (VIGS) resulted in increased drought sensitivity, whereas overexpression of PbrMYB21 in tobacco conferred enhanced tolerance to dehydration and drought stresses (Li et al. 2017). Park et al. (2017) generated CaWDP1-silenced pepper plants using VIGS and studied responses of CaWDP1-silenced pepper plants and CaWDP1-overexpressing (OX) *Arabidopsis* plants to ABA and drought. CaWDP1-silenced pepper plants showed enhanced tolerance to drought stress, and this was correlated with reduced water loss in the leaves. They found that in contrast to CaWDP1-silenced plants, CaWDP1-overexpressing plants exhibited an ABA-hyposensitive and drought-susceptible phenotype, which was accompanied by high levels of leaf water loss, low leaf temperatures, increased stomatal pore size, and low expression levels of stress-responsive genes. Ramegowda et al. (2017) carried out transcriptome analysis of finger millet (*Eleusine coracana*) under drought conditions by cDNA subtraction which identified drought-responsive genes that have a potential role in drought tolerance. They identified several genes including a *G-BOX BINDING FACTOR 3 (GBF3)* as candidate drought-stress response genes through VIGS in a related crop species, maize (*Zea mays*), and the role of GBF3 in drought tolerance was studied in *Arabidopsis thaliana*. Overexpression of both *EcGBF3* and *AtGBF3* in *A. thaliana* resulted in improved tolerance to osmotic stress, salinity stress, and drought stress. In contrast, downregulation of this gene increased the sensitivity of *A. thaliana* plants to drought stress. Guo et al. (2018) reported that silencing of *GhNAC79* using VIGS in cotton resulted in drought-sensitive phenotype. This result indicates that *GhNAC79* gene positively regulates drought stress and overexpression of *GhNAC79* gene results in an early flowering phenotype in *Arabidopsis* and it also improves drought tolerance in both *Arabidopsis* and cotton. Virus-induced gene silencing of *GhWRKY6*-like in cotton showed enhanced sensitivity compared to wild-type plants during salt and drought stresses (Ullah et al. 2018). Ullah et al. (2018) reported that *GhWRKY6*-like enhanced salt tolerance in *Arabidopsis* by scavenging ROS and regulating the ABA signaling pathway. This study suggested that overexpression of the *GhWRKY6*-like gene in cotton enhanced tolerance to salt, drought, and osmotic stresses. VIGS approach has also been used for identifying molecular factors involved in *Bacillus amyloliquefaciens* 5113 mediated abiotic stress tolerance in wheat (Abd El-Daim et al. 2018). Silencing of two abscisic acid-related TDFs revealed different effects upon heat and drought stress. Abd El-Daim et al. (2018) reported that treatment with *B. amyloliquefaciens* 5113 resulted in molecular modifications in wheat to induce heat, cold, and drought stress tolerance.

2.3 Advantages of Using VIGS to Study Abiotic Stress Tolerance Mechanisms in Crop Plants

VIGS has several advantages over other functional genomics tools (Burch-Smith et al. 2004; Purkayastha and Dasgupta 2009; Unver and Budak 2009; Stratmann and Hind 2011; Pflieger et al. 2013). VIGS technique is rapid and relatively easy to perform gene function studies in crop plants. VIGS can produce silencing phenotype of a specific gene in a short period resulting in rapid functional elucidation of genes (Dinesh-Kumar et al. 2003). Plant transformation is not required for VIGS. Functional elucidation of genes in difficult to transform species would be more easier once the VIGS system is established in that species (Burch-Smith et al. 2004). VIGS allows functional characterization of genes whose downregulation produces lethal phenotype. It can be used to study genes related to embryonic development and seedling emergence and vigor which are an important abiotic stress tolerance trait (Ratcliff et al. 2001; Burch-Smith et al. 2004; Liu et al. 2004). Functional redundancy can be overcome by VIGS. The multiple related genes or genes from gene families can be silenced together using conserved regions through VIGS (Ekengren et al. 2003; He et al. 2004). The complex signaling components associated with abiotic stresses such as drought, salinity, and heat stress can be deciphered by silencing two or more members of the gene family with redundant functions. Though other functional genomics tools like antisense RNAs, artificial miRNAs, or RNAi can also be used to uncover mechanisms underlying plants' response to various abiotic stresses, they are time consuming. VIGS enables rapid silencing of tissue-specific genes. For instance, plants being infected only at the time of flowering or panicle development will predominantly have genes silenced in that organ. Besides, VIGS can be used to quickly silence genes in a particular gene mutant, stable RNAi, or gene-overexpression plants. This will enable studying interactions of genes under abiotic stress condition in a shorter time. In addition VIGS is versatile tool for rapid characterization of gene function among species and works in different genetic backgrounds where genetic transformation is tedious and time consuming. VIGS as a high-throughput reverse genetics tool involves cloning of usually 300–500 base pair fragments from a large number of target genes into a suitable viral vector. Abiotic stress can be applied 2–3 weeks after inoculation, and silencing phenotype can be studied in the loss-of-function plants to attribute function for the target gene under abiotic stress.

2.4   Limitations of VIGS in Studying Abiotic Stress Tolerance Mechanisms and Possible Approaches to Overcome the Limitations

VIGS has been proved to be a wonderful technique for gene function studies. However, it also has some limitations. These limitations and ways to overcome the same are: (1) the virus may be accumulated to high levels in the silenced plant if the silenced target gene is associated with immunity of plants against the virus and such plants can become highly susceptible to abiotic stress. This will adversely influence studying the specific effect of gene silencing on abiotic stress tolerance. Quantification of virus titer in the silenced plants helps to decide whether the virus has accumulated higher than in the non-silenced control plant, and this information can be used to choose different region of the target gene for silencing (Senthil-Kumar and Mysore 2011b). (2) Virus-related symptoms itself can interfere with plants' response to abiotic stress. For example, infection of brome mosaic virus (BMV), cucumber mosaic virus (CMV), tobacco mosaic virus (TMV), and TRV delayed the appearance of drought stress-related phenotype in various plant species (Xu et al. 2008). The VIGS vector along with abiotic stress can create concurrent biotic and abiotic stress-related physiologies and phenotype. The phenotype produced under this situation might be different from abiotic stress alone (Suzuki et al. 2014). This can be overcome by including appropriate non-silenced empty vector control plants and comparing the results with specific gene silenced plants. (3) Silencing can be influenced by changes in environmental conditions during abiotic stress treatment. Temperature, relative humidity, and light can influence silencing (Fu et al. 2005; Kotakis et al. 2010). VIGS efficiency is reduced under high temperatures due to reduced virus replication (Chellappan et al. 2005). This can be overcome by maintaining the VIGS vector-inoculated plants under optimum environmental conditions until the loss of function followed by abiotic stress imposition.

2.5   Conclusion and Future Perspectives

VIGS-based functional characterization is an emerging option to initially analyze the large number of genes and to narrow the few promising genes that might have role in drought, salinity, heat, and cold stress tolerance. VIGS offers opportunities for rapid functional analysis of genes associated with abiotic stress tolerance in both dicotyledonous and monocotyledonous crop species. There is a possibility of extending VIGS as a tool to characterize genes involved in other abiotic stresses like low or high temperatures, salinity, mineral deficiency or toxicity, water-logging, and other general stresses. It can be effectively implemented as a reverse genetic tool if the genome sequence information is available, a gene identity is marked, and expression profiles are known. Currently, nearly 50 plant species have been shown to be amenable for VIGS (Lange et al. 2013), and VIGS is expected to be expanded to many other crop plants in the future. Stress imposition protocols for VIGS plants have been optimized for several abiotic stresses, including drought, salinity, and oxidative stress and extreme temperatures (Ramegowda et al. 2013).

Recently, both PTGS- and TGS-based VIGS were shown to be transmissible to progeny seedlings and persistent for a long duration. Therefore, the utility of VIGS has now become more versatile. However, studies related to heritable and long-duration VIGS have yet to be extended to other genes, apart from marker genes, in a wide range of plant species. The potential application of VIGS in crop improvement has yet to be realized. Taken into consideration of recent advances in VIGS technology, the prospects of using VIGS for various applications in modern plant biology are promising. PTGS achieved by VIGS vectors can be used for both genetic engineering and conventional or molecular breeding aimed at crop improvement for abiotic stress tolerance. The reduction or alteration of the flowering time of certain genotypes or of indeterminate cultivars, wild relatives, or inbred lines can be achieved by using viral vectors (Yamagishi and Yoshikawa 2011b; Yamagishi et al. 2011). Silencing of a negative regulator of flowering in a late-flowering genotype can help to match flowering time, enabling crossing with an early flowering genotype (Purwestri et al. 2009). This can also enable early and uniform flowering needed for crossing in indeterminate growth genotypes and reduce hurdles related to pollination time. A reduction in flowering time can speed up breeding programs by reducing generation time (Yamagishi and Yoshikawa 2011a).

References

Abd El-Daim IA, Bejai S, Fridborg I, Meijer J (2018) Identifying potential molecular factors involved in *Bacillus amyloliquefaciens* 5113 mediated abiotic stress tolerance in wheat. Plant Biol. https://doi.org/10.1111/plb.12680

Bao H, Chen X, Lv S, Jiang P, Feng J, Fan P, Nie L, Li Y (2015) Virus-induced gene silencing reveals control of reactive oxygen species accumulation and salt tolerance in tomato by γ-aminobutyric acid metabolic pathway. Plant Cell Environ 38:600–613

Baulcombe DC (1999) Fast forward genetics based on virus-induced gene silencing. Curr Opin Plant Biol 2:109–113

Becker A, Lange M (2010) VIGS genomics goes functional. Trends Plant Sci 15:1–4

Blair MW, Fernandez AC, Ishitani M, Moreta D, Seki M, Ayling S et al (2011) Construction and EST sequencing of full-length, drought stress cDNA libraries for common beans (*Phaseolus vulgaris* L.). BMC Plant Biol 11:171

Bohnert HJ, Gong Q, Li P, Ma S (2006) Unraveling abiotic stress tolerance mechanisms – getting genomics going. Curr Opin Plant Biol 9:180–188

Burch-Smith TM, Anderson JC, Martin GB, Dinesh-Kumar SP (2004) Applications and advantages of virus-induced gene silencing for gene function studies in plants. Plant J 39:734–746

Chellappan P, Vanitharani R, Ogbe F, Fauquet CM (2005) Effect of temperature on geminivirus-induced RNA silencing in plants. Plant Physiol 138:1828–1841

Chen RG, Jing H, Guo WL, Wang SB, Ma F, Pan BG, Gong ZH (2015) Silencing of *dehydrin CaDHN1* diminishes tolerance to multiple abiotic stresses in *Capsicum annuum* L. Plant Cell Rep 34:2189–2200

Cho SM, Kang BR, Han SH, Anderson AJ, Park JY, Lee YH et al (2008) 2R, 3R-butanediol, a bacterial volatile produced by *Pseudomonas chlororaphis* O6, is involved in induction of systemic tolerance to drought in *Arabidopsis thaliana*. Mol Plant-Microbe Interact 21:1067–1075

Choi HW, Hwang BK (2012) The pepper extracellular peroxidase CaPO₂ is required for salt, drought and oxidative stress tolerance as well as resistance to fungal pathogens. Planta 235:1369–1382

Dai F, Zhang C, Jiang X, Kang M, Yin X, Lu P et al (2012) *RhNAC2* and *RhEXPA4* are involved in the regulation of dehydration tolerance during the expansion of rose petals. Plant Physiol 160:2064–2082

Dinesh-Kumar SP, Anandalakshmi R, Marathe R, Schiff M, Liu Y (2003) Virus-induced gene silencing. In: Grotewold E (ed) Plant functional genomics. Humana Press, New York, pp 287–293

Ekengren SK, Liu Y, Schiff M, Dinesh-Kumar SP, Martin GB (2003) Two MAPK cascades, NPR1, and TGA transcription factors play a role in Pto-mediated disease resistance in tomato. Plant J 36:905–917

Fu DQ, Zhu BZ, Zhu HL, Jiang WB, Luo YB (2005) Virus-induced gene silencing in tomato fruit. Plant J 43:299–308

Gorantla M, Babu PR, Lachagari VB, Reddy AM, Wusirika R, Bennetzen JL et al (2007) Identification of stress-responsive genes in an indica rice (*Oryza sativa* L.) using ESTs generated from drought stressed seedlings. J Exp Bot 58:253–265

Govind G, Harshavardhan VT, Thammegowda HV, Patricia JK, Kalaiarasi PJ, Dhanalakshmi R et al (2009a) Identification and functional validation of a unique set of drought induced genes deferentially expressed in response to gradual water stress in peanut. Mol Gen Genomics 281:591–605

Govind G, Harshavardhan V, Patricia J, Dhanalakshmi R, Senthil-Kumar M, Sreenivasulu N, Udayakumar M (2009b) Identification and functional validation of a unique set of drought induced genes preferentially expressed in response to gradual water stress in peanut. Mol Gen Genomics 281:607–607

Guo Y, Huang C, Xie Y, Song F, Zhou X (2010) A tomato *Glutaredoxin* gene *SlGRX1* regulates plant responses to oxidative, drought and salts tresses. Planta 232:1499–1509

Guo Y, Pang C, Jia X, Ma Q, Dou L, Zhao F, Gu L, Wei H, Wang H, Fan S, Su J, Yu S (2018) An NAM Domain Gene, GhNAC79, Improves resistance to drought stress in Upland cotton. Front Plant Sci. https://doi.org/10.3389/fpls.2017.

Hasegawa PM, Bressan RA, Zhu JK, Bohnert HJ (2000) Plant cellular and molecular responses to high salinity. Annu Rev Plant Physiol Plant Mol Biol 51:463–499

He X, Anderson JC, Pozo OD, Gu YQ, Tang X, Martin GB (2004) Silencing of subfamily of protein phosphatase 2A catalytic subunits results in activation of plant defense responses and localized cell death. Plant J 38:563–577. x

He X, Jin C, Li G, You G, Zhou X, Zheng SJ (2008) Use of the modified viral satellite DNA vector to silence mineral nutrition-related genes in plants: silencing of the tomato ferric chelate reductase gene, FRO1, as an example. Sci China C Life Sci 51:402–409

Ingram J, Bartels D (1996) The molecular basis of dehydration tolerance in plants. Annu Rev Plant Physiol Plant Mol Biol 47:377–403

Jiang X, Zhang C, Lü P, Jiang G, Liu X, Dai F, Gao J (2014) RhNAC3, a stress-associated NAC transcription factor, has a role in dehydration tolerance through regulating osmotic stress-related genes in rose petals. Plant Biotechnology Journal 12(1):38–48.

Kang G, Li G, Ma H, Wang C, Guo T (2013) Proteomic analysis on the leaves of TaBTF3 gene virus-induced silenced wheat plants may reveal its regulatory mechanism. J Proteome 83:130–143

Kotakis C, Vrettos N, Kotsis D, Tsagris M, Kotzabasis K, Kalantidis K (2010) Light intensity affects RNA silencing of a transgene in Nicotiana benthamiana plants. BMC Plant Biol 10:220. 0

Kuzuoglu-Ozturk D, CebeciYalcinkaya O, Akpinar BA, Mitou G, Korkmaz G, Gozuacik D et al (2012) Autophagy-related gene, TdAtg8, in wild emmer wheat plays a role in drought and osmotic stress response. Planta 236:1081–1092

Lange M, Yellina A, Orashakova S, Becker A (2013) Virus-induced gene silencing (VIGS) in plants: an overview of target species and the virus-derived vector systems. In: Becker A (ed) Virus-induced gene silencing. Humana Press, New York, pp 1–14

Lee SC, Choi DS, Hwang IS, Hwang BK (2010) The pepper oxidoreductase CaOXR1 interacts with the transcription factor CaRAV1 and is required for salt and osmotic stress tolerance. Plant Mol Biol 73:409–424

Liang J, Deng G, Long H, Pan Z, Wang C, Cai P, Xu D, Nima Z-X, Yu M (2012) Virus-induced silencing of genes encoding LEA protein in Tibetan hulless barley (Hordeum vulgare ssp. vulgare) and their relationship to drought tolerance. Molecular Breeding 30(1):441–451.

Li C, Yan JM, Li YZ, Zhang ZC, Wang QL, Liang Y (2013) Silencing the SpMPK1, SpMPK2, and SpMPK3 genes in tomato reduces abscisic acid-mediated drought tolerance. Int J Mol Sci 14:21983–21996

Li K, Xing C, Yao Z, Huang X (2017) PbrMYB21, a novel MYB protein of Pyrus betulaefolia, functions in drought tolerance and modulates polyamine levels by regulating arginine decarboxylase gene. Plant Biotechnol J 15:1186–1203

Liu Y, Nakayama N, Schiff M, Litt A, Irish V, Dinesh-Kumar SP (2004) Virus induced gene silencing of a DEFICIENS ortholog in Nicotiana benthamiana. Plant Mol Biol 54:701–711

Liu D, Liu X, Meng Y, Sun C, Tang H, Jiang Y et al (2013) An organ-specific role for ethylene in rose petal expansion during dehydration and rehydration. J Exp Bot 64:2333–2344

Manmathan H, Shaner D, Snelling J, Tisserat N, Lapitan N (2013) Virus-induced gene silencing of Arabidopsis thaliana gene homologues in wheat identifies genes conferring improved drought tolerance. J Exp Bot 64:1381–1392

Marques MC, Alonso-Cantabrana H, Forment J, Arribas R, Alamar S, Conejero V, Perez-Amador M (2009) A new set of ESTs and cDNA clones from full-length and normalized libraries for gene discovery and functional characterization in citrus. BMC Genomics 10:428

Park C, Lim CW, Baek W, Kim JH, Lim S, Kim SH, Kim KN, Lee SC (2017) The pepper WPP domain protein, CaWDP1, acts as a novel negative regulator of drought stress via aba signaling. Plant Cell Physiol 58:779–788

Pflieger S, Richard MMS, Blanchet S, Meziadi C, Geffroy V (2013) VIGS technology: an attractive tool for functional genomics studies in legumes. Funct Plant Biol 40:1234–1248

Purkayastha A, Dasgupta I (2009) Virus-induced gene silencing: a versatile tool for discovery of gene functions in plants. Plant Physiol Biochem 47:967–976

Purwestri YA et al (2009) The 14-3-3 protein GF14c acts as a negative regulator of flowering in rice by interacting with the florigen Hd3a. Plant Cell Physiol 50:429–438

Ramegowda V, Senthil-Kumar M, Udayakumar M, Mysore KS (2013) A high-throughput virus-induced gene silencing protocol identifies genes involved in multi stress tolerance. BMC Plant Biol 13:193. 3

Ramegowda V, Gill US, Sivalingam PN, Gupta A, Gupta C, Govind G, Nataraja KN, Pereira A, Udayakumar M, Mysore KS, Senthil-Kumar M (2017) GBF3 transcription factor imparts drought tolerance in *Arabidopsis thaliana*. Sci Rep 7:9148. https://doi.org/10.1038/s41598-017-09542-1.

Rao SS, El-Habbak MH, Havens WM, Singh AK, Zheng D, Vaughn L, Haudenshield JS, Hartman GL, Korban SS, Ghabrial SA (2014) Overexpression of *GmCaM4* in soybean enhances resistance to pathogens and tolerance to salt stress. Mol Plant Pathol 15:145–160

Ratcliff F, Martin-Hernandez AM, Baulcombe DC (2001) Technical advance. Tobacco rattle virus as a vector for analysis of gene function by silencing. Plant J 25:237–245

Seki M, Narusaka M, Ishida J, Nanjo T, Fujita M, Oono Y, Kamiya A, Nakajima M, Enju A, Sakurai T et al (2002) Monitoring the expression profiles of 7000 Arabidopsis genes under drought, cold and high-salinity stresses using afull-length cDNA microarray. Plant J 31:279–292

Senthil-Kumar M, Mysore KS (2011a) New dimensions for VIGS in plant functional genomics. Trends Plant Sci 16:656–665

Senthil-Kumar M, Mysore KS (2011b) Virus-induced gene silencing can persist for more than 2 years and also be transmitted to progeny seedlings in *Nicotiana benthamiana* and tomato. Plant Biotechnol J 9:797–806

Senthil-Kumar M, Udayakumar M (2006) High-throughput virus induced gene silencing approach to assess the functional relevance of a moisture stress-induced cDNA homologous to Lea4. J Exp Bot 57:2291–2302

Senthil-Kumar M, Govind G, Kang L, Mysore KS, Udayakumar M (2007) Functional characterization of *Nicotiana benthamiana* homologs of peanut water deficit-induced genes by virus induced gene silencing. Planta 225:523–539

Senthil-Kumar M, Ramegowda HV, Hema R, Mysore KS, Udayakumar M (2008) Virus-induced gene silencing and its application in characterizing genes involved in water-deficit-stress tolerance. J Plant Physiol 165:1404–1421

Singh AK, Fu DQ, Mohamed A, Navarre D, Kachroo A, Ghabrial S (2011) Silencing genes encoding omega-3 fatty acid desaturase alters seed size and accumulation of bean pod mottle virus in soybean. Mol Plant Microbe Interact 24:506–515

Soares-Cavalcanti NM, Belarmino LC, Kido EA, Wanderley-Nogueira AC, Bezerra-Neto JP, Cavalcanti-Lira R et al (2012) In silico identification of known osmotic stress responsive genes from Arabidopsis in soybean and Medicago. Genet Mol Biol 35:315–321

Stratmann JW, Hind SR (2011) Gene silencing goes viral and uncovers the private life of plants. Entomol Exp Appl 140:91–102

Suzuki N, Rivero RM, Shulaev V, Blumwald E, Mittler R (2014) Abiotic and biotic stress combinations. New Phytol 203:32–43

Tran LS, Mochida K (2010) Identification and prediction of abiotic stress responsive transcription factors involved in abiotic stress signaling in soybean. Plant Signal Behav 5:255–257

Ullah A, Sun H, Hakim, Yang X, Zhang X (2018) A novel cotton *WRKY* gene, *GhWRKY6*-like, improves salt tolerance by activating the ABA signaling pathway and scavenging of reactive oxygen species. Physiol Plant. https://doi.org/10.1111/ppl.12651

Unver T, Budak H (2009) Virus-induced gene silencing, a posttranscriptional gene silencing method. Int J Plant Genomics 2009:198680. https://doi.org/10.1155/2009/198680

Virk N, Liu B, Zhang H, Li X, Zhang Y, Li D et al (2013) Tomato SlMPK4 is required for resistance against *Botrytis cinerea* and tolerance to drought stress. Acta Physiol Plant 35:1211–1221

Wang C, Lu W, He X, Wang F, Zhou Y, Guo X, Guo X (2016) The cotton mitogen-activated protein kinase kinase 3 functions in drought tolerance by regulating stomatal responses and root growth. Plant Cell Physiol 57:1629–1642

Wani SH, Singh NB, Saini HK, Devi LP, Monalisa P (2010) Expressed sequenced tags (ESTs)-a functional genomic approach for gene discovery. Int J Curr Res 5:74–79

Xu P, Chen F, Mannas JP, Feldman T, Sumner LW, Roossinck MJ (2008) Virus infection improves drought tolerance. New Phytol 180:911–921

Yamagishi N, Yoshikawa N (2011a) Expression of FLOWERING LOCUS T from Arabidopsis thaliana induces precocious flowering in soybean irrespective of maturity group and stem growth habit. Planta (in press). https://doi.org/10.1007/ s00425–010- 1318–3.

Yamagishi N, Yoshikawa N (2011b) Expression of FLOWERING LOCUS from *Arabidopsis thaliana* induces precocious flowering in soybean irrespective of maturity group and stem growth habit. Planta 233:561–568

Yamagishi N et al (2011) Promotion of flowering and reduction of a generation time in apple seedlings by ectopical expression of *Arabidopsis thaliana* gene using the Apple latent spherical virus vector. Plant Mol Biol 75:193–204

Yan Y, Jia H, Wang F, Wang C, Liu S, Guo X (2015) Overexpression of GhWRKY27a reduces tolerance to drought stress and resistance to *Rhizoctonia solani* infection in transgenic *Nicotiana benthamiana*. Front Plant Sci 6:265. https://doi.org/10.3389/fphys.00265

Chapter 3
RNA Interference: A Promising Approach for Crop Improvement

B. Mamta and Manchikatla V. Rajam

Abstract RNA interference (RNAi) is a naturally occurring biological process that regulates plant growth and development, defense against pathogens, and environmental stresses. It is a sequence-specific homology-based silencing mechanism in which the function of a gene is interfered or suppressed. Small interfering RNAs (siRNAs) and microRNAs (miRNAs) are produced inside the plant cell through the activation of RNAi machinery, which downregulates the expression of the target genes at transcriptional and translational levels. RNAi is more specific, precise in its action, and considered as a potential technology for functional genomics studies. In the last 15 years, it has emerged as a scientific breakthrough for crop improvement without affecting other agronomic traits. It has also been employed as a novel method in understanding the basic phenomenon of plant defense and metabolism. Several desirable traits have been improved in the crop varieties through RNAi, which include crop protection against biotic and abiotic stresses, enhancement of nutritional value, alteration in plant architecture for better adaptation to environmental conditions, overexpression or removal of secondary metabolites, enhancement of shelf life of fruits and vegetables, generation of male sterile lines, and development of seedless fruits. In this book chapter, we have discussed RNAi and its applications in crop improvement.

Keywords RNA interference · siRNAs · miRNAs · Gene silencing · Crop improvement · Transgenic plants · Stress tolerance · Plant architecture

B. Mamta · M. V. Rajam (✉)
Department of Genetics, University of Delhi South Campus, New Delhi, India
e-mail: rajam.mv@gmail.com

© Springer International Publishing AG, part of Springer Nature 2018
S. S. Gosal, S. H. Wani (eds.), *Biotechnologies of Crop Improvement, Volume 2*,
https://doi.org/10.1007/978-3-319-90650-8_3

3.1 Introduction

RNA interference is a conserved, naturally occurring gene regulatory mechanism. It is evolved to protect the organisms against the invading foreign nucleic acids. Besides, it is also required for maintaining genomic stability, regulation of transposon movement, epigenetic modification, and control of cellular processes at transcriptional and translational level (Ketting 2011; Castel and Martienssen 2013). Fire et al. (1998) coined the term RNAi for the unknown silencing mechanism observed upon exogenous supply of dsRNAs of sense and antisense transcripts in *Caenorhabditis elegans*. They have also detected that dsRNAs of target gene induced silencing even in low concentration and mentioned about the existence of amplification process in *C. elegans*. Earlier to RNAi, similar types of silencing phenomenon were also reported by scientists working on plant and fungal systems (Napoli et al. 1990; Romano and Macino 1992). They found that the introduction of transgene caused downregulation of transgene as well as the endogenous gene. The phenomenon was called as "co-suppression" in plants and "quelling" in fungi. Later, it was demonstrated that protein complexes involved in RNAi and related phenomena were conserved across the kingdoms (Baulcombe 2000; Matzke et al. 2001). So far, RNAi and related mechanisms have been described in prokaryotes such as bacteria (clustered regularly interspaced short palindromic repeats) (Wilson and Doudna 2013) and eukaryotes – algae (Cerutti et al. 2011), fungi (Romano and Macino 1992), moss (Bezanilla et al. 2003), plants (Napoli et al. 1990), nematodes (Fire et al. 1998), *Drosophila* (Hammond et al. 2000), and mammals (Elbashir et al. 2001).

RNAi involves homology-based sequence-specific degradation of target gene transcripts (Wilson and Doudna 2013). It is triggered by aberrant dsRNAs which can vary in length and origin. These aberrant or foreign dsRNAs are processed into small RNA duplexes of variable sizes ranging from 21 to 28 nucleotides (nt). Small RNA molecules are loaded on protein complex and then directed toward their cognate RNA where they cause cleavage of target gene or suppression of translation. They are also capable of inducing modification at DNA level through methylation or deacetylation (Molnar et al. 2010). RNAi and its executive molecules are, thus, responsible for gene regulation at transcriptional, posttranscriptional, and translational levels (Xie et al. 2003; Brodersen et al. 2008; Molnar et al. 2010; Khraiwesh et al. 2010). The discovery of RNAi gave a new tool in the hand of scientists to manipulate the plants through genetic engineering and to study the functional genomics. Nowadays, RNAi is extensively used for crop improvement through the alteration of desirable traits in plants. Steps involved in the development of effective RNAi-based strategies for crop improvement include the identification of a suitable target, preparation of an efficient RNAi construct, transformation of plants with the RNAi construct, and evaluation of RNAi lines for desirable characteristics (Saurabh et al. 2014). Different bioinformatic tools are used for initial screening of target genes so that the sequence used for preparation of RNAi construct do not bear any off-target effects on plant's development or nontargeted organisms (Saurabh et al. 2014). This chapter summarizes the various applications of RNAi for crop improvement.

3.2   RNA Interference (RNAi): siRNAs and miRNAs

The small noncoding RNAs (ncRNAs) mediate gene silencing at transcriptional and posttranscriptional levels. Both transcriptional gene silencing (TGS) and posttranscriptional gene silencing (PTGS) pathways start with dsRNAs but process through different machineries and mechanisms. PTGS is generally employed for host-induced gene silencing (HIGS), i.e., host plants engineered to produce siRNAs/miRNAs against the target gene. siRNAs and miRNA are the effector molecules for PTGS (Bartel 2004; Zamore and Haley 2005; Vazquez 2006). Both siRNAs and miRNAs are 20–24 nt long and are generated through the processing of long dsRNA. They vary in origin, initial precursor structure, biogenesis pathway, and mode of action (Axtell, 2013). Formation of siRNAs is triggered with the appearance of aberrant dsRNAs from endogenous or exogenous sources (Fire et al. 1998; Tuschl 2001). Plant cell recognizes these aberrant dsRNAs as foreign particles and cleaves them into 21–25 nt small siRNA duplexes with the help of dicer (DCL), an RNAse III endonuclease family member (Hamilton and Baulcombe 1999; Hammond et al. 2000; Bernstein et al. 2001). The generated siRNA duplexes have phosphate group at 5′ end and two nucleotide overhangs at 3′ end (Bernstein et al. 2001; Elbashir et al. 2001). The siRNA duplexes are then loaded onto RNA-induced silencing complex (siRISC) through the recognition at 3′ overhangs. Degradation of passenger/sense strand of siRNAs (strand which has same sequence as the target mRNA) activates the RISC complex and directs the remaining antisense siRNAs toward the cognate mRNA. Argonate (AGO) protein, the main component of RISC complex, then brings out the cleavage of target mRNA based on the sequence-specific homology between antisense siRNA and target mRNA. Plant miRNA biogenesis starts with the endogenous primary miRNA (pri-miRNA) precursor, which has partial double-stranded stem-loop structure and is transcribed by RNA polymerase II in the nucleus (Jones-Rhoades et al. 2006; Zhu 2008). The pri-miRNA is further processed into 70–110-nt-long precursor miRNA (pre-miRNA) by RNase III enzyme DCL1 (dicer-like 1) and other proteins (HYL1, SE, HEN1). The pre-miRNA is cleaved by DCL1 into 22–24-nt-long miRNA duplex, which is then moved to cytoplasm with the help of HASTY protein. The mature miRNA/miRNA* duplex is then recruited into RISC complex, and degradation of sense miRNA by SDN protein takes place leading to the activation of RISC complex. Mature miRNAs bind to the target mRNAs mostly at 3′ UTR (untranslated region) and mediate their cleavage or translational blockage (Bao et al. 2004; Khraiwesh et al. 2010). MicroRNAs are expressed during plant growth and development, synthesis of secondary metabolites, abiotic and biotic stress reactions, etc. Alteration in their expression and biosynthesis could be beneficial for the development of plants with valuable characteristics (Pareek et al. 2015) (Figs. 3.1 and 3.2).

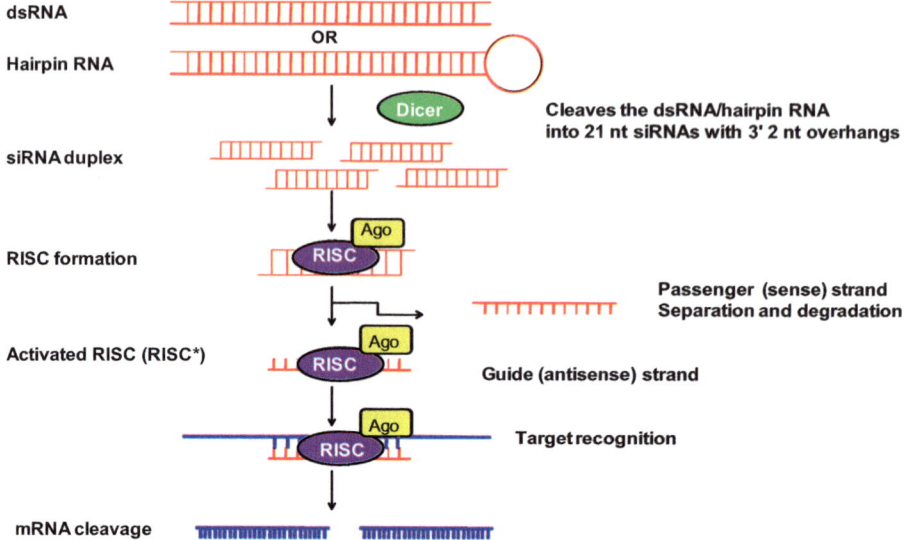

Fig. 3.1 Biogenesis and silencing mechanism of siRNAs (Reproduced from Yogindran and Rajam 2015)

3.3 RNAi for Crop Improvement

Plants are the major source of all kind of food to human being and livestock. Environmental changes, scarcity of land, and depletion of natural resources limit the crop productivity, thereby causing instability in food security and malnutrition across the world. The existing breeding and improvement programs are associated with various physiology, ecological, and biological drawbacks. In recent time, genetic engineering through RNAi has proven its potential for improving crop varieties for different useful agronomic traits.

3.3.1 Biotic Stress Resistance

The pathogens (viruses, bacteria, and fungi), insect pests, and nematode parasites are the biotic factors, which hinder the growth and development of crop plants and affect their quality and yield. Worldwide, biotic factors account for about 40% loss in six major food and cash crops (Oerke 2006). Geometrical elevation in world's population demands for novel techniques for effective management of biotic factors. RNAi-mediated crop protection against biotic factors opened up a new era in this direction.

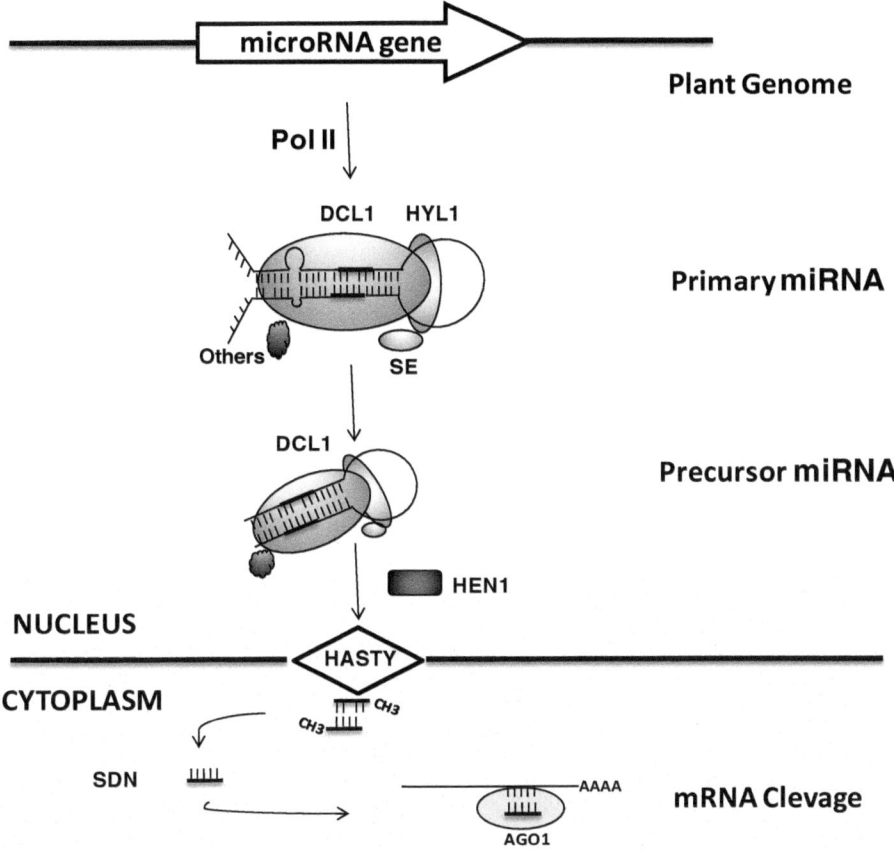

Fig. 3.2 Biogenesis and silencing mechanism of miRNAs in plants (Reproduced from Pareek et al. 2015)

Insect pests mostly damage the plants during reproductive stages. Insecticides offer a quick control for eradication of insect pests, but excessive use of insecticides and its persistence in environment and food crop makes it unsuitable for long use. The effectiveness of host-induced RNAi for control of insect pests was first demonstrated by Mao et al. (2007) and Baum et al. (2007). Mao et al. (2007) showed that sensitivity of insect pest toward phytotoxin (gossypol) can be increased by inhibiting the expression of insect P450 monoxygenase gene involved in the detoxification of gossypol through HI-RNAi in cotton. Transgenic maize plants were developed to produce siRNAs against the vital gene of an insect pest, Coleopteran western corn rootworm, V-ATPase for insect resistance. Recently, it has been shown that silencing of chitinase gene in *Helicoverpa armigera* through HI-RNAi caused downregulation of the target gene transcripts and induced mortality and developmental deformities at larval, pupal, and adult stages (Mamta et al. 2015). Previously, HI-RNAi was used to control different insect pests such as *H. armigera* (Zhu et al. 2012; Mao et al. 2013; Liu et al. 2015; Jin et al. 2015), *Manduca sexta* (Kumar et al. 2012), *Nilaparvata lugens* (Zha et al. 2011; Yu et al. 2014), and *Bemisia tabaci* (Thakur et al. 2014) through silencing of vital genes of the target pests.

Host genes involved in growth, development, and parasitism are found to be effective targets for control of nematodes through HI-RNAi (Huang et al. 2006; Yadav et al. 2006; Sindhu et al. 2009; Papolu et al. 2013; Tamilarasan and Rajam 2013; Xue et al. 2013; Banerjee et al. 2017). For example, the expression of *HgALD* dsRNA in soybean gave resistance against nematodes (Youssef et al. 2013). *HgALD* gene encodes for fructose-1,6-diphosphate aldolase enzyme required during gluconeogenesis. Silencing of *FMRF* amide-like peptide gene flp-14 and flp-18 through HI-RNAi led to the inhibition of invasion and reproduction pathways and thereby decreases the parasitic responses of root-knot nematode, *Meloidogyne incognita* (Papolu et al. 2013).

Fungal pathogens not only cause huge crop losses but also produce harmful mycotoxins in crop plants. Ingestion of mycotoxins even in low quantity leads to serious health problems in living beings. Control of fungal infections has been achieved through HI-RNAi in different crop plants (Nowara et al. 2010; Yin et al. 2011; Nunes and dean 2012; Koch et al. 2016). Tinoco et al. (2010) demonstrated HI-RNAi as a proof of concept in β-gluconidase (*GUS*) expressing necrotrophic fungi, *Fusarium verticillioides*, through silencing of *GUS* transgene expression by uptake of *GUS*-targeted siRNAs generated in transgenic tobacco. A recent study by Chen et al. (2016) showed the development of resistance in barley against *Fusarium culmorum* through downregulation of β-1, 3-glucan synthase (*FcGls1*) gene expression. Inhibition of *FcGls1* gene induced defects during hyphae growth and development and fungal invasion (Chen et al. 2016). In virus, mostly coat protein (CP) genes are reported to be the potential target for Hi-RNAi (Andika et al. 2005; Kamachi et al. 2007; Kumar et al. 2012; Zhou et al. 2012). For example, production of siRNAs against *CP* gene of cucumber green mottle mosaic virus (CGMMV) provided resistance against this virus in tobacco (Kamachi et al. 2007). Recently, resistance in tomato (*Solanum lycopersicum*) against yellow leaf curl virus was achieved through simultaneous silencing of different viral disease-responsive genes by expressing chimeric dsRNA (Chen et al. 2016). Viral resistance can also be achieved by targeting RNAi suppressor proteins. For example, downregulation of viral suppressor proteins (AC2 and AC4) by trans-acting siRNAs generated in tobacco provides high resistance against tomato leaf curl New Delhi virus (ToCNDV) (Singh et al. 2014).

Under biotic stress conditions, expression of various miRNAs gets enhanced or repressed (Khraiwesh et al. 2012; Li et al. 2012; Kumar 2014; Singh et al. 2014). Li et al. (2012) showed that 20 miRNAs were differentially expressed in susceptible soybean variety as compared to the resistant variety against soybean cyst nematodes (SCN – *Heterodera glycines*). Overexpression of osa-mi7696 provided immunity to rice (*Oryza sativa L.* ssp. *japonica*) against blast fungi *Magnaporthe oryzae* (Campo et al. 2013). This miRNA negatively regulates the expression of natural resistance-associated macrophage protein-6 (OsNramp6) in rice. Thus, RNAi is emerging as a potential alternate approach for gain of resistance under biotic stress conditions.

3.3.2 Abiotic Stress Tolerance

Under natural field conditions, plants are often not able to attain their full growth and development due to the continuous exposure to different abiotic stresses. Drought, salinity, and variation in temperature are the major abiotic conditions, which cause huge crop losses around the world. The changing environment and increasing food demands for growing population exerts great pressure on scientists for development of stress-tolerant crop varieties. Under stress conditions, plants synthesize different noncoding RNAs (ncRNAs) for gene regulation at transcriptional, posttranscriptional, and chromatin level. The ncRNAs and their targets can be utilized for generation of abiotic stress-tolerant variety through RNAi. Downregulation of receptor for activated C-kinase 1(*RACK1*) through RNAi resulted in development of drought tolerance in rice (Da-Hong et al. 2009). RACK1 is a conserved scaffold protein that regulates expression of antioxidant-related enzymes such as superoxide dismutases (SODs) in plants. Inhibition of *RACK*1 increases the accumulation of SODs and provides tolerance against drought as well as reactive oxygen species (ROS). Likewise, suppression of farnesyltransferase/squalene synthase (*SQS*) through siRNA generated from maize squalene synthase enhanced drought tolerance at vegetative and reproductive stages in rice (Manavalan et al. 2012). Increase in endogenous sterol level through silencing of *SQS* decreases the stomata density and prevents water loss through transpiration, thus preventing the plant from wilting under drought condition. *OsTZF1* is a CCCH-type zinc finger protein that gets expressed under drought, salinity, and ROS conditions. Silencing of *OsTZF1* gene enhances the tolerance of rice plants to high salt and low water conditions, indicating its role in abiotic stress tolerance (Jan et al. 2013). Low expression of *OsTZF1* gene maintains the internal homeostasis of plants through change in hormonal expression at cellular and molecular level under high salt condition. Similarly, suppression of the proline-rich proteins in *Poncirus trifoliata* through RNAi decreases the tolerance in plants against cold conditions. It was observed that PtrPRP protein gets accumulated at high level in cold condition and its inhibition disrupts the reactive oxidative species homeostasis and membrane permeability in plants (Peng et al. 2015).

Response to abiotic stress is regulated by different miRNAs in economically important crops such as rice, wheat, legumes, sugarcane, radish, etc. (Goswami et al. 2014; Kruszka et al. 2014; Naya et al. 2014; Gentile et al. 2015; Zhang 2015; Sun et al. 2015; Shriram et al. 2016). miRNAs mostly regulate transcription, detoxification, and development processes. Several miRNAs get upregulated under heat stress condition. For example, high expression of mir398 under heat stress condition suppresses copper/zinc super oxide dismutase (CSD) genes. It was observed that the overexpression of miRNA398 decreases the sustainability of *Arabidopsis thaliana* and common bean (*Phaseolus vulgaris*) plants under heat stress due to miRNA-mediated degradation of CSD mRNAs (Guan et al. 2013; Naya et al. 2014). Cold-tolerant rice plants can also be generated, without any developmental defect through downregulation of TF by Osa-miRNA319 (Yang et al. 2013). Similarly, a drought-responsive mir168 acts on nuclear factor Y (*NF-YA5*). Overexpression of soybean *NF-YA5* in *A. thaliana* increases its tolerance against drought condition (Ni et al. 2013). Gao et al. (2011) showed that high expression of Osa-miRNA393 decreases susceptibility of rice toward intolerable salt concentration present in soil. Recently, 22 novel miRNAs were identified in radish (*Raphanus sativus*) under high salt conditions, which regulate salt-responsive genes such as auxin response factors (ARFs), squamosa promoter-binding-like proteins (SPLs), and nuclear transcription factor Y (NF-Y) (Sun et al. 2015). *NAC* gene encodes for transcription factor required during plant development and environmental stresses. Thus, miRNAs maintain resistance to different abiotic stresses through up- and downregulation of the target gene transcripts.

3.3.3   Increasing Nutritional Value

Plants are the major source of various biologically active compounds required for overall growth of human. More than 2 billion people are found deficient in one or the other major nutrients and showed "hidden hunger" malnutrition (FAO 2013). Nowadays, various molecular biology and biotechnology techniques are being exploited in order to achieve the required level of nutrition in major staple foods. RNAi offers the new avenue for biofortification of nutrients in crop plants through modification of various physiological and biochemical pathways. Essential fatty acids found in oil are important for smooth functioning of the heart in human. Fatty acid composition in seeds can be easily modified through RNAi technology. The stability and oil quality of soybean oil were improved by downregulation of the expression of alpha-linolenic acid (18:3). Hairpin RNA-mediated tissue-specific suppression of omega-3 fatty acid desaturase enzyme significantly reduced the level of alpha-linolenic acid in transgenic soybean from 1 to 3% as compared to its level in untransformed soybean. Omega-3 fatty acid desaturase enzyme is responsible for the conversion of linoleic acid (18:2) to alpha-linolenic acid (18:3) in seeds (Flores et al. 2008). Opaque 2 gene (O2) encodes basic leucine zipper transcriptional factor, which regulates a storage protein. RNAi-mediated suppression of Opaque 2 gene resulted in increased lysine content in maize seeds without affecting the general function of O_2 (Angaji et al. 2010). Wheat expressing high level of amylase can be generated by downregulating the starch-branching enzymes (SBE) through RNAi (Regina et al. 2006). Low glutelin-containing rice is highly recommended for kidney patients due to its easy digestibility property. Kusaba et al. (2003) produced low in glutelin content rice variety, LGC-1 (low gluten content 1), through inhibition of *GluB* gene expression. When potatoes were grown at low temperature, starch was converted into sucrose and fructose, making the potato sweet, a phenomenon called "cold sweetening." RNAi-mediated downregulation of sucrose phosphatase gene (*SPP*) inhibited the cold sweetening phenomena in potato without significantly affecting its other agronomic parameters (Chen et al. 2008). Starch degradation is mediated by phosphorylation and dephosphorylation processes. Inhibition of starch degradation through alteration in the phosphate metabolic genes using RNAi enhanced the starch content in *A. thaliana* and *Zea mays* (Weise et al. 2012).

Tomatoes are known for their minerals, fibers, vitamins, and antioxidant property (Rajam et al. 2007). Overexpression of carotenoid or flavonoid synthetic genes or transcription factors increases either carotenoid or flavonoid content. RNAi has been employed in order to improve the level of both carotenoids and flavonoids in tomato fruit. *DET1* is a photomorphogenesis regulatory gene, which represses several light-mediated signaling pathways. Expression of dsRNA of *DET1* under fruit-specific promoter in tomato suppressed endogenous expression of *DET1* and resulted in high levels of flavonoids and carotenoids in tomato fruits (Davuluri et al. 2005). Similarly, downregulation of lycopene epsilon-cyclase (*ε-CYC*) gene expression through RNAi increased the carotenoid content in rapeseed (*Brassica napus*). High expression of β-carotene, lutein, zeaxanthin, and violaxanthin was observed in seeds obtained from these RNAi *Brassica* lines (Yu et al. 2007). Hence, RNAi has tremendous potential to eradicate the malnutrition across the world.

3.3.4 Increase in Shelf Life of Fruits

Fruits and vegetables are rich in various minerals and vitamins. They are harvested, stored, and transported for human consumption. The post-harvest crop losses include losses due to mishandling, spoilage, diseases, and pest infestation during storage and transportation. Delayed in ripening is one of the process through which post-harvest losses can be minimized. Climacteric fruits respond to ripening process according to the concentration of ethylene, where ethylene acts as ripening hormone which initiates, regulates, and coordinates the expression of various ripening-related genes. Blocking of ethylene biosynthesis, ethylene-mediated signaling, and ethylene response elements through RNAi has been shown to delay the ripening process and help in enhancement of shelf life of fruits and vegetables, a trait which is demanded in post-harvesting or transportation industry (Xiong et al. 2005; Gupta et al. 2013; Luo et al. 2013). 1-Aminocyclopropane-1-carboxylate (ACC) oxidase is an enzyme involved in synthesis of ethylene from its precursor ACC. Suppression of ACC oxidase through RNAi decreased the production of ethylene and delayed the ripening process in tomato (Xiong et al. 2005). Synthesis of ethylene precursor ACC is catalyzed by ACC synthase, a critical enzyme in ethylene biosynthetic pathway. Gupta et al. (2013) showed that simultaneous silencing of the three homologs of ACC synthase regulates the ethylene biosynthesis more efficiently. They expressed the chimeric dsRNA resulted from an off-target-free sequences of three tomato ACS homologs under the control of fruit-specific promoter 2A11 and observed a delay in ripening and increase in shelf life for about 45 days in transgenic tomato due to low production of ethylene. They also observed that the expression of ethylene-responsive genes gets affected by low expression of ethylene in RNAi plants. Ripening process also leads to accumulation of carotenoids in fruits. The key carotenoid biosynthesis gene (*SlPSY1*) is inhibited by STAY-GREEN (*SlSGR1*) protein. SlSGR1 also coordinates with ripening process through regulation of ethylene signaling and expression of ethylene-responsive genes. RNAi-mediated downregulation of *SlSGR1* enhances the shelf life of tomato up to 25–48 days. The low expression of SlSGR1 protein suppressed the production of H_2O_2 and alters the ethylene-mediated signal transduction in tomato, thus enhances ripening process through ethylene and carotenoid production (Luo et al. 2013).

Fruit development and ripening is a complex process, regulated by a variety of microRNAs (Moxon et al. 2008; Molesini et al. 2012; Karlova et al. 2013). The targets of miRNAs were found to be wide range of transcription factors (TFs), which act as negative or positive regulators for fruit ripening process. For example, tomato mir172 negatively regulates ripening process through downregulation of *APETALA2* (*SlAP2a*) gene (Chung et al. 2010; Karlova et al. 2011). Similarly, mir156 targets squamosa promoter-binding protein (SBP) and negatively regulates ripening process (Manning et al. 2006; Moxon et al. 2008). Inhibition of SBP through mir156 induced colorless never ripe (CNR) phenotype in tomato (Moxon et al. 2008). Karlova et al. (2013) proposed that a correlation exists between CNR and AP2 during ripening but these TFs are negatively regulated by miRNA 156 and 172, respectively. A recent study showed that the ripening inhibitor (RIN) transcription factor binds directly to the promoter sequence of mir172a and positively regulate its expression. Thus, fruit ripening process is coordinated by ripening inhibitors, miRNAs, and ethylene response elements (Gao et al. 2015)

3.3.5 Production of Seedless Fruits

Seedless fruits and vegetables are highly desirable in the market for fresh consumption as well as for production of processed food (Molesini et al. 2012). Various fruit characteristics also get improved through seedlessness, e.g., the absence of seed formation in watermelon and cucumber increases yield and shelf life (Pandolfini 2009). Seedless fruits are generally produced through parthenocarpy, a naturally occurring process which involves direct development of fruit from ovary without pollination or fertilization (Gorguet et al. 2005). It can also be artificially induced by disrupting the genes involved in the process of seed set and seed formation. The complex process of seed formation is temporally and spatially mediated by phytohormones. Generally, seedless fruits obtained through mutation and phytohormone alteration methods are generally found to bear pleiotropic effects such as reduced fruit size, effect on its taste, etc. (Varoquaux et al. 2000; Wang et al. 2005). Therefore, more efficient methods are now employed for generation of parthenocarpic fruits. RNAi-mediated downregulation of chalcone synthase, first gene in flavonoid biosynthesis, leads to development of parthenocarpy in tomato (Schijlen et al. 2007). Recent study showed that suppression of flavonol synthase through RNAi induced seedlessness in tobacco (*Nicotiana tabacum* cv *Xanthi*) (Mahajan et al. 2011). Flavonol synthase involved in formation of flavonols in flavonoid biosynthesis pathway is required for seed formation. Similarly, silencing of auxin-responsive element (*ARF7*) in tomato through RNAi causes seedlessness (De Jong et al. 2009). Parthenocarpy can also be induced through miRNA-mediated regulation of target genes. For example, miRNA167 regulates expression of auxin-responsive element (*ARF8*) and alteration in the expression of *ARF8* through aberrant expression of miRNA-induced parthenocarpy in both *Arabidopsis* and tomato (Molesini et al. 2009). Besides phytohormones, protein synthesis genes have also been utilized for generation of parthenocarpic fruits in tomato. Silencing of protein synthesis gene *AUCSIA* caused seedlessness in tomato through uncoupling of fruit formation with fertilization (Molesini et al. 2009). Thus, RNAi approach provides a good alternative to achieve parthenocarpy or seedlessness in fruits.

3.3.6 Modification of Flower Color

Various attributes of flower contribute to million dollar ornamental industry worldwide. Flower color is one of them, which is governed by combination of different pigments such as flavonoids, carotenoids, and betalains (He et al. 2013). These pigments also act as attractant to pollinators and protect the plants from harmful UV rays (He et al. 2013). RNAi-mediated genetic manipulation of pigment biosynthetic pathways offers a new platform for development of commercially valuable color pattern in flowers. Temporal and spatial regulation of gene expression in a highly specific manner through RNAi can induce desirable variation in flower color. Anthocyanins are the most prominent flavonoids, responsible for orange to red and purple to blue color in flowers. Anthocyanins are derived from a branch of flavonoid biosynthesis pathway. The first step in anthocyanin is mediated by chalcone synthase enzyme (CHS). Silencing of anthocyanin biosynthetic genes can be manifested into diverse flower colors. For example, downregulation of three anthocyanin biosynthetic genes – chalcone synthase (*CHS*), anthocyanidin synthase (*ANS*), and flavonoid 3' 5'-hydroxylase (*F 3' 5' H*) – through RNAi induced variable color patterns in *Gentiana* spp. (Nakatsuka et al. 2008). RNAi-CHS plants showed pure white to pale-blue color petals, and RNAi-ANS plants exhibited only pale-blue color, whereas RNAi plants expressing dsRNA of *F 3' 5' H* gene developed magenta flowers. The involvement of these genes at different steps in anthocyanin biosynthetic pathway is found to be responsible for observed variation in color patterns in three different gentian RNAi plants. Later, the same group demonstrated that more variation in flower color can be introduced by suppressing the target genes together. Silencing of anthocyanin biosynthesis gene 3'-aromatic acyltransferase (*5/3' AT*) and *F 3' 5' H* through chimeric RNAi produced variable colors in gentian flower ranging from liliac to pale blue (Nakatsuka et al. 2010). Downregulation of these enzymes caused accumulation of blue color dolphin dine pigment in flowers. Similarly, silencing of *CHS* gene in *Tricyrtis* sp., a monocotyledon plant by RNAi-induced alternation in flower color in sepals (Kamiishi et al. 2012). RNAi can also be used in combination of other techniques for improvement and change in color. For example, He et al. (2013) produced blue-colored *Chrysanthemum* flowers through RNAi-mediated silencing of *F3'H* and overexpression of the exogenous *Senecio cruentus F 3' 5' H* gene. Thus, commercial value of ornamental plant can be increased through production of desirable color variation in flowers using RNAi approach.

3.3.7 Development of Male Sterile Lines

Hybrids are contributing significantly to meet the future demand for increasing food worldwide. Hybridization leads to production of offsprings with superior characteristics in comparison to their parents, the process known as hybrid vigor or heterosis. Male sterility in female and its restoration in future generations is a prerequisite to produce hybrids in self-pollination plants. Male sterility is a naturally occurring phenomenon in some of the cross-pollinating plants such as grasses and safflower (Duvick 1999). To widen the scope for production of hybrid seeds in self-pollinating plants, male sterility can be artificially induced through various conventional and genetic engineering methods. RNAi-mediated suppression of genes involved in tapetum and pollen development was found to be more effective in producing male sterility (Nawaz-ul-Rehman et al. 2007; Tehseen et al. 2010; Sinha and Rajam 2013). Tapetum is a layer in the microsporangium which provides nutrition to the developing pollen grains. RNAi-mediated silencing of *TA29*, an anther-specific gene involved in pollen development in tobacco, resulted in male sterility (Nawaz-ul-Rehman et al. 2007). Similarly downregulation of another anther-specific gene *Bcp1* arrested the pollen development and induced male sterility in *A. thaliana*. The male sterile RNAi lines were found to be phenotypically normal and produced viable seeds when restored through cross-fertilization with male-fertile plants. *Bcp* 1 expressed during diploid tapetum and haploid microspore development and, thus, inhibition of its expression affected the pollen development adversely (Tehseen et al. 2010). S-adenosyl methionine decarboxylase (SAMDC) is a key enzyme in polyamine biosynthesis, required during pollen maturation and germination (Sinha and Rajam 2013). Thus, expression of chimeric (*SAMDC*) dsRNA under the control of tapetum-specific A9 promoter caused simultaneous silencing of three *SAMDC* isoforms in tapetum tissue, which resulted in formation of male sterile SAMDC-RNAi lines without affecting their female fertility.

MicroRNA-mediated regulation of male sterility has been reported in various plant species such as cotton (Yang et al. 2013; Wei et al. 2013; Yang et al. 2016), citrus (Fang et al. 2014), and radish (Zhang et al. 2016). For example, rsa-miR159a regulates the expression of transcription factor required during anther and pollen development. High expression of ras-mir159a decreases the expression of *MYB101* TF and, thus, induces male sterility through inhibition of normal pollen development in radish plants (Zhang et al. 2016). Yang et al. (2016) observed differential expression of 49 conserved and 51 novel miRNAs during male sterility in cotton. Hence, RNAi-mediated silencing of various genes involved in pollen development opens a new door for production of hybrid seeds in various plant systems.

3.3.8 Production of Secondary Metabolites

Plant secondary metabolites are major sources of pigments, fragrances, drugs, food additives, and pesticides. It is estimated that 70–80% of worldwide population fulfill their primary health requirements from the herbal medicines obtained from the plant secondary metabolites (Canter et al. 2005). Complex array of genes is responsible for synthesis of secondary metabolites. RNAi is recognized as an effective strategy for manipulation of secondary metabolites (Borgio 2009). Replacement of morphine with non-narcotic alkaloid reticuline in opium poppy (*Papaver somniferum*) presented a best example of metabolic engineering through RNAi. Allen et al. (2004) were the first to report RNAi-mediated silencing of multiple genes involved in different steps of a complex biochemical pathway. They designed hpRNA construct which caused simultaneous downregulation of all members of codeine reductase (COR) gene family. Silencing of *COR* gene family caused accumulation of (S)-reticuline, a non-narcotic alkaloid precursor in transgenic plants at the expense of morphine, codeine, and opium. Cassava (*Manihot esculenta*) is the third largest source of carbohydrates and major staple food in tropical countries. Presence of cyanogenic glucosides compound in cassava makes it unsuitable for food consumption at large scale. Silencing of cytochrome P450 through RNAi reduced the cyanogenic glucoside to a significant level in leaves as well as in tubers (Jørgensen et al. 2005). Potato (*Solanum tuberosum* L.) tubers have emerged as efficient bioreactors for production of recombinant human therapeutic glycoproteins. RNAi technology has been employed in inhibition of endogenous patatin expression at transcriptional and translational level. The developed potato tubers showed high accumulation of heterologous patatin glycoprotein, which has fastened the purification of recombinant protein (Kim et al. 2008).

Caffeine is a stimulant for the central nervous, respiratory, and circulatory system. It also gives protection against type 2 diabetes, Parkinson's disease, and liver disorders. However, high consumption of caffeine causes insomnia, restlessness, and palpitations. Decaffeinated coffee (DECAF) occupy only 10% of the world coffee market. Suppression of *CaMXMT1* (7-N-methylxanthine methyltransferase or theobromine synthase) by the RNAi resulted in reduction of caffeine content up to 70% in the silenced transgenic plant (Ogita et al. 2003). Similarly, low caffeine producing tea was generated through downregulation of caffeine synthase gene (*CS*) without affecting its stimulating property (Mohanpuria et al. 2011).

3.3.9 *Removal of Allergens from Food Crops*

Consumption of allergens containing food causes various health problems in humans, which even cannot be cured with the use of existing therapies. Allergens are naturally occurring compounds found in various food crops, capable of producing allergic response even if consumed in small quantities. Besides, they also cause hindrance in extraction of pure desirable products. Elimination of these unwanted compounds from plants is a costly and cumbersome process, requiring various chemical reactions and engineering processes, which reduces the nutritional value of food materials. RNAi has emerged as a powerful technology for removal of allergens through the alteration in their biosynthetic pathway or biochemical responses. It helps to enhance the edibility and food quality of crop plants without affecting their physiological processes. Major apple (*Malus domestica*) contains a pathogen-related protein PR10 allergens Mal d 1 which induce IgE-mediated hypersensitive response in organisms. Expression of Mal d1 dsRNA sequence reduces the expression of endogenous gene in developed RNAi apple plants and lowered the allergic response upon consumption (Gilisen et al. 2005). Le et al. (2006) provided a promising design for development of allergen-free tomato plants through RNAi. They identified a novel allergen Lyce3 in tomato which encodes a hydrophilic, non-specific lipid transfer protein (ns-LTP). Specific downregulation of *Lyce3* gene through RNAi resulted in suppression of Lyce 3 accumulation in tomato. Further, potential of allergens was tested with histamine test for the developed RNAi tomato plants. RNAi tomato plants showed reduced allergenicity and, thus, increased the edibility of tomato to allergen-sensitive population (Le et al. 2006). Consumption of heavy metals even in low concentration produces irreversible damaging effect on various physiological processes in human. Rice can accumulate cadmium (Cd) to a significant level in its seeds due to the presence of phytochelatin synthase (*PCS*) genes. RNAi-mediated suppression of phytochelatin synthase (*OsPCS1*) gene reduced the accumulation of Cd in rice (Li et al. 2007). Thus, accumulation of heavy metals in rice seeds can be regulated through RNAi even when plants are grown in heavy metal-polluted soil. Tearless onion can be generated through RNAi-mediated suppression of tear-inducing lachrymatory factor synthases gene (*LFS*). LFS is responsible for production of tear-inducing lachrymatory factor, Propanethial S-oxidase (LF), from 1-propenyle sulfenic acid. Inhibition of *LFS* resulted in generation of tearless onion due to low production of LF (Eady et al. 2008). RNAi can also be employed for removal of neurotoxic and carcinogenic compounds from food crops. Consumption of neurotoxin found in chickpeas causes lathyrism, a severe paralytic neurotoxic disease. RNAi-mediated downregulation of BOAA (β-N-oxalylamino-L-alanine) neurotoxin lowered its concentration in crop plants to a level, which is found safe for consumption. The production of carcinogenic compound in tobacco can also be minimized through RNAi-mediated silencing of nicotine demethylase gene (*DM*). DM is responsible for production of carcinogenic precursor from nicotine (Lewis et al. 2008). Cotton is a major cash crop, known for its fibers and oil worldwide. The cotton seeds are rich in proteins and calories, but they largely remain unutilized due to the presence of high amount of gossypol terpenoid. Gossypol is found in all parts of cotton plant and provides protection against herbivores. Downregulation of gossypol synthesis gene (δ-Cadinene synthase) in tissue-specific manner resulted in development of gossypol-free transgenic seeds, without affecting its expression in other parts of the plant (Sunilkumar et al. 2006). Even, ultralow gossypol-containing cotton seeds (ULGCS) can be produced by tight regulation of gossypol biosynthetic δ-Cadinene synthase gene through RNAi (Rathore et al. 2012).

3.3.10 Change in Plant Architecture

Plant architecture controls several important agronomic traits in plants. For example, plant height, pattern of shoot branching, plant morphology, inflorescence, crop yield, and resistance to environmental stresses (Khush 2001; Camp 2005;

Wang and Li 2006). Plant architecture can also be modified in order to minimize the negative effects of climate change on crop productivity. For example, plants are grown in drought and nutrient-deficient soil by manipulating its root architecture for maximum absorption of water and nutrients (de Dorlodot et al. 2007). Understanding of molecular basis of plant architecture has served as platforms for RNAi-mediated alternation in plant architecture (Wang and Li 2008). Shorter plants with erect leaf architecture were produced through RNAi-mediated silencing of *OsDWARF4* gene in rice (Feldmann 2006). RNAi-mediated downregulation of ornithine decarboxylase (*ODC*) gene, which is a key gene involved in polyamine biosynthesis resulted in significant physiological and morphological changes including reduced leaf size, decreased abiotic stress tolerance, delayed flowering, and early onset of senescence in tobacco (Choubey and Rajam 2017). Increase in biomass was observed in RNAi plants due to increase in rate of photosynthesis in lower erect leaves. Biofuel production can be enhanced through low lignin content in plant material. Lignin makes the plant material recalcitrant for conversion to ethanol. Low lignin-containing plants can be produced through downregulation of lignin biosynthetic gens by RNAi. For instance, RNAi-mediated downregulation of lignin-associated genes such as Cinnamate-4-hydroxylase, shikimate hydroxycinnamoyl transferase, and 4-coumarate-CoA ligase reduced the lignin content and increased its accessibility to cellulose to degradation (Hisano et al. 2009). RNAi technology is used to enhance the crop yield by manipulating the plant architecture. Taller rice variety QX1 was converted to semi-dwarf variety through RNAi-mediated suppression of GA 20-oxidase (*OsGA20ox2*) gene. The developed transgenic rice exhibited high yield due to the significant increase in panicle length, number of seeds per panicle, and weight of individual seeds (Qiao et al. 2007).

Plant architecture has been found to be regulated by miRNAs. The manipulation of miRNA expression directly or indirectly affected the plant architecture, biomass accumulation, and yield (Chuck et al. 2011; Wang et al. 2012; Rubinelli et al. 2013). Corngrass1 (Cg1) miRNA that belongs to the mir156 family regulates vegetative growth and flowering in plants. Overexpression of Cg1 miRNA caused prolongation of vegetative phase and delay in flowering time in maize (Chuck et al. 2011). Similarly, phenotype was also observed when Cg1 was overexpressed in other plant species, for example, overexpressing Cg1miRNA in *Populus* plants showed significant shortening of internode length, increase in the growth of axillary meristem, and ~30% reduction in stem lignin content as compared to the untransformed control (Rubinelli et al. 2013). The miRNA-mediated manipulation of plant architecture enhanced the grain yield in rice (Jiao et al. 2010; Miura et al. 2010; Wang et al. 2012). Osa-miR156 regulates the expression of *OsSPL14* (squamosa promoter-binding protein-like 14) and is found to have positive effects on plant architecture and yield in rice (Jiao et al. 2010; Miura et al. 2010). Overexpression of Osa-miR397 elevated grain production up to 25% in RNAi rice plants due to increase in panicle branching and grain size. Osa-miR397 downregulates a brassinosteroid-sensitive gene *OsLAC* (coding for a laccase-like protein) and, thus, directs the energy toward the growth of plants. Since miR397 is found to be highly conserved across different plant species, similar strategy can be used in other crops for increasing grain yield (Zhang et al. 2013). Thus, RNAi technology has a wide utility in manipulating the plant architecture for high yield, increase in biomass, flowering, and removal of undesirable phenotypes. Rose plant can be easily modified for its thorn characteristic or the plant architecture in mulberry, and tea plants can be manipulated for easy plucking of leaves.

3.4 Conclusions and Future Prospects

The main challenge for agriculture in the twenty-first century is to provide food security for existing and expanding population. Besides, malnutrition is also a major problem faced by people in developing countries. To ensure supply of balanced food to the world, it is necessary to develop biofortified staple food, vegetables, and fruits, enriched in essential compounds and elements such as fatty acids, vitamins, amino acids, and micro- or macronutrients. Increase in resistance toward the present technology, changing environment, increasing population, and pollution have put a high pressure on the existing natural resources for high crop productivity. The development of crop varieties resistant to pathogens and pests and tolerant to changing environmental conditions such as high temperature, drought, flood, oxidative stresses, high salt concentration, and heavy metal-polluted soil can be a blow for world food security, malnutrition, and famine problems. RNAi-based technology has proven its potential in development of crop varieties resistant to biotic and abiotic stresses. Besides, RNAi is creating a milestone in genetic manipulation of crop varieties for highly recommended agronomic traits. Crops have been engineered through RNAi for novel and commercially important agronomic traits including decaffeinated tea, coffee, nicotine-free tobacco, allergen-free cereals, low glutelin-containing wheat, healthy fatty acid-enriched oil crops, blue rose, browning-free apple, tear-free onion, easily packageable tomato, etc. Additionally, RNAi has been employed to remove carcinogenic, neurotoxin, and mycotoxin compounds from food crops. Most of the RNAi-based studies involve a single-gene silencing for the improvement of useful traits in crops. However, silencing of more than one gene can be achieved through chimeric RNAi constructs, which will be useful for improvement of several traits in crop plants simultaneously (Gupta et al. 2013; Sinha and Rajam 2013; Yogindran and Rajam 2015). RNAi-mediated crop improvement strategies hold a tremendous potential for enhancement of desirable traits and eradication of undesirable traits in crop plants (Fig. 3.3).

RNAi-mediated strategies are the most preferred and powerful alternative for crop improvement as compared to the existing approaches. It uses the existing conserved machinery for sequence-specific silencing at posttranscriptional level. So, there is no extra load on plant for production of effective proteins, and even, there are fewer chances of allergic responses. RNAi effector molecules can tolerate mutation without affecting the silencing efficiency; thus, it is very unlikely that the

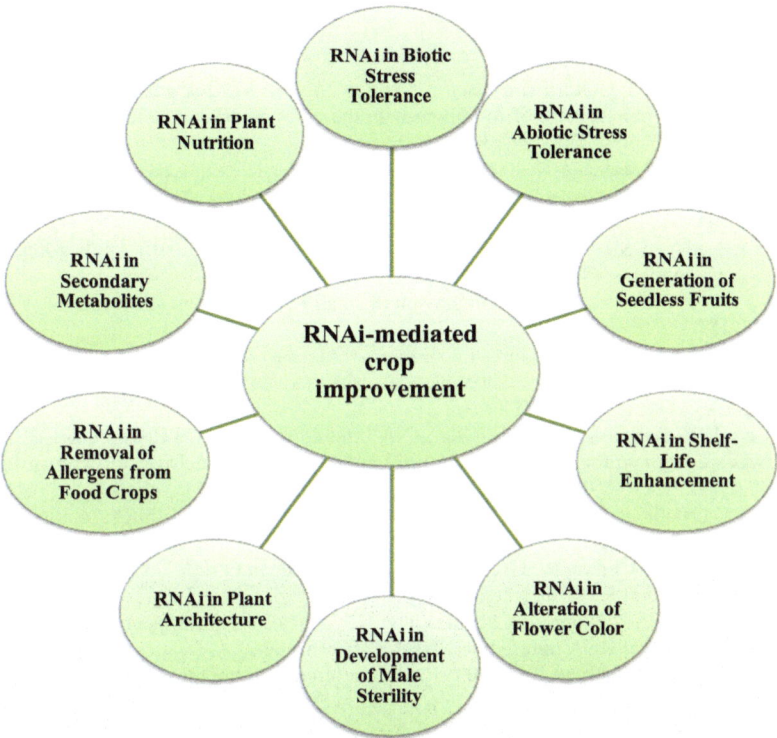

Fig. 3.3 Applications of RNAi for crop improvement

target pest or pathogen would gain resistance. However, RNA-based technology can result in off-target effects in the same or different plants and development of undesirable pleiotropic effects because of sequence homology. Therefore, RNAi-based transgenic approach should be planned in such a way that no or minimal off-target and pleiotropic effects will arise. For example, an in-depth in silico analysis should be performed before selecting the target gene sequence for preparation of hairpin RNAi construct. The applications of RNAi technology should move from lab to field and from model plant to crop plants (Jagtap et al. 2011; Katoch and Thakur 2013; Koch and Kogel 2014; Saurabh et al. 2014; Kamthan et al. 2015). In the near future, RNAi-mediated crop improvement programs in combination with other technologies will change the food security parameter across the world and improve the way of life.

Acknowledgments The financial assistance from the Department of Biotechnology (DBT) and Department of Science and Technology (DST), New Delhi, for RNAi work in the lab is acknowledged. We also thank the University Grants Commission (UGC) for Special Assistance Programme (DRS-III), DST for FIST (Level 2) program, and DU-DST PURSE (Phase II) grant.

References

Allen RS, Millgate AG, Chitty JA, Thisleton J, Miller JA, Fist AJ, Gerlach WL, Larkin PJ (2004) RNAi mediated replacement of morphine with the non-narcotic alkaloid reticuline in opium poppy. Nat Biotechnol 22:1559–1566

Andika IB, Kondo H, Tamada T (2005) Evidence that RNA silencing mediated resistance to beet necrotic yellow vein virus is less effective in roots than in leaves. Mol Plant-Microbe Interact 18:194–204

Angaji SA, Hedayati SS, Poor RH, Poor SS, Shiravi S, Madani S (2010) Application of RNA interference in plants. Plant Omics J 3:77–84

Axtell MJ (2013) Classification and comparison of small RNAs from plants. Annu Rev Plant Biol 64:137–159

Banerjee S, Banerjee A, Gill SS, Gupta OP, Dahuja A, Jain PK, Sirohi A (2017) RNA interference: a novel source of resistance to combat plant parasitic nematodes. Front Plant Sci 8:834. https://doi.org/10.3389/fpls.2017.00834

Bao N, Lye KW, Barton MK (2004) MicroRNA binding sites in *Arabidopsis* class III HD-ZIP mRNAs are required for methylation of the template chromosome. Dev Cell 7:653–662

Bartel DP (2004) MicroRNAs: genomics, biogenesis, mechanism, and function. Cell 116:281–297

Baulcombe D (2000) Unwinding RNA silencing. Science 290:1108–1109. https://doi.org/10.1126/science.290.5494.1108

Baum JA, Bogaert T, Clinton W, Heck GR, Feldmann P, Ilagan O (2007) Control of coleopteran insect pests through RNA interference. Nat Biotechnol 25:1322–1326

Bernstein E, Caudy AA, Hammond SM, Hannon GJ (2001) Role for a bidentate ribonuclease in the initiation step of RNA interference. Nature 409:363–366

Bezanilla M, Pan A, Quatrano RS (2003) RNA interference in the moss *Physcomitrella patens*. Plant Physiol 133:470–474

Borgio JF (2009) RNA interference (RNAi) technology: a promising tool for medicinal plant research. J Med Plant Res 3:1176–1183

Brodersen P, Sakvarelidze-Achard L, Bruun-Rasmussen M, Dunoyer P, Yamamoto YY, Sieburth L, Voinnet O (2008) Widespread translational inhibition by plant miRNAs and siRNAs. Science 320:1185–1190. https://doi.org/10.1126/science.1159151

Camp WV (2005) Yield enhancement genes: seeds for growth. Curr Opin Biotechnol 16:147–153

Campo S, Peris-Peris C, Sire C, Moreno AB, Donaire L, Zytnicki M, Notredame C, Llave C, San Segundo B (2013) Identification of a novel microRNA (miRNA) from rice that targets an alternatively spliced transcript of the Nramp6 (Natural resistance-associated macrophage protein 6) gene involved in pathogen resistance. New Phytol 199:212–227

Canter PH, Thomas H, Ernst E (2005) Bringing medicinal plants into cultivation: opportunities and challenges for biotechnology. Trends Biotechnol 23:180–185

Castel SE, Martienssen RA (2013) RNA interference in the nucleus: roles for small RNAs in transcription, epigenetics and beyond. Nat Rev Genet 14:100–112

Cerutti H, Ma X, Msanne J, Repas T (2011) RNA-mediated silencing in algae: biological roles and tools for analysis of gene function. Eukaryot Cell 10:1164–1172. https://doi.org/10.1128/EC.05106-11

Chen S, Hajirezaei MR, Zanor MI, Hornyik C, Debast S, Lacomme C, Fernie AR, Sonnewald U, Bornke F (2008) RNA interference-mediated repression of sucrose-phosphatase in transgenic potato tubers (*Solanum tuberosum*) strongly affects the hexose-to-sucrose ratio upon cold storage only minor effects on total soluble carbohydrate accumulation. Plant Cell Environ 31:165–176

Chen W, Kastner C, Nowara D, Oliveira-Garcia E, Rutten T, Zhao Y, Deising HB, Kumlehn J, Schweizer P (2016) Host-induced silencing of *Fusarium culmorum* genes protects wheat from infection. J Exp Bot 67:4979–4991. https://doi.org/10.1093/jxb/erw263

Choubey A, Rajam MV (2017) Transcriptome response and developmental implications of RNAi-mediated ODC knockdown in tobacco. Funct Integr Genomics DOI. https://doi.org/10.1007/s10142-016-0539-3

Chuck GS, Tobias C, Sun L, Kraemer F, Li C, Dibble D, Arora R, Bragg JN, Vogel JP, Singh S, Simmons BA, Pauly M, Hake S (2011) Overexpression of the maize corngrass1 microRNA prevents flowering, improves digestibility, and increases starch content of switchgrass. Proc Natl Acad Sci U S A 108:17550–17555. https://doi.org/10.1073/pnas.1113971108

Chung MY, Vrebalov J, Alba R, Lee J, McQuinn R, Chung JD, Klein P, Giovannoni J (2010) A tomato (Solanum lycopersicum) APETALA2/ERF gene, SlAP2a, is a negative regulator of fruit ripening. Plant J 64:936–947

Da-Hong L, Hui L, Yan-li Y, Ping-ping Z, Jian-sheng L (2009) Down-regulated expression of RACK1 gene by RNA interference enhances drought tolerance in rice. Rice Sci 16:14–20

Davuluri GR, Tuinen A, Fraser PD, Manfredonia A, Newman R, Burgess D, Brummell DA, King SR, Palys J, Uhlig J, Bramley PM, Pennings HMJ, Bowle C (2005) Fruit-specific RNAi-mediated suppression of DET1 enhances carotenoid and flavonoid content in tomatoes. Nat Biotechnol 23:890–895

de Dorlodot S, Forster B, Pages L, Price A, Tuberosa R, Draye X (2007) Root system architecture: opportunities and constraints for genetic improvement of crops. Trends Plant Sci 12:474–481

De Jong M, Wolters-Arts M, Feron R, Mariani C, Vriezen WH (2009) The Solanum lycopersicum 7 (auxin response factorSlARF7) regulates auxin signaling during tomato fruit set and development. Plant J 57:160–117

Duvick DN (1999) Heterosis: feeding people and protecting natural resources. In: Coors JG, Pandey S (eds) Genetics and exploitation of heterosis in crops. American Society of Agronomy/Crop Science Society of America, Madison, pp 19–29

Eady CC, Kamoi T, Kato M, Porter NG, Davis S, Shaw M, Kamoi A, Imai S (2008) Silencing onion lachrymatory factor synthase causes a significant change in the sulfur secondary metabolite profile. Plant Physiol 147:2096–2106. https://doi.org/10.1104/pp.108.123273

Elbashir SM, Harborth J, Lendeckel W, Yalcin A, Weber K, Tuschl T (2001) Duplexes of 21-nucleotide RNAs mediate RNA interference in cultured mammalian cells. Nature 411:494–508

Fang YN, Qiu WM, Wang Y, Wu XM, Xu Q, Guo WW (2014) Identification of differentially expressed microRNAs from a male sterile Ponkan mandarin (Citrus reticulata Blanco) and its fertile wild type by small RNA and degradome sequencing. Tree Genet Genome 10:1567–1581. https://doi.org/10.1007/s11295-014-0780-7

FAO (2013) The state of food insecurity in the world, executive summary. Rome, Italy, FAO

Feldmann KA (2006) Steroid regulation improves crop yield. Nat Biotechnol 24:46–47

Fire A, Xu S, Montgomery MK, Kostas SA, Driver SE, Mello CC (1998) Potent and specific genetic interference by double-stranded RNA in Caenorhabditis elegans. Nature 391:806–811

Flores T, Karpova O, Su X, Zeng P, Bilyeu K, Sleper DA, Nguyen HT, Zhang ZJ (2008) Silencing of GmFAD3 gene by siRNA leads to low alpha-linolenic acids (18:3) of fad 3-mutant phenotype in soybean [Glycine max (Merr.)]. Transgenic Res 17:839–850

Gao P, Bai X, Yang L, Lv D, Pan X, Li Y, Cai H, Ji W, Chen Q, Zhu Y (2011) osa-MIR393: a salinity and alkaline stress-related microRNA gene. Mol Biol Rep 38:237–242. https://doi.org/10.1007/s11033-010-0100-8

Gao C, Ju Z, Cao D, Zhai B, Qin G, Zhu H, Fu D, Luo Y, Zhu B (2015) MicroRNA profiling analysis throughout tomato fruit development and ripening reveals potential regulatory role of RIN on microRNAs accumulation. Plant Biotechnol J 13:370–382

Gentile A, Dias LI, Mattos RS, Ferreira TH, Menossi M (2015) MicroRNAs and drought responses in sugarcane. Front Plant Sci 6:58. https://doi.org/10.3389/fpls.2015.00058

Gilisen LJ, Bolhaar ST, Matos CI, Rouwendal GJ, Boone MJ, Krens FA et al (2005) Silencing of major apple allergen Mal d 1 by using the RNA interference approach. J Allergy Clin Immunol 115:364–369. https://doi.org/10.1016/j.jaci.2004.10.014

Gorguet B, van Heusden AW, Lindhout P (2005) Parthenocarpic fruit development in tomato. Plant Biol 7:131–139

Goswami S, Kumar RR, Rai RD (2014) Heat-responsive microRNAs regulate the transcription factors and heat shock proteins in modulating thermo-stability of starch biosynthesis enzymes in wheat (Triticum aestivum L.) under the heat stress. Aust J Crop Sci 8:697–705

Guan Q, Lu X, Zeng H, Zhang Y, Zhu J (2013) Heat stress induction of miR398 triggers a regulatory loop that is critical for thermo tolerance in *Arabidopsis*. Plant J 74:840–851. https://doi.
org/10.1111/tpj.12169

Gupta A, Pal RK, Rajam MV (2013) Delayed ripening and improved fruit processing quality
in tomato by RNAi-mediated silencing of three homologs of 1-aminopropane-1-carboxylate
synthase gene. J Plant Physiol 170:987–995

Hamilton AJ, Baulcombe DC (1999) A novel species of small antisense RNA in posttranscriptional
gene silencing. Science 286:950–952

Hammond SM, Bernstein E, Beach D, Hannon GJ (2000) An RNA-directed nuclease mediates
post-transcriptional gene silencing in *Drosophila* cells. Nature 404:293–296

He H, Ke H, Keting H, Qiaoyan X, Silan D (2013) Flower colour modification of Chrysanthemum
by suppression of F3'H and overexpression of the exogenous *Senecio cruentus F3' 5' H* gene.
PLoS One 8:e74395

Hisano H, Nandakumar R, Wang ZY (2009) Genetic modification of lignin biosynthesis for
improved biofuel production. In Vitro Cell Dev Biol Plant 45:306–313. https://doi.org/10.1007/
s11627-009-9219-5

Huang G, Allen R, Davis E, Baum T, Hussey R (2006) Engineering broad rootknot resistance in
transgenic plants by RNAi silencing of a conserved and essential root-knot nematode parasitism gene. Proc Natl Acad Sci U S A 103:14302–14306

Jagtap UB, Gurav RG, Bapat VA (2011) Role of RNA interference in plant improvement.
Naturwissenschaften 98:473–492

Jan A, Maruyama K, Todaka D, Kidokoro S, Abo M, Yoshimura E, Shinozaki K, Nakashima
K, Yamaguchi-Shinozaki K (2013) OsTZF1, a CCCH-tandem zinc finger protein, confers
delayed senescence and stress tolerance in rice by regulating stress-related genes. Plant Physiol
161:1202–1216

Jiao Y, Wang Y, Xue D, Wang J, Yan M, Liu G, Dong G, Zeng D, Lu Z, Zhu X, Qian Q, Li
J (2010) Regulation of OsSPL14 by OsmiR156 defines ideal plant architecture in rice. Nat
Genet 42:541–544

Jin S, Singh ND, Li L, Zhang X, Daniell H (2015) Engineered chloroplast dsRNA silences
cytochrome p450 monooxygenase, V-ATPase and chitin synthase genes in the insect gut and
disrupts *Helicoverpa armigera* larval development and pupation. Plant Biotech J 13:435–446

Jones-Rhoades MW, Bartel DP, Bartel B (2006) MicroRNAs and their regulatory roles in plants.
Annu Rev Plant Biol 57:19–53

Jørgensen K, Bak S, Busk PK, Sørensen C, Olsen CE, Puonti-Kaerlas J, Moller BL (2005)
Cassava plants with a depleted cyanogenic glucoside content in leaves and tubers. Plant Physiol
139:363–374

Kamachi S, Mochizuki A, Nishiguchi M, Tabei Y (2007) Transgenic *Nicotiana benthamiana*
plants resistant to cucumber green mottle mosaic virus based on RNA silencing. Plant Cell Rep
26:1283–1288. https://doi.org/10.1007/s00299-007-0358-z

Kamiishi Y, Otani M, Takagi H, Han DS, Mori S, Tatsuzawa F, Okuhara H, Kobayashi H, Nakano
M (2012) Flower color alteration in the liliaceous ornamental *Tricyrtis* sp. by RNA interference-
mediated suppression of the chalcone synthase gene. Mol Breed 30:671–680

Kamthan A, Chaudhuri A, Kamthan M, Datta A (2015) Small RNAs in plants: recent development and application for crop improvement. Front Plant Sci 6:208. https://doi.org/10.3389/
fpls.2015.00208

Karlova R, Rosin FM, Busscher-Lange J, Parapunova V, Do PT, Fernie AR, Fraser PD, Baxter
C, Angenent GC, de Maagd RA (2011) Transcriptome and metabolite profiling show that
APETALA2a is a major regulator of tomato fruit ripening. Plant Cell 23:923–941

Karlova R, van Haarst JC, Maliepaard C, van de Geest H, Bovy AG, Lammers M, Angenent GC,
de Maagd RA (2013) Identification of microRNA targets in tomato fruit development using
high-throughput sequencing and degradome analysis. J Exp Bot 64:1863–1878

Katoch R, Thakur N (2013) RNA interference: a promising technique for the improvement of
traditional crops. Int J Food Sci Nutr 64:248–259. https://doi.org/10.3109/09637486.2012.7
13918

Ketting RF (2011) The many faces of RNAi. Dev Cell 20:148–161

Khraiwesh B, Arif MA, Seumel GI, Ossowski S, Weigel D, Reski R, Frank W (2010) Transcriptional control of gene expression by microRNAs. Cell 140:111–122. https://doi.org/10.1016/j.cell.2009.12.023

Khraiwesh B, Zhu JK, Zhu J (2012) Role of miRNAs and siRNAs in biotic and abiotic stress responses of plants. Biochim Biophys Acta 1819:137–148

Khush GS (2001) Green revolution: the way forward. Nat Rev Genet 2:815–822

Kim YS, Lee YH, Kim HS, Kim MS, Hahn KW, Ko JH, Joung H, Jeon JH (2008) Development of patatin knockdown potato tubers using RNA interference (RNAi) technology, for the production of human-therapeutic glycoproteins. BMC Biotechnol 8:36. https://doi.org/10.1186/1472-6750-8-36

Koch A, Kogel KH (2014) New wind in the sails: improving the agronomic value of crop plants through RNAi-mediated gene silencing. Plant Biotechnol J 12:821–831. https://doi.org/10.1111/pbi.12226

Koch A, Biedenkopf D, Furch A, Weber L, Rossbach O, Abdellatef E, Linicus L, Johannsmeier J, Jelonek L, Goesmann A, Cardoza V, McMillan J, Mentzel T, Kogel KH (2016) An RNAi-based control of *Fusarium graminearum* infections through spraying of long dsRNAs involves a plant passage and is controlled by the fungal silencing machinery. PLoS Pathog 12:e1005901

Kruszka K, Pacak A, Swida-Barteczka A, Nuc P, Alaba S, Wroblewska Z, Karlowski W, Jarmolowski A, Szweykowska-Kulinska Z (2014) Transcriptionally and post-transcriptionally regulated microRNAs in heat stress response in barley. J Exp Bot 65:6123–6135. https://doi.org/10.1093/jxb/eru353

Kumar R (2014) Role of microRNAs in biotic and abiotic stress responses in crop plants. Appl Biochem Biotechnol 174:93–115

Kumar P, Pandit SS, Baldwin IT (2012) Tobacco rattle virus vector: a rapid and transient means of silencing *Manduca sexta* genes by plant mediated RNA interference. PLoS One 7:e31347

Kusaba M, Miyahara K, Iida S, Fukuoka H, Takano T, Sassa H, Nishimura M, Nishio T (2003) Low glutelin content 1: a dominant mutation that suppresses the glutelin multigene family via RNA silencing in rice. Plant Cell 15:1455–1467

Le LQ, Lorenz Y, Scheurer S, Fotisch K, Enrique E, Bartra J, Biemelt S, Vieths S, Sonnewald U (2006) Design of tomato fruits with reduced allergenicity by dsRNAi-mediated inhibition of ns-LTP (Lyc e 3) expression. Plant Biotechnol J 4:231–242. https://doi.org/10.1111/j.1467-7652.2005.00175.x

Lewis RS, Jack AM, Morris JW, Robert VJ, Gavilano LB, Siminszky B, Bush LP, Hayes AJ, Dewey RE (2008) RNA interference (RNAi)-induced suppression of nicotine demethylase activity reduces levels of a key carcinogen in cured tobacco leaves. Plant Biotechnol J 6:346–354. https://doi.org/10.1111/j.1467-7652.2008.00324.x

Li JC, Guo JB, Xu WZ, Ma M (2007) RNA interference-mediated silencing of phytochelatin synthase gene reduces cadmium accumulation in rice seeds. J Integr Plant Biol 49:1032–1037. https://doi.org/10.1111/j.1672-9072.2007.00473.x

Li X, Wang X, Zhang S, Liu D, Duan Y, Dong W (2012) Identification of soybean microRNAs involved in soybean cyst nematode infection by deep sequencing. PLoS One 7:e39650. https://doi.org/10.1371/journal.pone.0039650

Liu F, Wang XD, Zhao YY, Li YJ, Liu YC, Sun J (2015) Silencing the *HaAK* gene by transgenic plant-mediated RNAi impairs larval growth of *Helicoverpa armigera*. Int J Biol Sci 11:67–74. https://doi.org/10.7150/ijbs.10468

Luo Z, Zhang J, Li J, Yang C, Wang T, Ouyang B, Li H, Giovannoni J, Ye Z (2013) A STAY-GREEN protein SlSGR1 regulates lycopene and β-carotene accumulation by interacting directly with SlPSY1 during ripening processes in tomato. New Phytol 198:442–452. https://doi.org/10.1111/nph.12175

Mahajan M, Ahuja PS, Yadav SK (2011) Post-transcriptional silencing of flavonol synthase mRNA in tobacco leads to fruits with arrested seed set. PLoS One 6:e28315. https://doi.org/10.1371/journal.pone.0028315

Mamta, Reddy KR, Rajam MV (2015) Targeting chitinase gene of *Helicoverpa armigera* by host-induced RNA interference confers insect resistance in tobacco and tomato. Plant Mol Biol 90:281–292

Manavalan LP, Chen X, Clarke J, Salmeron J, Nguyen HT (2012) RNAi-mediated disruption of squalene synthase improves drought tolerance and yield in rice. J Exp Bot 63:163–175

Manning K, Tor M, Poole M, Hong Y, Thompson AJ, King GJ, Giovannoni J, Seymour GB (2006) A naturally occurring epigenetic mutation in a gene encoding an SBP-box transcription factor inhibits tomato fruit ripening. Nat Genet 38:948–952

Mao YB, Cai WJ, Wang JW, Hong GJ, Tao XY, Wang LJ, Huang YP, Chen X (2007) Silencing a cotton bollworm P450 monooxygenase gene by plant-mediated RNAi impairs larval tolerance of gossypol. Nat Biotechnol 25:1307–1313

Mao YB, Xue XY, Tao XY, Yang CQ, Wang LJ, Chen XY (2013) Cysteine protease enhances plant-mediated bollworm RNA interference. Plant Mol Biol 83:119–129

Matzke M, Matzke A, Pruss G, Vance V (2001) RNA-based silencing strategies in plants. Curr Opin Genet Dev 11:221–227. https://doi.org/10.1016/S0959-437X(00)00183-0

Miura K, Ikeda M, Matsubara A, Song XJ, Ito M, Asano K, Matsuoka M, Kitano H, Ashikari M (2010) OsSPL14 promotes panicle branching and higher grain productivity in rice. Nat Genet 42:545–549

Mohanpuria P, Kumar V, Ahuja PS, Yadav SK (2011) Producing low-caffeine tea through post-transcriptional silencing of caffeine synthase mRNA. Plant Mol Biol 76:523–234. https://doi.org/10.1007/s11103-011-9785-x

Molesini B, Pandolfini T, Rotino GL, Dani V, Spena A (2009) Aucsia gene silencing causes parthenocarpic fruit development in tomato[C][W]. Plant Physiol 149:534–548

Molesini B, Pii Y, Pandolfini T (2012) Fruit improvement using intragenesis and artificial microRNA. Trends Biotechnol 30:80–88. https://doi.org/10.1016/j.tibtech.2011.07.005

Molnar A, Melnyk CW, Bassett A, Hardcastle TJ, Dunn R, Baulcombe DC (2010) Small silencing RNAs in plants are mobile and direct epigenetic modification in recipient cells. Science 328:872–875. https://doi.org/10.1126/science.1187959

Moxon S, Jing R, Szittya G, Schwach F, Rusholme Pilcher RL, Moulton V, Dalmay T (2008) Deep sequencing of tomato short RNAs identifies microRNAs targeting genes involved in fruit ripening. Genome Res 18:1602–1609

Nakatsuka T, Mishiba K, Abe Y, Kubota A, Kakizaki Y, Yamamura S, Nishihara M (2008) Flower color modification of gentian plants by RNAi-mediated gene silencing. Plant Biotechnol TOKYO 25:61–68

Nakatsuka T, Mishiba KI, Kubota A, Abe Y, Yamamura S, Nakamura N, Tanaka Y, Nishihara M (2010) Genetic engineering of novel flower colour by suppression of anthocyanin modification genes in gentian. J Plant Physiol 167:231–237

Napoli C, Lemieux C, Jorgensen R (1990) Introduction of chimeric chalcone synthase gene into *Petunia* results in reversible cosuppression of homologous genes in trans. Plant Cell 2:279–289. https://doi.org/10.1105/tpc.2.4.279

Nawaz-ul-Rehman MS, Mansoor S, Khan AA, Zafar Y, Briddon RW (2007) RNAi-mediated male sterility of tobacco by silencing TA29. Mol Biotechnol 36:159–165

Naya L, Paul S, Valdés-López O, Mendoza-Soto AB, Nova-Franco B, Sosa-Valencia G, Reyes JL, Hernández G (2014) Regulation of copper homeostasis and biotic interactions by microRNA 398b in common bean. PLoS One 9:e84416. https://doi.org/10.1371/journal.pone.0084416

Ni Z, Hu Z, Jiang Q, Zhang H (2013) GmNFYA3, a target gene of miR169, is a positive regulator of plant tolerance to drought stress. Plant Mol Biol 82:113–129

Nowara D, Gay A, Lacomme C, Shaw J, Ridout C, Douchkov D, Hensel G, Kumlehn J, Schweizer P (2010) HIGS: host-induced gene silencing in the obligate biotrophic fungal pathogen *Blumeria graminis*. Plant Cell 22:3130–3141

Nunes CC, Dean RA (2012) Host – induced gene silencing: a tool for understanding fungal host interaction and for developing novel disease control strategies. Mol Plant Pathol 13:519–529. https://doi.org/10.1111/j.1364-3703.2011.00766.x

Oerke EC (2006) Crop losses to pests. J Agric Sci 144:31–43

Ogita S, Usefuji H, Yamaguchi Y, Koizumi N, Sano H (2003) Producing decaffeinated coffee plants. Nature 423:823. https://doi.org/10.1038/423823a

Pandolfini T (2009) Seedless fruit production by hormonal regulation of fruit set. Nutrients 1:168–177

Papolu PK, Gantasala NP, Kamaraju D, Banakar P, Sreevathsa R, Rao U (2013) Utility of host delivered RNAi of two FMRF amide like peptides, flp-14 and flp-18, for the management of root knot nematode, *Meloidogyne incognita*. PLoS One 8:e80603

Pareek M, Yogindran S, Mukherjee SK, Rajam MV (2015) Plant micro RNAs: biogenesis, functions and applications. In: Bahadur B, Rajam MV, Sahijram L, Krishnamurthy KV (eds) Plant biology and biotechnology, vol II.: Plant Genomics and Biotechnology. Springer, India, pp 639–661

Peng T, Jia MM, Liu JH (2015) RNAi-based functional elucidation of PtrPRP, a gene encoding a hybrid proline rich protein, in cold tolerance of *Poncirus trifoliata*. Front Plant Sci 29(6):808. https://doi.org/10.3389/fpls.2015.00808

Rajam MV, Madhulatha P, Pandey R, Hazarika PJ, Razdan MK (2007) Applications of genetic engineering in tomato. In: Razdan MK, Mattoo AK (eds) Genetic improvement of Solanaceae Crops: tomato, vol 2. Science Publishers, Enfield, pp 285–311

Rathore KS, Sundaram S, Sunilkumar G, Campbell LM, Puckhaber L, Marcel S, Palle SR, Stipanovic RD, Wedegaertner TC (2012) Ultra-low gossypol cottonseed: generational stability of the seed-specific, RNAi-mediated phenotype and resumption of terpenoid profile following seed germination. Plant Biotechnol J 10:174–183. https://doi.org/10.1111/j.1467-7652.2011.00652.x

Regina A, Bird A, Topping D, Bowden S, Freeman J, Barsby T, Kosar-Hashemi B, Li Z, Rahman S, Morell M (2006) High amylose wheat generated by RNA-Interference improves indices of large bowel health in rats. Proc Natl Acad Sci U S A 103:3546–3551

Romano N, Macino G (1992) Quelling: transient inactivation of gene expression in *Neurospora crassa* by transformation with homologous sequences. Mol Microbiol 6:3343–3353

Rubinelli PM, Chuck G, Li X, Meilan R (2013) Constitutive expression of the corn-grass1 microRNA in poplar affects plant architecture and stem lignin content and composition. Biomass Bioenergy 54:312–321

Saurabh S, Vidyarthi AS, Prasad D (2014) RNA interference: concept to reality in crop improvement. Planta 239:543–564. https://doi.org/10.1007/s00425-013-2019-5

Schijlen EGWM, de Vos RCH, Martens S, Jonker HH, Rosin FM, Molthoff JW, Tikunov YM, Angenent GC, van Tunen AJ, Bovy AG (2007) RNA interference silencing of chalcone synthase, the first step in the flavonoid biosynthesis pathway, leads to parthenocarpic tomato fruits. Plant Physiol 144:1520–1530

Shriram V, Kumar V, Devarumath RM, Khare TS, Wani SH (2016) MicroRNAs as potential targets for abiotic stress tolerance in plants. Front Plant Sci 7:817. https://doi.org/10.3389/fpls.2016.00817

Sindhu AS, Maier TR, Mitchum MG, Hussey RS, Davis EL, Baum TJ (2009) Effective and specific *in planta* RNAi in cyst nematodes: expression interference of four parasitism genes reduces parasitic success. J Exp Bot 60:315–324

Singh A, Taneja J, Dasgupta I, Mukherjee SK (2014) Development of plants resistant to tomato Gemini viruses using artificial trans-acting small interfering RNA. Mol Plant Pathol 16:725–734

Sinha R, Rajam MV (2013) RNAi silencing of three homologues of S-adenosylmethionine decarboxylase gene in tapetal tissue of tomato results in male sterility. Plant Mol Biol 82:169–180

Sun X, Xu L, Wang Y, Yu R, Zhu X, Luo X, Gong Y, Wang R, Limera C, Zhang K, Liu L (2015) Identification of novel and salt-responsive miRNAs to explore miRNA-mediated regulatory network of salt stress response in radish (*Raphanus sativus L.*). BMC Genomics 16:197. https://doi.org/10.1186/s12864-015-1416-5

Sunilkumar G, Campbell LM, Puckhaber L, Stipanovic RD, Rathore KS (2006) Engineering cottonseed for use in human nutrition by tissue-specifi c reduction of toxic gossypol. Proc Natl Acad Sci U S A 103:18054–18059

Tamilarasan S, Rajam MV (2013) Engineering crop plants for nematode resistance through host-derived RNA interference. Cell Dev Biol 2:114. https://doi.org/10.4172/2168-9296.1000114

Tehseen M, Imran M, Hussain M, Irum S, Ali S, Mansoor S, Zafar Y (2010) Development of male sterility by silencing Bcp1 gene of *Arabidopsis* through RNA interference. Afr J Biotechnol 9:2736–2741

Thakur N, Upadhyay SK, Verma PC, Chandrashekar K, Tuli R, Singh PK (2014) Enhanced white-fly resistance in transgenic tobacco plants expressing double stranded RNA of *v-ATPase A* gene. PLoS One 9:e87235

Tinoco ML, Dias BB, Dall'Astta RC, Pamphile JA, Aragao FJ (2010) In vivo trans-specific gene silencing in fungal cells by *in planta* expression of a double-stranded RNA. BMC Biol 31:27

Tuschl T (2001) RNA interference and small interfering RNAs. Chembiochem 2:239–245

Varoquaux F, Blanvillain R, Delseny M, Gallois P (2000) Less is better: new approaches for seedless fruit production. Trends Biotechnol 18:233–242

Vazquez F (2006) *Arabidopsis* endogenous small RNAs: highways and byways. Trends Plant Sci 11:460–468. https://doi.org/10.1016/j.tplants.2006.07.006

Wang Y, Li J (2006) Genes controlling plant architecture. Curr Opin Biotechnol 17:123–129

Wang H, Jones B, Li Z, Frasse P, Delalande C, Regad F, Chaabouni S, Latche A, Pech J-C, Bouzayen M (2005) The tomato Aux/IAA transcription factor IAA9 is involved in fruit development and leaf morphogenesis. Plant Cell 17:2676–2692

Wang S, Wu K, Yuan Q, Liu X, Liu Z, Lin X, Zeng R, Zhu H, Dong G, Qian Q, Zhang G, Fu X (2012) Control of grain size, shape and quality by OsSPL16 in rice. Nat Genet 44:950–954. https://doi.org/10.1038/ng.2327

Wei MM, Wei HL, Wu M, Song MZ, Zhang JF, Yu JW, Fan S, Yu S (2013) Comparative expression profiling of miRNA during anther development in genetic male sterile and wild type cotton. BMC Plant Biol 13:66. https://doi.org/10.1186/1471-2229-13-66

Weise SE, Aung K, Jarou ZJ, Mehrshahi P, Li Z, Hardy AC, Carr DJ, Sharkey TD (2012) Engineering starch accumulation by manipulation of phosphate metabolism of starch. Plant Biotechnol J 10:545–554. https://doi.org/10.1111/j.1467-7652.2012.00684.x

Wilson RC, Doudna JA (2013) Molecular mechanisms of RNA interference. Annu Rev Biophys 42:217–239

Xie Z, Kasschau KD, Carrington JC (2003) Negative feedback regulation of Dicer-Like1 in Arabidopsis by microRNA-guided mRNA degradation. Curr Biol 13:784–789. https://doi.org/10.1016/S0960-9822(03)00281-1

Xiong A, Yao Q, Peng R, Li X, Han P, Fan H (2005) Different effects on ACC oxidase gene silencing triggered by RNA interference in transgenic tomato. Plant Cell Rep 23:639–646

Xue B, Hamamouch N, Li C, Huang G, Hussey RS (2013) The 8D05 parasitism gene of *Meloidogyne incognita* is required for successful infection of host roots. Phytopathology 103:175–181

Yadav BC, Veluthambi K, Subramaniam K (2006) Host-generated double stranded RNA induces RNAi in plant-parasitic nematodes and protects the host from infection. Mol Biochem Parasitol 148:219–222

Yang C, Li D, Mao D, Liu X, Ji C, Li X, Zhao X, Cheng Z, Chen C, Zhu L (2013) Overexpression of microRNA319 impacts leaf morphogenesis and leads to enhanced cold tolerance in rice (*Oryza sativa* L.). Plant Cell Environ 36:2207–2218. https://doi.org/10.1111/pce.12130

Yang X, Zhao Y, Xie D, Sun Y, Zhu X, Esmaeili N, Yang Z, Wang Y, Yin G, Lv S, Nie L, Tang Z, Zhao F, Li W, Mishra N, Sun L, Zhu W, Fang W (2016) Identification and functional analysis of microRNAs involved in the anther development in cotton genic male sterile line Yu98-8A. Int J Mol Sci 17:1677. https://doi.org/10.3390/ijms17101677

Yin C, Jurgenson JE, Hulbert SH (2011) Development of a host-induced RNAi system in the wheat stripe rust fungus *Puccinia striiformis f. sp. tritici*. Mol Plant-Microbe Interact 24:554–561

Yogindran S, Rajam MV (2015) RNAi for crop improvement. In: Bahadur B, Rajam MV, Sahijram L, Krishnamurthy KV (eds) Plant biology and biotechnology: Volume II: Plant genomics and biotechnology. Springer, India, pp 623–637

Youssef RM, Kim KH, Haroon SA, Matthews BF (2013) Post-transcriptional gene silencing of the gene encoding aldolase from soybean cyst nematode by transformed soybean roots. Exp Parasitol 134:266–274

Yu B, Lydiate DJ, Young LW, Schafer UA, Hannoufa A (2007) Enhancing the carotenoid content of *Brassica napus* seeds by down regulating lycopene epsilon cyclase. Transgenic Res 17:573–585

Yu R, Xu X, Liang Y, Tian H, Pan Z, Jin S, Wang N, Zhang W (2014) The insect ecdysone receptor is a good potential target for RNAi-based pest control. Int J Biol Sci 10:1171–1180

Zamore PD, Haley B (2005) Ribo-genome: the big world of small RNAs. Science 309:1519–1524

Zha WJ, Peng XX, Chen RZ, Du B, Zhu LL, He GC (2011) Knockdown of midgut genes by dsRNA-transgenic plant-mediated RNA interference in the Hemipteran insect *Nilaparvata lugens*. PLoS One 6:e20504

Zhang B (2015) MicroRNA: a new target for improving plant tolerance to abiotic stress. J Exp Bot 66:1749–1761. https://doi.org/10.1093/jxb/erv013

Zhang YC, Yu Y, Wang CY, Li ZY, Liu Q, Xu J, Liao JY, Wang XJ, Qu LH, Chen F, Xin P, Yan C, Chu J, Li HQ, Chen YQ (2013) Over expression of microRNA OsmiR397 improves rice yield by increasing grain size and promoting panicle branching. Nat Biotechnol 31:848–852

Zhang W, Xie Y, Xu L, Wang Y, Zhu X, Wang R, Zhang Y, Muleke EM, Liu L (2016) Identification of microRNAs and their target genes explores miRNA-mediated regulatory network of cytoplasmic male sterility occurrence during anther development in radish (*Raphanus sativus* L.). Front Plant Sci 7:1054. https://doi.org/10.3389/fpls.2016.01054

Zhou Y, Yuan Y, Yuan F, Wang M, Zhong H, Gu M, Liang G (2012) RNAi-directed down-regulation of RSV results in increased resistance in rice (*Oryza sativa* L.). Biotechnol Lett 34:965–972. https://doi.org/10.1007/s10529-012-0848-0

Zhu JK (2008) Reconstituting plant miRNA biogenesis. Proc Natl Acad Sci U S A 105:9851–9852

Zhu JQ, Liu S, Ma Y, Zhang JQ, Qi HS, Wei ZJ, Yao Q, Zhang WQ, Li S (2012) Improvement of pest resistance in transgenic tobacco plants expressing dsRNA of an insect associated gene *EcR*. PLoS One 7:e3857

Chapter 4
RNAi for Resistance Against Biotic Stresses in Crop Plants

Pradeep Kumar Jain, Ramcharan Bhattacharya, Deshika Kohli, Raghavendra Aminedi, and Pawan Kumar Agrawal

Abstract RNA interference (RNAi)-based gene silencing has become one of the most successful strategies in not only identifying gene function but also in improving agronomical traits of crops by silencing genes of different pathogens/pests and also plant genes for improvement of desired trait. The conserved nature of RNAi pathway across different organisms increases its applicability in various basic and applied fields. Here we attempt to summarize the knowledge generated on the fundamental mechanisms of RNAi over the years, with emphasis on insects and plant-parasitic nematodes (PPNs). This chapter also reviews the rich history of RNAi research, gene regulation by small RNAs across different organisms, and application potential of RNAi for generating transgenic plants resistant to major pests. But, there are some limitations too which restrict wider applications of this technology to its full potential. Further refinement of this technology in terms of resolving these shortcomings constitutes one of the thrust areas in present RNAi research. Nevertheless, its application especially in breeding agricultural crops resistant against biotic stresses will certainly offer the possible solutions for some of the breeding objectives which are otherwise unattainable.

Keywords RNA interference · RNAi · Biotic stresses · Insect resistance · Disease resistance

4.1 Introduction

RNA interference (RNAi) is an invaluable technology for unraveling gene function in the area of functional genomics. It has been utilized in basic research ranging from functional studies to gene knockdown in plants and vertebrates and to suppression of cancer and viral diseases in medicine. Moreover, from application point of

P. K. Jain (✉) · R. Bhattacharya · D. Kohli · R. Aminedi
ICAR-NRC on Plant Biotechnology, IARI Campus, New Delhi, India

P. K. Agrawal
ICAR-NASF, KAB-I, IARI Campus, New Delhi 110012, India

© Springer International Publishing AG, part of Springer Nature 2018
S. S. Gosal, S. H. Wani (eds.), *Biotechnologies of Crop Improvement, Volume 2*,
https://doi.org/10.1007/978-3-319-90650-8_4

view, it is being used extensively for trait modification by selective inhibition of gene expression universally across the organisms. In agriculture, RNAi has been extensively employed particularly for imparting resistance against biotic stresses including insects, bacteria, nematodes, fungal infection, and viruses (Tan and Yin 2004; Yanagihara et al. 2006; Good and Stach 2011; Banerjee et al. 2017; Majumdar et al. 2017; Zhang et al. 2017). This chapter focuses on how RNAi has been extensively used in managing various biotic stresses which constitute serious impediments to crop productivity. Damage due to insects, fungus, parasitic weeds, and plant-parasitic nematodes is a major biotic constraint causing significant yield losses in agriculture year-round.

4.2   History of RNAi

The basic concept involves a double-stranded RNA (dsRNA) molecule which potentially silences the gene with complementary sequences post-transcriptionally. RNAi phenomenon was first discovered in a free-living nematode, *Caenorhabditis elegans* (Fire et al. 1998). They coined the term "RNAi" for describing effective silencing of gene expression by exogenously supplied sense and antisense RNAs in the model nematode, *Caenorhabditis elegans*. This phenomenon, conserved among eukaryotes, was described as post-transcriptional gene silencing (PTGS) (Carthew and Sontheimer 2009; Berezikov 2011). Historically the roots of this exciting development can be traced back to 1990 when chsA gene was overexpressed in transgenic petunia plants and the silencing of endogenous as well as transgene of chalcone synthase in the transgenic plants was observed (Napoli et al. 1990). Loss of endogenous as well as transgene-derived mRNAs was described as co-suppression, a term formulated by Napoli. Soon, importance of this technology was well understood by the scientific community, and since then, phenomenal growth in this technology has taken place. In fungi, this mechanism of PTGS is known as quelling (Agrawal et al. 2003). In nature, viruses mediate PTGS in plants, and the effect is amplified in cytoplasm or in the nucleus.

4.3   Biogenesis and Mechanism of RNAi Pathway

The major small noncoding RNAs (ncRNAs) include microRNAs (miRNAs), small interfering RNAs (siRNAs), and PIWI-interacting RNAs (piRNAs) which are all involved in downregulation of gene expression (Aalto and Pasquinelli 2012). Each class of small RNA is unique in its biogenesis and mechanism of action, but there are a few similarities too. Both miRNAs and siRNAs are processed from larger dsRNAs through cleavage by Dicer (a ribonuclease III enzyme). Both are associated with Argonaute proteins (AGO) (Ketting 2011) forming RNA-induced silencing complex (RISC). RISC basically is an Argonaute protein bound to a single strand of noncoding RNA. Varied ribonucleoprotein complexes arise due to several ncRNAs and Argonautes involved in formation of RISC (Darrington et al. 2017).

The RNAi-mediated gene silencing occurs basically in three stages (Siomi and Siomi 2009). First one involves processing of long dsRNA into small dsRNA by ribonuclease III; in the second stage, unwinding of these small RNAs leads to formation of one guide strand, which is loaded into the RISC, whereas the other strand known as passenger strand gets degraded. Finally, the RISC, directed by the guide strand, locates mRNAs containing sequences complementary to the guide, binds to these sequences, and either degrades the mRNA or blocks its translation (Winter et al. 2009). The mechanism of RNAi is emerging with all its complexity, but with clarity, as more and more players involved in the interference are getting identified and characterized.

The involvement of siRNA molecules as important intermediates of the RNAi process became evident through independent investigations carried out by researchers around the world. The first report of accumulation of siRNAs was confirmed by Hamilton and Baulcombe (1999) while studying tomato lines transformed with 1-aminocyclopropane-1-carboxyl oxidase (ACO) and later in *Drosophila* syncytial blastoderm embryo (Tuschl et al. 1999). Two other independent studies experimentally exhibited the 21–23 nucleotide small RNAs as intermediates for degradation of mRNA (Zamore et al. 2000; Elbashir et al. 2001). But how these small RNA molecules are excised from their precursor was yet to be discovered. As the role of RNase III enzymes had been recognized as dsRNA nucleases already, the RNase III domain-containing proteins were searched as one of the factors in siRNA biogenesis. Recently only, different experimental studies revealed the involvement of RNA-processing enzymes in chopping off the dsRNAs into siRNA molecules. One of the crucial enzymes, Dicer, was identified in *Drosophila*, by browsing its genome for the proteins dedicated for functioning like RNase III endonuclease activity (Bernstein et al. 2001). In another study, Dicer protein in *C. elegans* (a bidentate nuclease) was characterized revealing its functional role in small RNA regulatory pathways (Ketting et al. 2001). It was also deduced to be the ortholog of *Drosophila* DCR-1 protein. Ketting et al. (2001) in this study also showed the requirement of ATP for regulating the rate of siRNA synthesis. In yet another experiment reduction in ATP levels by 5000-fold in *Drosophila* revealed a decrease in the rate of siRNA production (Nykanen et al. 2001). It is now believed that Dicer acts as a complex of proteins with domains for dsRNA binding at its C terminus which are separable from motifs like helicase and PAZ. It was experimentally found to co-localize with an endoplasmic reticulum protein, calreticulin (Caudy et al. 2002). However, the role of ATP in the biogenesis of siRNA is abstruse due to its varied functions among different Dicer proteins in different organisms. An imperative involvement of ATPase in siRNA production was exhibited by *Drosophila* Dicer-2 and *C. elegans* Dcr-1 (Tomari and Zamore 2005) in contrast to human Dicer wherein an ATPase-defective mutant showed regular processing (Carthew and Sontheimer 2009). A comprehensive biochemical, molecular, genetic, and structural study revealed the presence of two main domains, namely, PAZ and RNaseIII, performing a crucial role in excising the siRNAs (Zhang et al. 2004; Macrae et al. 2006).

Once Dicer cuts off the dsRNA, synthesized siRNAs then enter the RISC complex. The double-stranded siRNAs act as a template for the RISC to recognize the complementary mRNA aided by Argonaute proteins. Agronaute proteins are required for the RISC assembly and have been biochemically characterized in *Drosophila*. Amplification of siRNAs has been reported in nematodes, fungus, plants and amoeba (Dykxhoorn et al. 2003). RNA-dependent RNA polymerase (RdRP) is proposed to be involved in augmenting the siRNA molecules on the basis of biochemical studies (Lipardi et al. 2001; Sijen et al. 2001). Sijen demonstrated the fundamental role of *rrf1* gene having sequence homology to RdRP for the production of secondary siRNAs in *C. elegans*. In this study, the concept of transitive RNAi pathway induced by secondary siRNAs came into the picture. Thus, catalytic nature of RNAi was proposed.

4.4   RNAi in Insect Resistance

The direct loss in crop productivity due to damage by insect pest and the input-cost accrued in agrochemical based protection amount to billions of dollars every year worldwide. In spite of alarming environmental hazard directly due to residual toxicity of insecticides in food chain, the consumption of insecticides has been ever incremental. This is primarily due to resistance development in insect-pest population and lack of awareness among the farming community. The worldwide consumption of insecticide increases by almost 30% in every 4 years. Therefore, insect-pest management, preferably through an integrative approach and without indiscriminate use of insecticide, has become a most sought-after area in research planning worldwide. Millions of dollars were granted for researching on sustainable and low-cost alternate avenues of pest control strategies in five most important agricultural crops. Development of resistant cultivars in crops seems to be the most acclaimed alternative for minimizing the application of insecticides. Unfortunately, for most of the major crop- insect damage, either such resistant cultivars are not available or the resistance has been broken down. Further insight into such examples reveals that lack of resistance source maneuverable either through classical breeding or through transgenesis has been the major constraint.

Accessing unrelated gene pool through development of transgenics has emerged as the most potential avenue for overcoming this bottleneck. Success of *Bacillus thuringiensis* (*Bt*) toxin-mediated protection of a large number of crops has been celebrated widely and in fact demonstrated for the first time the potential of biotechnological means in developing genetic resistance. However, applicability of Bt-mediated protection is limited as many of the insect pests are not affected by Bt toxin, and also this technology has faced second-generation challenge of some major insect species developing resistance to *Bt* (Tabashnik 2008; Tabashnik et al. 2008). It has been realized that lack of useful insecticidal transgenes is the major limitation in transgenic-based engineering of genetic resistance. In contrary, through RNAi, any important gene can be precisely targeted to elicit lethality in the insect species. Use of RNAi has rapidly progressed for gene function analysis in various insect orders, including Diptera (Lum et al. 2003; Dietzl et al. 2007), Lepidoptera (Tian et al. 2009; Terenius et al. 2011), Coleoptera (Baum et al. 2007; Zhu et al. 2011; Bolognesi et al. 2012), and Hymenoptera (Nunes and Simoes 2009; Meer and Choi 2013; Zhao and Chen 2013).

4.5   RNAi Pathway in Insects

Like in plants, RNAi is primarily involved in antiviral defense mechanisms of insects as a part of its innate immunity. However, a number of studies indicate several branches of RNAi involved in endogenous gene regulation in addition to silencing of genetic elements of pathogen invaders and transposons (Van Rij and Berezikov 2009). Gene silencing through RNAi is systemic and transitive as originally described in *C. elegans*. A host-derived RNA-dependent RNA polymerase (RdRp) amplifies the RNAi post-elicitation by dsRNA. In contrast to nematodes, in insects, there is no definite proof of the presence of RdRp. In the absence of RdRp-mediated amplification of dsRNA in insects, the silencing is expected to be more localized. Therefore, elicitation of an effective silencing will require delivery of the dsRNA directly to the target cells and tissues in a continuous manner. The administered dsRNA enters the insect cells via siRNA pathway in which a complex consisting of the RNAase III enzyme (Dicer-2) and TRBP cuts the dsRNA into small 21–23 bpsiRNAs. The RISC bound to AGO recognizes the guide strands of the siRNAs. This complex then binds to complementary sequences of target RNAs which are eventually degraded.

Two types of RNAi pathway are known to occur in insects: cell-autonomous and non-cell-autonomous RNAi. Cell-autonomous RNAi is limited to the cells in which the dsRNA is administered or delivered. In contrary, when the silencing occurs in cells different from the cells delivered with or producing the dsRNA, it is called non-cell-autonomous RNAi. Depending on how the dsRNA is acquired by the cell, non-cell-autonomous RNAi can be grouped in two kinds: environmental RNAi and systemic RNAi. In environmental RNAi, dsRNA is absorbed by a cell from the surrounding environment. Therefore, this is seen in unicellular organisms or any cell lines when administered with dsRNA. Environmental RNAi does not necessarily result into systemic spread of the response. In multicellular organisms, silencing signal is transported from one cell to another by systemic RNAi.

In case of transgenic host-mediated delivery of dsRNA, the dsRNA is delivered into the gut lumen of insects. For eliciting effective RNAi, dsRNA must be taken up by gut cells from the gut lumen which is known as environmental RNAi. If the transcripts of target genes are prevalently expressed in tissues outside the gut cells, the systemic RNAi has to occur for spreading of silencing signal. However, there is no definite study on assessing systemic RNAi in insects.

4.6   RNAi in Plant-Parasitic Nematodes (PPNs)

Plant-parasitic nematodes (PPNs) are grouped on the basis of different type of life-styles, i.e., sedentary, including root-knot nematode (RKN) and cyst nematodes, and migratory, including root-lesion nematodes. Sedentary endoparasites interact with the host through secretions which are vital cues for plant-nematode interactions. These secretory proteins are thus of major interest as targets for modulating the interaction. RNAi has been extensively used in functional genomics performed on *C. elegans* and opened up the possibility of deciphering the function of uncharacterized genes in other parasitic nematodes. Recent discoveries focused on unraveling the role of different components of RNAi in parasitic nematodes has eventually led to increasing our understanding of RNAi mechanism.

There are overwhelming reports on managing PPNs using RNAi. In nematodes, systemic RNAi can be observed resulting in a gene knockout that spreads throughout the organism. This is because RNA-dependent RNA polymerase (RdRP) is present in nematodes which interact with RISC and leads to production of new dsRNAs which are acted upon by Dicer enzymes and further produces new siRNAs (secondary siRNAs) in a well-coordinated amplification reaction. Therefore, the effect of dsRNA persists over development and also can be exported to neighboring cells thereby leading to silencing effect all over the organism (Daniel and John 2008). *C. elegans* displays systemic RNAi wherein the dsRNA/siRNAs entering from the environment can spread from one cell to another. Studies on identification of effectors of systemic RNAi revealed presence of protein SID-1 in *C. elegans* (Winston et al. 2002; Feinberg and Hunter 2003). Interestingly, *M. incognita* and *M. hapla*, along with other parasitic nematodes, despite exhibiting successful RNAi, were found deficient in SID-1 and other related proteins having a key role in dsRNA uptake and its spread. Several detailed comparative studies have postulated the presence of RNAi components in different PPNs and animal parasitic nematodes that were reported in *C. elegans* (Lendner et al. 2008; Dalzell et al. 2011; Haegeman et al. 2011). All these studies found rare proteins taking part in RNAi pathway. Seventy-seven orthologous effectors in *C. elegans* were searched in 13 nematode species, *Ancylostoma caninum, Oesophagostomum dentatum, Ascaris suum, Brugia malayi, C. brenneri, C. briggsae, C. japonica, C. remanei, Haemonchus contortus, Meloidogyne hapla, M. incognita, Pristionchus pacificus*, and *Trichinella spiralis*, using reciprocal BLAST followed by domain structure verification (Maule et al. 2011). It was concluded that effector deficiencies cannot, in any way, be associated with reduced susceptibility in parasitic nematodes. Surprisingly, minimum diversity was observed among these parasitic nematodes in most of the orthologous genes belonging to different functional groups (Table 4.1). Thus it was evident that all the species possess varied proteins from across the RNAi spectrum each with alternative proteins which are yet to be fully identified and characterized.

Table 4.1 RNAi effector components in selected nematodes[a]

Species	RNAi effectors – functional groupings					
	Small RNA biosynthesis	dsRNA uptake and spread	Amplification proteins	Argonautes and RISC components	RNAi inhibitors	Nuclear RNAi effectors
Free-living nematodes						
Caenorhabditis elegans	9	5	7	31	31	15
Caenorhabditis brenneri	9	4	6	21	9	15
Caenorhabditis briggsae	9	5	6	21	9	15
Caenorhabditis japonica	9	5	5	18	8	15
Caenorhabditis remanei	9	5	5	22	4	15
Pristionchus pacificus	6	2	4	14	4	5
Plant-parasitic nematodes						
Meloidogyne hapla	6	1	3	7	3	7
Meloidogyne incognita	7	1	3	9	2	6
Animal parasitic nematodes						
Trichinella spiralis	6	1	3	5	3	4
Ascaris suum	7	1	5	17	5	8
Brugia malayi	9	1	4	8	4	10
Haemonchus contortus	7	2	4	19	5	11
Oesophagostomum dentatum	6	2	3	14	5	6

[a]Data derived from Dalzell et al. (2011)

4.7   Mode of dsRNA Delivery

The efficacy of gene silencing substantially depends on the method of dsRNA uptake. In absence of systemic RNAi, gene silencing shall be limited to the cells that take up the dsRNA. Therefore, appropriate delivery system is pivotal (Terenius et al. 2011). Different delivery methods of dsRNA that have been used for successful RNAi in insects and nematodes include microinjection, feeding on either artificial diet (Table 4.2), and/or host-mediated delivery through transgenic plants (Fig. 4.1). Each of these methods has its own advantages and limitations.

Table 4.2 Summary of targeted genes silenced by RNAi approach in plant-parasitic nematodes

Target gene	Nematode	Host plant	Phenotype	Method of delivery	References
Hgctl	H. glycines		41% reduction in number of nematodes	Soaking	Urwin et al. (2002)
Hgcp-1	H. glycines		40% reduction in number of nematodes	Soaking	Urwin et al. (2002)
MiDuox1	M. incognita		70% reduction in number of nematodes	Soaking	Bakhetia et al. (2005)
Gr-eng-1 and Gr-ams-1	G. rostochiensis		Around 50% reduction in number of nematodes Reduced ability to locate and invade roots	Soaking	Chen et al. (2005)
Chitin synthase	M. artiellia		Delayed egg hatch	Soaking	Fanelli et al. (2005)
Hg-amp-1	H. glycines		61% decrease in number of female reproductive	Soaking	Lilley et al. (2005)
Integrase and splicing	M. incognita	Tobacco	>90% reduction in number of established nematodes	HD-RNAi	Yadav et al. (2006)
Secreted peptide 16D10	M. incognita M. arenaria M. javanica M. hapla	Arabidopsis	63–90% reduction in number of galls and gall size	Soaking and HD-RNAi	Huang et al. (2006)
Major sperm protein	H. glycines	Soybean	Up to 68% reduction in number of eggs	HD-RNAi	Steeves et al. (2006)
Putative transcription factor	M. javanica	Tobacco	None	HD-RNAi	Fairbairn et al. (2007)
Ribosomal protein 3a, ribosomal protein 4, spliceosomal SR protein	H. glycines	Soybean	87% reduction in number of female cysts 81% reduction in number of female cysts 88% reduction in number of female cysts	HD-RNAi	Klink et al. (2009)

Gene/target	Nematode	Plant	Effect	Method	Reference
4G06, ubiquitin-like, 3B05, cellulose-binding protein, 8H07, SKP1-like and 10AO6, zinc finger protein	H. schachtii	Arabidopsis	23–64% reduction in number of developing females; 12–47% reduction in number of developing females; >50% reduction in number of developing females; 42% reduction in number of developing females	HD-RNAi	Sindhu et al. (2009)
Y25, beta subunit of COPI complex	H. glycines	Soybean	81% reduction in number of nematode eggs	HD-RNAi	Li et al. (2010a, b)
Prp-17, pre-mRNA splicing factor and Cpn-1	H. glycines	Soybean	79% reduction in number of nematode eggs; 95% reduction in number of nematode eggs	HD-RNAi	Li et al. (2010b)
Fib-1	H. glycines	Soybean	24% and 37% reduction in cyst and eggs, respectively	HD-RNAi	Li et al. (2010a)
Rpn7	M. incognita	Tomato	Reduction in motility and infectivity of J2 s	Soaking and HD-RNAi	Niu et al. (2012)
AF531170, parasitism gene	M. incognita	Tomato	54–59% reduction in number of developing females	HD-RNAi	Choudhary et al. (2012)
8D05, parasitism gene	M. incognita	Arabidopsis	Reduction in number of galls	HD-RNAi	Xue et al. (2013)
flp-14 and flp-18, FMRF amide-like peptide	M. incognita	Tobacco	Reduction in parasitic ability from 67–86%; Reduction in parasitic ability from 53–82%	HD-RNAi	Papolu et al. (2013)
Mi-ser-1, serine protease, Mi-cpl-1, cysteine protease and Mi-asp-1 + Mi-ser-1 + Mi-cpl-1 (fusion)	M. incognita	Tobacco	Reduction in number of eggs per gram of root; Reduction in egg hatching ratio; Reduction in number of eggs per gram of root; Reduction in number of eggs per gram of root	HD-RNAi	Antonino de Souza Júnior et al. (2013)
Pv010	P. vulnus	Walnut	Reduction in number of nematodes	Feeding (bacterial) and HD-RNAi	Walawage et al. (2013)
Mc16D10L	M. chitwoodi	Potato	65–68% reduction in the number of egg masses	HD-RNAi	Dinh et al. (2014a)
Mc16D10L	M. chitwoodi	Arabidopsis	57 and 67% reduction in number of egg masses and eggs, respectively	HD-RNAi	Dinh et al. (2014b)

(continued)

Table 4.2 (continued)

Target gene	Nematode	Host plant	Phenotype	Method of delivery	References
Mi-cpl-1	*M. incognita*	Tomato	60–80% reduction in infection and multiplication	Soaking and HD-RNAi	Dutta et al. (2015)
Pp-pat-10 and Pp-unc-87	*P. penetrans*	Soybean	Up to 40% reduction in number of nematodes Up to 50% reduction in number of nematodes	Soaking and HD-RNAi	Vieira et al. (2015)
Rs-cb-1	*R. similis*	Tobacco	Reduced reproduction and pathogenicity	Soaking and HD-RNAi	Li et al. (2015)
HSP90, heat shock protein	*M. incognita*	Tobacco	Delayed gall formation and up to 46% reduction in the number of eggs	HD-RNAi	Lourenço-Tessutti et al. (2015)
ICL, isocitrate lyase			Up to 77% reduction in egg oviposition		
Unc-15	*Ditylenchus destructor*	Sweet potato	50% reduction in the infection area	HD-RNAi	Fan et al. (2015a, b)
MiMSP40	*M. incognita*	Arabidopsis	Reduction in the number of galls	HD-RNAi	Niu et al. (2016)
MeTCTP	*M. enterolobii*	Tomato	Reduction in number of nematodes	Tobacco rattle virus-mediated gene silencing	Zhuo et al. (2017)
Integrase and splicing	*M. incognita*	Arabidopsis	70% reduction in infection 60% reduction in infection	HD-RNAi	Kumar et al. (2017)
msp-18 and msp-20	*M. incognita*	Eggplant	43.64–69.68% and 41.74–67.30% reduction in nematode multiplication, respectively	HD-RNAi	Shivakumara et al. (2017)

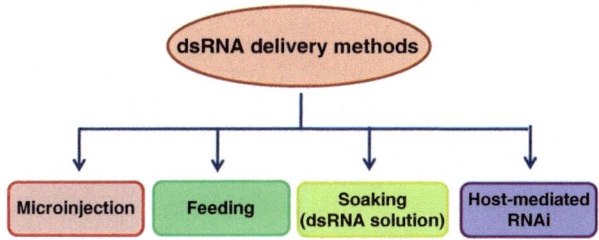

Fig. 4.1 Different delivery methods of dsRNA employed in RNAi strategy

4.7.1   Microinjection

Microinjection involves injection of dsRNA or siRNA directly into the body of an organism and has been demonstrated as one of the most successful delivery methods for RNAi to validate gene functions (Ober and Jockusch 2006). In this method, dsRNA is produced by in vitro transcription using T7 or Sp6 promoter sequences. It has been employed successfully for suppressing genes in both insects and nematodes.

4.7.1.1   In Insects

In *D. melanogaster*, microinjection has been successfully used for delivering dsR-NAs for two genes, viz., *frizzled* and *frizzled2*, into embryos. The silencing resulted in defects in embryonic patterning that was similar to loss of wingless (wg) function. This was the first study proving the function of *frizzle* through dsRNA microinjection in an insect (Kennerdell and Carthew 1998). Since then, microinjection-based delivery has been used in several insect species. A comprehensive list of Hemipteran insects subjected to microinjection for studying RNAi is presented in Table 4.3. Direct injection of dsRNA into the insect body leads to higher efficiency of gene expression attenuation compared to other methods. Nevertheless, there are several limitations in microinjection delivery method. In vitro synthesis of dsRNA is skill intensive and costly. Additionally, recovery of the insects, especially smaller insects, from aftershock of microinjection, is relatively low. The significant aftershock is due to damage of cuticle leading to adverse immune responses in the insect (Roxstrom-Lindquist et al. 2004). Therefore, microinjection is rarely used in functional analysis of large number of genes from the point of view of insect-pest control. It is evident from Table 4.3 that in the microinjection, mediated delivery has been carried out mostly in the case of hemipteran insects.

4.7.1.2   In Nematodes

After injecting dsRNAs into the worms, progeny is counted and recorded for the mutant phenotypes. Usually after 24 h of injection, good RNAi effect is observed (Fire et al. 1998). In *C. elegans*, dsRNAs of genes like *unc-22, unc-54, fem1*, and *hlh-1* were injected into the adult hermaphrodites, and the interference effect was observed. It was also proposed that in an antisense mechanism, interference of endogenous gene is due to the hybridization between the injected RNA and endogenous mRNA (Fire et al. 1998). It is a classical technique, and different target mRNAs can be used for injection simultaneously. However, microinjection has not been very successful in plant-parasitic nematodes in general and particularly in *M. incognita*. This is because of the small size of the infective stages and their inability to ingest fluid without host plant infection (Banerjee et al. 2017). In this process, although the range of dsRNA concentrations can be used, the success rate relies upon ample uptake or absorption by the worms (Hull and Timmons 2004).

4.7.2   Feeding on Artificial Diet

4.7.2.1   In Insects

dsRNA delivery through artificial diet has been the most popular method for delivering dsRNA into the insect gut especially for relatively smaller insects such as Hemipterans, which are sap-sucking. Several insect species of different taxa were studied for RNAi by the administration of dsRNA through artificial diet as presented in Table 4.3. Araujo et al. (2006) fed the blood-sucking *Rhodnius prolixus* with an artificial diet containing dsRNA of the *nitrophorin2* (*Np2*) gene and found that the saliva of control *R. prolixus* prolonged plasma coagulation by approximately fourfold compared with the saliva of *Np2*-knockdown *R. prolixus*. Feeding *A. pisum* with an artificial diet supplemented with dsRNA of the *A. pisum* aquaporin 1 (ApAQP1) gene caused attenuated expression of the target gene, which resulted in an increased osmotic pressure of the hemolymph in this insect (Shakesby et al. 2009).

4.7.2.2   In Nematodes

In a nematode, feeding involves ingestion of bacteria expressing dsRNA of the target gene against which RNAi is employed. Timmons et al. (2001) developed engineered bacteria deficient for RNaseIII producing high levels of dsRNA segments of a specific gene. *C. elegans* feeding on these engineered bacteria showed RNAi effect leading to loss-of-function phenotypes for the target genes. One of the advantages of this method is that it can be conducted for stage-specific RNAi experiments as worms of any stage can be fed with dsRNA (Kamath et al. 2001; Ahringer 2006).

Table 4.3 List of genes targeted for gene silencing in different insect orders

Organism	Target gene	Stage	Assay method	Conc. of dsRNA/siRNA	Phenotype/mRNA silencing	References
Coleoptera						
Diabrotica virgifera virgifera	Multiple targets	Neonates	Artificial diet	1–10 ppb	Larval stunting and mortality	Baum et al. (2007)
	Snf 7	Neonates	Artificial diet	4.3 ppb	Growth inhibition and mortality	Bolognesi et al. (2012)
Diabrotica undecimpunctata howardi	Snf 7	Neonates	Artificial diet	1.2 ppb	Mortality	Bolognesi et al. (2012)
	V- ATPase A and E	Neonates	Artificial diet	~0.1 ppm	Larval stunting and mortality	Baum et al. (2007)
	α-Tubulin	Neonates	Artificial diet	~0.1 ppm	Larval stunting and mortality	Baum et al. (2007)
Leptinotarsa decemlineata	V-ATPase A and E	Neonates	Artificial diet	~10 ppb	Larval stunting and mortality	Baum et al. (2007)
	Multiple targets	Neonates	Leaf tissue	ND	Reduced body weight and mortality	Zhu et al. (2011)
Phyllotreta striolata	Arginine kinase	Adults	Leaf tissue	0.8 ppb	Growth retardation, reduced fecundity, and increased mortality	Zhao et al. (2008)
Tribolium castaneum	V-ATPase E	Neonates	Artificial diet	2.5 ppm	Reduced growth and mortality	Whyard et al. (2009)
Diptera						
Aedes aegypti	V-ATPase A	Adults	Artificial diet	~1000 ppm	Significant transcript knockdown	Coy et al. (2012)
	Multiple targets	First instars	Water	200, 500 ppm	Reduced growth and mortality	Singh et al. (2013)
	ATP-dependent efflux pump	Second instars	Water	~30 ppm	Increased toxicity	Figueira-Mansur et al. (2013)
Anopheles gambiae	Chitin synthase 1, 2	Third instars	Artificial diet	–	Increased susceptibility to insecticides	Zhang et al. (2010)

(continued)

Table 4.3 (continued)

Organism	Target gene	Stage	Assay method	Conc. of dsRNA/siRNA	Phenotype/mRNA silencing	References
Anopheles stephensi	3-HKT	First instars	Transgenic Chlamydomonas	ND	>50% mortality	Kumar et al. (2013)
Bactrocera dorsalis	Multiple targets	Adults	Artificial diet	2000 ppm	Affected egg production and 20% mortality	Li et al. (2011b)
Glossina morsitans morsitans	Tsetse EP	Male adults	Blood meal	>400 ppm	No mortality	Walshe et al. (2009)
	Transferrin	Male adults	Blood meal	>400 ppm	No mortality	Walshe et al. (2009)
Hemiptera						
Acyrthosiphon pisum (pea aphid)	Aquaporin	6-day-old nymphs	Artificial diet	1000–5000 ppm	Elevated osmotic pressure of the hemolymph	Shakesby et al. (2009)
	V-ATPase E	First instars	Artificial diet	3.4 ppm	Reduced growth and mortality	Whyard et al. (2009)
	V-ATPase E	Neonates	Artificial diet	ND	dsRNA degradation in saliva hemolymph	Christiaens et al. (2014)
	Hunchback (hb)	Neonates	Artificial diet	750 ppm	Increased mortality	Mao and Zeng (2012)
	Salivary protein COO2	Adults	Injection	50 ng	Lethal	Mutti et al. (2006)
	Gut digestive enzyme cathepsin-L	Adults	Injection or feeding	92–460 ng 0.9–2.6 μg μL−1	Higher mortality and impaired molting	Sapountzis et al. (2014)
	Structural sheath protein SHP	Adults	Injection	50 ng	Impaired long-term feeding from sieve tubes and reduced fecundity	Will and Vilcinskas (2015)
	Angiotensin-converting enzymes ACE1 and ACE2	Adults	Injection	138 ng	Higher mortality of aphids feeding on plant	Wang et al. (2015)

	Peroxiredoxin 1 gene ApPrx1	Adults	Injection	285.2 ng	Decreased survival of aphids under oxidative stress	Zhang and Lu (2015)
	Macrophage migration inhibitory factor ApMIF1	Adults	Injection	100 ng	Decreased survival and fecundity of aphids feeding on their host plant	Naessens et al. (2015)
	Calreticulin, Cathepsin-L	Adults	Microinjection	5 nl, 23 nl and 46 nl of siRNA (6 µg/µl)	Insignificant RNAi effects	Possamai et al. (2007)
Aphis gossypii (cotton aphid)	Carboxylesterase gene CarE	Adults	Feeding	50–500 ng µL−1	Reduced resistance to organophosphorus insecticides	Gong et al. (2014)
	Cytochrome P450 monooxygenase gene CYP6A2	Adults	Feeding	100 ng µL−1	Increased sensitivity of the resistant aphids to spirotetramat and alpha-cypermethrin	Peng et al. (2016)
	Odorant-binding protein 2 AgOBP2	Adults	Feeding	62.5–250 ng µL−1	Impaired host-seeking and oviposition behavior of aphids	Rebijith et al. (2016)
Bactericera cockerelli	Multiple targets	Adults	Artificial diet	500–1000 ppm	Mortality	Wuriyanghan et al. (2011)
Bemisia tabaci	V-ATPase subunit A, rpL19	Adults	Artificial diet	3,11 ppm	Mortality	Upadhyay et al. (2011)
Nilaparvata lugens brown plant hopper	Trehalose PO4 synthase	Third instars	Artificial diet	500 ppm	Lethality	Chen et al. (2010)
	V-ATPase E	2nd instars	Artificial diet	50 ppm	Transcript knockdown and no mortality	Li et al. (2011a)

(continued)

Table 4.3 (continued)

Organism	Target gene	Stage	Assay method	Conc. of dsRNA/siRNA	Phenotype/mRNA silencing	References
Grain aphid (*Sitobion avenae*)	Catalase gene CAT	Third instar	Feeding	7.5 ng μL^{-1}	Reduced survival rate and ecdysis index	Deng and Zhao (2014)
	Acetylcholinesterase gene SaAce1	Adults	Injection	10 ng	Increased susceptibilities to pirimicarb and malathion and reduced fecundity	Xiao et al. (2015)
	Cytochrome c oxidase subunit VIIc precursor; zinc finger protein; three unknown proteins	Third instars	Feeding	3–7.5 ppm	Higher mortality and developmental stunting	Zhang et al. (2013)
	Secreted salivary peptide DSR32; salivary protein DSR33; serine protease 1 DSR48	Adults	Feeding	10 ng μL^{-1}	Higher mortality	Wang et al. (2015)
	Olfactory coreceptor gene SaveOrco	Adults	Feeding	20 ng μL^{-1}	Impaired response to behaviorally active odors	Fan et al. (2015a, b)
Bird cherry-oat aphid *Rhopalosiphum padi*	Acetylcholinesterase gene RpAce1	Adults	Injection	10 ng	Increased susceptibilities to pirimicarb and malathion and reduced fecundity	Xiao et al. (2015)
Greenbug *Schizaphis graminum*	Salivary protein C002	Adults	Feeding	20 ng μL^{-1}	Lethal	Zhang et al. (2015a, b)
Peregrinus maidis	V-ATPase B and D	Third instars	Artificial diet	500 ppm	Reduced fecundity and mortality	Yao et al. (2013)
Rhodnius prolixus	Nitrophorin 2	Second instars	Artificial diet	1000 ppm	Changes in aphid saliva content	Araujo et al. (2006)
Lygus lineolaris	Inhibitor of apoptosis	Neonates	Artificial diet	1000 ppm	Digestion of dsRNA	Allen and Walker (2012)

Hymenoptera						
Apis mellifera	Vitellogenin	Second instars	Natural diet	500–3000 ppm	Developmental stunting	Nunes and Simoes (2009)
Solenopsis invicta	PBAN/pyrokinin	Fourth instars	Artificial diet	1000 ppm	Mortality of pupae	Vander Meer and Choi (2013)
	Guanine nucleotide binding GNBP	Worker ants	Artificial diet	200 ppm	Mortality	Zhao and Chen (2013)
Isoptera						
Reticulitermes flavipes	Cellulase	Workers	Paper discs	5.1 μ g/cm²	Impact on molting and change in feeding behavior	Zhou et al. (2008)
	Hexamerin	Workers	Paper discs	2.2 μ g/cm²	Impact on molting and change in feeding behavior	Zhou et al. (2008)
Lepidoptera						
Chilo infuscatellus	CiHR3 molting factor	Third instars	Corn kernels	250 ppm	Yes	Zhang et al. (2012)
Epiphyas postvittana	Carboxylesterase	Third instars	Droplet	1000 ppm	Yes	Turner et al. (2006)
	Pheromone bp	Third instars	Droplet	1000 ppm	Yes	Turner et al. (2006)
Helicoverpa armigera	AchE receptor	Neonates	Artificial diet (Ec)	~0.35 ppm	Reduced fecundity, pupal weight reduction, mortality	Kumar et al. (2009)
	AchE receptor	Neonates	Leaf tissue	~0.35 ppm	Mortality	Kumar et al. (2009)
	Ecdysone receptor EcR	Third instars	Artificial diet (Ec)	ND	Molting defects and larval lethality	Zhu et al. (2012)
	HaHR3 molting factor	Third instars	Artificial diet (Ec)	ND	Developmental deformity and larval lethality	Xiong et al. (2013)

(continued)

Table 4.3 (continued)

Organism	Target gene	Stage	Assay method	Conc. of dsRNA/siRNA	Phenotype/mRNA silencing	References
	CYP6B6		Third instars	Artificial diet (Ec)	ND	Yes	Zhang et al. (2013b)
	Ultraspiracle protein, EcR		Third instars	Artificial diet	1000 ppm	Yes	Yang and Han (2014)
Manduca sexta	V-ATPase E		Neonates	Artificial diet	11 ppm	Yes	Whyard et al. (2009)
Ostrinia nubilalis	Chitinase		Neonates	Artificial diet	2500 ppm	Reduced body weight and mortality	Khajuria et al. (2010)
Plutella xylostella	CYP6BG1		Fourth instars	Droplet	800 ppm	Yes	Bautista et al. (2009)
	Rieske protein		Second instars	Leaf tissue	$3 \mu\ g/cm^2$	Mortality	Gong et al. (2011)
	AchE receptor		Second instars	Leaf tissue	$3 \mu\ g/cm^2$	Mortality	Gong et al. (2013)
Sesamia nonagrioides	JH esterase JHER		First to sixth instars	Artificial diet (Ec)	ND	Yes	Kontogiannatos et al. (2013)
Spodoptera exigua	Chitin synthase A		Neonates	Artificial diet (Ec)	ND	Yes	Tian et al. (2009)
	β1 integrin subunit		Fourth instars	Injection and leaf tissue	100–200 ppm	–	Surakasi et al. (2011)
Spodoptera litura	Aminopeptidase N		Neonates	Injection	ND	No	Rajagopal et al. (2002)

Spodoptera frugiperda	Allatostatin C	Fifth instars	Droplet	600 ppm	Yes	Griebler et al. (2008)
	Allatotropin 2	Fifth instars	Droplet	600 ppm	Yes	Griebler et al. (2008)
	SfT6 serine protease	Fourth instars	Droplet	600 ppm	Yes	Rodriguez-Cabrera et al. (2010)
Orthoptera						
Gryllus bimaculatus	Sulfakinins	Adults	Droplet	100–600 ppm	–	Meyering-Vos and Muller (2007)
Locusta migratoria	Multiple targets	Fourth instars	Artificial diet	~240 ppm	No	Luo et al. (2013)
Schistocerca gregaria	Tubulin, GAPDH	Adults	Artificial diet	ND	No	Wynant et al. (2012)

The feeding method has some major advantages over other methods of delivering dsRNA. These are as follows: (i) it is easy to perform; (ii) feeding dsRNA is less traumatic to the nymphs and juveniles than doing so via injections, the nymphs and juveniles remain healthier, and their mortality is comparatively lower (Shakesby et al. 2009); and (iii) perhaps most significantly, delivering dsRNAs in early stages of insects and nematodes is convenient by this method as compared to microinjection which needs special equipment and often causes high rate of mortality due to art effect. However, there are some challenges, viz., low efficiency of this method and requirement of large quantities of dsRNA, which need to be addressed. Moreover, a detailed study in understanding the mechanism of dsRNA delivery by ingestion for inhibiting gene expression is yet to be carried out.

4.7.2.3 Soaking Method for dsRNA Delivery in Nematodes

This method involves soaking of nematodes in concentrated dsRNA solution and subsequently scoring of worms or their progeny for phenotypes. RNAi by soaking is useful for treating a moderately large number of animals (e.g., 10–100). RNAi through soaking method was first employed in *C. elegans* as a tool for converting its genome sequence information into functional information (Tabara et al. 1998). Apart from *C. elegans*, silencing of genes in plant-parasitic nematodes (PPN) through soaking technique has been popularly used but with minor modifications. Other techniques like feeding and microinjection possess some limitations with respect to PPNs. In microinjection, successful recovery of injected juveniles is difficult and PPN juveniles do not take up dsRNA orally easily from the solutions. This was overcome by Urwin et al. (2002) by inducing oral uptake of dsRNA using octopamine, a neuroactive compound by cyst nematodes *Heterodera glycines* and *Globodera pallida*. This marked a revolution in imparting RNAi-mediated resistance in cyst and root-knot parasitic nematodes.

Since then many reports on successfully governing the nematode growth utilizing RNAi approach came into the picture. In later studies, compounds like resorcinol and serotonin were used for successful uptake of dsRNA in *M. incognita* (Rosso et al. 2005; Huang et al. 2006). Apart from neuroactive compounds, fluorescein isothiocyanate (FITC) as a marker for observing dsRNA uptake and as a mean of selecting affected individuals was used in many studies (Urwin et al. 2002; Rosso et al. 2005). Intestinal gene cysteine proteinase was suppressed through the soaking method in *G. pallida*, *H. glycines*, and *M. incognita* (Nakai and Horton 1999; Schmidt et al. 1999). Gene silencing by RNAi soaking has led to various abnormalities in processes like nematode hatching and molting and even resulted in reduced reproduction rates. Many genes, namely, chitin synthase, neuropeptides, msp, c-type lectin, and aminopeptidases, were targeted (Kennerdell and Carthew 1998; Schmidt et al. 1999; Dernburg and Karpen 2002; Ischizuka et al. 2002). But the efficiency and duration of the silencing effect were assessed for *M. incognita* calreticulin (Mi-crt) and polygalacturonase (Mi-pg-1) (Rosso et al. 2005). Other genes targeted by this approach are cellulases, pectate lyase, chorismate mutase, and glutathione-S transferase (Anandalakshmi et al. 1998; Cogoni and Macino 2000; Hammond et al. 2001; Matzke et al. 2001; Carmell et al. 2002). However, the silencing acquired by soaking in dsRNA solutions is often transient as duration of soaking and the concentration of dsRNAs affect the RNAi mechanism (Banerjee et al. 2017).

4.8 Resistance Via Transgenic Plants Expressing dsRNA

Another alternative method of dsRNA delivery is through host-delivered RNAi (HD-RNAi) where gene is silenced in target organism by the host plant. Since there is no synthesis of any gene product in HD-RNAi, it is likely to address the biosafety concerns more favorably.

4.8.1 In Insects

Genetic transformations of crop plants for expressing dsRNA homologous to important insect gene entail several advantages. It delivers the dsRNA to the target insect pest in a continuous fashion that leads to elicitation of RNAi throughout the life cycle of the insects. Host-mediated delivery of dsRNA was first demonstrated against two important agricultural pests, cotton bollworm, *Helicoverpa armigera*, and Western corn rootworm, *Diabrotica virgifera* (Baum et al. 2007; Mao et al. 2007). Transgenic rice was developed by delivering dsRNA targeting hexose transporter gene *NlHT1*, carboxypeptidase gene *Nlcar*, and the trypsin-like serine protease gene *Nltry* of *Nilaparvata lugens*. The study revealed reduced transcript levels of these three targeted genes in the insects that fed on these transgenic rice plants. However, insect lethality was not reported (Zha et al. 2011). Subsequently, several attempts have been made for attenuating key genes of the insects through transgenic host-mediated delivery of dsRNA as presented in Table 4.4. The gene construct for expression of the dsRNA essentially consists of 200–500 nucleotide tandem repeats of the target gene sequence under the control of a constitutive promoter. Such strategy also offers the scope of tissue specific expression of the dsRNA. For example, for targeting the phloem-feeding insect pests, phloem-specific expression of the dsRNA and their transport in phloem sieve elements would be more desirable. However, several attempts in this direction clearly indicated the effective level of protection would depend on targeting suitable target genes in addition to desired level of expression and delivery of intact dsRNA to the infesting insect pests (Price and Gatehouse 2008). Further understanding of the uptake process and elicitation of RNAi by dsRNA in insects will facilitate tailoring the gene expression cassette of dsRNA in order to achieve effective protection.

   Mao et al. (2007) used RNAi-mediated approach to reduce insect's ability to cope up when exposed to xenobiotic compounds, for example, gossypol. Transgenic cotton plants expressing a hairpin dsRNA targeting gossypol-inducible

Table 4.4 Summary of genes of the insects employed through transgenic host-mediated delivery of dsRNA

Insect	Target gene	Stage	Plant used	Phenotype	References
Coleoptera					
Diabrotica virgifera Virgifera	V-ATPase A	Neonates	Maize	Significant reduction in WCR feeding	Baum et al. (2007)
Western corn rootworm	Snf 7	Neonates	Maize	Stunted growth	Bolognesi et al. (2012)
Lepidoptera					
Helicoverpa armigera Cotton bollworm	CYP6AE14 GST	Third instars	Tobacco Arabidopsis	Retardation of larval growth	Mao et al. (2007)
	CYP6AE14	Third instars	Cotton	Increased tolerance and stunted larval growth	Mao et al. (2011, 2013)
	Ecdysone receptor (EcR)	Second instars	Tobacco	Growth reduction and mortality	Zhu et al. (2012)
	HaHR3 molting factor	Third instars	Tobacco	Reduced body weight and mortality	Xiong et al. (2013)
Spodoptera exigua	Ecdysone receptor (EcR)	Second instars	Tobacco	Increased mortality	Zhu et al. (2012)
Hemiptera					
Myzus persicae Green peach aphid	Salivary protein *MpC002* and receptor of activated kinase C gene *MpRack1*	Nymphs	Tobacco and Arabidopsis	Reduced fecundity	Pitino et al. (2011)
	Effector gene *MpC002, MpPIntO1*, and *MpPIntO2*		Tobacco and Arabidopsis	Reduced fecundity	Pitino and Hogenhout (2013)
	Receptor of activated kinase C gene *MpRack1*, effector gene *MpC002* and *MpPIntO2*		Arabidopsis	Reduced fecundity	Coleman et al. (2015)

Insect	Target gene	Life stage	Host plant	Effect	Reference
	Acetylcholinesterase 2 gene MpAChE2; V-ATPase E; tubulin folding cofactor D gene TBCD; 40S ribosomal protein S5-like isoform-1 Rps5; ribosomal protein S14 Rps14; mediator complex subunit 31 Med31; SWI/SNF-related matrix-associated actin-dependent regulator of chromatin subfamily D member 1-like gene SMARCD1		Tobacco	Reduced fecundity	Guo et al. (2014)
	Gap gene Hunchback (hb)	Neonates	Tobacco	Reduced fecundity	Mao and Zeng (2014)
	PEMV coat protein-Hv1a		Arabidopsis		Bonning and Chougule (2014)
	Serine proteinase gene MySP	Adults	Arabidopsis	Reduced fecundity and parthenogenetic population	Bhatia et al. (2012)
	Aquaporin gene MpAQP1; sucrase gene MpSUC1; sugar transporter gene MpSt4		Tobacco	Reduced fecundity	Tzin et al. (2015)
	Macrophage migration inhibitory factor MpMIF1	Adults	Tobacco	Decreased survival and fecundity of aphids feeding on their host plant	Naessens et al. (2015)
Sitobion avenae Grain aphid	Carboxylesterase gene CbE E4		Wheat	Impaired tolerance of phoxim insecticide	Xu et al. (2014)
	Structural sheath protein (shp)		Barley	Reduced fecundity and inhibited feeding behavior	Abdellatef et al. (2015)
Nilaparvata lugens Brown Plant Hopper	Multiple targets	Neonates	Transgenic plant		Zha et al. (2011)

cytochrome P450 gene CYP6AE14 of *H. armigera* showed increased tolerance to the cotton bollworm, *H. armigera* (Mao et al. 2011), but were not lethal to the larvae. Interestingly, when a cysteine proteinase which is supposed to damage larval peritrophic matrix leading to higher accumulation of gossypol in the midgut was co-delivered, the tolerance was further enhanced (Mao et al. 2013). The similar strategy may be applicable for restoring insecticide sensitivity among resistant insect species (Bautista et al. 2009; Tanget al. 2012; Figueira-Mansur et al. 2013).

The host-mediated RNAi for controlling insect pest has been considered to be particularly important for phloem-sucking hemipteran insect pests, viz., aphids. In green peach aphid, plant-mediated RNAi of several target insect-specific genes such as salivary proteins *MpC002*, *MpPIntO1*, and *MpPIntO2* and the gut-specific gene *Rack*-1 showed reduced fecundity (Table 4.3). In a similar study, stronger aphicidal activity of a hairpin RNA targeting V-ATPase E or the tubulin folding cofactor D (TBCD) was demonstrated (Guo et al. 2014). RNAi-mediated expression attenuation of a serine protease gene *MySP* in the green peach aphid, *Myzus persicae*, led to a remarkable decrease in their fecundity and parthenogeneticity (Bhatia et al. 2012). These studies on host-mediated delivery of dsRNA and elicitation of RNAi in infesting aphids demonstrated potential of RNAi approach for developing genetic resistance against aphids. Mao and Zeng (2014) reported reduced attack by aphids on transgenic tobacco plants expressing dsRNA against the gap gene *hunchback*, and reproduction rate of aphids was also retarded.

Interestingly, aphid nymphs parthenogenetically born from mothers reared on transgenic plants expressing dsRNA continued to show downregulation of the target gene even when transferred on normal plants. An assessment of RNAi effect over three generations of *M. persicae* revealed 60% reduction in aphid reproduction levels in transgenic Arabidopsis plants expressing ds*MpC002* compared to 40% decline on transgenics expressing dsRack1 and ds*MpPIntO2*. Such transgenerational RNAi was found to last over seven generations in *Sitobion avenae* reared on transgenic barley plants expressing shp-dsRNA (Abdellatef et al. 2015). Such parental transmission of RNAi effect adds to potential of the strategy.

4.8.2   In Nematodes

RNAi mechanism partly occurs in the host itself and partly in nematodes feeding on the transgenic host plant expressing dsRNA for the target gene. The plant RNAi machinery produces siRNAs which are ingested by nematodes feeding upon these plants through stylet (Li et al. 2011). By far HD-RNAi is the most successful methodology for developing resistance against nematodes in important crops. This technique exploits the capability of PPNs of ingesting macromolecules from the host plants. Specifically, the method involves producing dsRNA construct and developing transformed plants by *Agrobacterium*-mediated transformation. For generating dsRNA, a part of the target gene is cloned in sense and antisense orientation separated by an intron or spacer region and expressed under a constitutive or tissue-specific promoter. Majority of researchers have adopted this time-consuming methodology and have successfully developed transgenics resistant against nematodes. Another new approach with rapid screening system has been developed involving hairy root method for transformation of crops like soybean, tomato, and sugar beet.

Genes involved in various vital processes are mostly targeted by this approach being categorized into effector genes (most targeted), house-keeping genes, developmental genes, and genes associated with mRNA metabolism. Two genes encoding integrase and splicing factor were suppressed in *M. incognita* using host-delivered RNAi. It was the first report eliciting RNAi in *M. incognita* by developing transgenic tobacco lines (Yadav et al. 2006). The lethality of these genes as RNAi targets was further reconfirmed by Kumar et al. (2017) in *Arabidopsis* by utilizing this approach against *M. incognita*. Effective silencing of 16D10 effector genes leads to 63–90% reduction in the infectivity of *M. incognita* in *Arabidopsis* (Huang et al. 2006). Since 16D10 is highly conserved in *Meloidogyne* species, resistance against three other major species was also developed (Li et al. 2011). *M. chitwoodi* also showed a reduction in the number of nematodes and eggs on silencing 16D10L gene via HD-RNAi approach in transgenic Arabidopsis and potato plants (Dinh et al. 2014a, b).

Cyst nematodes also exhibited gene suppression by this technique successfully. The suppression of four parasitism genes, ubiquitin-like (4G06), cellulose-binding protein (3B05), SKP1-like (8H07), and zinc finger protein (10A06), in *Heterodera schachtii* resulted in the reduction of females in RNAi transgenic *Arabidopsis* lines (Sindhu et al. 2009). Silencing of esophageal proteins in *H. glycines* leads to the reduction in reproduction (Bakhetia et al. 2007). In another study, successful suppression of major sperm protein of *H. glycines* resulted in 68% decrease in eggs per gram root tissue when infected on transgenic soybean plants (Steeves et al. 2006). Transgenic tobacco lines expressing dsRNAs of two neuropeptides, flp-14 and flp-18, showed 50–80% decline in the infection of *M. incognita* (Papolu et al. 2013). Other genes silenced using this methodology are Mj-Tisll, Rpn7, tyrosine phosphotase, mitochondria stress 70 protein precursor and neuropepetides against Meloidogyne spp(s) (Hamann et al. 1993; Lindbo et al. 1993; Depicker and Montagu 1997; Pasquinelli 2002; Lim et al. 2003; Valdes et al. 2003). Host-mediated RNAi strategy is more successful in root-knot (RKN) nematodes as compared to cyst nematodes (CN) owing to factors like more RNAi sensitivity and larger size exclusion limit of RKNs than in CNs (Li et al. 2011). Host-delivered RNAi appears to be the most successful technique in controlling nematode infection.

Identification of appropriate target genes based on preliminary diet-based bioassay and ensuring adequate *in planta* expression of the dsRNA in the transgenic host are pivotal requirements for effective host-mediated RNAi. However, further understanding of the mechanisms on dsRNA uptake by insect and nematodes will facilitate the tailoring of dsRNA expression in HD-RNAi.

4.9 dsRNA Uptake Mechanisms

The dsRNA uptake mechanism in insects is known to be achieved by either of the two pathways, viz., a protein-mediated pathway and via endocytic pathway. The major component of protein-mediated pathway is a multi-pass transmembrane protein known as *systemic RNA interference deficient-1* (*Sid-1*) which exports the small interfering RNAs across neighboring cells (Bansal and Michel 2013). The second pathway is receptor-mediated pathway. In case of *C. elegans*, the endocytic pathway involves a *Sid-2* gene localized in intestinal cells. It encodes a membrane protein and is thought to import dsRNA from the intestinal lumen which are then exported to other cells with the help of sid-1 channels (Winston et al. 2007; McEwan et al. 2012). Hence, Sid-1 and Sid-2 proteins must work in conjunction to achieve environmental RNAi. *Sid-1* genes have been reported to be evolutionarily conserved among insects orders, but *Sid-2* gene is absent in insects. *Tribolium* is considered as the model insect for studying systemic RNAi with presence of *Sid-1* like proteins. However, the *Sid-1* gene of *Tribolium* was found orthologous to *Tag-130* gene of *C. elegans* and not *Ce-Sid-1* gene interestingly, where *Tag-130* has not been reported to be associated with systemic RNAi in nematodes (Tomoyasu et al. 2008). The presence of Sid-1-like channel proteins varies among different orders of insects. The involvement of Sid-1-like channel proteins in dsRNA uptake has been reported in brown plant hopper [BPH, *Nilaparvata lugens* (Xu et al. 2013)], the Colorado potato beetle [CPB, *Leptinotarsa decemlineata* (Cappelle et al. 2016)], and the red flour beetle *Tribolium castaneum* (Tomoyasu et al. 2008). In 2016, genes involved in RNAi pathway in insects were identified and classified. The study reveals absence of *Sid-1/Tag-130* orthologs in Diptera order (Dowling et al. 2016). It was suggested that in *Drosophila melanogaster*, dsRNA uptake is mediated via endocytic pathway along with pattern recognition receptors (PRRs) based on a study by Ulvila et al. (2006). This study reports more than 90% reduction in the uptake of double-stranded RNA on silencing of these two receptors by RNAi technology. Most of the studies examining dsRNA uptake so far focused on either the endocytic pathway or Sid-1-like dependent system. However, a clear understanding of the roles of these pathways on dsRNA uptake across the insect species is still lacking. Nevertheless, insects belonging to another order have been reported to have both the Sid-1-like channel proteins and receptor-mediated endocytosis pathways playing a role in dsRNA uptake (Cappelle et al. 2016).

However, the dsRNA uptake mechanism in worms is quite different. The components involved in dsRNA uptake have been well studied in *C. elegans*, and presence of *Sid-1* and *Sid-2* genes along with other components like *rsd-2*, *rsd-3*, and *rsd-6* has been well documented in the *C. elegans* genome. But surprisingly in a study, it was found these proteins were not evolutionary conserved (Dalzell et al. 2011). The dataset recognizes *sid-1*orthologs in two parasitic nematodes, viz., in *Haemonchus contortus* and *Oesophagostomum dentatum* only. The *Sid-2* protein was not found to be present in other nematode species. Intriguingly, the plant-parasitic nematodes such as *Meloidogyne* and *Globodera* spp. despite the absence of Sid-1 and Sid-2 genes exhibit systemic RNAi when subjected to silencing technology indicating a presence of similar receptor-mediated endocytic process for dsRNA uptake as reported in insects (Dalzell et al. 2011). Though lot of information has been generated over past few years, a clear understanding on dsRNA uptake mechanism(s) in worms is still elusive

4.10   RNAi Resistance in Other Agricultural Pests

Other than insects and nematodes, there are agricultural pests belonging to phylum Arthropoda that affect the crop productivity worldwide, and RNAi-based strategy to control these pests has shown some success. These pests are fire ants, mites, locusts (order Orthoptera), and many more. Systemic RNAi has already been demonstrated in these pests via microinjection. On feeding the worker ants, *Solenopsis invicta*, with 1000 ppm dsRNA targeting PBAN/pyrokinin gene, increased mortality rate of the fourth instar larvae. Direct toxic effect was also observed even when the dsRNA concentration was reduced to 200 ppm (Zhao and Chen 2013). In spider mite, gene silencing and increased mortality rate was observed when 160 ppm of dsRNA, targeting several genes, was employed via permeated leaf disc assay (Kwon et al. 2013). In another mite, *Varroa destructor*, an ectoparasite of the honey bee, *Apis mellifera*, both the delivery methods of dsRNA, i.e., by immersing mites in a dsRNA solution or by host-mediated RNAi, wherein dsRNA was fed to the honey bees and eventually delivered to mites, were found to attenuate the target gene expression through environmental RNAi (Campbell et al. 2010; Garbian et al. 2012).

Interestingly, locust species displayed systemic RNAi response but were refractory to environmental RNAi. Even a considerate concentration of 15 pg of dsRNA per mg body mass (~10 ng/insect) was enough to silence a gene in the desert locust, *Schistocerca gregaria* (Wynant et al. 2012). In case of *Tribolium castaneum*, the systemic response continued to increase over time in a dose-dependent manner and furthermore led to mortality 7 days postinjection. A similar dose-dependent response was also exhibited by the migratory locust, *Locusta migratoria*, leading to target gene suppression and lethality, but was unresponsive to environmental RNAi (Luo et al. 2013).

4.11   RNAi for Fungus Resistance

Fungi are classified as a separate eukaryotic kingdom from plants and animals. The vital RNAi components (RNA-dependent RNA polymerase (RdRP), Dicer, and Argonaute) have been found in different fungi indicating the presence of functional RNAi pathway (Dang et al. 2011). The RNAi phenomenon is termed as "quelling" in fungi which was first demonstrated in ascomycete *Neurospora crassa* (Romano and Macino 1992). Silencing of fungal genes by RNAi has shown to be desirable for many fungal species like Ascomycota, Basidiomycota, Zygomycota, and Phytophthora species (Nunes and Dean 2012). Several studies have been published reporting the successful use of host-induced gene silencing (HIGS) to control fungal diseases (Table 4.5) (Koch and Kogel 2014). Suppression of GUS transcripts in a GUS-expressing strain of *Fusarium verticillioides* (phytopathogenic filamentous fungi) while colonizing transgenic tobacco plants expressing GUS gene-interfering cassette was reported (Tinoco et al. 2010).

In vitro feeding of dsRNA complementary to three genes involved in ergosterol biosynthetic pathway, viz., CYP51A, CYP51B, and CYP51C, showed reduced growth of *Fusarium graminearum* (Koch et al. 2013). In wheat, mycotoxin-specific genes were silenced in *F. graminearum* and resulted in inhibition of virulence (McDonald et al. 2005). Fungal pathogenicity genes have shown to be an appropriate target for controlling fungal infection. A complete loss of pathogenicity was reported on targeting two of the host-selective ACT-toxin

Table 4.5 List of genes targeted in fungus through RNAi

Species	Target gene	Host plant	Effect/comments	References
Blumeria graminis f. sp. *tritici*	MLO	Wheat	Resistance	Riechen (2007)
Phytophthora parasitica var. *nicotianae*	GST (glutathione S-transferase gene)	Tobacco	Resistance; GST negative regulator of defense response	Hernandez et al. (2009)
Blumeria graminis	Avra10 (effector gene)	Barley and wheat	Reduced fungal development in the absence of the matching resistance gene Mla10	Nowara et al. (2010)
Fusarium verticillioides (F. moniliforme)	GUS (reporter gene)	Tobacco	GUS silencing; proof of concept	Tinoco et al. (2010)
Puccinia striiformis f. sp. *tritici* or *P. graminis* f. sp. *tritici*	PSTha12J12 (haustorial Pst transcript)	Barley and wheat	No obvious reductions in rust development or sporulation	Yin et al. (2011)
Phytophthora parasitica	PnPMA1 (H + -ATPase) and GFP (reporter gene)	Arabidopsis	Not sufficient; no reduction in GFP and PnPMA1 transcripts	Zhang et al. (2011)
P. triticina, P. graminis, and *P. striiformis*	PtMAPK1 (MAP kinase), PtCYC1 (cyclophilin), and PtCNB (calcineurin B)	Wheat	Disease suppression, compromising fungal growth and sporulation	Panwar et al. (2013)
Fusarium graminearum	CYP51A, CYP51B, and CYP51C	Arabidopsis and barley	Resistance	Koch et al. (2013)

genes in the fungus *Alternaria alternata* (Miyamoto et al. 2008; Ajiro et al. 2010). Similar reports on silencing of pathogenicity gene or avirulent gene proved successful in inhibiting the fungal growth and development. In *Magnaporthe oryzae*, silencing of 37 genes involved in calcium signaling process adversely affected hyphal growth, sporulation, and pathogenicity (Nunes and Dean 2012). HIGS-mediated silencing of effector gene *Avra10* showed a reduction in the number of haustoria in powdery mildew-susceptible barley cultivar (Koch and Kogel 2014).

To date, there are several successful reports of gene silencing in fungi with varied silencing efficiency. For instance, in *Moniliophthora perniciosa*, the silencing efficiency varied depending upon the targeted gene with reduction rates ranging from 18% to 97% in case of hydrophobin transcripts and 23% to 87% in peroxiredoxin transcripts (Santos et al. 2009), while when RNA hairpin precursor used to transform the Ascomycota *Ophiostoma novo-ulmi*, the expression of 6%, 22%, and 31% relative to the wild type was reported (Carneiro et al. 2010) in three transformants. Although usage of RNAi for managing fungus growth is nowadays a favored approach by researchers, RNAi silencing also leads to some off-target effects as observed by Lacroix and Spanu (2009) on silencing various genes in *C. fulvum*. These off-targets can be avoided by using specific silencing trigger sequence in RNAi vector, by tissue-specific and inducible silencing (Senthil-Kumar and Mysore 2011).

4.12 Barriers Limiting RNAi

The potential of RNAi technology for controlling various pests has been well documented over the past decade. However, there are many limitations which need to be taken care of for successful deployment of RNAi technology. There are several factors which need to be carefully looked into while designing RNAi experiments, including the off-target effects, dsRNA design, length and concentration of dsRNA, and many more. Therefore, to ensure a successful and effective RNAi-based silencing, these factors need to be balanced optimally. In case of insects, persistency of RNAi is a major problem due to which an optimum amount of dsRNA needs to be determined for an effective silencing. Interestingly, it is not true for every order of insect which is to be managed. For instance, about 60% (or lower) of gene knockdown was reported in certain recalcitrant insect species, while in coleopterans, 90% knockdown of gene was successfully achieved ensuing a long-lasting hereditary (Baum et al. 2007; Huvenne and Smagghe 2010; Zhu et al. 2011; Bolognesi et al. 2012; Rangasamy and Siegfried 2012; Li et al. 2013). Not only in insects but in nematodes also barriers like off-target effects have been reported while performing RNAi technology based management approaches. Designing an effective siRNA sequence is a major limitation in RNAi technology-based silencing. The following are some major barriers.

4.12.1   Off-Target Effects

Off-target effects result from the knockdown of unintended genes other than the target gene. Therefore, one of the most important aspects is avoidance of nonspecific target effects. It is the sequence used that determines possible off-target effects in the target organism and also in other species. Other than sequence, off-target effects can arise due to wide range of siRNAs being produced from a single dsRNA which increases the chance of nontarget effects. There are many reports of off-target effects, for instance, in triatomid bug *R. prolixus*, two homologous nitroprin genes were silenced other than the targeted gene (Araujo et al. 2006). Thus, selecting a sequence for synthesizing dsRNA is a crucial and limiting step in RNAi technology.

4.12.2   The Design of dsRNA

Selection of target gene is the first step in decision-making process for successful induction of RNAi in an organism. The gene selected should have a crucial role in the concerned organism, and genes involved in parasitism or development are likely candidate genes fulfilling all such requirements. Moreover, it should be highly specific and not conserved across different genera (Danchin et al. 2013) especially in pollinators. Next stage is to choose a suitable target site from the selected target gene. It is necessary to ensure the designing of a species-specific dsRNA. For identifying potential target sites for eliciting effective RNAi, bioinformatic tools are available online. Specificity of the dsRNAs could be conferred by either targeting conserved domain or variable region depending on the candidate gene with the aim to minimize possibility of affecting any unintended genes or organisms. This is particularly important to ensure that dsRNAs targeting agricultural pests should not possess any overlapping similarity to the genes of beneficial pollinators. By targeting the UTR regions, even closely related homologous genes can be selectively silenced through RNAi as demonstrated in *D. melanogaster*, *T. castaneum*, *A. pisum*, and tobacco hornworms, *Manduca sexta*, with respect to vATPase gene (Whyard et al. 2009).The concept of dsRNAs being used as tailor-made pesticides is emerging wherein highly specific dsRNAs are employed against havoc-creating pests and are also eco-friendly to the environment.

4.12.3   Length and Concentration of dsRNA

In general, longer RNA molecules tend to have longer half-life and therefore may be considered desirable while designing dsRNAs. However, size of the dsRNA molecule could be a limiting factor toward efficient uptake by the organisms. In nematodes, 28–140 kDa dsRNA could be efficiently ingested by *Meloidogyne* species (Urwin et al. 1997; Li et al. 2007; Zhang et al. 2012), though the limit is not known for other pests. In red flour beetle, the length and concentration of dsRNA had profound effect on efficiency as well as persistence of the RNAi effect, for example, 60- and 30-bp dsRNAs induced 70 and 30% of gene knockdown, respectively (Miller et al. 2012). In the same study, it has been also shown that multiple dsRNAs, when injected together, led to competitive inhibition influencing the effectiveness of RNAi. In contrary, dsRNA longer than 200 nucleotides and likely to generate multiple siRNAs contribute efficient RNAi response (Andrade and Hunter 2016). Multiple siRNAs will help in overcoming the target resistance that may arise due to polymorphism in the target. However, more studies are warranted to understand unambiguously the effect of length and concentration of dsRNAs on the initial efficiency and persistence of the RNAi effect.

4.12.4 Screening of Target Genes

For realizing RNAi-mediated gene silencing as an applicable strategy of pest control in agriculture, it remains imperative to achieve significant mortality or growth arrest of the pest population. Therefore, any attenuation of the target gene must be indispensible for the pest organism. This in turn underlines the importance of identifying appropriate target gene for the target pest. Though most of the studies have used limited set of target genes reported earlier, more emphasis should be given on identification of novel candidate genes (Pitino et al. 2011; Zhu et al. 2011). The upcoming genomics and bioinformatics tools, like genome search (Bai et al. 2009), cDNA library (Mao et al. 2007; Baum et al. 2007), RNA-seq and digital gene expression tag profile (DGE-tag) (Wang et al. 2011), and RIT-seq (Alsford et al. 2011), have been used for identification of new target genes.

4.12.5 Persistence of the Silencing Effect

The persistence of silencing signal determines the effectiveness of RNAi. Studies on low persistence of silencing effect have been reported in *A. pisum* wherein silencing effect on aquaporin persisted for 5 days of delivery before subsiding (Shakesby et al. 2009) indicating transient nature of RNAi effect. Thus, continuous supply of dsRNA seems to be essential for effective RNAi. It lends support for the transgenic host-mediated expression of the dsRNA for persistent and effective silencing. Persistent RNAi will also be useful in manifesting desired effect on the target organism even in case of inefficient and partial downregulation of the target gene.

4.12.6 Life Stage of the Target Organism

Selecting a life stage for larger silencing effects is species dependent that is to be targeted. In most cases, younger stage is preferred despite the efficient handling of older stages. In plant-parasitic nematodes, selecting the pre-parasitic juvenile stage for delivering dsRNAs shows better silencing effect. Similar observation was reported in insects, for example, in case of *R. prolixus*, no silencing effect was observed after treating its fourth instars compared to 42% silencing when using second instars (Araujo et al. 2006).

4.12.7 Methods of Delivery and Uptake Mechanisms

Various methods of dsRNA delivery have been used across the organisms. Such methods include microinjection, feeding with bacteria expressing dsRNA, feeding through diet supplementation, and host-mediated ingestion. The efficiency of RNAi varies significantly among different organisms and when using different delivery methods. In insects, either microinjection or diet supplementation has been the method of choice, though the aftershock effect of microinjection remains a concern in many species. Microinjection-mediated direct delivery bypasses the exposure of the dsRNA molecule to the nucleases present in the digestive tract. However, for realizing true efficacy of the dsRNA, it is desired to deliver through oral delivery that mimics the host-mediated delivery through ingestion.

4.12.8 Nucleases and Viruses

Limited success in RNAi in some of the insects has been attributed to rapid degradation of dsRNA by saliva of the insects. The saliva of *Lygus lineolaris* was found to contain RNases which interact with plant material prior to ingestion (Allen and Walker 2012). Presence of nucleases in the saliva and viruses in the hemolymph of insects also limits the silencing efficiency by degrading dsRNAs (Thompson et al. 2012; Christensen et al. 2013).

4.13 Improving RNAi

4.13.1 Large Throughput Screening for Selection of Target Genes

An ample number of studies in insect orders of Coleoptera, Diptera, Lepidoptera, Hemiptera, and others comprising of several insect pests have shown that RNAi targeting insect genes can affect growth and development of insects, often leading to insect death (Tables 4.3 and 4.4). The kind of genes for which a relatively high RNAi efficiency could be achieved included genes encoding detoxification enzymes, metabolism and cytoskeleton structure, cell synthesis, nutrition, etc. Alternative pathways of many of these genes in insects as well as relative importance of a particular pathway in an insect species are not known with certainty. Therefore, use of RNAi as a strategy for pest control will require an essential step of target selection. If an indispensible gene has to be identified for an insect species, it will involve large throughput screening rather than going for homologous genes, effective for other insect species.

Chitin covers the exoskeleton of insect body, and the insect midgut lined by peritrophic membrane (PM) constitutes the major channel for absorption of nutrients as well as orally administered dsRNA. Therefore genes expressed and functioning in the insect midgut have been screened by many researchers (Wang and Granados 2001). For example, a chitinase gene (*OnCht*) and a chitin synthase gene (*OnCHS2*) were identified from gut-specific EST of European corn borer (*Ostrinia nubilalis*) (Khajuria et al. 2010). Chitin content of the PM is regulated by *OnCht* as demonstrated in feeding experiment with dsRNA- and RNAi-based suppression which led to reduced growth and development of European corn borer larvae (Khajuria et al. 2010). In a similar study, Mao et al. (2007) identified several gossypol inducible genes, including a putative P450 monooxygenase, CYP6AE14, from a midgut-specific cDNA library from fifth-instar larvae exposed to gossypol. Similarly, for screening targets for RNAi in coleopteran insects, a large number of cDNAs from the cDNA libraries of Western corn rootworm (*Diabrotica virgifera virgifera*) were in vitro transcribed and used in feeding-based bioassays (Baum et al. 2007).

A rapid method of cDNA screening was demonstrated by Wang et al. (2011) by combining Illumina's RNA-seq and digital gene expression tag profile (DGE-tag) in Asian corn borer (ACB) (*Ostrinia furnacalis*). In addition to being a rapid and cost-effective method, this method allows monitoring expression of the genes throughout the insect body and thus broadening the base of target selection. Using Illumina parallel sequencing technology, abundance of >90,000 transcripts from trypanosome libraries was scored before and after induction of RNAi. The results led to constitution of non-redundant set of protein-coding sequences (CDS) comprising ~7500 genes (Alsford et al. (2011). Thus these methods can derive core set of essential gene loci if genome sequence of the organism is known. RNAi-mediated attenuation of these core loci is most likely to significantly retard survival and fitness of the insect pests.

In recent years, several modifications and methods for effective delivery and uptake of dsRNA have been proposed. Such methods include chemical modifications of siRNA duplex delivery through nanoparticles and liposomes, sprayable RNAi-based products, root absorption and trunk injection, and bacteria- or virus-based delivery. A few of them with much potentiality have been described below.

4.13.2   Nanoparticles

Synthetic, nontoxic nanoparticles could be generated from natural as well as synthetic polymers. Nanoparticles offer ease of surface modifications and biodegradability in addition to more penetration ability, thus an effective vehicle for delivery of dsRNA (Vauthier et al. 2003; Herrero-Vanrell et al. 2005). In mosquito dsRNA encapsulated in polymer, chitosan was used to achieve RNAi (Zhang et al. 2010). The encapsulation process used the electrostatic forces between the negative charges of the RNA backbone and positive charges of the amino groups of chitosan. Zhang et al. (2015a, b) demonstrated effective knockdown of *AgCHS1* and *AgCHS2* in *A. gambiae* and *A. aegypti* (*sema1a*) during larval development by using chitosan nanoparticles. He et al. (2013) fed lepidopteran pest, Asian corn borer (*Ostrinia furnacalis*), with diet containing the mixture of fluorescent nanoparticle (FNP) and CHT10-dsRNA, naked CHT10-dsRNA, FNP and GFP-dsRNA, and GFP-dsRNA. RNAi-mediated gene silencing occurred only in the larvae fed on the diet containing the mixture of FNP and CHT10-dsRNA leading to retarded growth and eventually death of the larvae.

4.13.3   Liposomes

Liposome vesicles composed of nontoxic natural lipids are already being used in drug delivery. Liposomes can cross the cell membrane effectively and deliver the exogenous molecules. Whyard et al. (2009) used cationic liposomes for encapsulating and delivering dsRNA targeting 3′-UTR of the g-tubulin gene in four different species of *Drosophila* (*D. melanogaster*, *D. sechellia*, *D. yakuba*, and *D. pseudoobscura*) and demonstrated mortality of the insects only in case of encapsulated dsRNA. In *Drosophila*, presence of *sid1* homologues has never been confirmed, and the uptake of dsRNA is likely to be by receptor-mediated endocytosis (Ulvila et al. 2006). Higher efficiency of RNAi in case of liposome-mediated delivery in certain cases could be attributed to the fact that it bypasses the gut nucleases which reduces the efficacy of orally delivered dsRNA.

4.13.4   Chemical Modifications

Chemical modifications are known to increase the stability of RNA molecules. In case of siRNA also such modifications have been proposed to improve half-life and pharmacokinetic properties of the siRNA duplexes, target-binding affinity, and delivery (Kurreck 2003; Manoharan 2003; Dorsett and Tuschl 2004). Interestingly a couple of examples have demonstrated that such modifications may increase the specificity of dsRNA. For example, methylation at 2′-position of the ribosyl ring of the second base of the siRNA could decrease off-target effects (Jackson et al. 2003), siRNA duplex with 3′-overhangs at each end was more effective in gene silencing compared to blunt-ended duplex (Elbashir et al. 2001), and addition of 3′-TT overhangs (the "Tuschl design") on both strands of duplex siRNA has been preferred in many cases. A few other designs, for instance, siRNAs without 3′-overhangs and single 3′-overhang structures in the guide strand, have been active in gene silencing (Czauderna et al. 2003; Lorenz et al. 2004).

4.14 Future Perspective and Conclusion

Despite few limitations, the applicability of RNAi in improving crop resistance especially against biotic stresses is expected to be the most reliable and significant approach in the future as evident from a plethora of studies. Certain products based on RNAi-mediated resistance such as Monsanto's SmartStax Pro, for control of Western corn earworm, and DuPont Pioneer's Plenish high oleic acid soybean (Majumdar et al. 2017) are likely to be commercialized soon. However, efficacy of these plants remains to be proven in actual field situations. Diverse classes of biotic factors, affecting crop production worldwide, have shown varied levels of susceptibility toward RNAi, which warrants need for modified and improved versions of dsRNA delivery methods. The better understanding of host-pest interaction and the genetic basis of parasitism are likely to generate more potential target genes for effective HD-RNAi. CRISPR/Cas system has come up as a powerful technique in creating knockout mutants to unravel complex mechanism of parasitism and thus paves the way for identification of the key pest genes. Transplastomic expression of dsRNA in the plants would be a further improvement for achieving higher expression. Applying dsRNA through methods with low environmental risks, for instance, irrigation water, root drench, or trunk injection, would obviate the need for genetic transformation. These methods result in localized application along with rapid breakdown of dsRNA and therefore likely to be more acceptable from a biosafety point of view (Joga et al. 2016). Successful demonstration of using layered double hydroxide clay nanosheets for topical application of dsRNA against viruses (Mitter et al. 2017) opens up possibilities of applying dsRNA like any other protective agrochemicals.

To conclude, RNAi has emerged as one of the most potential control mechanisms for pests like insects, nematodes, fungus, etc. Although still a lot remains to be explored and understood about the molecular process of RNAi in plants and their pests, the present available knowledge and the studies reviewed in this chapter have proved RNAi technology as an important tool in identifying gene functions and targeting vital genes for controlling pest development. RNAi-mediated loss-of-function phenotypes not only determine functions of unknown genes but also lead to identification of new specific targets for managing pest or improving agricultural traits. But understanding RNAi mechanism is of utmost importance as RNAi machinery varies from genus to genus. There are several shortcomings that need to be addressed, for instance, persistence of silencing effects, off-target effects of silencing, etc. Not only this, the biosafety, risk assessment, and government regulations related to commercialization of RNAi-based transgenics still have to be developed. The revelation of RNAi technology has revolutionized the area of research in biotechnology. Not only in pest management, the wide range of RNAi application includes modification of agronomic traits, eliminating mycotoxin contamination, improving nutritional value of crops, etc. It is also proving its worth in RNAi-based therapeutics research for human welfare. In toto, this technology is a potential boon in the arsenal of the scientific community to address the challenges associated with climatic changes, burgeoning population, and sustainability of human race.

References

Aalto AP, Pasquinelli AE (2012) Small non-coding RNAs mount a silent revolution in gene expression. Curr Opin Cell Biol 24:333–340

Abdellatef E, Will T, Koch A et al (2015) Silencing the expression of the salivary sheath protein causes transgenerational feeding suppression in the aphid *Sitobion avenae*. Plant Biotechnol J B13:849–857

Agrawal N, Dasaradhi PVN, Mohmmed A et al (2003) RNA interference: biology, mechanism and applications. Microbiol Mol Biol Rev 67:657–685

Ahringer J (ed) (2006) The C elegans research community. Reverse genetics J. http://www.wormbook.org

Ajiro N, Miyamoto Y, Masunaka A et al (2010) Role of the host-selective ACT-toxin synthesis gene ACTTS2 encoding an enoylreductase in pathogenicity of the tangerine pathotype of *Alternaria alternata*. Phytopathology 100:120–126

Allen ML, Walker WB (2012) Saliva of *Lygus lineolaris* digests double stranded ribonucleic acids. J Insect Physiol 58:391–396

Alsford S, Turner DJ, Obado SO et al (2011) High-throughput phenotyping using parallel sequencing of RNA interference targets in the African trypanosome. Genome Res 21:915–924. https://doi.org/10.1101/gr.115089.110

Anandalakshmi R, Pruss GJ, Ge X et al (1998) A viral suppressor of gene silencing in plants. Proc Natl Acad Sci U S A 95:13079–13084

Andrade CE, Hunter WB (2016) RNA interference– natural gene-based technology for highly specific pest control (HiSPeC). In: Abdurakhmonov IY (ed) RNA interference. InTech, Croatia, pp 391–409

Antonino JD, Coelho R, Lourenço T et al (2013) Knocking- down *Meloidogyne incognita* proteases by plant-delivered dsRNA has negative pleiotropic effect on nematode vigor. PLoS One 8:e85364. https://doi.org/10.1371/journal.pone.0085364

Araujo RN, Santos A, Pinto FS et al (2006) RNA interference of the salivary gland nitrophorin 2 in the triatomine bug *Rhodnius prolixus* (Hemiptera: Reduviidae) by dsRNA ingestion or injection. Insect Biochem Mol Biol 36:683–693

Bai J, Sepp KJ, Perrimon N (2009) Culture of Drosophila primary cells dissociated from gastrula embryos and their use in RNAi screening. Nat Protoc 4:1502–1512

Bakhetia M, Charlton W, Atkinson HJ et al (2005) RNA interference of dual oxidase in the plant nematode *Meloidogyne incognita*. Mol Plant Microbe Interact 18:1099–1106. https://doi.org/10.1094/MPMI-18-1099

Bakhetia M, Urwin PE, Atkinson HJ (2007) qPCR analysis and RNAi define pharyngeal gland cell-expressed genes of *Heterodera glycines* required for initial interactions with the host. Mol Plant Microbe Interact 20:306–312. https://doi.org/10.1094/mpmi-20-3-0306

Banerjee S, Banerjee A, Gill SS et al (2017) RNA interference: a novel source of resistance to combat plant parasitic nematodes. Front Plant Sci 8:834. https://doi.org/10.3389/fpls.2017.00834

Bansal R, Michel AP (2013) Core RNAi machinery and *Sid1*, a component for systemic RNAi, in the Hemipteran insect, *Aphis glycines*. Int J Mol Sci 14:3786–3801. https://doi.org/10.3390/ijms14023786

Baum JA, Bogaert T, Clinton W et al (2007) Control of coleopteran insect pests through RNA interference. Nat Biotechnol 25:1322–1326. https://doi.org/10.1038/nbt1359

Bautista MA, Miyata T, et a MK (2009) RNA interference-mediated knockdown of a cytochrome P450, CYP6BG1, from the diamondback moth, *Plutella xylostella* , reduces larval resistance to permethrin. Insect Biochem Mol Biol 39:38–46

Berezikov E (2011) Evolution of microRNA diversity and regulation in animals. Nat Rev Genet 12(12):846–860

Bernstein E, Caudy AA, Hammond SM et al (2001) Role for a bidentate ribonuclease in the initiation step of RNA interference. Nature 409:363–366

Bhatia V, Bhattacharya R, Uniyal PL et al (2012) Host generated siRNAs attenuate expression of serine protease gene in *Myzus persicae*. PLoS One 7(10):e46343. https://doi.org/10.1371/journal.pone.0046343

Bolognesi R, Ramaseshadri P, Anderson J et al (2012) Characterizing the mechanism of action of double-stranded RNA activity against western corn rootworm (*Diabrotica virgifera virgifera* LeConte). PLoSONE 7:e47534. https://doi.org/10.1371/journal.pone.0047534

Bonning BC, Chougule NP (2014) Delivery of intrahemocoelic peptides for insect pest management. Trends Biotechnol 32:91–98

Campbell ME, Budge GE, Bowman AS (2010) Gene-knockdown in the honey bee mite *Varroa destructor* by a non-invasive approach: studies on a glutathione S-transferase. Parasit Vectors 3:73. https://doi.org/10.1186/1756-3305-3-73

Cappelle K, de Oliveira CFR, Eynde BV et al (2016) The involvement of clathrin-mediated endocytosis and two Sid-1-like transmembrane proteins in double-stranded RNA uptake in the Colorado potato beetle midgut. Insect Mol Biol 25:315–323

Carmell MA, Xuan Z, Zhang MQ et al (2002) The Argonaute family: tentacles that reach into RNAi, developmental control, stem cell maintenance, and tumorigenesis. Genes Dev 16:2733–2742

Carneiro JS, Bastide PY, Chabot M et al (2010) Suppression of polygalacturonase gene expression in the phytopathogenic fungus *Ophiostoma novo-ulmi* by RNA interference. Fungal Genet Biol 47:399–405

Carthew RW, Sontheimer EJ (2009) Origins and mechanisms of miRNAs and siRNAs. Cell 136:642–655

Caudy AA, Myers M, Hannon GJ et al (2002) Fragile X-related protein and VIG associate with RNA interference machinery. Genes Dev 16:2491–2496

Chen Q, Rehman S, Smant G et al (2005) Functional analysis of pathogenicity proteins of the potato cyst nematode *Globodera rostochiensis* using RNAi. Mol. Plant Microb. Interact. 18:621–625

Chen J, Zhang D, Yao Q et al (2010) Feeding based RNA interference of a trehalose phosphate synthase gene in the brown plant hopper, *Nilaparvata lugens*. Insect Mol Biol 19:777–786

Choudhary D, Koulagi R, Rohatagi D et al (2012) Engineering resistance against root-knot nematode, Meloidogyne incognita, by host delivered RNAi. In: Abstracts of international conference on plant biotechnology for food security: new frontiers. National Agricultural Science Centre, New Delhi, pp 21–24

Christensen J, Litherland K, Faller T et al (2013) Metabolism studies of unformulated internally [3H]- labeled short interfering RNAs in mice. Drug Metab Dispos 41:1211–1219. https://doi.org/10.1124/dmd.112.050666

Christiaens O, Sweveres L, Smagghe G (2014) DsRNA degradation in the pea aphid(*Acyrthosiphon pisum*) associated with lack of response in RNAi feeding and injection assay. Peptides 53:307–314

Cogoni C, Macino G (2000) Post-transcriptional gene silencing across kingdoms. Curr Opin Genet Dev 10:638–643

Coleman AD, Wouters RH, Mugford ST et al (2015) Persistence and transgenerational effect of plant-mediated RNAi in aphids. JExp Bot 66:541–548

Coy MR, Sanscrainte ND, Chalaire KC et al (2012) Gene silencing in adult *Aedes aegypti* mosquitoes through oral delivery of double-stranded RNA. J Appl Entomol 136:741–748

Czauderna F, Fechtner M, Dames S et al (2003) Structural variations and stabilizing modifications of synthetic siRNAs in mammalian cells. Nucleic Acids Res 31:2705–2716. https://doi.org/10.1093/nar/gkg393

Dalzell JJ, McVeigh P, Warnock et al (2011) RNAi effector diversity in nematodes. PLoS Negl Trop Dis 5:e1176

Danchin EGJ, Arguel MJ, Campan-Fournier A et al (2013) Identification of novel target genes for safer and more specific control of root-knot nematodes from a pan-genome mining. PLoS Pathog 9:e1003745. https://doi.org/10.1371/journal.ppat.1003745

Dang Y, Yang Q, Xue Z et al (2011) RNA interference in fungi: pathways, functions, and applications. Eukaryot Cell 10(9):1148–1155. https://doi.org/10.1128/EC.05109-11

Daniel RGP, John AG (2008) RNAi-mediated crop protection against insects. Trends Biotechnol 26:393–400

Darrington M, Dalmay T, Morrison NI et al (2017) Implementing the sterile insect technique with RNA interference – a review. Entomol Exp Appl 164:155–175. https://doi.org/10.1111/eea.12575

Deng F, Zhao Z (2014) Influence of catalase gene silencing on the survivability of *Sitobion avenae*. Arch Insect Biochem Physiol 86:46–57

Depicker A, Montagu MV (1997) Post-transcriptional gene silencing in plants. Curr Opin Cell Biol 9:373–382

Dernburg AF, Karpen GH (2002) A chromosome RNAissance. Cell 111:159–162

Dietzl G, Chen D, Schnorrer F et al (2007) A genome-wide transgenic RNAi library for conditional gene inactivation in Drosophila. Nature 448:151–156

Dinh PTY, Brown CR, Elling AA (2014a) RNA interference of effector gene Mc16D10L confers resistance against *Meloidogyne chitwoodi* in Arabidopsis and Potato. Phytopathology 104:1098–1106. https://doi.org/10.1094/PHYTO-03-14-0063-R

Dinh PTY, Zhang L, Brown CR et al (2014b) Plant mediated RNA interference of effector gene Mc16D10L confers resistance against *Meloidogyne chitwoodi* in diverse genetic backgrounds of potato and reduces pathogenicity of nematode offspring. Nematology 6:669–682. https://doi.org/10.1163/15685411-00002796

Dorsett Y, Tuschl T (2004) siRNAs: applications in functional genomics and potential as therapeutics. Nat Rev Drug Discov 3:318–329. https://doi.org/10.1038/nrd1345

Dowling D, Pauli T, Donath A et al (2016) Phylogenetic origin and diversification of RNAi pathway genes in insects. Genome Biol Evol 8:3784–3793. https://doi.org/10.1093/gbe/evw281

Dutta TK, Papolu PK, Banakar P et al (2015) Tomato transgenic plants expressing hairpin construct of a nematode protease gene conferred enhanced resistance to root-knot nematodes. Front Microbiol 6:260. https://doi.org/10.3389/fmicb.2015.00260

Dykxhoorn DM, Novina CD, Sharp PA (2003) Killing the messenger: short RNAs that silence gene expression. Nat Rev Mol Cell Biol 4:457–467

Elbashir SM, Lendeckel W, Tuschl T (2001) RNA interference is mediated by 21- and 22-nucleotide RNAs. Genes Dev 15:188–200

Fairbairn DJ, Cavallaro AS, Bernard M et al (2007) Host-delivered RNAi: an effective strategy to silence genes in plant parasitic nematodes. Planta 226:1525–1533. https://doi.org/10.1007/s00425-007-0588-x

Fan W, Wei Z, Zhang M et al (2015a) Resistance to Ditylenchus destructor infection in sweet potato by the expression of small interfering RNAs targeting unc-15, a movement-related gene. Phytopahol 105:1458–1465. https://doi.org/10.1094/PHYTO-04-15-0087-R

Fan J, Zhang Y, Francis F et al (2015b) Orco mediates olfactory behaviors and winged morph differentiation induced by alarm pheromone in the grain aphid, *Sitobion avenae*. Insect Biochem Mol Biol 64:16–24

Fanelli E, Di Vito M, Jones JT et al (2005) Analysis of chitin synthase function in a plant parasitic nematode, Meloidogyne artiellia, using RNAi. Gene 349:87–95

Feinberg EH, Hunter CP (2003) Transport of dsRNA into cells by the transmembrane protein SID-1. Science 301:1545–1547

Figueira-Mansur J, Ferreira-Pereira A, Mansur JF et al (2013) Silencing of P-glycoprotein increases mortality in temephos-treated *Aedes aegypti* larvae. Insect Biochem Mol Biol 22:648–658

Fire A, Xu S, Montgomery MK et al (1998) Potent and specific genetic interference by double-stranded RNA in *C. elegans*. Nature 391:806–811

Garbian Y, Maori E, Kalev H et al (2012) Bidirectional transfer of RNAi between honey bee and *Varroa destructor*: *Varroa* gene silencing reduces *Varroa* population. PLoS Pathog 8(12):e1003035

Gong L, Yang X, Zhang B et al (2011) Silencing of Rieske iron-sulfur protein using chemically synthesized siRNA as a potential biopesticide against *Plutella xylostella*. Pest Manag Sci 67:514–520

Gong L, Chen Y, Hu Z et al (2013) Testing insecticidal activity of novel chemically synthesized sirna against *Plutella xylostella* under laboratory and field conditions. PLoSOne 8:e62990

Gong YH, Yu XR, Shang QL et al (2014) Oral delivery mediated RNA interference of a carboxylesterase gene results in reduced resistance to organophosphorus insecticides in the cotton aphid, *Aphis gossypii* glover. PLoS One 9:e102823

Good L, Stach JEM (2011) Synthetic RNA silencing in bacteria antimicrobial discovery and resistance breaking. Front Microbiol 2:185

Griebler M, Westerlund SA, Hoffmann KH et al (2008) RNA interference with the allato regulating neuropeptide genes from the fall armyworm *Spodoptera frugiperda* and its effects on the JH titer in the hemolymph. J Insect Physiol 54:997–1007

Guo H, Song X, Wang G et al (2014) Plant-generated artificial small RNAs mediated aphid resistance. PLoS One 9:e97410

Haegeman A, Joseph S, Gheysen G et al (2011) Analysis of the transcriptome of the root lesion nematode Pratylenchus coffeae generated by 454 sequencing technology. Mol Biochem Parasitol 178:7–14

Hamann L, Jensen K, Harbers K (1993) Consecutive inactivation of both alleles of the gb110 gene has no effect on the proliferation and differentiation of mouse embryonic stem cells. Gene 126:279–284

Hamilton AJ, Baulcombe DC (1999) A species of small antisense RNA in posttranscriptional gene silencing in plants. Science 286:950–952

Hammond SM, Caudy AA, Hannon GJ (2001) Post-transcriptional gene silencing by double-stranded RNA. Nat Rev Genet 2:110–119

He B, Chu Y, Yin M et al (2013) Fluorescent nanoparticle delivered dsRNA toward genetic control of insect pests. Adv Mater Weinheim 25:4580–4584. https://doi.org/10.1002/adma.201301201

Herrero-Vanrell R, Rincón AC, Alonso M et al (2005) Self-assembled particles of an elastin-like polymer as vehicles for controlled drug release. J Control Release 102:113–122. https://doi.org/10.1016/j.jconrel.2004.10.001

Hernandez I, Chacon O, Rodriguez R et al. (2009) Black shank resistant tobacco by silencing of glutathione S- transferase. Biochem Biophys Res Commun 387:300–304

Huang G, Allen R, Davis EL et al (2006) Engineering broad root-knot resistance in transgenic plants by RNAi silencing of a conserved and essential root-knot nematode parasitism gene. Proc Natl Acad Sci U S A 103:14302–14306. https://doi.org/10.1073/pnas.0604698103

Hull D, Timmons L (2004) Methods for delivery of double-stranded RNA into Caenorhabditis elegans. Methods Mol Biol 265:23–58

Huvenne H, Smagghe G (2010) Mechanisms of dsRNA uptake in insects and potential of RNAi for pest control: a review. J Insect Physiol 56:227–235. https://doi.org/10.1016/j.jinsphys.2009.10.004

Ischizuka A, Siomi MC, Siomi H (2002) A *Drosophila* fragile X protein interacts with components of RNAi and ribosomal proteins. Genes Dev 16:2497–2508

Jackson AL, Bartz SR, Schelter J et al (2003) Expression profiling reveals off-target gene regulation by RNAi. Nat Biotechnol 21:635–637. https://doi.org/10.1038/nbt831

Joga MR, Zotti MJ, Smagghe G et al (2016) RNAi efficiency, systemic properties, and novel delivery methods for pest insect control: what we know so far. Front Physiol 7:553. https://doi.org/10.3389/fphys.2016.00553

Kamath RS, Martinez-Campos M, Zipperlen P et al (2001) Effectiveness of specific RNA-mediated interference through ingested double-stranded RNA in *Caenorhabditis elegans*. Genome Biol 2:2.1–2.10. https://doi.org/10.1186/gb-2000-2-1-research0002

Kennerdell JR, Carthew RW (1998) Use of dsRNA-mediated genetic interference to demonstrate that *frizzled* and *frizzled 2* Act in the wingless pathway. Cell 95:1017–1026

Ketting RF, Fischer SE, Bernstein E et al (2001) Dicer functions in RNA interference and in synthesis of small RNA involved in developmental timing in *C. elegans*. Genes Dev 15:2654–2659

Ketting RF (2011) The many faces of RNAi. Dev Cell 15:148-161. https://doi.org/10.1016/j.devcel.2011.01.012

Khajuria C, Buschman LL, Chen MS et al (2010) A gut-specific chitinase gene essential for regulation of chitin content of peritrophic matrix and growth of *Ostrinia nubilalis* larvae. Insect Biochem Mol Biol 40:621–629. https://doi.org/10.1016/j.ibmb.2010.06.003

Klink VP, Kim KH, Martins V et al (2009) A correlation between host-mediated expression of parasite genes as tandem inverted repeats and abrogation of development of female *Heterodera glycines* cyst formation during infection of *Glycine max*. Planta 230:53–71. https://doi.org/10.1007/s00425-009-0926-2

Koch A, Kogel KH (2014) New wind in the sails: improving the agronomic value of crop plants through RNAi-mediated gene silencing. Plant Biotechnol J 12:821–831

Koch A, Kumar N, Weber L et al (2013) Host-induced gene silencing of cytochrome P450 lanosterol C14α-demethylase-encoding genes confers strong resistance to Fusarium species. Proc Natl Acad Sci U S A 110(48):19324–19329. pmid:24218613

Kontogiannatos D, Swevers L, Maenaka K et al (2013) Functional characterization of a juvenile hormone esterase related gene in the moth *Sesamia nonagrioides* through RNA interference. PLoS One 8:e73834

Kumar M, Gupta GP, Rajam MV (2009) Silencing of acetyl cholinesterase gene of *Helicoverpa armigera* by siRNA affects larval growth and its life cycle. J Insect Physiol 55:273–278. https://doi.org/10.1016/j.jinsphys.2008.12.005

Kumar A, Wang S, Ou R et al (2013) Development of an RNAi based microalgal larvicide to control mosquitoes. Malaria World J 4:6

Kumar A, Kakrana A, Sirohi A et al (2017) Host-delivered RNAi-mediated root-knot nematode resistance in Arabidopsis by targeting splicing factor and integrase genes. J Gen Plant Pathol 83:91–97. https://doi.org/10.1007/s10327-017-0701-3

Kurreck J (2003) Antisense technologies: improvement through novel chemical modifications. Eur J Biochem 270:1628–1644. https://doi.org/10.1046/j.1432-1033.2003.03555.x

Kwon DH, Park JH, Lee SH (2013) Screening of lethal genes for feeding RNAi by leaf disc-mediated systematic delivery of dsRNA in *Tetranychus urticae*. Pestic Biochem Physiol 105:69–75. https://doi.org/10.1016/j.pestbp.2012.12.001

Lacroix H, Spanu PD (2009) Silencing of six hydrophobins in *Cladosporium fulvum*: complexities of simultaneously targeting multiple genes. Appl Environ Microbiol 75:542–546

Lendner M, Doligalska M, Lucius R et al. (2008) Attempts to establish RNA interference in the parasitic nematode Heligmosomoides polygyrus. Mol Biochem Parasitol 161:21–31

Li J, Todd TC, Oakley TR et al (2010a) Host-derived suppression of nematode reproductive and fitness genes decreases fecundity of Heterodera glycines Ichinohe. Planta 232:775–785. https://doi.org/10.1007/s00425-010-1209-7

Li J, Todd TC, Trick HN (2010b) Rapid in planta evaluation of root expressed transgenes in chimeric soybean plants. Plant Cell Rep 29:113–123. https://doi.org/10.1007/s00299-009-0803-2

Li J, Chen Q, Lin Y et al (2011a) RNA interference in *Nilaparvata lugens* (Homoptera, Delphacidae) based on dsRNA ingestion. Pest Manag Sci 67:852–859

Li X, Zhang M, Zhang H (2011b) RNA interference of four genes in adult Bactrocera dorsalis by feeding their dsRNAs. PLoS One 6:e17788

Li J, Todd TC, Lee J et al (2011c) Biotechnological application of functional genomics towards plant parasitic nematode control. Plant Biotech J 9:936–944. https://doi.org/10.1111/j.1467-7652.2011.00601.x

Li J, Wang XP, Wang MQ et al (2013) Advances in the use of the RNA interference technique in Hemiptera. Insect Sci 20:31–39. https://doi.org/10.1111/j.1744-7917.2012.01550.x

Li XQ, Wei JZ, Tan A et al. (2007) Resistance to root-knot nematode in tomato roots expressing a nematicidal Bacillus thuringiensis crystal protein. Plant Biotechnol J 5:455–464. https://doi.org/10.1111/j.1467-7652.2007.00257.x

Li Y, Wang K, Xie H et al (2015) Cathepsin B cysteine proteinase is essential for the development and pathogenesis of the plant parasitic nematode *Radopholus similis*. Int J Biol Sci 11:1073–1087. https://doi.org/10.7150/ijbs.12065

Lilley CJ, Goodchild SA, Atkinson HJ et al (2005) Cloning and characterisation of a *Heterodera glycines* aminopeptidase cDNA. Int J Parasitol 35:1577–1585

Lim LP, Glasner ME, Yekta S et al (2003) Vertebrate micro-RNA genes. Science 299:1540

Lindbo JA, Silva-Rosales L, Proebsting WM et al (1993) Induction of a highly specific antiviral state in transgenic plants: implications for regulation of gene expression and virus resistance. Plant Cell 5:1749–1759

Lipardi C, Wei Q, Paterrson BM (2001) RNAi as random degradation PCR: siRNA primers convert mRNA into dsRNA that are degraded to generate new siRNAs. Cell 101:297–307

Lorenz C, Hadwiger P, John M et al (2004) Steroid and lipid conjugates of siRNAs to enhance cellular uptake and gene silencing in liver cells. Bioorg Med Chem Lett 14:4975–4977. https://doi.org/10.1016/j.bmcl.2004.07.018

Lourenço-Tessutti IT, Souza JDA, Martins-de-Sa D et al (2015) Knockdown of heat-shock protein 90 and isocitrate lyase gene expression reduced root-knot nematode reproduction. Phytopathology 105:628–637. https://doi.org/10.1094/PHYTO-09-14-0237-R

Lum L, Yao S, Mozer B et al (2003) Identification of Hedgehog pathway components by RNAi in *Drosophila* cultured cells. Science 299:2039–2045

Luo Y, Wang X, Wang X et al (2013) Differential responses of migratory locusts to systemic RNA interference via double-stranded RNA injection and feeding. Insect Mol Biol 22:574–583. https://doi.org/10.1111/imb.12046

Macrae IJ, Zhou K, Li F et al (2006) Structural basis for double-stranded RNA processing by Dicer. Science 311:195–198

Majumdar R, Rajasekaran K, Cary JW (2017) RNA interference (RNAi) as a potential tool for control of mycotoxin contamination in crop plants: concepts and considerations. Front Plant Sci 8:200. https://doi.org/10.3389/fpls.2017.00200

Manoharan M (2003) RNA interference and chemically modified siRNAs. Nucleic Acids Res Suppl 3:115–116. https://doi.org/10.1093/nass/3.1.115

Mao J, Zeng F (2012) Feeding-based RNA interference of a gap gene is lethal to the pea aphid, *Acyrthosiphon pisum*. PLoS One 7:e48718

Mao J, Zeng F (2014) Plant-mediated RNAi of a gap gene-enhanced tobacco tolerance against the *Myzus persicae*. Transgenic Res 23:145–152. https://doi.org/10.1007/s11248-013-9739-y

Mao YB, Cai WJ, Wang JW et al (2007) Silencing a cotton bollworm P450 monooxygenase gene by plant-mediated RNAi impairs larval tolerance of gossypol. Nat Biotechnol 25:1307–1313. https://doi.org/10.1038/nbt1352

Mao YB, Tao XY, Xue XY et al (2011) Cotton plants expressing *CYP6AE14* double-stranded RNA show enhanced resistance to bollworms. Transgenic Res 20:665–673. https://doi.org/10.1007/s11248-010-9450-1

Mao YB, Xue XY, Tao XY et al (2013) Cysteine protease enhances plant-mediated bollworm RNA interference. Plant Mol Biol 83:119–129. https://doi.org/10.1007/s11103-013-0030-7

Matzke M, Matzke AJ, Kooter JM (2001) RNA: guiding gene silencing. Science 293:1080–1083

Maule AG, McVeigh P, Dalzell JJ et al (2011) An eye on RNAi in nematode parasites. Trends Parasitol 27:505–513. https://doi.org/10.1016/j.pt.2011.07.004

McDonald T, Brown D, Keller NP et al (2005) RNA silencing of mycotoxin production in *Aspergillus* and *Fusarium* species. Mol Plant-Microbe Interact 18:539–545

McEwan DL, Weisman AS, Hunter CP (2012) Uptake of extracellular double-stranded RNA by SID-2. Mol Cell 47:746–754. https://doi.org/10.1016/j.molcel.2012.07.014

Meer VRK, Choi MY (2013) Formicidae (ant) control using double-stranded RNA constructs. US Patent No. 8,575,328

Meyering-Vos M, Muller A (2007) RNA interference suggests sulfakinins as satiety effectors in the cricket *Gryllus bimaculatus*. J Insect Physiol 53:840–848

Miller SC, Miyata K, Brown SJ et al (2012) Dissecting systemic RNA interference in the red flour beetle *Tribolium castaneum*: parameters affecting the efficiency of RNAi. PLoS One 7:e47431. https://doi.org/10.1371/journal.pone.0047431

Mitter N, Worrall EA, Robinson KE et al (2017) Clay nanosheets for topical delivery of RNAi for sustained protection against plant viruses. Nature Plants 3:16207

Miyamoto Y, Masunaka A, Tsuge T et al (2008) Functional analysis of a multicopy host-selective ACT-toxin biosynthesis gene in the tangerine pathotype of *Alternaria alternata* using RNA silencing. Mol Plant-Microbe Interact 21:1591–1599

Mutti NS, Park Y, Reese JC et al (2006) RNAi knockdown of a salivary transcript leading to lethality in the pea aphid, *Acyrthosiphon pisum*. J Insect Sci 6:38

Naessens E, Dubreuil G, Giordanengo P et al (2015) A secreted MIF cytokine enables aphid feeding and represses plant immune responses. Curr Biol 25:1898–1903

Nakai K, Horton P (1999) PSORT: a program for detecting sorting signals in proteins and predicting their subcellular localization. Trends Biochem Sci 24:34–36

Napoli C, Lemieux C, Jorgensen R (1990) Introduction of a chimeric chalcone synthase gene into Petunia results in reversible cosuppression of homologous genes in trans. Plant Cell 2:279–289

Niu JH, Jian H, Xu J et al (2012) RNAi silencing of the Meloidogyne incognita Rpn7 gene reduces nematode parasitic success. Euro J Plant Pathol 134:131–144. https://doi.org/10.1007/s10658-012-9971-y

Niu J, Liu P, Liu Q et al (2016) Msp40 effector of root-knot nematode manipulates plant immunity to facilitate parasitism. Sci Rep 6:19443. https://doi.org/10.1038/srep19443

Nowara D, Gay A, Lacomme C et al (2010) HIGS: host-induced gene silencing in the obligate biotrophic fungal pathogen *Blumeria graminis*. Plant Cell 22:3130–3141. https://doi.org/10.1105/tpc.110.077040

Nunes CC, Dean RA (2012) Host-induced gene silencing: a tool for understanding fungal host interaction and for developing novel disease control strategies. Mol Plant Pathol 13:519–529. https://doi.org/10.1111/J.1364-3703.2011.00766.X

Nunes FMF, Simoes ZLP (2009) A non-invasive method for silencing gene transcription in honeybees maintained under natural conditions. Insect Biochem Mol Biol 39:157–160

Nykanen A, Haley B, Zamore PD (2001) ATP requirement and small interfering RNA structure in the RNA interference pathway. Cell 107:309–321

Ober KA, Jockusch EL (2006) The roles of wingless and decapentaplegic in axis and appendage development in the red flour beetle, *Tribolium castaneum*. Dev Biol 294:391–405

Panwar V, McCallum B, Bakkeren G (2013) Host-induced gene silencing of wheat leaf rust fungus *Puccinia triticina* pathogenicity genes mediated by the Barley stripe mosaic virus. Plant Mol Biol 81:595–608. https://doi.org/10.1007/s11103-013-0022-7

Papolu PK, Gantasala NP, Kamaraju D et al (2013) Utility of host delivered RNAi of two FMRF amide like peptides, flp-14 and flp-18, for the management of root knot nematode, *Meloidogyne incognita*. PLoS One 8:e80603. https://doi.org/10.1371/journal.pone.0080603

Pasquinelli AE (2002) MicroRNAs: deviants no longer. Trends Genet 18:171–173

Peng T, Pan Y, Yang C et al (2016) Over-expression of CYP6A2 is associated with spirotetramat resistance and cross-resistance in the resistant strain of *Aphis gossypii* glover. Pestic Biochem Physiol 126:64–69

Pitino M, Hogenhout SA (2013) Aphid protein effectors promote aphid colonization in a plant species-specific manner. Mol Plant-Microbe Interact 26:130–139

Pitino M, Coleman AD, Maffei ME et al (2011) Silencing of aphid genes by dsRNA feeding from plants. PLoSONE 6:e25709. https://doi.org/10.1371/journal.pone.0025709

Possamai SJ, Trionnaire GL, Bonhomme J et al (2007) Gene knockdown by RNAi in the pea aphid *Acyrthosiphon pisum*. BMC Biotechnol 7:63

Price DR, Gatehouse JA (2008) RNAi-mediated crop protection against insects. Trends Biotechnol 26:393–400. https://doi.org/10.1016/j.tibtech.2008.04.004

Rajagopal R, Sivakumar S, Agrawal N et al (2002) Silencing of midgut aminopeptidase N of *Spodoptera litura* by double-stranded RNA establishes its role as *Bacillus thuringiensis* toxin receptor. J Biol Chem 277:46849–46851

Rangasamy M, Siegfried BD (2012) Validation of RNA interference in western corn rootworm *Diabrotica virgifera virgifera* LeConte (Coleoptera, Chrysomelidae) adults. Pest Manag Sci 68:587–591. https://doi.org/10.1002/ps.2301

Rebijith KB, Asokan R, Hande HR et al (2016) RNA interference of odorant-binding protein 2 (OBP2) of the cotton aphid, *Aphis gossypii* (glover), resulted in altered electrophysiological responses. Appl Biochem Biotechnol 178:251–266

Riechen J (2007) Establishment of broad-spectrum resistance against *Blumeria graminis* f. sp. *tritici* in *Triticum aestivum* by RNAi-mediated knock-down of MLO. J Verbrauch Lebensm 2:120. https://doi.org/10.1007/s00003-007-0282-8

Rodriguez-Cabrera L, Trujillo-Bacallao D, Borra's-Hidalgo O et al (2010) RNAi-mediated knock-down of a *Spodoptera frugiperda* trypsin-like serine-protease gene reduces susceptibility to a *Bacillus thuringiensis* Cry1Ca1 protoxin. Environ Microbiol 12:2894–2903

Romano N, Macino G (1992) Quelling: transient inactivation of gene expression in Neurospora crassa by transformation with homologous sequences. Mol Microbiol 6:3343–3353

Rosso MN, Dubrana MP, Cimbolini N et al (2005) Application of RNA interference to root-knot nematode genes encoding esophageal gland proteins. Mol. Plant Microb. Interact. 18:615–620. https://doi.org/10.1094/MPMI-18-0615

Roxström-Lindquist K, Terenius O, Faye I (2004) Parasite-specific immune response in adult *Drosophila melanogaster:* a genomic study. Scientific Report 5:207–212. https://doi.org/10.1038/sj.embor.7400073

Santos AC, Sena JAL, Santos SC et al (2009) dsRNA induced gene silencing in *Moniliophthora perniciosa*, the causal agent of witches' broom disease of cacao. Fungal Genet Biol 46:825–836

Sapountzis P, Duport G, Balmand S et al (2014) New insight into the RNA interference response against cathepsin-L gene in the pea aphid, *Acyrthosiphon pisum*: molting or gut phenotypes specifically induced by injection or feeding treatments. Insect Biochem Mol Biol 51:20–32

Schmidt A, Palumbo G, Bozzetti MP et al (1999) Genetic and molecular characterization of sting, a gene involved in crystal formation and meiotic drive in the male germ line of *Drosophila melanogaster*. Genetics 151:749–760

Senthil-Kumar M, Mysore KS (2011) Caveat of RNAi in plants: the off-target effect. In: Kodama H, Komamine A (eds) RNAi and plant gene function analysis. Methods in molecular biology (methods and protocols), vol 744. Humana Press

Shakesby AJ, Wallace IS, Isaacs HV et al (2009) A water-specific aquaporin involved in aphid osmoregulation. Insect Biochem Mol Biol 39:1–10. https://doi.org/10.1016/j.ibmb.2008.08.008

Shivakumara TN, Sonam C, Divya K et al (2017) Host-induced silencing of two pharyngeal gland genes conferred transcriptional alteration of cell wall-modifying enzymes of *Meloidogyne incognita* vis-à-vis perturbed nematode infectivity in eggplant. Front Plant Sci 8:473. https://doi.org/10.3389/fpls.2017.00473

Sijen T, Fleenor J, Simmer F et al (2001) On the role of RNA amplification in dsRNA-triggered gene silencing. Cell 107:465–476

Sindhu A, Maier TR, Mittchum MG et al (2009) Effective and specific in planta RNAi in cyst nematodes: expression interference of four parasitism genes reduces parasitic success. J Exp Bot 1:315–324. https://doi.org/10.1093/jxb/ern289

Singh AD, Wong S, Ryan CP et al (2013) Oral delivery of double-stranded RNA in larvae of the yellow fever mosquito, *Aedes aegypti*: implications for pest mosquito control. J Insect Sci 13:69

Siomi H, Siomi MC (2009) On the road to reading the RNA-interference code. Nature 457:396–404. https://doi.org/10.1038/nature07754

Steeves RM, Todd TC, Essig JS et al (2006) Transgenic soybeans expressing siRNAs specific to a major sperm protein gene suppress Heterodera glycines reproduction. Func Plant Biol 33:991–999. https://doi.org/10.1071/FP06130

Surakasi VP, Mohamed AAM, Kim Y (2011) RNA interference of β1 integrin subunit impairs development and immune responses of the beet armyworm, *Spodoptera exigua*. J Insect Physiol 57:1537–1544

Tabara H, Grishok A, Mello CC (1998) RNAi in C. elegans: soaking in the genome sequence. Science 282:430–431. https://doi.org/10.1126/science.282.5388.430

Tabashnik BE (2008) Delaying insect resistance to transgenic crops. PNAS 105:19029–19030. https://doi.org/10.1073/pnas.0810763106

Tabashnik BE, Gassmann AJ, Crowder DW et al (2008) Insect resistance to Bt crops: evidence versus theory. Nat Biotechnol 26:199–202. https://doi.org/10.1038/nbt1382

Tan FL, Yin JQ (2004) RNAi, a new therapeutic strategy against viral infection. Cell Res 14:460–466

Terenius O, Papanicolaou A, Garbutt JS et al (2011) RNA interference in Lepidoptera: an overview of successful and unsuccessful studies and implications for experimental design. J Insect Physiol 57:231–245. https://doi.org/10.1016/j.jinsphys.2010.11.006

Thompson JD, Kornbrust DJ, Foy JW et al (2012) Toxicological and pharmacokinetic properties of chemically modified siRNAs targeting p53 RNA following intravenous administration. Nucleic Acid Ther 22:255–264. https://doi.org/10.1089/nat.2012.0371

Tian H, Peng H, Yao Q et al (2009) Developmental control of a lepidopteran pest *Spodoptera exigua* by ingestion of bacteria expressing dsRNA of a non-midgut gene. PLoS One 4:e6225

Timmons L, Court DL, Fire A (2001) Ingestion of bacterially expressed dsRNAs can produce specific and potent genetic interference in Caenorhabditis elegans. Gene 263:103–112. https://doi.org/10.1016/S0378-1119(00)00579-5

Tinoco ML, Dias BB, Astta RCD et al (2010) In vivo trans-specific gene silencing in fungal cells by in planta expression of a double-stranded RNA. BMC Biol 8:1–11

Tomari Y, Zamore PD (2005) MicroRNA biogenesis: drosha can't cut it without a partner. Curr Biol 15:R61–R64

Tomoyasu Y, Miller SC, Tomita S et al (2008) Exploring systemic RNA interference in insects: a genome-wide survey for RNAi genes in Tribolium. Genome Biol 9:R10. https://doi.org/10.1186/gb-2008-9-1-r10

Turner CT, Davy MW, MacDiarmid RM et al (2006) RNA interference in the light brown apple moth, *Epiphyas postvittana* Walker induced by double-stranded RNA feeding. Insect Mol Biol 15:383–391

Tuschl T, Zamore PD, Lehmann R et al (1999) Targeted mRNA degradation by double-stranded RNA *in vitro*. Genes Dev 13:3191–3197

Tzin V, Yang X, Jing X et al (2015) RNA interference against gut osmoregulatory genes in phloem-feeding insects. J Insect Physiol 79:105–112

Ulvila J, Parikka M, Kleino A et al (2006) Double-stranded RNA is internalized by scavenger receptor-mediated endocytosis in Drosophila S2 cells. J Biol Chem 281:14370–14375

Upadhyay SK, Chandrashekar K, Thakur N et al (2011) RNA interference for the control of whiteflies (*Bemisia tabaci*) by oral route. J Biosci 36:153–161

Urwin PE, Lilley CJ, Atkinson HJ (2002) Ingestion of double-stranded RNA by pre-parasitic juvenile cyst nematodes leads to RNA interference. Mol Plant Microb Interact 15:747–752. https://doi.org/10.1094/MPMI.2002.15.8.747

Urwin PE, Lilley CJ, McPherson MJ et al. (1997) Resistance to both cyst- and root-knot nematodes conferred by transgenic Arabidopsis expressing a modified plant cystatin. Plant J 12:455–461

Valdes VJ, Sampieri A, Sepulveda J et al (2003) With double stranded RNA to prevent in vitro and in vivo viral infections by recombinant baculovirus. J Biol Chem 278:19317–19324

Van Rij RP, Berezikov E (2009) Small RNAs and the control of transposons and viruses in *Drosophila*. Trends Microbiol 17:163–171

Vauthier C, Dubernet C, Chauvierre C et al (2003) Drug delivery to resistant tumors: the potential of poly (alkyl cyanoacrylate) nanoparticles. J Control Release 93:151–160. https://doi.org/10.1016/j.jconrel.2003.08.005

Vieira P, Akker EDS, Verma R et al (2015) The Pratylenchus penetrans transcriptome as a source for the development of alternative control strategies: mining for putative genes involved in parasitism and evaluation of in planta RNAi. PLoS One 10:e0144674. https://doi.org/10.1371/journal.pone.0144674

Walawage SL, Britton MT, Leslie CA et al (2013) Stacking resistance to crown gall and nematodes in walnut rootstocks. BMC Genomics 14:668. https://doi.org/10.1186/1471-2164-14-668

Walshe DP, Lehane SM, Lehane MJ (2009) Prolonged gene knockdown in the tsetse fly Glossina by feeding double stranded RNA. Insect Mol Biol 18:11–19

Wang P, Granados RR (2001) Molecular structure of the peritrophic membrane (PM): identification of potential PM target sites for insect control. Arch Insect Biochem Physiol 47:110–118. https://doi.org/10.1002/arch.1041

Wang Y, Zhang H, Li H et al (2011) Second-generation sequencing supply an effective way to screen RNAi targets in large scale for potential application in pest insect control. PLoS One 6:e18644. https://doi.org/10.1371/journal.pone.0018644

Wang W, Luo L, Lu H et al (2015) Angiotensin-converting enzymes modulate aphid–plant interactions. Sci Reports 5:8885

Whyard S, Singh AD, Wong S (2009) Ingested double-stranded RNAs can act as species-specific insecticides. Insect Biochem Mol Biol 39:824–832. https://doi.org/10.1016/j.ibmb.2009.09.007

Will T, Vilcinskas A (2015) The structural sheath protein of aphids is required for phloem feeding. Insect Biochem Mol Biol 57:34–40

Winston WM, Molodowitch C, Hunter CP et al (2002) Systemic RNAi in C. elegans requires the putative transmembrane protein SID-1. Science 295:2456–2459

Winston WM, Sutherlin M, Wright AJ et al (2007) Caenorhabditis elegans SID-2 is required for environmental RNA interference. PNAS 104:10565–10570. https://doi.org/10.1073/pnas.0611282104

Winter J, Jung S, Keller S et al (2009) Many roads to maturity: microRNA biogenesis pathways and their regulation. Nat Cell Biol 11:228–234

Wuriyanghan H, Rosa C, Falk BW (2011) Oral delivery of double-stranded RNAs and siRNAs induces RNAi effects in the potato/tomato psyllid, Bactericera cockerelli. PLoS One 6:e27736

Wynant N, Verlinden H, Breugelmans B et al (2012) Tissue-dependence and sensitivity of the systemic RNA interference response in the desert locust, Schistocerca gregaria. Insect Biochem Mol Biol 42:911–971

Xiao D, Lu YH, Shang QL et al (2015) Gene silencing of two acetylcholinesterases reveals their cholinergic and non-cholinergic functions in Rhopalosiphum padi and Sitobion avenae. Pest Manag Sci 71:523–530

Xiong Y, Zeng H, Zhang Y et al (2013) Silencing the HaHR3 gene by transgenic plant-mediated RNAi to disrupt Helicoverpa armigera development. Int J Biol Sci 9:370–381

Xu HJ, Chen T, Ma XF et al (2013) Genome-wide screening for components of small interfering RNA (siRNA) and micro-RNA (miRNA) pathways in the brown planthopper, Nilaparvata lugens (Hemiptera: Delphacidae). Insect Mol Biol 22:635–647. https://doi.org/10.1111/imb.12051

Xu L, Duan X, Lv Y et al (2014) Silencing of an aphid carboxylesterase gene by use of plant-mediated RNAi impairs Sitobion avenae tolerance of Phoxim insecticides. Transgenic Res 23:389–396

Xue B, Hamamouch N, Li C et al (2013) The 8D05 parasitism gene of Meloidogyne incognita is required for successful infection of host roots. Phytopathology 103:175–181. https://doi.org/10.1094/PHYTO-07-12-0173-R

Yadav BC, Veluthambi K, Subramaniam K (2006) Host generated double stranded RNA induces RNAi in plant parasitic nematodes and protects the host from infection. Mol Biochem Parasitol 148:219–222. https://doi.org/10.1016/j.molbiopara.2006.03.013

Yanagihara K, Tashiro M, Fukuda Y et al (2006) Effects of short interfering RNA against methicillin-resistant Staphylococcus aureus coagulase in vitro and in vivo. J Antimicrob Chemother 57:122–126

Yang J, Han Z (2014) Efficiency of different methods for dsRNA delivery in cotton bollworm (Helicoverpa armigera). J Integr Agric 13:115–123

Yao J, Rotenberg D, Afsharifar A et al (2013) Development of RNAi methods for Peregrinus maidis, the corn planthopper. PLoS One 8:e370243

Yin C, Jurgenson JE, Hulbert SH (2011) Development of a host-induced RNAi system in the wheat stripe rust fungus *Puccinia striiformis* f. sp. *tritici*. Mol Plant-Microbe Interact 24:554–561. https://doi.org/10.1094/MPMI-10-10-0229

Zamore PD, Tuschl T, Sharp PA (2000) RNAi: double-stranded RNA directs the ATP-dependent cleavage of mRNA at 21- to 23-nucleotide intervals. Cell 101:25–33

Zha W, Peng X, Chen R et al (2011) Knockdown of midgut genes by dsRNA-transgenic plant-mediated RNA interference in the hemipteran insect *Nilaparvata lugens*. PLoS One 6:e20504. https://doi.org/10.1371/journal.pone.0020504

Zhang Y, Lu Z (2015) Peroxiredoxin 1 protects the pea aphid *Acyrthosiphon pisum* from oxidative stress induced by *Micrococcus luteus* infection. J Invertebr Pathol 127:115–121

Zhang H, Kolb F, Jaskiewicz L et al (2004) Single processing center models for human Dicer and bacterial RNase III. Cell 118:57–68

Zhang X, Zhang J, Zhu KY (2010) Chitosan/double-stranded RNA nanoparticle-mediated RNA interference to silence chitin synthase genes through larval feeding in the African malaria mosquito (*Anopheles gambiae*). Insect Mol Biol 19:683–693. https://doi.org/10.1111/j.1365-2583.2010.01029.x

Zhang M, Wang Q, Xu K et al (2011) Production of dsRNA sequences in the host plant is not sufficient to initiate gene silencing in the colonizing oomycete pathogen *Phytophthora parasitica*. PLoS One 6:e28114. https://doi.org/10.1371/journal.pone.0028114

Zhang Y, Zhang SZ, Kulye M et al (2012) Silencing of molt-regulating transcription factor gene, CiHR3, affects growth and development of sugarcane stem borer, *Chilo infuscatellus*. J Insect Sci 12:1–12

Zhang H, Li HC, Miao XX (2013a) Feasibility, limitation and possible solutions of RNAi-based technology for insect pest control. Insect Sci 20:15–30

Zhang X, Liu X, Ma J et al (2013b) Silencing of cytochrome P450 CYP6B6 gene of cotton bollworm (*Helicoverpa armigera*) by RNAi. Bull Entomol Res 103:584–591

Zhang J, Khan SA, Hasse C et al (2015a) Full crop protection from an insect pest by expression of long double-stranded RNAs in plastids. Science 347:991–994

Zhang X, Mysore K, Flannery E et al (2015b) Chitosan/interfering RNA nanoparticle mediated gene silencing in disease vector mosquito larvae. J Vis Exp 97:52523. https://doi.org/10.3791/52523

Zhang J, Khan SA, Heckel DG et al (2017) Next-generation insect-resistant plants: RNAi-mediated crop protection. Trends Biotechnol 35:871–882

Zhao L, Chen J (2013) Double stranded RNA constructs to control ants. US Patent Application Publication No. 2013/0078212

Zhao Y, Yang G, Wang-Pruski G et al (2008) Phyllotreta striolata (Coleoptera, Chrysomelidae): arginine kinase cloning and RNAi-based pest control. Eur J Biochem 105:815–822

Zhou X, Wheeler MM, Oi FM et al (2008) RNA interference in the termite *Reticulitermes flavipes* through ingestion of double-stranded RNA. Insect Biochem Mol Biol 38:805–815

Zhu F, Xu J, Palli R et al (2011) Ingested RNA interference for managing the populations of the Colorado potato beetle, *Leptinotarsa decemlineata*. Pest Manag Sci 67:175–182. https://doi.org/10.1002/ps.2048

Zhu JQ, Liu S, Ma Y et al (2012) Improvement of pest resistance in transgenic tobacco plants expressing dsRNA of an insect-associated gene EcR. PLoS One 7:e38572

Zhuo K, Chen J, Lin B et al (2017) A novel *Meloidogyne enterolobii* effector MeTCTP promotes parasitism by suppressing programmed cell death in host plants. Mol Plant Pathol 18:45–54. https://doi.org/10.1111/mpp.12374

Chapter 5
RNAi Approach: A Powerful Technique for Gene Function Studies and Enhancing Abiotic Stress Tolerance in Crop Plants

Ajay Kumar Singh, Mahesh Kumar, Deepika Choudhary, Lalitkumar Aher, Jagadish Rane, and Narendra Pratap Singh

Abstract RNA interference (RNAi) is a versatile tool frequently used for gene function studies in plants. RNAi phenomenon involves small interfering RNA (siRNA) or short hairpin or microRNA (miRNA) to suppress the expression of sequence-specific gene at posttranscriptional or translational level. This technology has been used to study functional relevance of genes, enhancing crop yield, improving nutritional quality, and increasing crop productivity through suppression of expression of genes responsive to abiotic stress, involved in biomass and grain yield. Here, we describe mechanism of RNAi-mediated gene silencing and application of RNAi technique involving siRNA, shRNA, and microRNA for elucidating function of genes responsive to abiotic stress in crops and also for improving abiotic stress tolerance in crop plants.

Keywords siRNA- small interfering RNA · miRNA- microRNA · RNAi- RNA interference · Drought stress · Salinity stress · Abiotic stress tolerance · Dicer · Argonaute · Gene silencing · Posttranscriptional gene silencing

5.1 Introduction

Crop yield is adversely affected by various kinds of abiotic stresses such as drought, salinity, and heat and cold stresses. Therefore, in the future, there will be huge demand of genetically improved crops with ability to maintain yield stability under adverse environmental conditions. Drought, salinity, heat, and cold stress tolerance and adaptation of crop plants to these stresses have been improved through RNAi approach for manipulating expression of transcription factor genes, genes associated with signaling and biosynthetic pathways and accumulation of antioxidants

A. K. Singh (✉) · M. Kumar · D. Choudhary · L. Aher · J. Rane · N. P. Singh
ICAR-National Institute of Abiotic Stress Management, Malegaon,
Baramati, Pune, India
e-mail: ajay.singh4@icar.gov.in

© Springer International Publishing AG, part of Springer Nature 2018
S. S. Gosal, S. H. Wani (eds.), *Biotechnologies of Crop Improvement, Volume 2*,
https://doi.org/10.1007/978-3-319-90650-8_5

(Gupta et al. 2014; Pradhan et al. 2015; Wang et al. 2015; Li et al. 2017a; Meena et al. 2017). Several genes associated with metabolic pathways have been functionally characterized to understand stress tolerance mechanisms and to improve abiotic stress tolerance in crop plants (Zhou et al. 2015; Guo et al. 2016; Ji et al. 2016; Li et al. 2017a; Ma et al. 2017; Cai et al. 2018; Huang et al. 2018). It is utmost important to elucidate the role of transcription factors or genes by genetic manipulation for higher yield and also for maintaining yield stability under various abiotic stress conditions. Several researchers are trying to identify and characterize various genes responsive to drought and salinity stress by using genomics, transcriptomics, proteomics, and metabolomics approaches (Wang et al. 2015; Ji et al. 2016; Qin et al. 2016; Tripathi et al. 2016; Li et al. 2017a; Huang et al. 2018). Therefore, it is essential to know the exact role of specific small RNA followed by genetic manipulation for improvement of abiotic stress tolerance in crop plants.

RNA interference (RNAi) phenomenon involves suppression of the gene expression by degrading the specific messenger RNAs. The RNAi technology involves small noncoding RNAs, viz., small interfering RNA (siRNA), short hairpin RNA (shRNA), and microRNA (miRNA) that are the cleavage product of dsRNA. The mRNA degradation process is triggered by the introduction of double-stranded RNA (dsRNA) which is further cleaved by the enzyme dicer (Kumar et al. 2012). In addition to small noncoding RNAs, the RNAi phenomenon also involves an RNA-induced silencing complex (RISC) (Redfern et al. 2013; Wilson and Doudna 2013) and Argonaute proteins (AGOs) (Ender and Meister 2010; Riley et al. 2012). The phenomenon of gene silencing was discovered accidentally in petunia flowers where expression of pigment-producing gene *chalcone synthase* resulted in variegated flowers instead of expected deep purple color. Since, the expression of both the transgene and the homologous endogenous gene was suppressed, the phenomenon was termed co-suppression (Napoli et al. 1990; Campbell 2005). RNAi technology can be used to identify and functionally characterize numerous genes within any genome which can be exploited for crop improvement (Pradhan et al. 2015; Li et al. 2017a; Meena et al. 2017). The RNAi technology has been successfully used for improvement of several plant species in terms of enhancing abiotic stress tolerance (Wang et al. 2015; Li et al. 2017a; Srivastava et al. 2017; Mao et al. 2018).

5.2 Mechanisms of RNAi-Mediated Gene Silencing

There are two classes of small RNA in the RNAi pathway, a small interfering RNA (siRNA) and microRNA (miRNA) (Figs. 5.1, 5.2, and 5.3) The miRNAs are similar to siRNA in many aspects as they originate from double-stranded structure, the size of the miRNA is 20 to 24 bp, and both are generated by dicer or dicer-like enzyme (DCL1, DCL2). The miRNA is derived from genomic DNA, while siRNA is generated by cleavage of dsRNA into smaller segment. Active miRNA has two phases including primary miRNA (primiRNA) and pre-miRNA. Both pri- and pre-miRNAs are characterized by a hairpin structure.

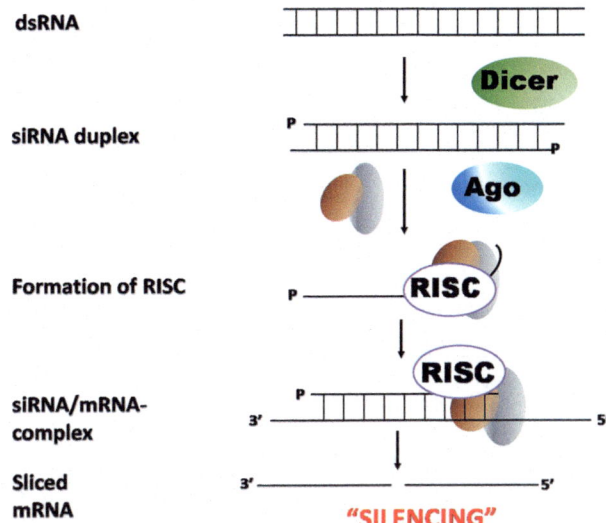

Fig. 5.1 siRNA-mediated RNAi gene silencing

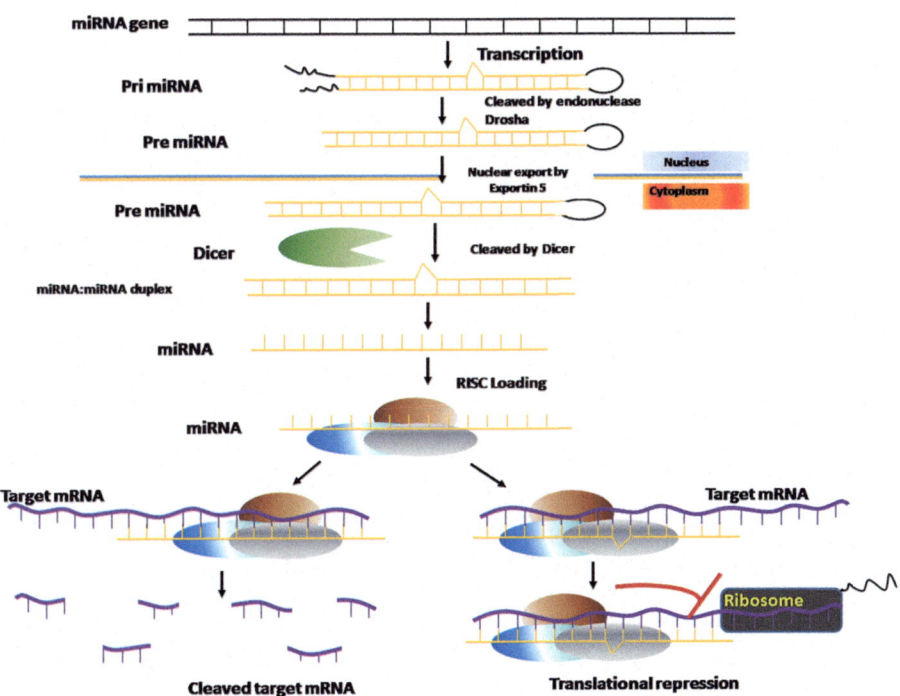

Fig. 5.2 miRNA-mediated RNAi gene silencing

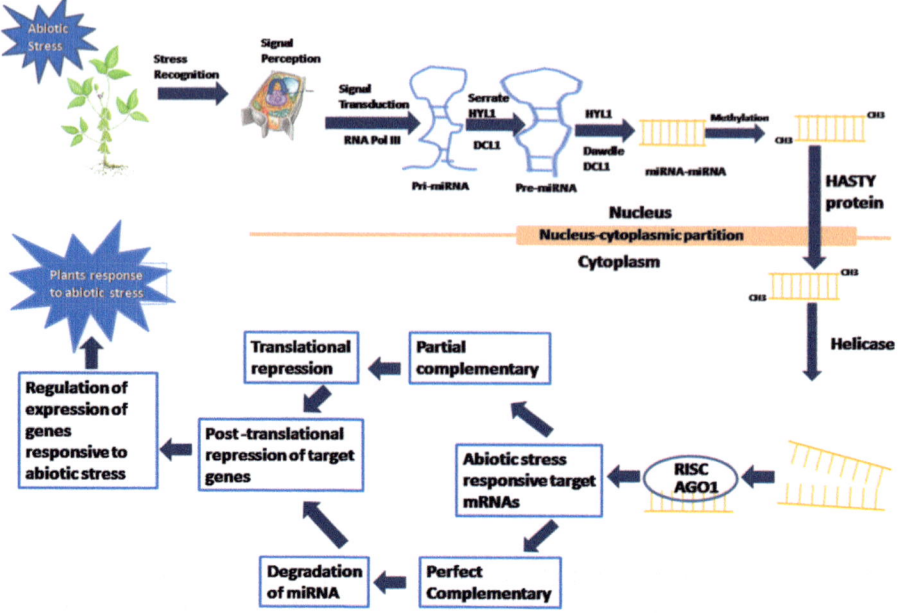

Fig. 5.3 Role of RNAi in plants response to abiotic stresses

Processing of miRNA occurs at the nuclear and cytoplasmic levels (Williams et al. 2004). Once a miRNA gene is transcribed, the transcript forms a roughly 42 to 60 bp long hairpin structure with two arms of approximately the same length. One of the strands produces active miRNA via dicer. RNA interference pathway involves four common steps: cleavage of dsRNA by dicer, entry of SiRNA into RISC complex, silencing complex activation, and mRNA degradation (Ali et al. 2010). The first step of RNAi involves the introduction of dsRNA in the cell which is recognized by dicer enzyme (Fig. 5.1). Dicer enzyme further processes the dsRNA into dsSiRNA of 21–25 nucleotides. Then, the siRNA produced by the dicer is loaded onto multicomponent nuclear complex into the RNA-induced silencing complex, which is inactive in this form to conduct RNAi. The next step involves unwinding of the siRNA by a helicase and further remodeling of the complex to create an active form of the RISC. RISC is a ribonucleoprotein complex, and its two important components are the single-stranded siRNA and the Argonaute protein (Kumar et al. 2012). The next step is degradation of mRNA. The active component of RISC is an endonuclease called Argonaute protein which cleaves the target mRNA strand complementary to their bound siRNA; therefore Argonaute contributes "silencer" activity to RISC. When the dsRNA is cleaved by the dicer, it produces the small siRNA in which one strand is known as guide strand that binds the Argonaute protein and directs gene silencing. After the cleavage is complete, the RISC departs, and the siRNA can be reused in a new cycle of mRNA recognition and cleavage (Figs. 5.1 and 5.2).

5.3   Functional Elucidation of Genes Responsive to Abiotic Stress Employing RNAi Technology

Considerable progress has been made in developing genomic resources for plants such as soybean, chickpea, pigeon pea, peanut, rice, wheat, maize, barley, grape and sorghum. A large number of genes have been identified through transcriptome profiling under various abiotic stress conditions, but most of them with unknown function. Therefore, a major research priority in the post-genomic sequencing era is determining the function of these genes (Wesley et al. 2001). The primary tool for dissecting a genetic pathway is the screen for the loss of gene function and disrupting the target pathway. Modern biotechnology has enabled the elucidation of gene function through the systematic modification of gene expression followed by quantitative and qualitative analyses of the gene expression products. The modulation of gene expression can be achieved by the integration of foreign DNA sequences in the plant genome, leading to either overexpression or gene silencing. Gene silencing is currently achieved through RNA interference (RNAi), a process of sequence-specific, posttranscriptional gene silencing initiated by double-stranded RNA that is homologous in sequence to the target gene (Figs. 5.1 and 5.2). Overexpression and silencing are complementary strategies which have been used to functionally characterize genes responsive to abiotic stresses in many crop plants (Table 5.1) (Guo et al. 2016; Ma et al. 2017; Srivastava et al. 2017; Mao et al. 2018).

5.4   Application of RNAi Technique in Elucidating Function of Genes Associated with Abiotic Stress Tolerance and Understanding Mechanisms of Plants' Response to Abiotic Stresses

Stress is usually defined as an external factor that exerts a disadvantageous effect or harmful effect on the plant. Abiotic stress causes serious damages to the plant by negatively affecting its growth and yield potential. It has been estimated that nearly 60–70% of crop yield is reduced due to the abiotic stress (Younis et al. 2014). Plants are subjected to many types of fluctuations in the physical environment. Plants have adapted numerous physiological, biochemical, and metabolic approaches for tolerating abiotic stresses. Abiotic stresses are classified into drought, salinity, heat, cold and oxidative stress. Classical techniques of breeding crop plants with enhanced tolerance to abiotic stress have until now achieved inadequate success. Therefore, transgenic technology is one of the numerous tools that offered improvement in modern plant breeding program. Identification of candidate gene through functional genomics programs discovered multiple gene families which regulate the abiotic stress tolerance phenomena and high production. Therefore, plant biologists are trying to incorporate the candidate gene or multiple numbers of genes to express ectopically for crop improvements (Younis et al. 2014). Nowadays, RNAi technology has been evolved as a modern approach for gene function analysis and in translational research program. Recent findings suggest the RNAi is playing an important role in abiotic stress stimulation in different crops (Fig. 5.3). RNAi technology could be a substitute of complex molecular techniques because it contains several benefits, its specificity and sequence-based gene silencing. Due to this property, RNAi has been effectively utilized for incorporating desired trait for abiotic stress tolerance in various plant species (Table 5.1) (Jagtap et al. 2011; Pradhan et al. 2015; Meena et al. 2017; Li et al. 2017a). This ability of RNAi has been efficaciously utilized for incorporating desired traits for abiotic stress tolerance in various plants species.

The first evidence that siRNAs are involved in abiotic stress responses in plants was provided by Sunkar and Zhu (2004). In recent years, RNA approach has been used to describe functional relevance of several genes responsive to various kinds of abiotic stresses (Pradhan et al. 2015; Li et al. 2017a; Meena et al. 2017). Zhou et al. (2015) functionally characterized Glossy1 (GL1)-homologous gene OsGL1–3 in rice using overexpression and RNAi transgenic rice plants. OsGL1–3 gene was ubiquitously expressed at different levels in rice plants except root, and its expression was upregulated under ABA and PEG treatments. Overexpressing rice plants exhibited stunted growth, more wax crystallization on leaf surface, and significantly increased total cuticular wax load due to the prominent changes of C30–C32 aldehydes and C30 primary alcohols compared to wild-type plants. While OsGL1–3 RNAi-silenced plants exhibited no significant difference in plant height, there was less wax crystallization and decreased total cuticular wax accumulation on leaf surface. Based on all these evidences, together with the effects of OsGL1–3 on the expression of some wax synthesis-related genes, Zhou et al. (2015) suggested that OsGL1–3 plays an important role in drought tolerance. Wang et al. (2015) studied the interaction of GmWRKY27 with GmMYB174 and reported that these two cooperatively inhibit transcription of GmNAC29 by binding to the core sequences in its promoter. The downregulation of expression of GmNAC29 leads to reduced intracellular ROS levels. GmWRKY27 may also increase proline content by indirectly suppressing the transcription of PDH which ultimately led to improvement in stress tolerance in soybean (Wang et al. 2015). Basic helix-loop-helix (bHLH) leucine zipper transcription factors regulate plants' response to abiotic stress. However, the exact role of bHLH in abiotic stress tolerance is not fully known. Ji et al. (2016) functionally characterized a *bHLH* gene, *ThbHLH1*, from *Tamarix hispida* in abiotic stress tolerance. *T. hispida* plantlets with transiently overexpressed ThbHLH1 and RNAi-silenced ThbHLH1 were generated for gain- and loss-of-function analysis. Transgenic overexpressing ThbHLH1 *Arabidopsis thaliana* lines were generated to confirm the gain- and loss-of-function analysis. Overexpression of ThbHLH1 increases glycine betaine and proline levels, increases Ca^{2+} concentration, and enhances peroxidase (POD) and superoxide dismutase (SOD) activities to decrease reactive oxygen species (ROS) accumulation. Additionally, ThbHLH1 regulates the expression of the genes including P5CS, BADH, CaM, POD, and SOD, to activate ROS scavenging process, and also induces the expression of stress tolerance-related genes LEAs and HSPs. Ji et al. (2016) reported that ThbHLH1 induces the expression of stress tolerance-related genes to improve abiotic stress tolerance by increasing osmotic potential, improving ROS scavenging capability, and enhancing second messenger in stress signaling cascades. Salinity is a severe environmental stress that greatly reduces crops' productivity worldwide. The GMPase plays an important role in tolerance of plants to salt stress at vegetative stage. The OsVTC1–1 expression was suppressed using RNAi-mediated gene silencing to elucidate the function of GMPase in response of rice to salt stress (Qin et al. 2016). OsVTC1–1 RNAi lines of rice accumulated more ROS under salt stress, and supplying exogenous ascorbic acid restored salt tolerance of OsVTC1–1 RNAi lines, suggesting that OsVTC1–1 is involved in salt tolerance of rice through the biosynthesis regulation of ascorbic acid (Qin et al. 2016). Qin et al. (2016) demonstrated that rice GMPase gene OsVTC1–1 plays a critical role in salt tolerance of rice at both vegetative and reproductive stages through AsA scavenging of excess ROS. Guo et al. (2016) reported that MID1 (MYB Important for Drought Response1), encoding a putative R-R-type MYB-like transcription factor, improves rice yield under drought. The MID1 transcription factor was functionally enumerated by developing overexpressing plants and RNAi lines in rice and found that MID1 play an important role in response to drought stress during reproductive stage. Guo et al. (2016) demonstrated that MID1 acts as a transcriptional regulator that promotes rice male development under drought by modulating the expressions of drought-related and anther developmental genes. Ma et al. (2017) elucidated signal transduction involving MAPK cascades in rice by developing RNAi and overexpressing plants. MPKK10.2-overexpressing plants showed enhanced resistance to drought, whereas MPKK10.2-RNA interference (RNAi) plants had increased sensitivity to drought. Ma et al. (2017) reported that MAPKK10.2 is associated with abiotic stress responses by functioning in the cross-point of two MAPK cascades leading to drought tolerance. Li et al. (2017b) studied physiological significance of glycosyltransferase genes, UGT79B2 and UGT79B3, strongly induced by various abiotic stresses such as cold, salt, and drought stresses. Overexpression of UGT79B2/B3 enhanced plant tolerance to low temperatures as well as drought and salt stresses, whereas the ugt79b2/b3 double mutants generated by RNAi (RNA interference) were more susceptible to adverse environmental conditions. Li et al. (2017b) identified enzyme activities of UGT79B2/B3 in adding UDP-rhamnose to cyanidin and cyanidin 3-O-glucoside. Ectopic expression of UGT79B2/B3 increased anthocyanin accumulation and enhanced antioxidant activity in coping with abiotic stresses, whereas the ugt79b2/b3 double mutants showed reduced anthocyanin levels. Li et al. (2017b) demonstrated that UGT79B2 and UGT79B3 are regulated by CBF1 and confer abiotic stress tolerance by modulating anthocyanin accumulation. Srivastava et al. (2017) revealed the role of the SUMO protease, OsOTS1, in mediating tolerance to drought in rice. They demonstrated that OsOTS1RNAi lines accumulated more ABA and exhibited more productive agronomic traits during drought, while OsOTS1 overexpressing lines were drought sensitive but ABA insensitive. Srivastava et al. (2017) reported that OsOTS1 SUMO protease directly targets the ABA and drought-responsive transcription factor OsbZIP23 for de-SUMOylation affecting its stabil-

Table 5.1 Functional characterization of genes responsive to abiotic stress using RNAi approach and response of plants to abiotic stress in response to silencing/overexpression of genes

Plant species	Target gene	Plant response to abiotic stress		Abiotic stresses	Reference
		Silencing of gene	Overexpression of gene		
Soybean	GmNAC29	Reduced intracellular ROS levels	–	Abiotic stress	Wang et al. (2015)
Rice	Glossy1 (GL1)-homologous gene OsGL1-3	No alteration in plant height, less wax crystallization, and decreased total cuticular wax accumulation on leaf surface	Stunted growth, more wax crystallization on leaf surface, and significantly increased total cuticular wax load due to the prominent changes of C_{30}–C_{32} aldehydes and C_{30} primary alcohols	Drought stress	Zhou et al. (2015)
Arabidopsis Tamarix hispida	Basic helix-loop-helix (bHLH) leucine zipper transcription factor	Increase in ROS accumulation, decrease in glycine betaine and proline levels and decrease in Ca^{2+} concentration	Increase in glycine betaine and proline levels. Increase in Ca^{2+} concentration and enhancement of peroxidase (POD) and superoxide dismutase (SOD) activities lead to decreased reactive oxygen species (ROS) accumulation	Abiotic stress	Ji et al. (2016)
Rice	OsVTC1-1	More accumulation of ROS under salt stress	–	Salinity stress	Qin et al. (2016)
Rice	MAPK	Increased sensitivity to drought	Enhanced resistance to drought	Drought stress	Ma et al. (2017)
Arabidopsis	UDP-glycosyltransferases UGT79B2 and UGT79B3	Susceptible response to adverse environmental conditions	Enhanced plant tolerance to low temperatures as well as drought and salt stresses	Cold, salinity, and drought	Li et al. (2017a, b)
Rice	OsOTS1	Accumulated more ABA and exhibited more productive agronomic traits during drought	Drought sensitive but ABA insensitive		Srivastava et al. (2017)

Tomato	Fructose 1,6-bisphosphate aldolase (FBAs)	Decrease in FBA activity reduces plant growth and tolerance to chilling stress in tomato seedlings	–	Cold stress/heat stress	Cai et al. (2018)
Rice	DRAP1	Reduced drought tolerance with negative effects on development and yield	Improved drought tolerance	Drought stress	Huang et al. (2018)
Rice	OsNAC2	Markedly decreased cell death in response to severe salt stress	Enhancement of salt-induced cell death accompanied by the loss of plasma membrane integrity and nuclear DNA fragmentation	Salt stress	Mao et al. (2018)

ity. OsOTS-RNAi lines showed increased abundance of OsbZIP23 and increased drought-responsive gene expression, while OsOTS1 overexpressing lines show reduced levels of OsbZIP23 leading to suppressed drought-responsive gene expression (Srivastava et al. 2017). Srivastava et al. (2017) uncovered a mechanism in which rice plants govern ABA-dependant drought-responsive gene expression by controlling the stability of OsbZIP23 by SUMO conjugation through manipulating specific SUMO protease levels. It is a well-known fact that plant development and adaptations to environmental stresses are closely associated with programmed cell death (PCD). Mechanisms regulating PCD phenomenon such as accumulation of reactive oxygen species (ROS) are common among responses to different abiotic stresses. Recently, the pathways mediating salt-induced PCD were characterized by Mao et al. (2018). Mao et al. (2018) demonstrated that overexpressing OsNAC2 transcription factor enhances salt-induced cell death accompanied by the loss of plasma membrane integrity and nuclear DNA fragmentation. In OsNAC2-knockdown lines, cell death was markedly decreased in response to severe salt stress. Findings of Mao et al. (2018) revealed that OsNAC2 accelerates NaCl-induced PCD and provides new insights into the mechanisms affecting ROS accumulation, plant caspase-like activity, and K+ efflux. Huang et al. (2018) functionally characterized a DREB2-like gene, transcription factor gene OsDRAP1 in rice. Overexpressing OsDRAP1 transgenic plants exhibited improved drought tolerance, while OsDRAP1 RNA interfering plants exhibited reduced drought tolerance with negative effects on development and yield. OsDRAP1 interacted with several genes/proteins and activate many downstream DT-related genes, including transcription factors such as OsCBSX3. These findings can provide a basis for further understanding OsDRAP1-mediated gene networks and their related phenotypic effects. Cai et al. (2018) studied the role of fructose 1, 6-bisphosphate aldolase in photosynthesis and in regulating cold stress responses in tomato using RNA interference (RNAi) vector containing SlFBA7 reverse tandem repeat sequence. They reported that decrease in FBA activity reduces plant growth and tolerance to chilling stress in tomato seedlings.

5.5   Role of MicroRNAs in Elucidating Function of Genes Associated with Abiotic Stress Tolerance

MicroRNAs (miRNAs) are small RNAs that regulate expression of target genes posttranscriptionally and play major roles in development and responses to abiotic stress (Gupta et al. 2014). In abiotic stress condition, the plant after signal perception, the abiotic stress responsible miRNA gene undergoes transcription by RNA polymerase II enzyme into primary miRNA (primiRNA), and the miRNA is preceded by dicer-like DCL 1 into a miRNA duplex. The miRNA is then exported into the cytoplasm from the nucleus. The mature miRNAs are incorporated into RNA-induced silencing complex (RISC), where the mature single-stranded miRNA guides the RNA silencing activity of AGO1 to partially complementary mRNA

(Fig. 5.2). The microRNA then targets the abiotic stress-responsive mRNA and that causes translation repression and mRNA degradation (Ding et al. 2009). The function of miRNAs (microRNA) in relation to abiotic stress like oxidative, cold, drought, and salinity stresses was reported by Sunkar and Zhu (2004). MicroRNA have been characterized for their role in abiotic stress tolerance in several crop plants (Ma et al. 2015; Yuan et al. 2015; Chung et al. 2016; Ding et al. 2017). miR396, miR394, miR164, miR408, and miR2118 are a group of drought-inducible miRNAs. Overexpression of these miRNAs enhances drought resistance. Such an effect is likely to be caused by changes in development or oxidative status related to target repression (Song et al. 2013; Fang et al. 2014; Chen et al. 2015; Hajyzadeh et al. 2015; Wu et al. 2015a, b).

Overexpression of salinity-inducible miR393 reduces the levels of TIR1 and AFB2, and causes hypersensitivity to salinity stress, whereas expression of an miR393-resistant TIR1 transgene increases plant tolerance to salinity (Chen et al. 2011; Iglesias et al. 2014). The miRNVL5 from cotton and miR417 from *Arabidopsis* also negatively influence plant responses to salinity stress (Jung and Kang 2007). Conversely, two salinity-inducible miRNAs, miR319, and miR528, can positively affect plant response to salinity stress through the downregulation of their targets (Zhou et al. 2013; Yuan et al. 2015). The miR408 is a highly conserved miRNA in plants that responds to the availability of copper and targets genes encoding copper-containing proteins. Expression of miR408 is significantly affected by a variety of developmental and environmental conditions. Involvement of miR408 in the abiotic stress response was investigated in *Arabidopsis* (Ma et al. 2015). Ma et al. (2015) investigated the expression of miR408 and its target genes in response to salinity, cold, oxidative, drought, and osmotic stresses. Ma et al. (2015) generated transgenic plants with modulated miR408 expression and demonstrated that higher miR408 expression leads to improved tolerance to salinity, cold, and oxidative stresses, but enhanced sensitivity to drought and osmotic stresses. Cellular antioxidant capacity was enhanced in plants in response to elevated miR408 expression, and it was associated with reduced levels of reactive oxygen species and induced expression of genes associated with antioxidative functions, including Cu/Zn superoxide dismutases (CSD1 and CSD2) and glutathione-S-transferase (GST-U25), as well as auxiliary genes such as the copper chaperone CCS1 and the redox stress-associated gene SAP12. MicroRNA528 (miR528) is a conserved monocot-specific small RNA capable of mediating multiple stress responses. Yuan et al. (2015) reported that overexpression of a rice (*Oryza sativa*) miR528 (Osa-miR528) in transgenic creeping bent grass (*Agrostis stolonifera*) altered plant development and improved plant salt-stress tolerance. Morphologically, miR528-overexpressing transgenic plants displayed shortened internodes, increased tiller number, and upright growth. Improved salt-stress resistance was associated with increased water retention, cell membrane integrity, chlorophyll content, capacity for maintaining potassium homeostasis, and *catalase* activity and reduced *ascorbic acid oxidase* (AAO) activity (Yuan et al. 2015). In addition, AsAAO and *copper ion-binding protein 1* are identified as two putative targets of miR528 in creeping bent grass. Both of them respond to salinity and N starvation and are significantly downregulated in miR528-overexpressing transgenics (Yuan et al. 2015). Yuan et al. (2015) demonstrated that miR528 plays an important role in modulating plant growth and development and in the plant response to salinity and indicate the potential of manipulating miR528 in improving plant abiotic stress resistance. Timing of flowering is not only an interesting topic in developmental biology, but it also plays a significant role in agriculture for its effects on the maturation time of seed, and it is associated with abiotic stress tolerance capability of plants. The hexaploid wheat (*Triticum aestivum*) is one of the most important crop species whose flowering time, i.e., heading time, greatly influences yield under water stress and non-stressed condition. Zhao et al. (2016) identified the tae-miR408 in wheat and its targets in vivo, including *Triticum aestivum timing of CAB expression-A1* (TaTOC-A1), TaTOC-B1, and TaTOC-D1. The tae-miR408 levels were reciprocal to those of TaTOC1s under long-day and short-day conditions. Wheat plants with a knockdown of TaTOC1s via RNA interference and overexpression of tae-miR408 showed early-heading phenotype (Zhao et al. 2016). TaTOC1s expression was downregulated by the tae-miR408 in the hexaploid wheat. In addition, other important agronomic traits in wheat, such as plant height and flag leaf angle, were regulated by both tae-miR408 and TaTOC1s (Zhao et al. 2016). Zhao et al. (2016) demonstrated that the tae-miR408 functions in the wheat heading time by mediating TaTOC1s expression and findings of Zhao et al. (2016) provide important new information on the mechanism underlying heading time regulation in wheat. Expression of many plant microRNAs is responsive to hormone and environmental stimuli. *Arabidopsis* (*Arabidopsis thaliana*) miR163 is 24 nucleotides in length and targets mRNAs encoding several S-adenosyl-Met-dependent carboxyl methyltransferase family members. Chung et al. (2016) reported that miR163 is highly induced by light during seedling de-etiolation as well as seed germination. Under the same condition, its target PXMT1, encoding a methyltransferase that methylates 1,7-paraxanthine, was downregulated. During seed germination, miR163 and its target PXMT1 were predominantly expressed in the radicle, and the expression patterns of the two genes are inversely correlated. The mir163 mutant or PXMT1 overexpression line shows delayed seed germination under continuous light, and seedlings develop shorter primary roots with an increased number of lateral roots under long-day condition compared to the wild type. Chung et al. (2016) demonstrated that miR163 targets PXMT1 mRNA to promote seed germination and modulate root architecture during early development of *Arabidopsis* seedlings. Male sterility caused by long-term high-temperature (HT) stress occurs widely in crops. A total of 112 known miRNAs, 270 novel miRNAs, and 347 target genes were identified from anthers of HT-insensitive (84021) and HT-sensitive (H05) cotton cultivars under normal temperature and HT conditions through small RNA and degradome sequencing to understand the role of miRNAs in male sterility under high-temperature stress (Ding et al. 2017). Ding et al. (2017) reported that miR156 was suppressed by HT stress in both 84,021 and H05; miR160 was suppressed in 84,021 but induced in H05. Correspondingly, SPLs (target genes of miR156) were induced both in 84,021 and H05; ARF10 and ARF17 (target genes of miR160) were induced in 84,021 but suppressed in H05. Overexpressing miR160 increased cotton sensitivity to HT stress seen as anther indehiscence, associated with the suppression of ARF10 and ARF17 expression, thereby activating the auxin response that leads to anther indehiscence (Ding et al. 2017). Supporting this role for auxin, exogenous indole-3-acetic acid (IAA) leads to a stronger male sterility phenotype both in 84,021 and H05 under HT stress. Cotton plants overexpressing miR157 suppressed the auxin signal, and also exhibited enhanced sensitivity to HT stress, with microspore abortion and anther indehiscence.

5.6   Conclusion and Future Perspectives

RNA interference technology involving siRNA and miRNA has emerged as an attractive tool used by plant biologists not only to elucidate the function of genes responsive to abiotic stresses but also to improve novel agronomic traits by manipulation of both desirable and undesirable genes. Identification of entire sets of miRNAs and siRNAs and their targets will lay the foundation that is need of hour to unravel the complex miRNA and siRNA-mediated regulatory networks associated with various physiological processes that may contribute abiotic stress tolerance and maintenance of yield stability under abiotic stress conditions. Since miRNAs and siRNAs are crucial components of gene regulatory networks, a complete understanding of mechanisms and functions of miRNAs and siRNAs will greatly increase our understanding of plants response to abiotic stresses. The regulatory role of miRNAs in plants is definitely a subject that will require much more investigation in plant biology. Several miRNAs have been functionally characterized to be commonly involved in drought and salinity stress responses and also plant performance under adverse environmental conditions. The miRNAs regulate numerous transcription factors in response to different stresses. For many drought and salinity stress-related genes, miRNAs function as critical posttranscription modulator for their expression. Although a number of drought-associated miRNAs have been identified, their precise role remains to be verified. Additional strategies need to be employed to investigate the functions of miRNAs and their associated signaling pathways and gene networks under various abiotic stress conditions.

References

Ali N, Datta KS, Datta K (2010) RNA interference in designing transgenic crops. Landes Biosci 1:207–213

Cai B, Li Q, Liu F, Bi H, Ai X (2018) Decreasing fructose 1,6-bisphosphate aldolase activity reduces plant growth and tolerance to chilling stress in tomato seedlings. Physiol Plant. https://doi.org/10.1111/ppl.12682

Campbell TN (2005) Choy FYM, RNA interference: past, present and future. Mol Biol 7:1–6

Chen ZH, Bao ML, Sun YZ, Yang YJ, Xu XH, Wang JH, Han N, Bian HW, Zhu MY (2011) Regulation of auxin response by miR393-targeted transport inhibitor response protein 1 is involved in normal development in Arabidopsis. Plant Mol Biol 77:619–629

Chen L, Luan Y, Zhai J (2015) Sp-miR396a-5p acts as a stress-responsive genes regulator by conferring tolerance to abiotic stresses and susceptibility to Phytophthora nicotianae infection in transgenic tobacco. Plant Cell Rep 34:2013–2025

Chung PJ, Park BS, Wang H, Liu J, Jang IC, Chua NH (2016) Light-inducible MiR163 targets PXMT1 transcripts to promote seed germination and primary root elongation in Arabidopsis. Plant Physiol 170:1772–1782

Ding D, Zhang L, Wang H, Liu Z, Zhang Z, Zheng Y (2009) Differential expression of miRNAs in response to salt stress in maize roots. Ann Bot 103:29–38

Ding Y, Ma Y, Liu N, Xu J, Hu Q, Li Y, Wu Y, Xie S, Zhu L, Min L, Zhang X (2017) microRNAs involved in auxin signalling modulate male sterility under high-temperature stress in cotton (*Gossypium hirsutum*). Plant J 91:977–994

Ender C, Meister G (2010) Argonaute proteins at a glance. J Cell Sci 123(11):1819–1823

Fang Y, Xie K, Xiong L (2014) Conserved miR164-targeted NAC genes negatively regulate drought resistance in rice. J Exp Bot 65:2119–2135

Guo C, Yao L, You C, Wang S, Cui J, Ge X, Ma H (2016) MID1 plays an important role in response to drought stress during reproductive development. Plant J 88:280–293

Gupta K, Sengupta A, Saha J, Gupta B (2014) The attributes of RNA interference in relation to plant abiotic stress tolerance. Gene Technol 3:110. https://doi.org/10.4172/2329-6682.1000110

Hajyzadeh M, Turktas M, Khawar KM, Unver T (2015) miR408 over-expression causes increased drought tolerance in chickpea. Gene 555:186–193

Huang L, Wang Y, Wang W, Zhao X, Qin Q, Sun F, Hu F, Zhao Y, Li Z, Fu B, Li Z (2018) Characterization of transcription factor gene OsDRAP1 conferring drought tolerance in rice. Front Plant Sci 9. https://doi.org/10.3389/fpls.2018.00094

Iglesias MJ, Terrile MC, Windels D, Lombardo MC, Bartoli CG, Vazquez F, Estelle M, Casalongue CA (2014) MiR393 regulation of auxin signaling and redox-related components during acclimation to salinity in Arabidopsis. PLoS One 9:e107678

Jagtap UB, Gurav RG, Bapat VA (2011) Role of RNA interference in plant improvement. Naturwissenschaften 98:473–492

Ji X, Nie X, Liu Y, Zheng L, Zhao H, Zhang B, Huo L, Wang Y (2016) A bHLH gene from Tamarix hispida improves abiotic stress tolerance by enhancing osmotic potential and decreasing reactive oxygen species accumulation. Tree Physiol 36:193–207

Jung HJ, Kang H (2007) Expression and functional analyses of microRNA417 in Arabidopsis thaliana under stress conditions. Plant Physiol Biochem 45:805–811

Kumar P, Kamle M, Pandey A (2012) RNAi: new era of functional genomics for crop improvement. Frontiers Recent Dev Plant Sci 1:24–38

Li S, Castillo-González C, Yu B, Zhang X (2017a) The functions of plant small RNAs in development and in stress responses. Plant J 90:654–670

Li P, Li YJ, Zhang FJ, Zhang GZ, Jiang XY, Yu HM, Hou BK (2017b) The Arabidopsis UDP-glycosyltransferases UGT79B2 and UGT79B3, contribute to cold, salt and drought stress tolerance via modulating anthocyanin accumulation. Plant J 89:85–103

Ma H, Chen J, Zhang Z, Ma L, Yang Z, Zhang Q, Li X, Xiao J, Wang S (2017) MAPK kinase 10.2 promotes disease resistance and drought tolerance by activating different MAPKs in rice. Plant J 92:557–570

Ma C, Burd S, Lers A (2015) miR408 is involved in abiotic stress responses in Arabidopsis. Plant J 84:169–187

Mao C, Ding J, Zhang B, Xi D, Ming F (2018) OsNAC2 positively affects salt-induced cell death and binds to the OsAP37 and OsCOX11 promoters. Plant J. https://doi.org/10.1111/tpj.13867

Meena AK, Verma LK, Kumhar BL (2017) RNAi, It's mechanism and potential use in crop improvement : a review. Int J Pure App Biosci 5:294–311

Napoli C, Lemieux C, Jorgensen R (1990) Introduction of chimeric chalcone synthase gene into Petunia results in reversible co suppression of homologous genes intrans. Plant Cell 2:279–289

Pradhan A, Naik N, Sahoo KK (2015) RNAi mediated drought and salinity stress tolerance in plants. Am J Plant Sci 6:1990–2008

Qin H, Wang Y, Wang J, Liu H, Zhao H, Deng Z et al (2016) Knocking down the expression of GMPase gene OsVTC1-1 decreases salt tolerance of Rice at seedling and reproductive stages. PLoS ONE 11(12):e0168650. https://doi.org/10.1371/journal.pone.0168650

Redfern AD, Colley SM, Beveridge DJ, Ikeda N, Epis MR, Li X et al (2013) RNA-induced silencing complex (RISC) proteins PACT, TRBP, and Dicer are SRA binding nuclear receptor coregulators. Proc Natl Acad Sci USA 110:6536–6541

Riley KJ, Yario TA, Steitz JA (2012) Association of Argonaute proteins and microRNAs can occur after cell lysis. RNA 18(9):1581–1585

Song JB, Gao S, Sun D, Li H, Shu XX, Yang ZM (2013) miR394 and LCR are involved in Arabidopsis salt and drought stress responses in an abscisic acid-dependent manner. BMC Plant Biol 13:210

Srivastava AK, Zhang C, Caine RS, Gray J, Sadanandom A (2017) Rice SUMO protease overly tolerant to salt 1 targets the transcription factor, OsbZIP23 to promote drought tolerance in rice. Plant J 92:1031–1043

Sunkar R, Zhu JK (2004) Novel and stress-regulated microRNAs and other small RNAs from Arabidopsis. Plant Cell 16:2001–2019

Tripathi AK, Pareek A, Singla-Pareek SL (2016) A NAP-family histone chaperone functions in abiotic stress response and adaptation. Plant Physiol 171:2854–2868

Wang F, Chen HW, Li QT, Wei W, Li W, Zhang WK, Ma B, Bi YD, Lai YC, Liu XL, Man WQ, Zhang JS, Chen SY (2015) GmWRKY27 interacts with GmMYB174 to reduce expression of GmNAC29 for stress tolerance in soybean plants. Plant J 83:224–236

Wesley SV, Helliwell CA, Smith NA, Wang MB, Rouse DT, Liu Q et al (2001) Construct design for efficient, effective and high through put gene silencing in plants. Plant J 27:581–590

Williams M, Clark G, Sathasivan K, Islam AS (2004) RNA interference and its application in crop improvement. Plant Tissue Cult 14(1):1–18

Wilson RC, Doudna JA (2013) Molecular mechanisms of RNA interference. Annu Rev Biophys 42:217–239

Wu BF, LiWF XHY, Qi LW, Han SY (2015a) Role of cinmiR2118 in drought stress responses in Caragana intermedia and Tobacco. Gene 574:34–40

Wu J, Yang Z, Wang Y et al (2015b) Viral-inducible Argonaute18 confers broad-spectrum virus resistance in rice by sequestering a host microRNA. Elife 4:1–19

Younis A, Siddique MI, Kim CK, Lim KB (2014) RNA interference (RNAi) induced gene silencing: a promising approach of hi-tech plant breeding. Int J Biol Sci 10:1150–1158

Yuan S, Li Z, Li D, Yuan N, Hu Q, Luo H (2015) Constitutive expression of rice MicroRNA528 alters plant development and enhances tolerance to salinity stress and nitrogen starvation in creeping bentgrass. Plant Physiol 169:576–593

Zhao XY, Hong P, Wu JY, Chen XB, Ye XG, Pan YY, Wang J, Zhang XS (2016) The tae-miR408-mediated control of TaTOC1 genes transcription is required for the regulation of heading time in wheat. Plant Physiol 170:1578–1594

Zhou M, Li D, Li Z, Hu Q, Yang C, Zhu L, Luo H (2013) Constitutive expression of a miR319 gene alters plant development and enhances salt and drought tolerance in transgenic creeping bentgrass. Plant Physiol 161:1375–1391

Zhou X, Li L, Xiang J, Gao G, Xu F, Liu A, Zhang X, Peng Y, Chen X, Wan X (2015) OsGL1-3 is involved in cuticular wax biosynthesis and tolerance to water deficit in rice. PLoS ONE 10(1):e116676. https://doi.org/10.1371/journal.pone.0116676

Chapter 6
Antifungal Plant Defensins: Insights into Modes of Action and Prospects for Engineering Disease-Resistant Plants

Jagdeep Kaur, Siva LS Velivelli, and Dilip Shah

Abstract Defensins are small, cysteine-rich peptides that are ubiquitously present in all plants. They are important components of the plant immune system and serve as first line of defense against invading pathogens. Plant defensins share conserved tetradisulfide connectivity but vary in their sequence, net charge, and hydrophobicity. A number of plant defensins with potent broad-spectrum antifungal activity have been identified and characterized. Studies conducted during the past decade have highlighted the diverse modes of action (MOA) of a few antifungal defensins. Constitutive expression of these defensins has been demonstrated to confer *in planta* resistance to several economically important fungal and oomycete pathogens in transgenic crops. Here, we provide a brief review of recent findings that have contributed to our current understanding of the MOA of these peptides and their deployment for disease resistance in crops.

Keywords Plant defensins · Antifungal activity · Mode of action · Fungal resistance · Genetic engineering

6.1 Introduction

Plants are continuously exposed to a plethora of potentially harmful pathogens. Fungal and oomycete pathogens of considerable economic importance impose major constraints globally on agricultural production and pose a clear threat to food security (Collinge et al. 2010). Plants lack a somatic adaptive immune system to protect themselves from pathogen attack and therefore must rely on their

J. Kaur (✉)
Donald Danforth Plant Science Center, St. Louis, MO, USA

Monsanto Company, Chesterfield, MO, USA
e-mail: jagdeep.kaur@monsanto.com

S. L. Velivelli · D. Shah (✉)
Donald Danforth Plant Science Center, St. Louis, MO, USA
e-mail: dshah@danforthcenter.org

© Springer International Publishing AG, part of Springer Nature 2018
S. S. Gosal, S. H. Wani (eds.), *Biotechnologies of Crop Improvement, Volume 2*,
https://doi.org/10.1007/978-3-319-90650-8_6

sophisticated innate immune system for defense against these pathogens (Jones and Dangl 2006). The innate immunity of plants comprises fortification of cell wall, hypersensitive response, and production of antimicrobial compounds and antimicrobial peptides (AMPs). AMPs serve as one of the first lines of defense against pathogen invasion and make up the crucial effector arm of the plant's immune system (van der Weerden et al. 2013).

Defensins represent a large family of AMPs in all higher plants and are at the forefront of their defense against pathogens. Plant defensins are cysteine-rich cationic peptides of 45–54 amino acids. First isolated from wheat and barley seeds in 1990 (Colilla et al. 1990; Mendez et al. 1990), plant defensins have since been discovered in several phyla of plant kingdom. They have been identified in a variety of plant tissues and are either ubiquitously or conditionally expressed in response to various biological or environmental cues. Based on their subcellular localization, plant defensins have been designated either as class I or class II. Class I defensins are secreted to the apoplast and synthesized by plant cells as precursor proteins comprising of the secretory signal peptide followed by the mature peptide. Class II defensins, localized to the vacuole and expressed in floral tissue of plants from Solanaceae and Poaceae families, are synthesized containing an additional carboxy-terminal propeptide (CTPP) (Lay et al. 2003).This C-terminal propeptide may serve dual function, i.e., it protects against autocytotoxicity by neutralizing the deleterious cationicity of the peptides during export to the vacuole and acts as a chaperone to assist in folding (Lay et al. 2014). Plant defensins share a conserved 3D structure consisting of one α-helix and three antiparallel β-strands that are connected by four disulfide bonds forming a cysteine-stabilized αβ (CSαβ) motif (Broekaert et al. 1997; Thomma et al. 2002). The structure of each plant defensin is also characterized by the occurrence of a functionally important signature γ-core motif $GXCX_{3-9}C$ (where G is glycine, C is cysteine, and X is any amino acid) that is conserved among all antimicrobial peptides containing disulfide bonds. Despite their structural similarity, plant defensins exhibit very low sequence homology outside the eight conserved cysteines. This divergence in primary sequences may account for the multi-functionality of plant defensins including antifungal and antibacterial activity, proteinase inhibitor activity, pollen tube guidance and discharge of male gametes, zinc tolerance, and plant development (Carvalho Ade and Gomes 2009). During the past decade, significant inroads into understanding the structure-activity relationships and MOA of a few antifungal plant defensins have been made. This chapter highlights current knowledge of their MOA and their deployment for improving plant resistance to fungal and oomycete pathogens.

6.2 MOA of Antifungal Plant Defensins

To fully comprehend the roles of defensins in plant defense and to harness their potential for engineering disease-resistant crops, it is important to unravel the MOA of antifungal plant defensins. First studies aimed at unraveling the MOA of

Sphingolipid binding defensins

```
MsDef1    -RTCENLADKYRGPCFS--GCDTHCTTKENAVSGRCRDDF---RCWCTKRC
Psd1      -KTCEHLADTYRGVCFTNASCDDHCKNKAHLISGTCH-NW---KCFCTQNC
RsAFP2    QKLCQRPSGTWSGVCGNNNACKNQCIRLEKARHGSCNYVFPAHKCICYFPC
DmAMP1    -ELCEKASKTWSGNCGNTGHCDNQCKSWEGAAHGACHVRNGKHMCFCYFNC
```

Phospholipid binding defensins

```
MtDef4    ---RTCESQSHKFKGPCASDHNCASVCQ-TERFS-GGRC--RGFRRRCFCTTHC
NaD2      ---RTCESQSHRFKGPCARDSNCATVCL-TEGFS-GGDC--RGFRRRCFCTRPC
TPP3      -AQQICKAPSQTFPGLCFMDSSCRKYCI-KEKFT-GGHC--SKLQRKCLCTKPC
NsD7      ---KDCKRESNTFPGICITKPPCRKACI-REKFT-DGHC--SKILRRCLCTKPC
NaD1      ---RECKTESNTFPGICITKPPCRKACI-SEKFT-DGHC--SKILRRCLCTKPC
HsAFP1    DGVKLCDVPSGTWSGHCGSSSKCSQQCKDREHFAYGGACHYQFPSVKCFCKRQC
MtDef5    ---KLCQKRSTTWSGPCLNTGNCKRQCINVEHAT-FGACHRQGFGFACFCYKKCAPKKVEP
          KLCERRSKTWSGPCLISGNCKRQCINVEHAT-SGACHRQGIGFACFCKKKC
```

Fig. 6.1 Amino acid sequence comparison of sphingolipid (MsDef1, Psd1, RsAFP2, and DmAMP1)- and phospholipid (MtDef4, NaD2, TPP3, NsD7, NaD1, HsAFP1, and MtDef5)-binding defensins. The presence of eight cysteines and the γ-core motif (GXCX3-9C, where X is any amino acid) in each defensin is shown in red and blue, respectively. Each defensin comprises one α-helix and three β-strands as shown below the amino acid sequences

defensins revealed interactions with fungus-specific membrane components (Thevissen et al. 1997, 2000, 2004). Defensins permeabilize fungal plasma membrane, induce Ca^{2+} influx, and disrupt a tip-focused Ca^{2+} gradient essential for polar growth of hyphal tips (Thevissen et al. 1996, 1997, 1999). In 2004, we presented evidence that defensin MsDef1 from *Medicago sativa* blocks the L-type calcium channel in mammalian cells (Spelbrink et al. 2004). It remains to be determined, however, if a fungal calcium channel plays a functional role in the antifungal action of MsDef1. Using *Neurospora crassa* expressing the Ca^{2+} reporter aequorin, MsDef1 and the cognate peptides containing its γ-core motif were each shown to perturb Ca^{2+} homeostasis in a highly specific and distinct manner (Munoz et al. 2014). Recently, *Arabidopsis thaliana* defensin AtPDF2.3 has been shown to block voltage-gated potassium channels expressed in frog oocytes indicating the role for potassium transport and/or homeostasis in the antifungal action of this defensin (Vriens et al. 2016).

Some defensins bind with high affinity to specific sphingolipids present in the fungal cell wall and/or plasma membrane of their target fungi (Thevissen et al. 2003, 2005, 2007; Aerts et al. 2008). Sphingolipids are important structural components of the fungal cell wall and plasma membrane and serve as second messengers regulating delicate balance between cell death and survival (Thevissen et al. 2006). Plant defensins RsAFP2 from *Raphanus sativus*, MsDef1 from *M. sativa*, and Psd1 from *Pisum sativum* bind specifically to fungal cell wall localized glucosylceramide (GlcCer) (Fig. 6.1). Importantly, RsAFP2 does not interact with soybean or human GlcCer, suggesting that its preferential binding to yeast GlcCer may be due to structural differences. RsAFP2/GlcCer interaction has been shown to result in the induction of cell wall stress, accumulation of ceramides and reactive oxygen

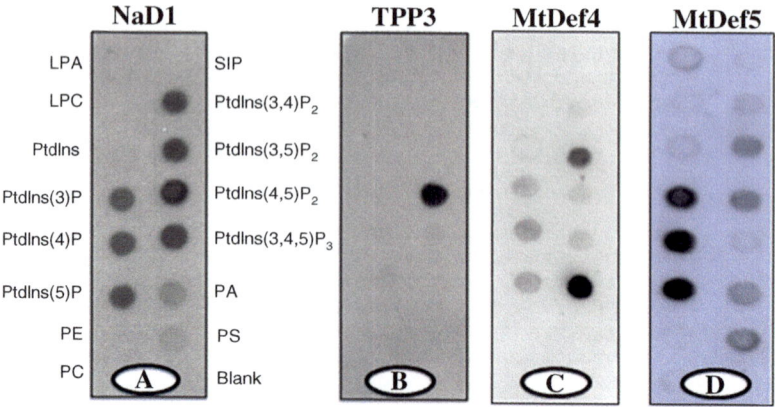

Fig. 6.2 Phospholipid-binding profile of defensins using protein-lipid overlay assay. (**A**) NaD1 binds to a broad range of bioactive phospholipids with a strong preference for phosphatidylinositol 4,5-bisphosphate PI(4,5)P2. (**B**) TPP3 binds specifically to PI(4,5)P2. (**A**, **B**, reproduced from Baxter et al. 2015). (**C**) MtDef4 bound preferentially to phosphatidic acid and PI(3,5)P2. (Reproduced from Sagaram et al. 2013). (**D**) MtDef5 binds to a range of bioactive phospholipids with a strong preference to PI3P, PI4P, and PI5P. (Reproduced from Islam et al. 2017)

species (ROS), and ultimately cell death (De Coninck et al. 2013; van der Weerden et al. 2013). The ability of MsDef1 to disrupt a Ca^{2+} gradient in *N. crassa* is also dependent on its ability to interact with GlcCer (Munoz et al. 2014). *Dahlia merckii* defensin DmAMP1 targets a different sphingolipid mannose-(inositol phosphate)$_2$-ceramide in fungal cells (Thevissen et al. 2000). At present, very little is known regarding the biochemical steps downstream of defensin/sphingolipid interaction that lead to fungal growth arrest or cell death.

During the past decade, evidence has emerged that some antifungal plant defensins bind to a variety of bioactive plasma membrane resident phospholipids, induce membrane disruption, and gain entry into fungal cells (Lobo et al. 2007; van der Weerden et al. 2008, 2010). The phospholipid-binding plant defensins shown in Fig. 6.1 have received much attention lately, for studies aimed at unraveling their MOA. These include NaD1 from *Nicotiana alata*, TPP3 from tomato (Baxter et al. 2015), NsD7 from *N. suaveolens* (Kvansakul et al. 2016), and MtDef4 and MtDef5 from *M. truncatula* (Sagaram et al. 2013; Islam et al. 2017). These defensins bind to different membrane phospholipids (Fig. 6.2). MtDef4 and NsD7 target phosphatidic acid (PA), a precursor for the biosynthesis of other phospholipids and a regulator of membrane-cytoskeleton interactions and membrane curvature. They also bind to a lesser extent to other bioactive phospholipids, in particular, phosphatidylinositol mono- and bisphosphates. MtDef4 mutants that fail to bind PA also fail to gain entry into fungal cells and show much reduced or complete loss of antifungal activity. Recently, HsAFP1 from *Heuchera sanguinea* has also been shown to bind PA (Cools et al. 2017). HsAFP1 mutant that exhibits much reduced PA binding also exhibits loss of antifungal activity greater than twofold. Structural analysis of the NsD7-PA complex has revealed a double helix forming right-handed coiled

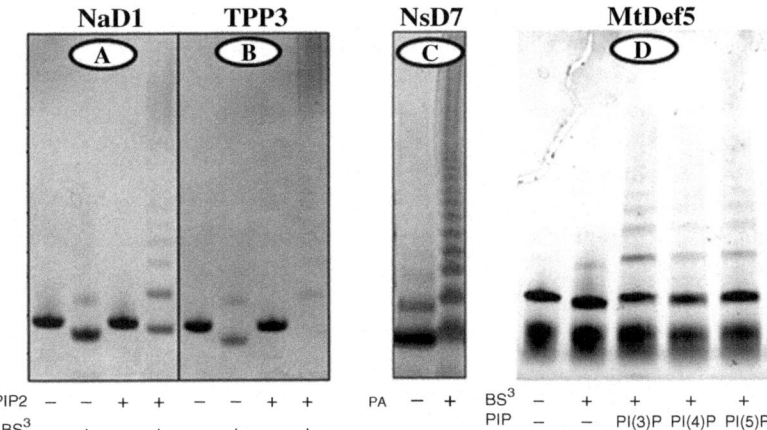

Fig. 6.3 Oligomerization profile of defensins using protein cross-linking analysis. (**A, B**) NaD1 and TPP3 form oligomers in the presence of PIP2. (Reproduced from Baxter et al. 2015). (**C**) MtDef4 forms oligomers in the presence of phosphatidic acid. (Reproduced from Sagaram et al. 2013). (**D**) MtDef5 forms oligomers in the presence of PI3P, PI4P, and PI5P. (Reproduced from Islam et al. 2017)

oligomeric defensin fibril, and PA is required for oligomerization. It remains to be determined if MtDef4 forms oligomeric complexes in the presence of PA. Interaction with PA is important for the antifungal activity of these defensins.

NaD1 and TPP3 bind to plasma membrane-localized phosphoinositides, in particular, phosphatidylinositol 4,5-bisphosphate (PIP_2), a key mediator of cytoskeletal-membrane interactions (Fig. 6.2a, b). NaD1 also binds to other phospholipids and thus appears to be more promiscuous than TPP3 (Poon et al. 2014; Baxter et al. 2015). In absence of phospholipids, NaD1 has been shown to form dimers in solution, and dimerization enhances its antifungal activity (Lay et al. 2012). However, in presence of PIP_2, structural analysis has shown that 7 dimers cooperatively bind to the anionic head groups of 14 molecules of PIP2 and form arch-shaped "cationic grip" configuration, and this PIP_2-mediated oligomerization is important for fungal plasma membrane permeabilization. NaD1 thus employs a unique PIP_2-dependent mechanism to disrupt fungal plasma membrane. NaD1 also forms oligomers in solution in presence of PIP_2 (Fig. 6.3A), and these oligomers lead to the formation of fibrils as observed by transmission electron microscopy. TPP3, which shares 63% sequence identity with NaD1, binds specifically to PIP_2, and structural analysis has shown that it too forms a dimeric cationic grip only in presence of this phospholipid (Fig. 6.3B). This interaction with PIP_2 also leads to higher order oligomerization of this defensin and formation of string-like fibrils (Poon et al. 2014; Baxter et al. 2015). Whether TPP3 and NaD1 form fibrils in contact with the fungal plasma membrane in vivo is not known.

Another phospholipid-binding antifungal defensin MtDef5 has been recently studied in our lab (Islam et al. 2017). It is a novel bi-domain defensin which consists of 2 defensin monomers, 50 amino acids each, linked by a 7-amino acid peptide sequence APKKVEP (Fig. 6.1). It carries a net charge of +16 and exhibits broad-spectrum antifungal activity against filamentous fungi at submicromolar concentrations. MOA studies have shown that MtDef5 is a highly promiscuous defensin which binds to a number of phospholipids but with a strong preference for phosphatidylinositol 3-phosphates (PI3P), PI4P, and PI5P, substrates for synthesis of PIP_2 and contributors to polar tip growth in fungi (Fig. 6.2D).The phospholipid-binding profile of MtDef5 is different from those of MtDef4 and NsD7 which bind to PA and also different from those of NaD1and TPP3 which bind preferentially to PIP_2. MtDef5 forms oligomers in presence of PIP, but surprisingly, it also oligomer-izes in presence of PI and PA to which it has relatively weak binding (Fig. 6.3D). Similarly, both NaD1 and NsD7 have also been reported to oligomerize in presence of PA and PIP2 (Poon et al. 2014; Baxter et al. 2015; Kvansakul et al. 2016) (Fig. 6.3A, C). In preliminary studies, MtDef5/PIP complexes form nanonet-like structures when observed under the transmission electron microscope. Mutagenesis studies have revealed that cationic amino acids present in the γ-core motif are involved in PIP binding and oligomerization of this defensin and facilitating mem-brane disruption and fungal killing by this protein (Islam et al. 2017).

From the studies described above, it is becoming increasingly clear that plant defensins which gain entry into fungal cells utilize a broad "phospholipid code" to identify and attack fungal membranes as part of the first line of plant defense (Baxter et al. 2017). However, one outstanding issue which needs to be addressed is whether phospholipid binding is required for their antifungal activity. Bleackley and col-leagues have addressed this issue by analyzing the phospholipid binding and anti-fungal activity of a series of NaD1 chimeras with NaD2 that exhibits poor antifungal activity (Bleackley et al. 2016). These chimeras were produced by replacing the sequence between the two neighboring cysteine residues of NaD1 with the corre-sponding sequence of NaD2. Surprisingly, some of the chimeras that lost PIP_2 bind-ing retained their ability to inhibit fungal growth suggesting mechanisms other than phospholipid binding exist for antifungal activity.

Defensin NaD1 has been shown to permeabilize the plasma membrane of *Fusarium oxysporum* f. sp. *vasinfectum* ultimately leading to granulation of the cytoplasm and cell death (van der Weerden et al. 2008) (Fig. 6.4A). Similarly, MtDef4 rapidly permeabilizes plasma membrane of *F. graminearum* where it accu-mulates in the cytoplasm that eventually leads to death (Sagaram et al. 2013) (Fig. 6.4B). The important question related to the antifungal action of a specific plant defensin is whether its MOA is conserved in different fungi. Surprisingly, we have found that mechanisms used by MtDef4 to inhibit the growth of *F. gra-minearum* and *N. crassa* are not the same even though these two fungi belong to the same phylum *Ascomycota*, subphylum *Pezizomycotina*, and order *Sordariomycetes* (El-Mounadi et al. 2016).When used at minimal inhibitory concentration, MtDef4 permeabilizes the plasma membrane of *F. graminearum* but not *N. crassa* (Fig. 6.5). After its internalization, MtDef4 is localized to vesicular bodies in the conidia and germlings of *N. crassa* but shows diffuse cytoplasmic localization in those of *F. graminearum*. Further, cellular uptake of MtDef4 into *N. crassa* is energy

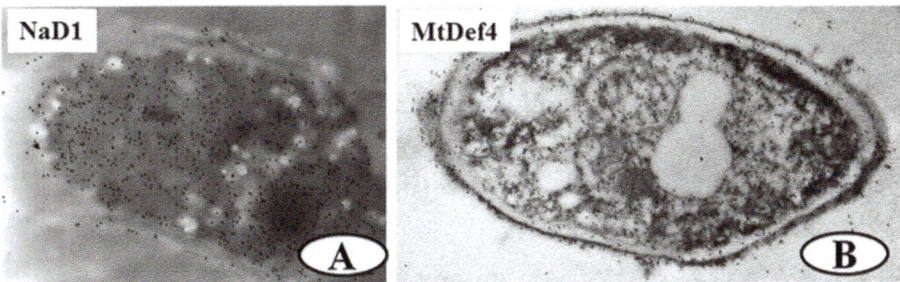

Fig. 6.4 Immunogold labeling of defensins (**A**) Micrograph of NaD1-treated hyphae of *F. oxysporum* f. sp. *vasinfectum*. NaD1 has been internalized at a high concentration inside a treated hypha. (Reproduced from van der Weerden et al. 2008). (**B**) Micrograph of MtDef4-treated *F. graminearum* hypha. MtDef4 internalized at a high concentration inside the fungal cell. (Reproduced from Sagaram et al. 2013)

Fig. 6.5 Permeabilization of fungal plasma membrane by plant defensins. (**A, B**) MtDef4 permeabilizes the plasma membrane of *F. graminearum* but not of *N. crassa* as revealed by SYTOX green uptake assay. (Reproduced from El-Mounadi et al. 2016). (**C–D**) MtDef5, in contrast, permeabilizes the plasma membrane of both fungi. (Reproduced from Islam et al. 2017)

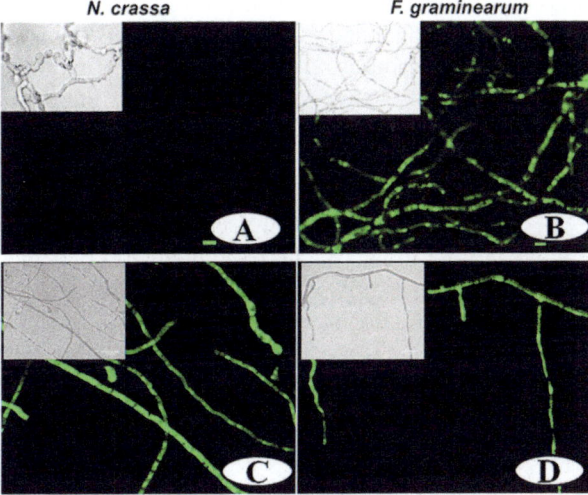

dependent and involves endocytosis, whereas it is only partially energy dependent in *F. graminearum*. Brefeldin A (an ER to Golgi transport inhibitor) and filipin (a lipid raft-mediated endocytosis inhibitor) significantly inhibit internalization of MtDef4 in *N. crassa* but not in *F. graminearum*. In fungi, PA is generated mainly through the action of phospholipase D (PLD). *N. crassa* and *F. graminearum* each express three PLDs, namely, PLD1, PLD, and PLDA. Surprisingly, the plasma membrane-localized PLD1 is required for entry of this defensin in *N. crassa*, but not in *F. graminearum* (El-Mounadi et al. 2016). These findings indicate that the cell wall and plasma membrane compositions are different even in closely related fungi and markedly influence the antifungal activity of plant defensins. They also raise the possibility that pathogenic and saprophytic fungi respond differently to challenge by a specific antifungal plant defensin.

Deeper knowledge of different MOA employed by defensins for fungal killing highlighted here will undoubtedly enable rational design and exploitation of these peptides for engineering disease-resistant crops, a topic discussed below.

6.3 Deployment of Plant Defensins to Engineer Disease-Resistant Plants

Combatting plant fungal and oomycete diseases by varietal genetic resistance, management practices, and fungicide application is the current norm. Novel technologies that could augment the above disease control strategies, however, will be required to stay ahead of fast-evolving pathogens and changing climate. Plant defensins with their potent broad-spectrum antifungal and anti-oomycete activity combined with nontoxicity to humans hold potential for use as antifungal agents in transgenic crops. Several labs including ours have reported enhanced resistance to various plant fungal and oomycete pathogens in transgenic plants expressing plant defensins. The reader is referred to excellent reviews on this topic (Kaur et al. 2011; DeConinck et al. 2013). While majority of these studies demonstrated *in planta* efficacy of defensins in controlled environment of a growth chamber or a greenhouse, few have shown resistance to fungal and/or oomycete pathogens in the field. One of the pioneer studies, first reported in 2000, demonstrated that constitutive expression of alfAFP (MsDef1) controlled Verticillium wilt caused by fungus *V. dahliae* in field-grown potato (Gao et al. 2000). Almost a decade later, constitutive expression of *N. megalosiphon* defensin NmDef02 was shown to confer resistance to an oomycete pathogen *Phytophthora infestans*, causal agent of potato late blight, in the field (Portieles et al. 2010). In both studies, defensin peptides were secreted to the apoplast in transgenic lines and field efficacy of these lines correlated with the peptide expression levels. In another exciting study, expression of vacuole-targeted NaD1 was shown to provide substantial field level resistance to *F. oxysporum* f. sp. *vasinfectum* and *V. dahliae* in transgenic cotton (Fig. 6.6A). When compared with non-transgenic control lines, transgenic cotton lines had increased survival rate and produced two- to fourfold increase in lint yield under disease pressure. In non-diseased soil, transgenic lines showed no negative impact on agronomic characteristics relative to non-transgenic lines (Gasper et al. 2014). Collectively, these studies demonstrate that antifungal defensins can be used to engineer resistance to economically important fungal and oomycete pathogens.

Targeting a specific defensin to the appropriate subcellular compartment to match the lifestyle of a fungal or oomycete pathogen is the key for design of effective disease control strategies. We have shown that targeting MtDef4 to the apoplast, but not to the intracellular compartments, is necessary to control an obligate biotrophic oomycete *Hyaloperonospora arabidopsidis* causing downy mildew in transgenic *Arabidopsis thaliana* (Kaur et al. 2012). The efficacy of apoplast-targeted MtDef4 to confer resistance to an obligate biotroph *Puccinia triticina*, causal fungal pathogen of leaf rust, was also demonstrated in transgenic wheat (Kaur et al. 2017)

Fig. 6.6 Disease resistance in transgenic lines overexpressing defensins. (**A**) Transgenic cotton line D1 expressing NaD1 showed resistance to *F. oxysporum* f. sp. *vasinfectum* compared to non-transgenic Coker parent line. (Reproduced from Gasper et al. 2014). (**B**) MtDef4 overexpressing transgenic wheat lines conferred resistance to *Puccinia triticina*. Transgenic lines BW-A-11, BW-B-4, BW-F-10, and XC9–104-1 in comparison to their respective non-transgenic controls BW and XC9. (Reproduced from Kaur et al. 2017)

(Fig. 6.6B). It is proposed that antifungal defensin optimally expressed in the extracellular milieu makes direct contact with the biotrophic pathogen and impedes its growth. In a recently published study, transgenic peanuts overexpressing apoplast-targeted MsDef1 or MtDef4 exhibit near immunity to *Aspergillus flavus* and accumulate drastically reduced levels of aflatoxins (Sharma et al. 2017). Aflatoxin contamination caused by *A. flavus* infection of peanuts poses a major food safety problem for people in sub-Saharan Africa and South Asia. This finding clearly indicates that the defensin technology when employed strategically could have important implications to mitigate the toxin levels.

Expression of defensins using strong constitutive promoters in transgenic plants provides yield advantage under epidemic conditions but can also result in deleterious side effects in the absence of a disease. In such cases, expression of defensins using tissue-specific or pathogen-inducible promoters will be crucial for commercially viable deployment of defensin-mediated resistance. A number of such tissue-specific and pathogen-inducible promoters are available to choose from to match the target pathogen's lifestyle. For root-colonizing pathogens, for example, expression of defensins using root-specific promoters might be sufficient to confer optimal resistance without the deleterious effects of their constitutive expression. Recently, kernel-specific zein promoter was used to express RNA interference gene cassette directed against the aflatoxin biosynthesis pathway for development of *A. flavus*-resistant maize with much reduced aflatoxin accumulation (Thakare et al. 2017). This promoter might prove useful for expression of *A. flavus* inhibitory defensins in transgenic maize kernels to reduce aflatoxin levels. The *Lem2* promoter known to be expressed in lemma and palea of florets in wheat and barley might prove useful for expression of antifungal defensins and conferring resistance to *Fusarium graminearum*. Interestingly, *Lem2* promoter is also induced by *Fusarium* infection (Abebe et al. 2005, 2006). Pathogen-inducible promoters such as *OsPR10a* (Hwang et al. 2008) and *GER4c* (Himmelbach et al. 2010) induced by pathogens and defense hormones can also be tested for resistance to economically important fungal and oomycete pathogens in transgenic cereals. In addition, synthetic designer promoters that are responsive to a number of phytohormones (Liu et al. 2011) could also be used for appropriate targeting of defensins. With much greater understanding of the MOA of sequence divergent antifungal plant defensins in recent years and the availability of tools for their pathogen-inducible or tissue-specific expression and subcellular localization, we are in an excellent position to engineer durable, agronomically useful level of fungal resistance in transgenic crops.

References

Abebe T, Skadsen RW, Kaeppler HF (2005) A proximal upstream sequence controls tissue-specific expression of Lem2, a salicylate-inducible barley lectin-like gene. Planta 221:170–183

Abebe T, Skadsen R, Patel M, Kaeppler H (2006) The Lem2 gene promoter of barley directs cell- and development-specific expression of gfp in transgenic plants. Plant Biotechnol J 4:35–44

Aerts AM, François IEJA, Cammue BPA, Thevissen K (2008) The mode of antifungal action of plant, insect and human defensins. Cell Mol Life Sci 65:2069–2079

Baxter AA, Richter V, Lay FT, Poon IKH, Adda CG, Veneer PK, Phan TK, Bleackley MR, Anderson MA, Kvansakul M, Hulett MD (2015) The tomato defensin TPP3 binds phosphatidylinositol (4,5)-bisphosphate via a conserved dimeric cationic grip conformation to mediate cell lysis. Mol Cell Biol 35:1964–1978

Baxter AA, Poon IKH, Hulett MD (2017) The lure of the lipids: how defensins exploit membrane phospholipids to induce cytolysis in target cells. Cell Death Dis 8:e2712

Bleackley MR, Payne JAE, Hayes BME, Durek T, Craik DJ, Shafee TMA, Poon IKH, Hulett MD, van der Weerden NL, Anderson MA (2016) *Nicotiana alata* defensin chimeras reveal differences in the mechanism of fungal and tumour cell killing and an enhanced antifungal variant. Antimicrob Agents Chemother 60:6302–6312

Broekaert WF, Cammue BPA, De Bolle MFC, Thevissen K, De Samblanx GW, Osborn RW, Nielson K (1997) Antimicrobial peptides from plants. Crit Rev Plant Sci 16:297–323

Carvalho Ade O, Gomes VM (2009) Plant defensins-prospects for the biological functions and biotechnological properties. Peptides 30:1007–1020

Colilla FJ, Rocher A, Mendez E (1990) Gamma-purothionins: amino acid sequence of two polypeptides of a new family of thionins from wheat endosperm. FEBS Lett 270:191–194

Collinge DB, Jorgensen HJ, Lund OS, Lyngkjaer MF (2010) Engineering pathogen resistance in crop plants: current trends and future prospects. Annu Rev Phytopathol 48:269–291

Cools TL, Vriens K, Struyfs C, Verbandt S, Ramada MHS, Brand GD, Block C Jr, Koch B, Traven A, Drijfhout JW, Demuyser L, Kucharikova S, Van Dijck P, Spasic D, Lammertyn J, Cammue BPA, Thevissen K (2017) The antifungal plant defensin HsAFP1 is a phosphatidic acid-interacting peptide inducing membrane permeabilization. Front Microbiol 8:2295. https://doi.org/10.3389/fmicb.2017.02295

De Coninck B, Cammue BPA, Thevissen K (2013) Modes of antifungal action and in planta functions of plant defensins and defensin-like peptides. Fungal Biol Rev 26:109–120

El-Mounadi K, Islam KT, Hernández-Ortiz P, Read ND, Shah DM (2016) Antifungal mechanisms of a plant defensin MtDef4 are not conserved between the ascomycete fungi *Neurospora crassa* and *Fusarium graminearum*. Mol Microbiol 100:542–559

Gao AG, Hakimi SM, Mittanck CA, Wu Y, Woerner BM, Stark DM, Shah DM, Liang J, Rommens CM (2000) Fungal pathogen protection in potato by expression of a plant defensin peptide. Nat Biotechnol 18:1307–1310

Gaspar YM, McKenna JA, McGinness BS, Hinch J, Poon S, Connelly AA, Anderson MA, Heath RL (2014) Field resistance to *Fusarium oxysporum* and *Verticillium dahliae* in transgenic cotton expressing the plant defensin NaD1. J Exp Bot 65:1541–1550

Himmelbach A, Liu L, Zierold U, Altschmied L, Maucher H, Beier F, Müller D, Hensel G, Heise A, Schützendübel A, Kumlehn J, Schweizer P (2010) Promoters of the barley germin-like GER4 gene cluster enable strong transgene expression in response to pathogen attack. Plant Cell 22:937–952

Hwang S-H, Lee IA, Yie SW, Hwang D-J (2008) Identification of an OsPR10a promoter region responsive to salicylic acid. Planta 227:1141–1150

Islam KT, Velivelli SLS, Berg RH, Oakley B, Shah DM (2017) A novel bi-domain plant defensin MtDef5 with potent broad-spectrum antifungal activity binds to multiple phospholipids and forms oligomers. Sci Rep 7:16157

Jones JDG, Dangl JL (2006) The plant immune system. Nature 444:323

Kaur J, Sagaram US, Shah D (2011) Can plant defensins be used to engineer durable commercially useful fungal resistance in crop plants? Fungal Biol Rev 25:128–135

Kaur J, Thokala M, Robert-Seilaniantz A, Zhao P, Peyret H, Berg H, Pandey S, Jones J, Shah D (2012) Subcellular targeting of an evolutionarily conserved plant defensin MtDef4.2 determines the outcome of plant-pathogen interaction in transgenic Arabidopsis. Mol Plant Pathol 13:1032–1046

Kaur J, Fellers J, Adholeya A, Velivelli SL, El-Mounadi K, Nersesian N, Clemente T, Shah D (2017) Expression of apoplast-targeted plant defensin MtDef4.2 confers resistance to leaf rust pathogen *Puccinia triticina* but does not affect mycorrhizal symbiosis in transgenic wheat. Transgenic Res 26:37–49

Kvansakul M, Lay FT, Adda CG, Veneer PK, Baxter AA, Phan TK, Poon IKH, Hulett MD (2016) Binding of phosphatidic acid by NsD7 mediates the formation of helical defensin-lipid oligomeric assemblies and membrane permeabilization. Proc Natl Acad Sci 113:11202–11207

Lay FT, Brugliera F, Anderson MA (2003) Isolation and properties of floral defensins from ornamental tobacco and petunia. Plant Physiol 131:1283–1293

Lay FT, Mills GD, Poon IK, Cowieson NP, Kirby N, Baxter AA, van der Weerden NL, Dogovski C, Perugini MA, Anderson MA, Kvansakul M, Hulett MD (2012) Dimerization of plant defensin NaD1 enhances its antifungal activity. J Biol Chem 287:19961–19972

Lay FT, Poon S, McKenna JA, Connelly AA, Barbeta BL, McGinness BS, Fox JL, Daly NL, Craik DJ, Heath RL, Anderson MA (2014) The C-terminal propeptide of a plant defensin confers cytoprotective and subcellular targeting functions. BMC Plant Biol 14:41

Liu W, Mazarei M, Rudis MR, Fethe MH, Stewart CN (2011) Rapid in vivo analysis of synthetic promoters for plant pathogen phytosensing. BMC Biotechnol 11:108–108

Lobo DS, Pereira IB, Fragel-Madeira L, Medeiros LN, Cabral LM, Faria J, Bellio M, Campos RC, Linden R, Kurtenbach E (2007) Antifungal *Pisum sativum* defensin 1 interacts with *Neurospora crassa* Cyclin F related to the cell cycle. Biochemistry 46:987–996

Mendez E, Moreno A, Colilla F, Pelaez F, Limas GG, Mendez R, Soriano F, Salinas M, de Haro C (1990) Primary structure and inhibition of protein synthesis in eukaryotic cell-free system of a novel thionin, gamma-hordothionin, from barley endosperm. Eur J Biochem 194:533–539

Munoz A, Chu M, Marris PI, Sagaram US, Kaur J, Shah DM, Read ND (2014) Specific domains of plant defensins differentially disrupt colony initiation, cell fusion and calcium homeostasis in *Neurospora crassa*. Mol Microbiol 92:1357–1374

Poon IKH, Baxter AA, Lay FT, Mills GD, Adda CG, Payne JAE, Phan TK, Ryan GF, White JA, Veneer PK, van der Weerden NL, Anderson MA, Kvansakul M, Hulett MD (2014) Phosphoinositide-mediated oligomerization of a defensin induces cell lysis. elife 3:e01808

Portieles R, Ayra C, Gonzalez E, Gallo A, Rodriguez R, Chacon O, Lopez Y, Rodriguez M, Castillo J, Pujol M, Enriquez G, Borroto C, Trujillo L, Thomma BP, Borras-Hidalgo O

(2010) NmDef02, a novel antimicrobial gene isolated from *Nicotiana megalosiphon* confers high-level pathogen resistance under greenhouse and field conditions. Plant Biotechnol J 8:678–690

Sagaram US, El-Mounadi K, Buchko GW, Berg HR, Kaur J, Pandurangi RS, Smith TJ, Shah DM (2013) Structural and functional studies of a phosphatidic acid-binding antifungal plant defensin MtDef4: identification of an RGFRRR motif governing fungal cell entry. PLoS One 8(12):e82485

Sharma KK, Pothana A, Prasad K, Shah D, Kaur J, Bhatnagar D, Chen ZY, Raruang Y, Cary JW, Rajasekaran K, Sudini HK, Bhatnagar-Mathur P (2017) Peanuts that keep aflatoxin at bay: a threshold that matters. Plant Biotechnol J 16:1024. https://doi.org/10.1111/pbi.12846

Spelbrink RG, Dilmac N, Allen A, Smith TJ, Shah DM, Hockerman GH (2004) Differential antifungal and calcium channel-blocking activity among structurally related plant defensins. Plant Physiol 135:2055–2067

Thakare D, Zhang J, Wing RA, Cotty PJ, Schmidt MA (2017) Aflatoxin-free transgenic maize using host-induced gene silencing. Sci Adv 3:e1602382

Thevissen K, Ghazi A, De Samblanx GW, Brownlee C, Osborn RW, Broekaert WF (1996) Fungal membrane responses induced by plant defensins and thionins. J Biol Chem 271:15018–15025

Thevissen K, Osborn RW, Acland DP, Broekaert WF (1997) Specific, high affinity binding sites for an antifungal plant defensin on *Neurospora crassa* hyphae and microsomal membranes. J Biol Chem 272:32176–32181

Thevissen K, Terras FRG, Broekaert WF (1999) Permeabilization of fungal membranes by plant defensins inhibits fungal growth. Appl Environ Microbiol 65:5451–5458

Thevissen K, Cammue BP, Lemaire K, Winderickx J, Dickson RC, Lester RL, Ferket KK, Van Even F, Parret AH, Broekaert WF (2000) A gene encoding a sphingolipid biosynthesis enzyme determines the sensitivity of *Saccharomyces cerevisiae* to an antifungal plant defensin from dahlia (*Dahlia merckii*). Proc Natl Acad Sci U S A 97:9531–9536

Thevissen K, François IEJA, Takemoto JY, Ferket KKA, Meert EM, Cammue BPA (2003) DmAMP1, an antifungal palnt defensins from dahlia (Dahlia merckii), interats with sphingolipids from *Saccharomyces cerevisiae*. FEMS Microbiol Lett 226:169–173

Thevissen K, Warnecke DC, François IEJA, Leipelt M, Heinz E, Ott C, Zähringer U, Thomma BPHJ, Ferket KKA, Cammue BPA (2004) Defensins from insects and plants interact with fungal glucosylceramides. J Biol Chem 279:3900–3905

Thevissen K, Francois IE, Aerts AM, Cammue BP (2005) Fungal sphingolipids as targets for the development of selective antifungal therapeutics. Curr Drug Targets 6:923–928

Thevissen K, Francois IE, Winderickx J, Pannecouque C, Cammue BP (2006) Ceramide involvement in apoptosis and apoptotic diseases. Mini Rev Med Chem 6:699–709

Thevissen K, Kristensen H-H, Thomma BPHJ, Cammue BPA, François IEJA (2007) Therapeutic potential of antifungal plant and insect defensins. Drug Discov Today 12:966–971

Thomma BP, Cammue BP, Thevissen K (2002) Plant defensins. Planta 216(2):193–202

van der Weerden NL, Lay FT, Anderson MA (2008) The plant defensin, NaD1, enters the cytoplasm of *Fusarium oxysporum* hyphae. J Biol Chem 283:14445–14452

van der Weerden NL, Hancock REW, Anderson MA (2010) Permeabilization of fungal hyphae by the plant defensin NaD1 occurs through a cell wall-dependent process. J Biol Chem 285:37513–37520

van der Weerden NL, Bleackley MR, Anderson MA (2013) Properties and mechanisms of action of naturally occurring antifungal peptides. Cell Mol Life Sci CMLS 70:3545–3570

Vriens K, Peigneur S, De Coninck B, Tytgat J, Cammue BPA, Thevissen K (2016) The antifungal plant defensin AtPDF2.3 from Arabidopsis thaliana blocks potassium channels. Sci Rep 6:32121

Chapter 7
Transgenic Plants for Improved Salinity and Drought Tolerance

Saikat Paul and Aryadeep Roychoudhury

Abstract Salinity and drought are the two most brutal environmental stresses that greatly affect plant growth and productivity. The worldwide increase in human population has made such stress assume a more disastrous form. In order to provide sufficient food and mitigate global hunger, a more sustainable and sufficient means of crop production is of urgent necessity. In the last decade, scientists have carried out extensive research to develop salt- and drought-tolerant crops through conventional breeding, but the outcome of these programs was not found to be convincing, as indicated by the limited number of salt- and drought-tolerant genotypes released so far. This is because hybridization is time-consuming and labor intensive. Whole genome sequencing, proteomic and metabolomic analysis of different crop plants under salt and drought stress has led scientists to identify different groups of genes involved in stress tolerance. Genetic engineering approach provides a comprehensive and more promising or practical tool to clone single gene or gene clusters and precisely characterize their function by introgression into other crop species, as compared to traditional crossing technique. The present chapter highlights the recent developments in transgenic research through incorporation and overexpression of single or multiple genes, either in homologous or heterologous background, thereby enhancing tolerance to salt and drought stress.

Keywords Abiotic stress · Salinity · Drought · Transgenics · Gene overexpression

7.1 Introduction

Plants are continually exposed to harsh environmental conditions which is life-threatening for their survival. Soil salinity and drought are the two major environmental constraints that highly affect plant growth and productivity worldwide. Osmotic stress due to limited availability of water during drought and high Na^+

S. Paul · A. Roychoudhury (✉)
Post Graduate Department of Biotechnology, St. Xavier's College (Autonomous),
Kolkata, West Bengal, India

© Springer International Publishing AG, part of Springer Nature 2018   141
S. S. Gosal, S. H. Wani (eds.), *Biotechnologies of Crop Improvement, Volume 2*,
https://doi.org/10.1007/978-3-319-90650-8_7

concentration during salt stress lead to the inhibition of photosynthesis which ultimately affect plant growth, yield, and productivity. As sessile in nature, plants cannot escape from such adverse situations (Gosal et al. 2009). Hence, to cope up with these adverse situations, plants have developed a complex array of adaptive strategies including intricate regulation of cellular, physiological, biochemical, and metabolic processes to avoid or tolerate cellular dehydration (Yamaguchi-Shinozaki and Shinozaki 2006; Roychoudhury and Chakraborty 2013). Under limited water availability, stomata plays an essential role to check water loss due to transpiration (Dow et al. 2014). In addition, upon perception of stress signal, a wide range of signaling cascade has been activated which ultimately initiates the expression of stress-responsive genes in a timely and coordinated manner. Abscisic acid (ABA), the universal stress hormone, highly accumulated under stress condition, also plays an important role in stress adaptation including stomatal closure and expression of stress-responsive genes (Hirayama and Shinozaki 2007; Osakabe et al. 2014). In recent times, whole genome sequencing analysis of different plants reveals that a large family of genes is expressed under different types of abiotic stresses that are involved in defense-related pathways. These genes can be grouped into three categories (Fig. 7.1), genes involving recognition of osmotic stress, signal perception, and transduction and production of stress-adaptive components for physiological responses (Li et al. 2013).

It has been estimated that by 2050, the world population will be increased from 6.2 billion to 9 billon (http://www.unfpa.org/swp/2001/). In order to provide sufficient and sustainable means of food production to this ever-increasing growing human population, it is essential to improve crop production and develop stress-tolerant plants. The conventional breeding program has been greatly implemented in the twentieth century for tackling food security worldwide. However, limited success has been achieved so far in terms of generating salt- and drought-tolerant genotypes through breeding. Such limitation is due to the fact that most of the crop species contain low magnitude of genetic variation in their gene pool so that wild varieties of crops are generally used as donor for developing better salt- and drought-tolerant crops. Because of the reproductive barriers between species, it is very difficult to transfer desirable alleles from wild relatives to domesticated crops. Breeding strategy also takes extensive time and intensive labor and does not have any control in transferring undesirable genes along with the desirable ones, Therefore, it is difficult to transfer selectively the favorable alleles from interspecific and intergeneric sources (Ashraf and Akram 2009; Ashraf 2010). With the advancement of marker-assisted breeding (MAB) and genetic engineering (transgenic approach), a desirable gene can be tagged and easily selected within breeding population and can be manipulated and transferred from one species to the other without transferring undesirable ones from the donor species. Genetic engineering has been implemented worldwide as a potential technology for developing abiotic stress-tolerant plants (Wani et al. 2016). In this chapter, we have focused upon the recent advances in crop improvement by genetic manipulation of different groups of genes (Fig. 7.2) to combat against salinity and drought stress (Fig. 7.2).

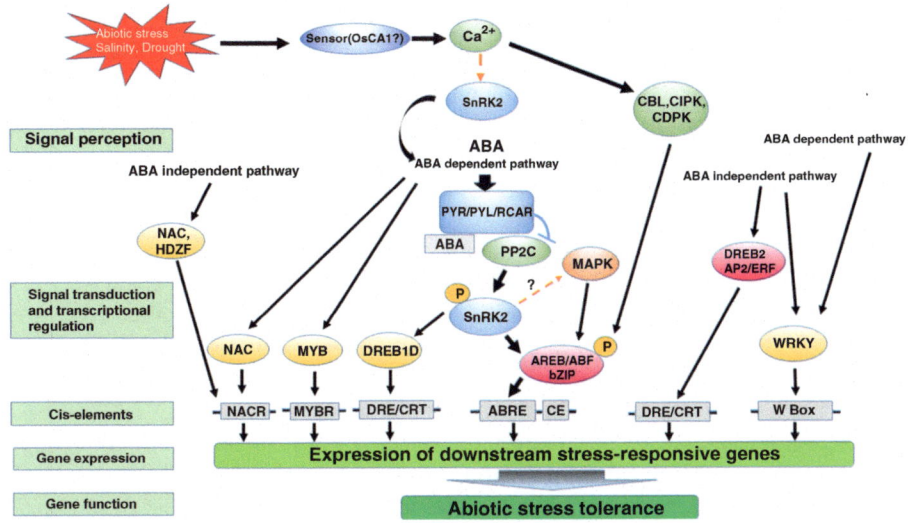

Fig. 7.1 An integrated circuit of the overall signal transduction mechanism during salinity and drought stress. Ca^{2+} channel OSCA1 acts as an osmosensor and increases cytosolic-free Ca^{2+} in response to abiotic stress which eventually activates CIPKs, CBLs, and CDPKs. Under osmotic stress, SnRK2s are activated by Ca^{2+}, a prerequisite to ABA accumulation, and control osmotic adjustment during stress conditions. In the presence of ABA, the ABA receptors PYR/PIL/RCAR bind to ABA and inhibit PP2C activity, resulting in autoactivation of SnRK2s to phosphorylate their downstream targets, such as the bZIP transcription factors, which recognize and bind to the ABRE of their target downstream genes and modulate their expression. The products of these genes confer abiotic stress tolerance. In ABA-dependent pathway, along with bZIP group TFs, NAC, MYB, and DREB1D TFs are also functional in ABA-inducible gene expression. In ABA-independent pathway, DREB2, WRKY, NAC, and HDZF TFs are involved in salinity and drought responses and possibly mediating the cross talk between ABA-dependent and ABA-independent pathway

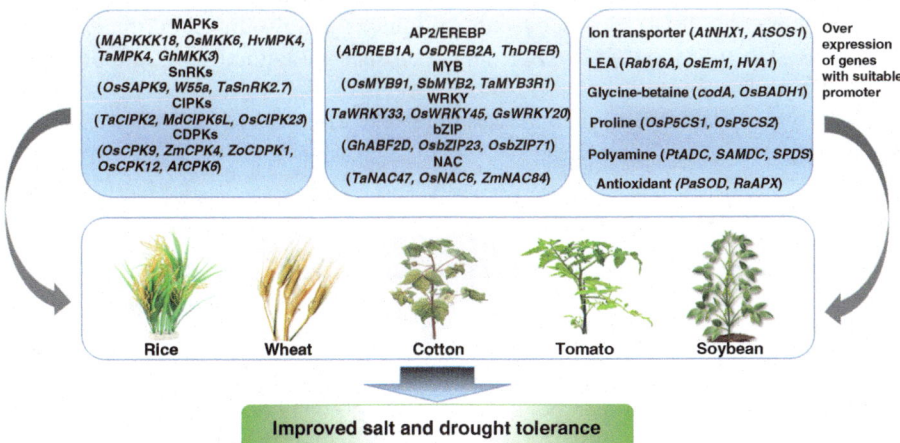

Fig. 7.2 Transgenic approaches through overexpression of different groups of genes, including regulatory and functional genes under the control of suitable promoters (preferably stress-inducible) to improve salinity and drought tolerance in different crop plants

7.2 Genes Involved in Signal Perception

In bacteria and yeast, the two-component signaling cascade acts as the osmosensor that transduces extracellular signal to the cytoplasm. *Arabidopsis thaliana Histidine kinase 1 (AtHK1)* has been first reported to function as a plant osmosensor (Urao et al. 1999) analogous to yeast SLN1. *AtHK1* was transcriptionally regulated and was accumulated more in roots than other tissues under high or low osmotic condition. Overexpression of the *AtHK1* complemented and suppressed the lethality of temperature-sensitive, osmosensing-defective yeast mutants, *sln1-ts* and *sln1Δ*. They have also demonstrated that *AtHK1* function as osmosensor in saline condition and transmitted signal through mitogen-activated protein kinase (MAPK) pathway. Transgenic *Arabidopsis* overexpressing *AtHK1/AHK1* enhances drought tolerance and acts as a positive regulator in osmotic stress signaling (Tran et al. 2007). Loss-of-function analysis of *ahk1* mutant indicated that *AHK1* is an important positive regulator of the ABA signal transduction pathway and enhances osmotic stress tolerance through ABA-dependent pathway. Recent study has shown that *Arabidopsis ahk1* mutants increased stomatal density and stomatal index which was consistent with greater transpirational water loss, thereby suggesting the role of *AtHK1* in drought tolerance by decreasing the stomatal density and preventing water loss during soil drying (Kumar et al. 2013). Calcium (Ca^{2+}) ion is a well-known intracellular secondary messenger molecule involved in signal transduction pathways in plants (Tuteja and Mahajan 2007). It has been well established that the cytosolic-free Ca^{2+} increases in response to various abiotic and biotic stress by activating Ca^{2+} channels in plants (Monshausen and Gilroy 2009). Previous studies have shown that osmotic/mechanical stimuli-gated Ca^{2+}-permeable channels serve as osmosensor in bacteria and animals (Árnadóttir and Chalfie 2010). In plants, the first genetically identified Ca^{2+}-permeable channels which acts as a osmosensor is OSCA1 from *Arabidopsis* which provide potential targets for genetic engineering to generate drought-resistant crops (Yuan et al. 2014). Several groups of kinases are also involved in perceiving the stress stimuli, along with ABA, and transduce the signal downstream to the transcription factor and their target genes (Fig. 7.1). Below, we present a comprehensive overview of several such classes of protein kinases and their involvement in tolerance mechanism (Table 7.1).

7.2.1 Mitogen-Activated Protein Kinase (MAPK)

Protein kinases and phosphatases play a central role in signal transduction through the phosphorylation and dephosphorylation mechanism. This not only leads to the activation of defense responses but also to the activation of developmental processes like cell growth and differentiation. Mitogen-activated protein kinase (MAPK) cascades are conserved and ubiquitous signaling modules found in unicellular and multicellular eukaryotes, linking the signal perception from external stimuli to cellular

Table 7.1 Kinase families overexpressed to generate salt- and drought-tolerant transgenic plants

Family	Gene	Source	Target	Enhanced tolerance	Reference
MAPK	*MAPKKK18*	*Arabidopsis*	*Arabidopsis*	Drought	Li et al. (2017b)
	GhMKK3	*Gossypium hirsutum*	Tobacco	Drought	Wang et al. (2016a)
	GhRaf19	*Gossypium hirsutum*	Tobacco	Cold	Jia et al. (2016)
	AtMKK5	*Arabidopsis*	*Arabidopsis*	Salinity	Xing et al. (2015)
	TaMPK4	*Triticum aestivum*	Tobacco	Salinity	Hao et al. (2015)
	Raf43	*Arabidopsis*	*Arabidopsis*	Salinity, drought	Virk et al. (2015)
	ZmSIMK1	*Zea mays*	Tobacco	Drought	Wang et al. (2014a)
	SpMPK3	*Solanum pimpinellifolium*	*Arabidopsis*	Osmotic stress	Li et al. (2014a)
	ZmMPK5	*Zea mays*	Tobacco	Salinity	Zhang et al. (2014a)
	ZmMKK1	*Zea mays*	*Arabidopsis*	Salinity, drought	Cai et al. (2014)
	GhMPK17	*Gossypium hirsutum*	*Arabidopsis*	Salinity, osmotic stress	Zhang et al. (2014b)
	OsMKK6	*Oryza sativa*	Rice	Salinity	Kumar and Sinha (2013)
	HvMPK4	*Hordeum vulgare*	Barley	Salinity	Abass and Morris (2013)
	GhMKK1	*Gossypium hirsutum*	Tobacco	Salinity, drought	Lu et al. (2013)
	GhMPK2	*Gossypium hirsutum*	Tobacco	Salinity, drought	Zhang et al. (2011b)
	ZmMKK4	*Zea mays*	*Arabidopsis*	Salinity, cold	Kong et al. (2011)
	CsNMAPK	*Cucumis sativus*	Tobacco	Salinity, osmotic stress	Xu et al. (2010a)
	DSM1	*Oryza sativa*	Rice	Drought	Ning et al. (2010)
	ZmSIMK1	*Zea mays*	*Arabidopsis*	Salinity	Gu et al. (2010)

(continued)

Table 7.1 (continued)

Family	Gene	Source	Target	Enhanced tolerance	Reference
SnRK	PtSnRK2	Populus trichocarpa	Arabidopsis	Salinity	Song et al. (2016)
	SAPK9	Oryza rufipogon	Rice	Drought	Dey et al. (2016b)
	JcSnRK2	Jatropha curcas	Arabidopsis	Salinity, drought	Chun et al. (2014)
	GsAPK	Glycine soja	Arabidopsis	Salinity	Yang et al. (2012c)
	TaSnRK2.7	Triticum aestivum	Arabidopsis	Salinity, drought, cold	Zhang et al. (2011a)
	TaSnRK2.8	Triticum aestivum	Arabidopsis	Salinity, drought, cold	Zhang et al. (2010b)
	TaSnRK2.4	Triticum aestivum	Arabidopsis	Salinity, drought, cold	Mao et al. (2010)
	W55a	Triticum aestivum	Arabidopsis	Drought	Xu et al. (2009)
	SRK2C	Arabidopsis	Arabidopsis	Drought	Umezawa et al. (2004)
CIPK	MdSOS2L1	Malus domestica	Apple, tomato	Salinity	Hu et al. (2016)
	TaCIPK2	Triticum aestivum	Tobacco	Drought	Wang et al. (2016e)
	PeCIPK26	Populus euphratica	Arabidopsis	Salinity	Lv et al. (2014)
	ZmCIPK21	Zea mays	Arabidopsis	Salinity	Chen et al. (2014a)
	TaCIPK29	Triticum aestivum	Tobacco	Salinity	Deng et al. (2013a)
	TaCIPK14	Triticum aestivum	Tobacco	Salinity, cold	Deng et al. (2013b)
	AtCIPK6	Arabidopsis	Arabidopsis	Salinity	Chen et al. (2013c)
	MdSOS2	Malus domestica	Arabidopsis	Salinity	Hu et al. (2012)
	MdCIPK6L	Malus domestica	Apple, tomato, Arabidopsis	Salinity, drought, cold	Wang et al. (2012)
	CaCIPK6	Cicer arietinum	Tobacco	Salinity	Tripathi et al. (2009)
	ZmCIPK16	Zea mays	Arabidopsis	Salinity	Zhao et al. (2009)
	OsCIPK23	Oryza sativa	Rice	Drought	Yang et al. (2008)

(continued)

Table 7.1 (continued)

Family	Gene	Source	Target	Enhanced tolerance	Reference
CDPK	OsCPK17	Oryza sativa	Rice	Cold	Almadanim et al. (2017)
	VaCPK20	Vitis amurensis	Arabidopsis	Drought, cold	Dubrovina et al. (2015)
	AtCPK8	Arabidopsis	Arabidopsis	Drought	Zou et al. (2015)
	OsCPK9	Oryza sativa	Rice	Drought	Wei et al. (2014)
	ZmCPK4	Zea mays	Arabidopsis	Drought	Jiang et al. (2013)
	PeCPK10	Populus euphratica	Arabidopsis	Drought, cold	Chen et al. (2013a)
	ZoCDPK1	Zingiber officinale	Tobacco	Salinity, drought	Vivek et al. (2013)
	OsCPK12	Oryza sativa	Rice	Salinity	Asano et al. (2012)
	OsCPK21	Oryza sativa	Rice	Salinity	Asano et al. (2011)
	AtCPK6	Arabidopsis	Arabidopsis	Salinity, drought	Xu et al. (2010b)
	AtCPK10	Arabidopsis	Arabidopsis	Drought	Zou et al. (2010)

processes together with changes in gene expression or cell organization. MAPK cascade is composed of three essential kinases: MAPK kinase kinase (MAPKKK), MAPKK, and MAPK which are activated in a sequential phosphorylation-dependent manner (Zelicourt et al. 2016). Upon perception of external stimuli, MAPKKK can selectively phosphorylate downstream MAPKK(s) on Ser/Thr residue in a conserved (S/T)X3/5(S/T) motif, which in turn activates specific MAPK(s) by phosphorylating the Thr and Tyr in the TXY motif, eventually leading to the activation of transcription factors, phospholipases, cytoskeletal and microtubule-associated proteins, and the expression of specific sets of genes in response to various stimuli (Taj et al. 2014; Li et al. 2017b). Traditional genetic and biochemical methods have identified MKKK/MKK/MPK signaling modules with overlapping roles in controlling cell division, development, hormone signaling and synthesis, and response to abiotic stress. Transgenic analysis with MAPK genes from different plants has been extensively studied to generate salt- and drought-tolerant varieties. Transgenic Arabidopsis overexpressing MAPKKK18 significantly enhanced drought tolerance by canonical ABA signaling pathway and exerted its regulatory roles via downstream of MAPKK3 (Li et al. 2017b). Wang et al. (2016a) isolated and characterized cotton (Gossypium hirsutum) group B MAPKK gene GhMKK3. Transgenic tobacco overexpressing GhMKK3 conferred tolerance to drought. RNA sequencing (RNA-seq) analysis of transgenic plant showed that GhMKK3 plays an important role in abiotic stress tolerance by regulating stomatal responses and root hair growth. It was also shown that in cotton, ABA-induced MAPK kinase cascade was composed of GhMKK3, GhMPK7, and GhPIP1, which played an important role in water transport in roots and stomata. Wheat MAPK member, *TaMPK4*, conferred tolerance to various abiotic stresses. Overexpression of *TaMPK4* in tobacco improved salt tolerance by modifying root architecture through transcriptional regulation of the auxin transport-associated genes and also by modulating ROS metabolism and nutrient acquisition (Hao et al. 2015). *Arabidopsis* Raf-like MAPKKK gene *Raf43* has been identified and genetically characterized as a positive regulator against multiple abiotic stresses including salinity and drought (Virk et al. 2015). In mutant plants, the expression of *RD17* and *DREB2A*, the two stress-responsive genes, was downregulated, suggesting the possible role of *Raf43* in stress-responsive signaling pathway. The DSM1 protein belongs to a B3 subgroup of plant Raf-like MAPKKKs that conferred tolerance to dehydration stress at the seedling stage in rice. Microarray analysis of *dsm1* mutants showed that the expression of two *peroxidase* (*POX*) genes, *POX22.3* and *POX8.1*, were inhibited, as compared to wild type, suggesting the involvement of *DSM1* in drought tolerance through the reactive oxygen species (ROS) signaling pathway (Ning et al. 2010). Transgenic *Arabidopsis* overexpressing *mitogen-activated protein kinase kinase 5* (*MKK5*) showed enhanced tolerance to salinity and the expression of iron *superoxide dismutase* (*SOD*) gene was mediated by MAP kinase cascade (*MEKK1, MKK5,* and *MPK6*) under salinity stress (Xing et al. 2015). A maize MAPKK gene, *ZmMKK1*, has been reported to be upregulated by diverse stresses and ABA signaling molecule in maize root. Transgenic *Arabidopsis* overexpressing *ZmMKK1* conferred tolerance to salt and drought stress by upregulating the expression of genes encoding ROS scavenging enzyme and ABA-related genes, such as *POX, catalase* (*CAT*), *RAB18*, and *RD29A* (Cai et al. 2014). Transgenic *Arabidopsis* overexpressing cotton MAPK, *GhMPK17*, showed salt and osmotic stress tolerance by elevating higher seed germination rate, root elongation, and cotyledon greening/expansion under stress condition (Zhang et al. 2014b).

7.2.2   Sucrose Non-fermenting-1 (Snf1)-Related Protein Kinase (SnRK)

The second family of plant protein kinase is SnRK orthologs of the budding yeast SNF1 (sucrose non-fermenting-1). Mammalian AMPK (AMP-activated protein kinase) represents subfamily of serine/threonine kinases which are involved in acclimation of plants to various abiotic stresses and ABA-dependent plant development (Lovas et al. 2003; Baena-González et al. 2007). The SnRK could be classified into three subgroups: SnRK1, SnRK2, and SnRK3. Earlier studies suggest that the members of SnRK2 and SnRK3 family are unique to plant and are involved in ABA-mediated stress-signaling pathway (Coello et al. 2011). Rice has ten members of SnRK2 subfamily; the expression of all the members are induced by hyperosmotic stress, and three of them are also activated by ABA (Kobayashi et al. 2004). In *Arabidopsis*, 10 *SnRK2* and 25 *SnRK3* genes have been reported so far

(Halford et al. 2003). Transgenic *Arabidopsis* expressing *SRK2C*, a member of *Arabidopsis* SnRK2 family protein kinase, has been reported to be significantly enhanced during drought by controlling the expression of many stress-responsive genes (Umezawa et al. 2004). The *SAPK9*, one of the members of ten rice SnRK2 family protein, has been functionally characterized by analyzing the gain-of-function and loss-of-function transgenic plants. Transgenic rice line overexpressing *SAPK9* from *Oryza rufipogon* exhibited enhanced drought tolerance and grain yield by altering stress-responsive gene expression, stomatal closure, and cellular osmotic potential (Dey et al. 2016b). Heterologous overexpression of poplar (*Populus trichocarpa*) SnRK2, *PtSnRK2.5*, and *PtSnRK2.7* genes in *Arabidopsis thaliana* enhanced salt tolerance through the activation of cellular signaling pathways (Song et al. 2016). Functional analysis of wheat *TaSnRK2.7* through transgenic approach showed that its overexpression in *Arabidopsis* enhanced tolerance potential to multiple abiotic stresses including salt and polyethylene glycol (PEG)-mediated dehydration stress through promoting root growth, enhancing photosystem II activity, and lowering osmotic potential of cell (Zhang et al. 2011a). Transgenic *Arabidopsis* overexpressing *Jatropha* SnRK2, *JcSnRK2*, improved tolerance to high salt and drought stress (Chun et al. 2014). *GsAPK*, an ABA-inducible and calcium-independent SnRK2-type kinase from *Glycine soja*, enhanced tolerance to salinity and was presumably involved in ABA-mediated signal transduction (Yang et al. 2012b). SnRK3 family was also known to be involved in salt tolerance. *Arabidopsis SnRK3.11* or *Salt Overly Sensitive 2* (*SOS2*), one of the members of SnRK3 subfamily, is involved in enhancing salt tolerance by maintaining cellular Na^+/H^+ homeostasis under salt stress (Liu et al. 2000). Transgenic analysis of activated *Arabidopsis SOS2* mutants revealed that C-terminal domain is critical for salt tolerance and its kinase activity (Guo et al. 2004). SnRK3 (also called CIPKs) interact with calcineurin B-like (CBL) calcium-binding protein (Guo et al. 2001). Transgenic *Arabidopsis* and tall fescue overexpressing *Arabidopsis* SOS pathway genes showed enhanced salt tolerance by decreasing Na^+/K^+ ratio and increasing antioxidative enzyme activity along with higher accumulation of proline in transgenic plants (Yang et al. 2009; Ma et al. 2014).

7.2.3 Calcium-Dependent Protein Kinase (CDPK)

As mentioned earlier, calcium (Ca^{2+}) plays an important role as secondary messenger in plant responses under abiotic stresses. Calcium-dependent protein kinase (CDPK), containing a CaM-like domain as well as a catalytic Ser/Thr kinase domain, acts as a Ca^{2+} sensor and mediates various cellular processes (Kudla et al. 2010). In *Arabidopsis thaliana*, 34 CDPKs have been reported and are involved in plant development and abiotic stress responses. Transgenic analysis revealed that *CPK8*, a *Arabidopsis* CDPK, conferred drought tolerance and functions in ABA- and Ca^{2+}-mediated plant responses through the regulation of catalase 3 (CAT3) activity under drought stress (Zou et al. 2015). Zhou et al. (2013) reported that

Arabidopsis CYCLIN H; 1 (CYCH; 1) interacts with and activates CYCLIN-DEPENDENT KINASE Ds (CDKDs) and regulates blue light-mediated stomatal opening by controlling the ROS homeostasis, thus reducing transpiration and enhancing drought tolerance. Another CDPK, *CPK10*, from *Arabidopsis* has been functionally characterized through transgenic analysis in relation to drought tolerance (Zou et al. 2010). Genetic analysis of *cpk10* mutant and overexpressed plant revealed that *CPK10* conferred drought tolerance possibly by interacting with *HSP1* (*heat shock protein 1*) and plays a vital role in ABA- and Ca^{2+}-mediated regulation of stomatal movements.

7.3 Transcription Factors in Salt and Drought Tolerance

Transcription factors (TFs) are terminal transducer in a signaling cascade and are able to regulate the expression of downstream genes involved in stress responses by recognizing and binding to conserved cis-acting sequences in their promoter (Joshi et al. 2016; Banerjee and Roychoudhury 2017). Considerable research in the last few decades has led to the identification and characterization of several groups of TFs, and numerous studies have been made to improve stress tolerance by genetic manipulation of these TFs (Fig. 7.1). The most important TF families involved in abiotic stress tolerance are AP2/EREBP (Apetala2/Ethylene-Responsive Element-Binding Protein), MYB, WRKY, NAC (NAM, ATAF, and CUC), and bZIP (basic leucine zipper domain) (Golldack et al. 2014). In this section, recent advances in genetic manipulation of TFs through transgenic approaches in relation to salinity and drought tolerance have been summarized (Table 7.2).

7.3.1 AP2/EREBP (Apetala2/Ethylene-Responsive Element-Binding Protein) Transcription Factors

AP2/EREBP is a large family of TFs involved in a wide range of physiological processes including development and abiotic and biotic stress responses (Sharoni et al. 2011). AP2/EREBP TFs contain highly conserved AP2/ERF DNA-binding domain which recognizes and binds to GCC box and/or dehydration-responsive element (DRE)/C-repeat element (CRT) at the promoter of their downstream target genes (Riechmann and Meyerowitz 1998). The AP2/EREBP gene family has been divided into four subgroups on the basis of the number of AP2/ERF domain. These are AP2, RAV (related to ABI3/VP1), dehydration-responsive element-binding protein (DREB), and ERF (ethylene-responsive element-binding factor) (Sharoni et al. 2011). Previous reports have shown that among AP2/EREBP gene family, DREB and ERF are involved in biotic and abiotic stress tolerance. Whole genome sequencing analysis revealed that *Arabidopsis* genome encodes at least eight *DREB2* genes, of which *DREB2A* and *DREB2B* are highly induced by drought and salinity stress (Nakashima et al. 2000; Sakuma et al. 2002). Transgenic *Arabidopsis* constitutively overexpressing *DREB2A* significantly increased drought tolerance by altering the expression of many stress-inducible genes (Sakuma et al. 2006). Transgenic rice harboring *OsDREB2A* enhanced stress tolerance against osmotic, salt, and dehydration stress with enhanced growth performance (Mallikarjuna et al. 2011). Recent study has shown that overexpression of *ThDREB* gene form *Tamarix hispida* in transgenic tobacco and *T. hispida* resulted in an increased tolerance to salt and drought stress by increasing ROS-scavenging enzyme activity (Yang et al. 2017). Liao et al. (2016) isolated and characterized apple *MsDREB6.2* with respect to drought tolerance through overexpression analysis and chimeric repressor gene-silencing technology (CRES-T). They have shown that in transgenic plant, the expression of cytokinin catabolism gene, *MdCKX4a*, increased which led to decreased endogenous cytokinin levels and caused a decrease in shoot/root ratio in

Table 7.2 Transcription factor families overexpressed to generate salt- and drought-tolerant transgenic plants

Family	Gene	Source	Target	Enhanced tolerance	Reference
AP2/ EREBP	*ThDREB*	*Tamarix hispida*	Tobacco	Salinity, drought	Yang et al. (2017)
	DREB1A	*Arabidopsis*	*Arabidopsis*	Drought	Kudo et al. (2017)
	MsDREB6.2	*Malus sieversii*	Apple	Drought	Liao et al. (2016)
	AtDREB1A	*Arabidopsis*	*Salvia miltiorrhiza*	Drought	Wei et al. (2016)
	VrDREB2A	*Vigna radiata*	*Arabidopsis*	Salinity, drought	Chen et al. (2016)
	SsDREB	*Suaeda salsa*	Tobacco	Salinity, drought	Zhang et al. (2015c)
	AtDREB1A	*Arabidopsis*	Rice	Drought	Ravikumar et al. (2014)
	TaERF3	*Triticum aestivum*	Wheat	Salinity, drought	Rong et al. (2014)
	EaDREB2	*Erianthus arundinaceus*	Sugarcane	Salinity, drought	Augustine et al. (2015)
	LcERF054	*Lotus corniculatus*	*Arabidopsis*	Salinity	Sun et al. (2014b)
	LcDREB2	*Leymus chinensis*	*Arabidopsis*	Salinity	Peng et al. (2013)
	GmERF7	*Glycine max*	Tobacco	Salinity	Zhai et al. (2013)
	StDREB1	*Solanum tuberosum*	Potato	Salinity	Bouaziz et al. (2013)
	OsERF4a	*Oryza sativa*	Rice	Drought	Joo et al. (2013)
	OsDREB2A	*Oryza sativa*	Rice	Salinity, drought	Mallikarjuna et al. (2011)
	JERF3	*Solanum lycopersicum*	Rice	Drought	Zhang et al. (2010a)

(continued)

Table 7.2 (continued)

Family	Gene	Source	Target	Enhanced tolerance	Reference
MYB	PbrMYB21	Pyrus betulaefolia	Tobacco	Drought	Li et al. (2017a)
	LpMYB1	Lablab purpureus	Arabidopsis	Drought	Yao et al. (2016)
	SbMYB15	Salicornia brachiata	Tobacco	Salinity, drought	Shukla et al. (2015)
	OsMYB91	Oryza sativa	Rice	Salinity	Zhu et al. (2015)
	SRM1	Arabidopsis	Arabidopsis	Salinity	Wang et al. (2015a)
	SbMYB2	Scutellaria baicalensis	Tobacco	NaCl, mannitol, ABA	Qi et al. (2015)
	SbMYB7	Scutellaria baicalensis	Tobacco	NaCl, mannitol, ABA	Qi et al. (2015)
	TaMYB3R1	Triticum aestivum	Arabidopsis	Salinity, drought	Cai et al. (2015)
	LeAN2	Lycopersicum esculentum	Tomato	Heat	Meng et al. (2015)
	TaMYB19-B	Triticum aestivum	Arabidopsis	Salinity, drought	Zhang et al. (2014c)
	OsMYB48–1	Oryza sativa	Rice	Salinity, drought	Xiong et al. (2014)
	MdSIMYB1	Malus domestica	Tobacco, apple	Salinity, drought, cold	Wang et al. (2014b)
	SpMYB	Solanum pimpinellifolium	Tobacco	Salinity, drought	Li et al. (2014b)
	PtsrMYB	Poncirus trifoliata	Tobacco	Drought	Sun et al. (2014a)
	GmMYBJ1	Glycine max	Arabidopsis	Drought, cold	Su et al. (2014)
	MdoMYB121	Malus domestica	Tomato, apple	Salinity, drought, cold	Cao et al. (2013)
	OsMYB2	Oryza sativa	Rice	Salinity, drought, cold	Yang et al. (2012b)
	TaMYB73	Triticum aestivum	Arabidopsis	Salinity	He et al. (2012)
	TaMYB33	Triticum aestivum	Arabidopsis	Salinity, drought	Qin et al. (2012)
	StMYB1R-1	Solanum tuberosum	Potato	Drought	Shin et al. (2011)
	AtMYB15	Arabidopsis	Arabidopsis	Salinity, drought	Ding et al. (2009)
	Osmyb4	Oryza sativa	Apple	Drought, cold	Pasquali et al. (2008)

(continued)

Table 7.2 (continued)

Family	Gene	Source	Target	Enhanced tolerance	Reference
WRKY	JcWRKY	Jatropha curcas	Tobacco	Salinity	Agarwal et al. (2016)
	TaWRKY33	Triticum aestivum	Arabidopsis	Drought, heat	He et al. (2016)
	TaWRKY1	Triticum aestivum	Arabidopsis	Drought, heat	He et al. (2016)
	OsWRKY71	Oryza sativa	Rice	Cold	Kim et al. (2016)
	MtWRKY76	Medicago truncatula	Medicago truncatula	Salinity, drought	Liu et al. (2016a)
	GhWRKY25	Gossypium Hirsutum	Tobacco	Salinity	Liu et al. (2016b)
	GhWRKY41	Gossypium hirsutum	Tobacco	Salinity, drought	Chu et al. (2015)
	FcWRKY70	Fortunella crassifolia	Tobacco, Lemon	Drought	Gong et al. (2015)
	GhWRKY68	Gossypium hirsutum	Tobacco	Salinity, drought	Jia et al. (2015)
	SpWRKY1	Solanum pimpinellifolium	Tobacco	Salinity, drought	Li et al. (2015b)
	TaWRKY93	Triticum aestivum	Arabidopsis	Salinity, drought, cold	Qin et al. (2015)
	TaWRKY44	Triticum aestivum	Tobacco	Salinity, drought	Wang et al. (2015b)
	GhWRKY34	Gossypium hirsutum	Arabidopsis	Salinity	Zhou et al. (2015)
	BdWRKY36	Brachypodium distachyon	Tobacco	Drought	Sun et al. (2015)
	GsWRKY20	Glycine soja	Arabidopsis	Drought	Luo et al. (2013)
	TaWRKY10	Triticum aestivum	Tobacco	Salinity, drought	Wang et al. (2013)
	ThWRKY4	Tamarix hispida	Arabidopsis	Salinity	Zheng et al. (2013)
	TaWRKY2	Triticum aestivum	Arabidopsis	Salinity, drought	Niu et al. (2012)
	TaWRKY19	Triticum aestivum	Arabidopsis	Salinity, drought, cold	Niu et al. (2012)
	VvWRKY11	Vitis vinifera	Arabidopsis	Drought	Liu et al. (2011)
	OsWRKY08	Oryza sativa	Arabidopsis	Salinity	Song et al. (2010)
	OsWRKY45	Oryza sativa	Arabidopsis	Salinity, drought	Qiu and Yu (2009)

(continued)

Table 7.2 (continued)

Family	Gene	Source	Target	Enhanced tolerance	Reference
NAC	VaNAC26	Vitis amurensis	Arabidopsis	Salinity, drought	Fang et al. (2016)
	TaNAC47	Triticum aestivum	Arabidopsis	Salinity, drought, cold	Zhang et al. (2015a)
	ZmNAC84	Zea mays	Tobacco	Drought	Zhu et al. (2016)
	MusaNAC042	Musa acuminata	Banana	Salinity, drought	Tak et al. (2016)
	MlNAC9	Miscanthus lutarioriparius	Arabidopsis	Drought, cold	Zhao et al. (2016)
	GhNAC2	Gossypium hirsutum	Arabidopsis	Drought	Gunapati et al. (2016)
	ONAC095	Oryza sativa	Rice	Cold	Huang et al. (2016a)
	EcNAC67	Eleusine coracana	Rice	Salinity, drought	Rahman et al. (2016)
	AaNAC1	Artemisia annua	Artemisia Arabidopsis	Drought	Lv et al. (2016)
	TaNAC2D	Triticum aestivum	Arabidopsis	Salinity	Huang and Wang (2016)
	OsNAC6	Oryza sativa	Rice	Drought	Lee et al. (2016)
	ONAC022	Oryza sativa	Rice	Salinity, drought	Hong et al. (2016)
	TaNAC29	Triticum aestivum	Arabidopsis	Salinity, drought	Huang et al. (2015)
	MLNAC5	Miscanthus lutarioriparius	Arabidopsis	Drought, cold	Yang et al. (2015)
	OsNAC	Oryza sativa	Rice	Salinity, drought, cold	Chen et al. (2014b)
	TaNAC67	Triticum aestivum	Arabidopsis	Salinity, drought, cold	Mao et al. (2014)
	AhNAC3	Arachis hypogaea	Tobacco	Drought	Liu et al. (2013)
	SNAC1	Oryza sativa	Wheat	Salinity, drought	Saad et al. (2013)
	OsNAC5	Oryza sativa	Rice	Drought	Jeong et al. (2013)
	OsNAC9	Oryza sativa	Rice	Drought	Redillas et al. (2012)
	TaNAC2a	Triticum aestivum	Tobacco	Drought	Tang et al. (2012b)
	GmNAC20	Glycine max	Arabidopsis	Salinity, cold	Hao et al. (2011)
	GmNAC11	Glycine max	Arabidopsis	Salinity	Hao et al. (2011)
	OsNAC10	Oryza sativa	Rice	Drought	Jeong et al. (2010)
	ONAC063	Oryza sativa	Arabidopsis	Salinity, Osmotic stress	Yokotani et al. (2009)

(continued)

Table 7.2 (continued)

Family	Gene	Source	Target	Enhanced tolerance	Reference
bZIP	PtabZIP1L	Populus tremula x P. alba	Poplar	Drought	Dash et al. (2017)
	GhABF2D	Gossypium hirsutum	Cotton	Drought	Kerr et al. (2017)
	OsbZIP23	Oryza sativa	Rice	Drought	Dey et al. (2016a)
	TabZIP174	Triticum aestivum	Arabidopsis	Drought	Li et al. (2016)
	GhABF2	Gossypium hirsutum	Arabidopsis, cotton	Salinity, drought	Liang et al. (2016a)
	VlbZIP36	Vitis vinifera	Arabidopsis	Drought	Tu et al. (2016)
	TaAREB3	Triticum aestivum	Arabidopsis	Drought, freezing	Wang et al. (2016d)
	AtABF3	Arabidopsis	Alfalfa	Salinity, drought	Wang et al. (2016f)
	TabZIP60	Triticum aestivum	Arabidopsis	Salinity, drought, cold	Zhang et al. (2015b)
	OsbZIP71	Oryza sativa	Rice	Salinity, drought, osmotic stress	Liu et al. (2014)
	LrbZIP	Nelumbo nucifera	Tobacco	Salinity	Cheng et al. (2013)
	OsbZIP46	Oryza sativa	Rice	Drought	Tang et al. (2012a)
	ZmbZIP72	Zea may	Arabidopsis	Salinity, drought	Ying et al. (2012)
	ABP9	Zea may	Arabidopsis	Salinity, drought, cold	Zhang et al. (2011c)
	GmbZIP1	Glycine max	Arabidopsis	Salinity, drought, cold	Gao et al. (2011)
	PtrABF	Poncirus trifoliata	Tobacco	Drought	Huang et al. (2010)
	OsbZIP72	Oryza sativa	Rice	Drought	Lu et al. (2009)
	OsbZIP23	Oryza sativa	Rice	Salinity, drought	Xiang et al. (2008)

transgenic apple plants. Transgenic plants exhibited better drought tolerance by influencing stomatal density and opening and root growth and modulating aquaporin (*AQP*) gene expression. Kudo et al. (2017) applied a gene stacking approach using two transcription factors, *DREB1A* and rice *PHYTOCHROME-INTERACTING FACTOR- LIKE 1* (*OsPIL1*) to enhance plant growth under drought condition. The transgenic *Arabidopsis* overexpressing two TFs showed improved drought tolerance. Transcriptomic and metabolomic analysis revealed that double overexpressor plant accumulated more compatible solutes, such as sugars and amino acids with higher expression of abiotic stress-responsive genes and cell elongation-related *OsPIL1* downstream genes.

7.3.2   MYB TFs

MYB family is one of the largest groups of TFs characterized by the presence of conserved MYB domain at N-terminal for DNA binding. MYB domain comprises one to four imperfect repeats (R) with about 52 amino acid residues in each repeat, forming a helix-turn-helix structure (Dubos et al. 2010). Based on the presence of number of repeats, MYB family can be divided into four different classes, 1R-MYB (MYB-related type), R2R3-MYB, R1R2R3-MYB, and 4R-MYB, containing one, two, three, and four MYB repeats, respectively. Over two decades ago, the first MYB gene, *COLOURED1* (*C1*), was isolated and characterized as a transcriptional activator involved in synthesis of anthocyanins in the aleurone of maize kernels (Paz-Ares et al. 1987). Since then, numerous studies have been conducted to identify and characterize function of *MYB* genes in important plants like *Arabidopsis*, maize, rice, petunia (*Petunia hybrida*), snapdragon (*Antirrhinum majus*), grapevine (*Vitis vinifera* L.), poplar (*Populus tremuloides*), and apple (*Malus domestica*), using both genetic and molecular analyses. Accumulation of polyamine (PA) is considered as a metabolic stress marker under drought stress in various plant groups (Saha et al. 2015). In *Arabidopsis*, accumulation of endogenous PA with higher expression of *ADC (arginine decarboxylase)* transcript in relation to drought tolerance has been clearly demonstrated (Alcázar et al. 2010). Recent study has shown that a novel stress-responsive MYB gene, *PbrMYB21*, from *Pyrus betulaefolia* was induced by various abiotic stresses, particularly dehydration stress. Transgenic tobacco overexpressing *PbrMYB21* conferred drought tolerance with higher expression of *ADC* and greater accumulation of endogenous PAs, as compared to wild-type plant. On the other hand, virus-induced gene silencing (VIGS) of *PbrMYB21* in *Pyrus betulaefolia* suppressed the expression of *ADC* and decreased PA concentration with decreased level of drought tolerance. Transgenic analysis suggested that *PbrMYB21* played a positive role in drought tolerance by modulating the *ADC* expression with subsequent PA synthesis (Li et al. 2017a). Yao et al. (2016) isolated and characterized the function of a novel R2R3-MYB transcription factor, *LpMYB1* from *Lablab purpureus,* a multipurpose leguminous plant. Transgenic *Arabidopsis* overexpressing *LpMYB1* enhanced salt and drought tolerance along with improved germination potential of transgenic seeds under NaCl and ABA, suggesting that *LpMYB1* acts as a positive regulator under stress condition. *SbMYB15*, a R2R3-type transcription factor from the extreme halophyte *Salicornia brachiata*, has been functionally characterized for involvement in abiotic stress tolerance. The *SbMYB15* overexpression in tobacco greatly improved salt and dehydration tolerance by lowering lipid peroxidation, H_2O_2 production, and Na^+/K^+ ratio, together with higher accumulation of proline, reducing sugar and total amino acid, and increased expression of stress-responsive genes in transgenic plants, suggesting its role as a positive regulator in salt and dehydration tolerance (Shukla et al. 2015). In *Arabidopsis, Salt-Related MYB1 (SRM1)*, a MYB-like R2R3 TF has been identified as an important transcriptional regulator that directly targets *NCED3/ STO1*, the ABA biosynthesis genes and *RESPONSIVE TO DESICCATION 26 (RD26), NAC DOMAIN CONTAINING PROTEIN19 (ANAC019)* genes involved in stress signaling, thereby influencing vegetative growth under salt stress (Wang et al. 2015a). In rice, very few *MYB* genes have been characterized that are involved in salt and drought tolerance. Yang et al. (2012a) isolated and functionally characterized a R2R3-MYB TF, *OsMYB2*, in response to salt, cold, and dehydration stress. Overexpressed *OsMYB2* rice showed higher tolerance to salt, cold, and dehydration stresses with higher accumulation of proline and soluble sugar, along with upregulated expression of *Δ¹-pyrroline-5-carboxylate synthase* (*P5CS*) and proline transporter genes as compared to the wild type. Overexpressed plants also showed higher antioxidant enzyme activity along with increased expression of genes encoding antioxidative enzymes and lowered H_2O_2 and malondialdehyde (MDA) level. Microarray and real-time PCR analysis also showed that *OsMYB2*-overexpressed rice showed altered expression of stress-responsive genes like *OsLEA3*, *OsRab16A*, and *OsDREB2A*, suggesting a regulatory role of *OsMYB2* in tolerance of rice to salt, cold, and dehydration stress. The *OsMYB48-1* has been functionally characterized as a novel MYB-related TF involved in drought and salinity tolerance by modulating ABA biosynthesis and expression of stress-responsive *lea* genes (Xiong et al. 2014). Recently, the function of rice *OsMYB91*, R2R3-type MYB TF, has been characterized in relation to plant growth and survival under salt stress (Zhu et al. 2015). Salt stress induced the expression of *OsMYB91* and also led to histone modifications in the promoter as well as in the transcribed region. Interestingly, transgenic rice overexpressing *OsMYB91* showed reduced growth and endogenous ABA level under control condition. Upon salt stress, overexpressed plant showed enhanced tolerance with decreased content of H_2O_2 and MDA and higher accumulation of proline along with increased expression of *P5CS*, *OsLEA3*, *Rab16A*, *OsNHX1*, and *OsSOS1* genes. The expression of *Slender Rice1*(*SLR1*), the rice homolog of *Arabidopsis DELLA*, involved in coordinating plant growth under stress, was highly induced in *OsMYB91* overexpressed line, while the salt-induced *SLR1* expression was suppressed by RNA interference (RNAi) approach (Zhu et al. 2015).

7.3.3 WRKY TFs

WRKY belongs to the large family of TFs characterized by the presence of WRKY domain of about 60 amino acids in length, containing WRKY sequence at the amino terminal and a putative C_2H_2 or C_2HC zinc finger motif at carboxyl terminal ends. WRKY TFs bind to the conserved cis-elements called W box, characterized by (T) (T)TGAC(C/T) sequence at the promoter region of their target genes (Rushton et al. 2010; Banerjee and Roychoudhury 2015). Based on the number of WRKY domain and type of zinc finger-like motif, WRKY TFs can be divided into three groups. Group I contains two WRKY domain with C_2H_2 motif, and group II and III contain one WRKY domain with C_2H_2 or C2HC zinc finger-like motif. The large group II can be further divided into five subgroups (IIa, IIb, IIc, IId, and IIe) on the basis of peptide sequence (Rushton et al. 2010). *WRKY* genes are known to be involved in multiple biological processes including abiotic stress tolerance (Wang et al. 2016c).

Recent study has shown that *JcWRKY*, a group II WRKY TF from biofuel crop *Jatropha curcas*, enhanced salt tolerance in transgenic tobacco via salicylic acid (SA)-induced ROS homeostasis (Agarwal et al. 2016). Transgenic *Arabidopsis* overexpressing wheat *WRKY* genes, *TaWRKY1* and *TaWRKY33* enhanced drought and high-temperature tolerance; however, drought tolerance potential of *TaWRKY33* was found to be higher than *TaWRKY1* in terms of water retention ability compared to wild type under stress condition (He et al. 2016). Qin et al. (2015) identified *TaWRKY93*, a wheat group II *WRKY* gene as a new positive regulator of abiotic stress, as it increased multiple abiotic stress tolerance like salinity, drought, and low-temperature stress via modulating osmotic adjustment, maintaining membrane stability, and increasing the expression of stress-responsive genes. They have also shown that *TaWRKY93* conferred stress tolerance by providing superior agricultural traits like forming longer primary roots or more lateral roots in transgenic plants. Wang et al. (2015b) reported that a group I WRKY TF of wheat, *TaWRKY44*, was upregulated by treatments with PEG6000, NaCl, cold (4 °C), and abscisic acid (ABA) and acted as a positive regulator in drought/salt/osmotic stress responses via modulating cellular antioxidative system and expression of stress-associated genes. Another study has shown that transgenic *Arabidopsis* plants overexpressing wheat *TaWRKY2* and *TaWRKY19* conferred tolerance to salt, drought, and freezing tolerance by influencing the downstream stress-responsive gene expression when compared with the wild-type plant (Niu et al. 2012). Chu et al. (2015) isolated and characterized the function of cotton (*Gossypium hirsutum*) *GhWRKY41* in drought stress tolerance. Transgenic tobacco overexpressing *GhWRKY41* conferred drought tolerance by enhancing stomatal closure and ROS scavenging mechanism with the increased expression of antioxidant genes. Gong et al. (2015) isolated and characterized the function of group III WRKY gene (*FcWRKY70*) from *Fortunella crassifolia*. The expression of *FcWRKY70* was highly induced by drought and ABA but slightly by salt stress. They have also shown that transgenic tobacco and lemon conferred enhanced tolerance to dehydration and drought stress via modulating putrescine production by regulating *ADC* expression. Shen et al. (2012) reported that rice *OsWRKY30* interacted with MAPKs and were phosphorylated by *OsMPK3*, *OsMPK7*, and *OsMPK14*. Transgenic rice overexpressing *OsWRKY30* showed increased drought tolerance. The overexpression of the mutant *OsWRKY30AA* (all serine residue followed by proline residue were replaced by alanine) was unable to improve drought tolerance, suggesting that the biological activity of *OsWRKY30* was dependent on phosphorylation status by MAPKs.

7.3.4   NAC TFs

NAC family consists of large number of members and specific to higher plants. NAC TF is named after three letter of genes, petunia hybrid *No apical meristem* (*NAM*), *Arabidopsis transcription activation factor 1/2* (*ATAF 1* and *ATAF 2*), and *CUP-SHAPED COTYLEDON 2* (*CUC 2*). The protein of this family harbor conserved NAC domain at the N-terminal region for DNA binding and variable transcriptional activator domain at C-terminal site. Whole genome sequencing have identified *NAC* genes in a number of plant species like rice, *Arabidopsis*, foxtail millet (*Setaria italica*), and soybean (*Glycine max*) (Nuruzzaman et al. 2010; Le et al. 2011; Puranik et al. 2013). NAC TFs are known to be associated with a number of biological processes like development, fruit ripening, hormone signaling, and biotic and abiotic stress responses (Puranik et al. 2012). NAC TFs have also been genetically modified for crop improvement under abiotic stresses. In rice, seven *NAC* genes have been extensively studied and characterized through transgenic analysis for involvement in salt and drought tolerance (Jeong et al. 2010; Redillas et al. 2012; Jeong et al. 2013; Chen et al. (2014b). *OsNAP*, member of the NAC TF family, was highly induced by salt, drought, and low temperature, conferred salt and drought tolerance in overexpressed rice lines without growth retardation, and improved yield at reproductive stage by modulating many stress-responsive genes, like *OsPP2C06/ OsABI2*, *OsPP2C09*, *OsPP2C68*, and *OsSalT*, and some stress-inducible transcription factors, viz., *OsDREB1A*, *OsMYB2*, *OsAP37*, and *OsAP59* Chen et al. (2014b). Recently, a rice *NAC* gene, *ONAC022*, was identified as a positive regulator of salt and drought stress tolerance. Transgenic rice lines overexpressing *ONAC022* increased salt and drought tolerance by accumulating less Na$^+$ in roots than shoots, lowering water loss by decreasing transpiration rate, causing more accumulation of proline and soluble sugar and increased expression of *OsRAB21*, *OsLEA3*, *OsP5CS1*, and other stress-responsive genes (Hong et al. 2016). Lee et al. (2016) identified *OsNAC6* as positive regulator of drought tolerance by influencing root architecture including increased root number and root diameter under drought stress and also via nicotianamine biosynthesis pathway. Root-specific *OsNAC6* overexpressed rice line showed higher yield as compared to wild type under drought stress.

7.3.5 bZIP TFs

bZIP (basic leucine zipper) group of TFs is characterized by the presence conserved bZIP domain of basic amino acids at the N-terminal site for DNA binding and with leucine zipper at the C-terminal for dimerization of TFs. bZIP group members are also involved in abiotic stress responses like high salinity, drought, and cold stress. Studies have shown that such genes are induced by ABA and regulate the expression of many stress-responsive genes by recognizing and binding to conserved ABRE (abscisic acid-responsive elements) at their promoter region (Roychoudhury et al. 2013). Many members of bZIP TFs have been identified in different plant species, like 75 in *Arabidopsis* (Jakoby et al. 2002), 89 in rice (Nijhawan et al. 2008), 125 in maize (Wei et al. 2012), 96 in *Brachypodium distachyon* (Liu and Chu 2015), 119 in *Brassica oleracea* (Hwang et al. 2016), and 69 in *Solanum lycopersicum* (Li et al. 2015a). bZIP TFs can be divided into 11 groups (I–XI) (Nijhawan et al. 2008) or 13 groups (A, B, C, D, E, F, G, H, I, J, K, L, and S) (Guedes Corrêa et al. 2008). Group A members are also known as ABRE-binding factors (ABFs/AREBs), involved in plant responses to dehydration and salt stress (Banerjee and Roychoudhury 2017). Among group A members, rice *TRAB1*, *OsbZIP23*, *OsbZIP46*, *OsbZIP72*, *OsbZIP12/OsABF1*, and *Arabidopsis ABI5* have been found to be involved in salinity and dehydration stress tolerance (Hobo et al. 1999; Kang et al. 2002; Xiang et al. 2008; Lu et al. 2009; Hossain et al. 2010; Tang et al. (2012a). Transgenic analysis through overexpression and RNAi knockdown approach revealed that rice *OsbZIP71*, a group S1 bZIP TF, enhanced salt and drought tolerance by interacting with group C bZIP members, viz., *OsbZIP15*, *OsbZIP20*, *OsbZIP33*, and *OsbZIP88*, or forming heterodimer with *OsMyb4*. These complexes then bind to the promoters of *OsNHX1* and *COR413-TM* containing G-box elements which lead to enhanced tolerance to drought and salt stresses (Liu et al. 2014). Transgenic alfalfa (*Medicago sativa* L.) overexpressing *Arabidopsis ABSCISIC ACID-RESPONSIVE ELEMENT-BINDING FACTOR 3* (*ABF3*), under the control of sweet potato oxidative stress-inducible *SWPA2* promoter, increased drought tolerance by lowering transpiration rate and endogenous ROS content in transgenic plants (Wang et al. 2016d).

7.4 Late Embryogenesis Abundant Protein

Late Embryogenesis Abundant (LEA) proteins, as their name suggests, are highly expressed during later phase of embryo development. This group of proteins is not only induced under water-limiting condition but also accumulated during seed or pollen development or some stages of shoot and root development when water is limiting. LEA proteins are hydrophilic in nature and contain high proportion of glycine or small amino acids, forming a hydrophilic structure with high stability against heat. Compiling evidence indicates that they are involved in various functions, including protection of cellular structures from the effects of water loss and desiccation, protection of proteins from stress-induced damage, sequestration of ions, and folding of denatured proteins (Banerjee and Roychoudhury 2016). According to recent classification based on the presence of specific motifs conserved across species, LEA family can be divided into seven subfamilies (Battaglia et al. 2008). Genome-wide analysis has led to the identification of *LEA* genes in different plant species, like 51 in *Arabidopsis*, 34 in rice, 108 in *Brassica napus*, and 23 in Moso bamboo (*Phyllostachys edulis*) (Wang et al. 2007; Hundertmark and Hincha 2008; Huang et al. 2016b; Liang et al. 2016b). Transgenic rice overexpressing wheat group I *LEA* gene, *PMA1959*, and group II gene *PMA80* conferred salinity and dehydration tolerance by enhancing cell membrane integrity and growth performance under stress condition (Chen et al. 2002). Dalal et al. (2009) isolated and characterized group IV *LEA* gene, *BnLEA4-1*, from *Brassica napus* for involvement in salt and drought tolerance. Overexpression of *BnLEA4-1*, either constitutively under CaMV35S promoter or stress-inducible *RD29A* promoter, enhanced tolerance to salt and drought stresses at vegetative stages of development. Xiao et al. (2007) isolated and functionally characterized rice *OsLEA3-1* during drought stress under field condition. Increased grain yield under drought condition was noted in transgenic rice when *OsLEA3-1* was overexpressed either constitutively or under stress-inducible promoter. Yu et al. (2016) identified a group I *LEA* gene, *OsEm1*, in rice and found that the expression of *OsEm1* was induced by multiple abiotic stresses like ABA, polyethylene glycol (PEG), cold, and NaCl. Moreover, transgenic rice line over-expressing *OsEm1* conferred drought tolerance with increased expression of *LEA* genes, including *RAB16A*, *RAB16C*, and *LEA3*, leading to better survival rate of transgenic plants at the vegetative stage. Rice *RAB16A*, a group II *LEA* gene also known as dehydrins, rendered salt tolerance when overexpressed in rice and tobacco along with increased accumulation of osmolytes like proline and polyamines and efficient antioxidative machinery (RoyChoudhury et al. 2007; Ganguly et al. 2012). Recent study has shown that four dehydrin genes of *Prunus mume* conferred toler-ance to drought stress when overexpressed in tobacco plant by lowering MDA level and increased relative water content and cell membrane integrity (Bao et al. 2017). *SmLEA2*, a group II *LEA* gene of *Salvia miltiorrhiza*, has been isolated and charac-terized through overexpression and RNAi approach. Transgenic *S. miltiorrhiza* over-expressing *SmLEA2* enhanced salt and drought tolerance by significantly increasing SOD activity and reduced level of lipid peroxidation and efficient growth perfor-mance compared to RNAi and wild-type plants (Wang et al. 2017). Overexpression of barley group III *LEA* gene, *HVA1* in rice, maize, oat, and mulberry (*Morus indica*) enhanced prolonged drought and salt tolerance in transgenic lines by retain-ing more water status in cell and minimizing cell membrane injury under drought stress (Xu et al. 1996; Babu et al. 2004; Oraby et al. 2005; Checker et al. 2012; Nguyen and Sticklen 2013). Ke et al. (2016b) identified four rice group III *LEA* genes, *OsG3LEA-47.3*, *OsG3LEA-41.9*, *OsG3LEA-20.5*, and *OsG3LEA-24.5*, which showed tissue-specific differential expression pattern during developmental stages and under abiotic stresses. Transgenic analysis revealed that *OsG3LEA-47.3* played an important role in conferring cross-tolerance for alleviating the detrimental effects of drought and heat treatment. The *OsG3LEA-41.9* and *OsG3LEA-24.5* conferred resistance to drought and heat stress, respectively. However, *OsG3LEA-20.5* was unable to confer tolerance in transgenic *Arabidopsis* under heat or drought treat-ment. The atypical hydrophobic group V LEA protein, *AdLEA*, from *Arachis diogoi* rendered tolerance to salt, dehydration, and oxidative stress when ectopically over-expressed in tobacco plants showing higher chlorophyll content and reduced lipid peroxidation as compared to wild-type plants (Sharma et al. 2016).

7.5 Ion Transporter

Soil salinity causes higher accumulation of Na^+ and Cl^- ions into plant cell which is responsible for membrane depolarization and interferes with cellular metabolism including suppression of enzymatic function and protein biosynthesis. Ion homeo-stasis by concerted action of ion transporter is one of the important adaptive responses of a plant under salt stress (Fig. 7.3). In plant cell, tonoplast-localized NHX1 and plasma membrane-localized SALT OVERLY SENSITIVE 1 (SOS1), also known as Na$^+$/H$^+$ antiporter, have played a pivotal role in cellular ion homeostasis. In plants, NHX1 antiporters sequester Na$^+$ ions into the vacuole, thus minimizing cellular Na$^+$ ion toxicity and subsequently increasing osmotic potential for water uptake. Numerous studies have been made to generate salt-tolerant transgenic plants like *Arabidopsis*, *Brassica*, rice, maize, tomato, *Arachis hypogaea*, and tobacco using *Arabidopsis* vacuolar antiporter gene *AtNHX1* (Apse et al. 1999; Zhang et al. 2001; Yin et al. 2004; Chen et al. 2007; Leidi et al. 2010; Ahsan et al. 2011; Zhou et al. 2011; Banjara et al. 2012). *Arabidopsis SOS1* that encode a plasma membrane-localized Na$^+$/H$^+$ antiporter can transport Na$^+$ ion across the plasma membrane from cytoplasm to apoplastic space in exchange of H$^+$ ion (Qiu et al. 2002). Overexpression of *AtSOS1* in *Arabidopsis* and tobacco enhanced salt tolerance by maintaining higher K$^+$/Na$^+$ ratio (Shi et al. 2003; Yue et al. 2012). However, in all earlier studies, the transgenic plants overexpressing either *NHX1* or *SOS1* were not able to tolerate more than 200 mM NaCl stress. Recently, gene stacking approach has been implemented for co-expressing *AtNHX1* and *AtSOS1* in *Arabidopsis* for improving salt tolerance. Transgenic *Arabidopsis* could tolerate up to 250 mM NaCl concentration without compromising yield as compared to wild-type plant (Pehlivan et al. 2016).

7.6   Osmolytes in Salt and Drought Tolerance

Osmotic stress induced by high salinity and drought is one of the major causes for cellular and metabolic dysfunction which ultimately affects plant growth and productivity. To cope up with such harsh environmental conditions, plants alter physiological, molecular, and metabolic function to produce low molecular weight, electrochemically neutral small molecules, maintaining cellular osmotic homeostasis by lowering osmotic potential of a cell under stress (Fig. 7.3). Such compounds are known as compatible solutes or osmolytes, which are nontoxic and do not interfere with cellular metabolism when accumulated in cytosol. The common forms of osmolytes include proline, glycine betaine, and polyamines (Khan et al. 2015). Numerous studies have been conducted to engineer plants with genes for osmolyte biosynthesis (Sah et al. 2016). In this section, we have discussed the recent findings of transgenic research using genes encoding biosynthesis of important osmolytes for improving plant growth and yield under abiotic stresses.

7.6.1   Proline

Proline (Pro), one of the essential amino acids in plants, accumulates at higher level during salt and drought. The main function of Pro is to stabilize protein structure and scavenge free radicals (Biedermannova et al. 2008). In plants, Pro is synthesized from glutamatic acid via the intermediate Δ^1-*pyrroline*-5-carboxylate (P5C),

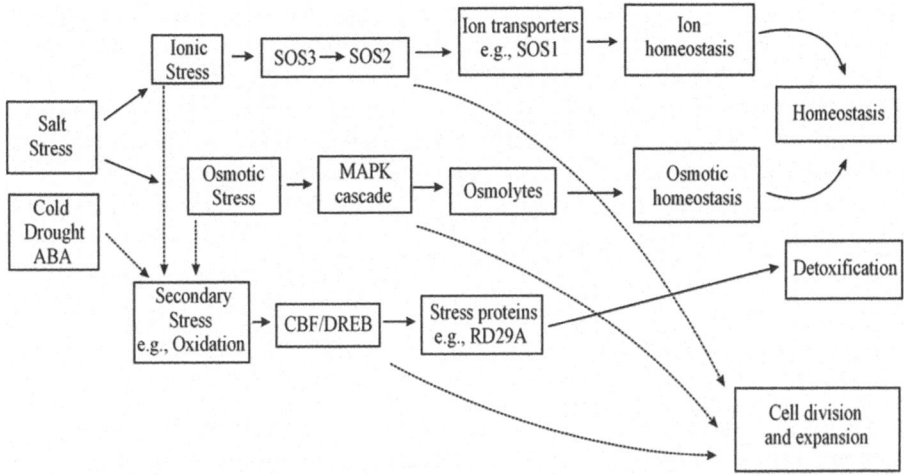

Fig. 7.3 The three aspects of salt tolerance in plants, homeostasis, detoxification, and growth control, and the pathways that interconnect them

catalyzed by Δ^1-*pyrroline*-5-carboxylate synthetase (P5CS) and Δ^1-*pyrroline*-5-carboxylate reductase (P5CR). An alternative precursor for Pro biosynthesis is ornithine, which can be transaminated to P5C by ornithine-D-aminotransferase (OAT), a mitochondrially located enzyme (Hayat et al. 2012). Pro homeostasis is also maintained by Pro-catabolizing enzyme proline dehydrogenase (PDH) which is repressed during dehydration and activated with rehydration. Pro is sequentially oxidized to P5C and then to glutamate by the action of enzymes PDH and pyrroline-5-carboxylate dehydrogenase (P5CDH), respectively (Szabados and Savouré 2010; Roychoudhury et al. 2015). Genetic analysis revealed that P5CS is the rate-limiting enzyme in Pro biosynthesis. Transgenic tobacco overexpressing moth bean *P5CS* produced higher concentration of P5CS enzyme and more Pro than wild-type plant which led to increased tolerance against drought, together with increased root biomass and flower development in transgenic lines under drought stress (Kishor et al. 1995). Recently, a study has been conducted to manipulate Pro biosynthesis by introducing a mutation at Glu155 residue of P5CS that relieves feedback inhibition by Pro. In this experiment, E155G mutation (glutamate at position 155 changed to glycine) was introduced in *Arabidopsis* P5CS and the mutated gene overexpressed under the control of CaMV35S promoter in wild-type *Nicotiana plumbaginifolia* (P2) and in the RNa mutant (over produced Pro and have mutated *P5CS* with reduced feedback inhibition of Pro biosynthesis). Enhanced salt tolerance was noted in transgenic plants with higher expression of *P5CS1* gene along with greater accumulation of Pro and exhibiting better growth performance compared to control P2 plants. However, in transgenic RNaP5CS lines, growth rate and Pro content did not significantly differ under stress as compared to RNa control plant, possibly because RNa plants already have one copy of mutated *P5CS* with higher endogenous Pro level. Introduction of second mutated

P5CS was not able to increase Pro level beyond their threshold level in transgenic plants (Ahmed et al. 2015). Another recent study showed that transgenic tobacco co-expressing two rice *P5CS* ortholog, *OsP5CS1* and *OsP5CS2*, conferred abiotic stress tolerance with higher accumulation of Pro and reduced oxidative damage as compared to wild-type plant (Xia et al. 2014). Several transgenic approaches have been undertaken to enhance osmotic stress tolerance in different plant species by genetic manipulation of *P5CS* gene (Kumar et al. 2010; Kim and Nam 2013; Chen et al. 2013b; Ibragimova et al. 2015).

7.6.2 Glycine Betaine

In plants, glycine betaine (GB) is one of the most important osmolytes which provide protection to osmotic stress by stabilizing membrane and photosynthetic machinery under salt, drought, and cold stress (Chen and Murata 2011; Wani et al. 2013). GB is synthesized from choline in two-step oxidation processes. In the first step, choline is oxidized to betaine aldehyde, catalyzed by the action of choline monooxygenase (CMO), which is then oxidized to GB by NAD^+-dependent betaine aldehyde dehydrogenase (BADH). Another enzyme choline oxidase A (codA), from soil bacterium *Arthrobacter globiformis*, converts choline into GB in single step (Khan et al. 2015). Some plant species like *Arabidopsis*, tobacco, tomato, and rice do not contain GB under normal or stress condition and are known as non-accumulators of GB. Therefore, they are the potential targets for genetic engineering to improve stress tolerance. Among cereal crops, rice do not accumulate GB naturally due to the truncated transcripts for GB biosynthesizing enzyme (BADH) (Niu et al. 2007; Roychoudhury and Banerjee 2016). Transgenic approach has been implemented to overproduce GB in rice by overexpressing *codA* gene from *A. globiformis*. Genetically engineered rice with targeted expression of *codA* to chloroplast (ChlCOD plants) and to the cytosol (CytCOD plants) was developed, both exhibiting higher accumulation of GB; however, ChlCOD plants showed enhanced tolerance to salt stress by providing better protection to photosynthetic machinery. This suggests that GB provides protection to osmotic stress by stabilizing the structure and function of photosystem II complex against inactivation caused by salt stress (Sakamoto et al. 1998; Mohanty et al. 2002; Su et al. 2006). Recently, transgenic poplar plants were generated by overexpressing *codA* gene under the control of stress-inducible promoter SWPA2. Transgenic lines conferred higher tolerance against salt and drought stress with increased efficiency of photosystem II at vegetative stage (Ke et al. 2016a). Transgenic tobacco and Japonica rice variety Nipponbare, overexpressing *OsBADH1* from *indica* rice under control of the maize ubiquitin promoter, conferred salt tolerance in transgenic lines by increasing the expression of *OsBADH1* gene and OsBADH1 enzyme production, resulting in the higher accumulation of GB (Hasthanasombut et al. 2010, 2011). On the other hand, transgenic *OsBADH1* RNAi knockdown line showed reduced tolerance to salt, drought, and cold stress with higher MDA and H_2O_2 level which greatly affected crop productivity. Transgenic *Arabidopsis* overexpressing *ScBADH* gene from halophyte *Suaeda corniculata* conferred salt and drought tolerance over wild type by increased accumulation of GB and Pro with efficient antioxidative enzyme activity (Wang et al. 2016b).

7.6.3   Polyamines

Polyamines (PAs), including tetraamine spermine (Spm^{4+}), triamine spermidine (Spd^{3+}), and their diamine precursor, putrescine (Put^{2+}), are polycationic, ubiquitous molecules that play an important role in abiotic stress tolerance (Liu et al. 2015a). PA biosynthesis is initiated by the synthesis of Put via decarboxylation of arginine and ornithine, catalyzed by arginine decarboxylase (ADC) and ornithine decarboxylase (ODC), respectively. Following Put synthesis, Spd and Spm are synthesized by sequential addition of amino-propyl group to PA biosynthetic pathway, catalyzed by the enzymatic action of spermidine synthase (SPDS) and spermine synthase (SPMS), respectively. The amino-propyl group is generated from decarboxylated S-adenosylmethionine (dcAdoMet) which is synthesized from S-adenosylmethionine (SAM) by the enzymatic action of S-adenosylmethionine decarboxylase (SAMDC). Like other osmolytes, PA biosynthetic pathway has been genetically manipulated to increase PA level in transgenic crops which led to increased abiotic stress tolerance (Roychoudhury and Das 2014). Capell et al. (2004) modulated PA biosynthetic pathway to investigate their response during drought stress in transgenic rice. *Datura stramonium ADC* gene was overexpressed in rice under the control of maize *ubiquitin 1* (*Ubi-1*) promoter. Transgenic rice exhibited better tolerance to drought stress with higher steady state of rice *SAMDC* mRNA level, correlated with increased accumulation of PAs (Spm, Spd, and Put) in overexpressed lines. Transgenic tobacco and *Arabidopsis* overexpressing *PtADC* from *Poncirus trifoliata* enhanced salt, drought, and cold tolerance with increased Put level, thereby modulating ROS scavenging pathway and increasing root length (Wang et al. 2011a, b). Transgenic *Arabidopsis* overexpressing oat *ADC* gene under the control of a stress-inducible promoter (pRD29A) conferred drought and cold tolerance with increased level of endogenous Put, which also directly or indirectly affected ABA metabolism (Alet et al. 2011). *SAMDC* is one of the key regulatory genes in PA biosynthetic pathway, influencing endogenous PAs under stress. To characterize its function on abiotic stress tolerance in different plant species like rice, tobacco, *Arabidopsis*, and tomato, transgenic approaches have been implemented to overexpress *SAMDC* from different sources, and overexpressed lines showed enhanced tolerance to multiple stresses compared to wild-type plants (Wi et al. 2006; Cheng et al. 2009; Liu et al. 2017). Recently *SAMDC* gene from *Leymus chinensis* has been isolated and functionally characterized. Transgenic *Arabidopsis* overexpressing *LcSAMDC1* exhibited increased accumulation of Spm, Pro, and chlorophyll content and tolerance to salt and cold stress in transgenic lines (Liu et al. 2017). Transgenic *Arabidopsis*, potato, and pear (*Pyrus communis*), overexpressing *SPDS*, showed increased accumulation of Spd and attenuated susceptibility to salt and drought stress by modulating the activity of enzymatic and nonenzymatic antioxidants (Kasukabe 2004; Kasukabe et al. 2006; Wen et al. 2008, 2010; Pathak et al. 2014).

7.7   Antioxidants in Salt and Drought Tolerance

Oxidative stress caused by high salinity and drought leads to the generation of ROS which is detrimental for plant growth and survival. In order to maintain cellular redox homeostasis, plants have developed a complex system of ROS scavenging enzymes. The common antioxidant enzymes are superoxide dismutase (SOD), ascorbate peroxidase (APX), catalase (CAT), and glutathione reductase (GR) (Das and Roychoudhury 2014; Anjum et al. 2016). The SOD converts superoxide anion $(O_2^{\cdot-})$ to hydrogen peroxide (H_2O_2) and molecular oxygen. H_2O_2 is then finally scavenged by CAT and APX and converted into water and oxygen. Several transgenic studies involving single-gene modification of antioxidants or gene stacking strategies have been implemented to improve abiotic stress tolerance of plant. Transgenic potato overexpressing *Cu/Zn SOD* and *APX* under the control of stress-inducible promoter *SWPA2* conferred tolerance to salt stress (Yan et al. 2016). Simultaneous overexpression of *PaSOD* and *RaAPX* genes from *Potentilla atrosanguinea* and *Rheum australe*, respectively, in *Arabidopsis* enhanced salt tolerance with increased lignin biosynthesis, yield, and biomass production in transgenic plants under salt stress (Shafi et al. 2015). Similar gene stacking approach with antioxidant genes like *SOD*, *CAT*, and *APX* has been applied in different plant species for generating multiple abiotic stress-tolerant plants (Tang et al. 2006; Tseng et al. 2007; Ahmad et al. 2010; Diaz-Vivancos et al. 2013; Xu et al. 2013). Kaouthar et al. (2016) isolated and functionally characterized a novel manganese superoxide dismutase, *TdMnSOD*, from durum wheat (*Triticum turgidum*). The expression of *TdMnSOD* was induced by salt, osmotic, and H_2O_2-induced oxidative stress. Transgenic *Arabidopsis* overexpressing *TdMnSOD* conferred tolerance to multiple abiotic stresses with increased accumulation of antioxidative enzymes as compared to non-transformed plants. Overexpression of *Cu/Zn SOD* from different plant sources in transgenic *Arabidopsis*, tobacco, and *Arachis hypogaea* exhibited tolerance to high salt and drought stress (Jing et al. 2015; Negi et al. 2015; Liu et al. 2015b). Transgenic *Arabidopsis* overexpressing *OsAPXa* or *OsAPXb* and two *CAT* genes from rice enhanced salt tolerance by modulating the activity of ROS scavenging enzymes (Lu et al. 2007).

7.8   Conclusion

This chapter briefly summarizes the recent developments in transgenic research in rendering salt and drought tolerance in diverse plant species. High-throughput technologies, including whole genome sequencing, and genomics or proteomics approaches have led to the identification of specific groups of genes which are expressed differentially in timely and coordinated manner to enhance stress tolerance. Such novel genes can be used as candidates for overexpression or downregulation by RNAi or microRNA (miRNA) approach (Shriram et al. 2016). Although several transgenic plants have been developed so far, most of them have been tested either in controlled growth condition or greenhouse facility, and very few studies have provided a quantitative basis of stress tolerance. When such transgenic lines are transferred and tested under field conditions, the function of these genes often becomes suppressed so that the transgenics show decreased tolerance level. Under field condition, the plants in reality are also encountered by a multitude of stress factors, apart from salt and drought stress, so that the overall performance may not be satisfactory. Overexpression of a single gene governing tolerance to multiple stresses, gene stacking and gene pyramiding strategy through overexpression of multiple genes from diverse pathways, and overexpression of genes involved in biotic-abiotic cross talk appear to be more logical approaches in improving the overall performance of the transgenics in a more holistic manner. Tolerance to a combination of different stress conditions, particularly those that mimic the field environment, should therefore be the focus of future research programs aimed at developing transgenic crops and plants with enhanced tolerance to environmental conditions.

Acknowledgment The financial support from the Council of Scientific and Industrial Research (CSIR), Government of India, through the project [38(1387)/14/EMR-II] to Dr. Aryadeep Roychoudhury is gratefully acknowledged. The authors are thankful to University Grants Commission (UGC), Government of India, for providing Senior Research Fellowship to Saikat Paul.

References

Abass M, Morris PC (2013) The *Hordeum vulgare* signalling protein MAP kinase 4 is a regulator of biotic and abiotic stress responses. J Plant Physiol 170:1353–1359

Agarwal P, Dabi M, Sapara KK et al (2016) Ectopic expression of *JcWRKY* transcription factor confers salinity tolerance via salicylic acid signaling. Front Plant Sci 7:1541

Ahmad R, Kim YH, Kim MD et al (2010) Simultaneous expression of choline oxidase, superoxide dismutase and ascorbate peroxidase in potato plant chloroplasts provides synergistically enhanced protection against various abiotic stresses. Physiol Plant 138:520–533

Ahmed AAM, Roosens N, Dewaele E et al (2015) Overexpression of a novel feedback-desensitized Δ^1-pyrroline-5-carboxylate synthetase increases proline accumulation and confers salt tolerance in transgenic *Nicotiana plumbaginifolia*. Plant Cell Tissue Organ Cult 122:383

Ahsan M, Zafar AY, Iqbal J et al (2011) Enhanced expression of *AtNHX1*, in transgenic groundnut (*Arachis hypogaea* L.) improves salt and drought tolerence. Mol Biotechnol 49:250–256

Alcázar R, Altabella T, Marco F et al (2010) Polyamines: molecules with regulatory functions in plant abiotic stress tolerance. Planta 231:1237–1249

Alet AI, Sanchez DH, Cuevas JC et al (2011) Putrescine accumulation in *Arabidopsis thaliana* transgenic lines enhances tolerance to dehydration and freezing stress. Plant Signal Behav 6:278–286

Almadanim MC, Alexandre BM, Rosa MTG et al (2017) Rice calcium-dependent protein kinase *OsCPK17* targets plasma membrane intrinsic protein and sucrose-phosphate synthase and is required for a proper cold stress response. Plant Cell Environ. https://doi.org/10.1111/pce.12916

Anjum NA, Sharma P, Gill SS et al (2016) Catalase and ascorbate peroxidase-representative H_2O_2-detoxifying heme enzymes in plants. Environ Sci Pollut Res 23:19002–19029

Apse MP, Aharon GS, Snedden WA, Blumwald E (1999) Salt tolerance conferred by overexpression of a vacuolar Na^+/H^+ antiport in *Arabidopsis*. Science 285:1256–1258

Árnadóttir J, Chalfie M (2010) Eukaryotic mechanosensitive channels. Annu Rev Biophys 39:111–137

Asano T, Hakata M, Nakamura H et al (2011) Functional characterisation of *OsCPK21*, a calcium-dependent protein kinase that confers salt tolerance in rice. Plant Mol Biol 75:179–191

Asano T, Hayashi N, Kobayashi M et al (2012) A rice calcium-dependent protein kinase *OsCPK12* oppositely modulates salt-stress tolerance and blast disease resistance. Plant J 69:26–36

Ashraf M (2010) Inducing drought tolerance in plants: recent advances. Biotechnol Adv 28:169–183

Ashraf M, Akram NA (2009) Improving salinity tolerance of plants through conventional breeding and genetic engineering: an analytical comparison. Biotechnol Adv 27:744–752

Augustine SM, Ashwin Narayan J, Syamaladevi DP et al (2015) Overexpression of *EaDREB2* and pyramiding of *EaDREB2* with the pea DNA helicase gene (*PDH45*) enhance drought and salinity tolerance in sugarcane (*Saccharum* spp. hybrid). Plant Cell Rep 34:247–263

Babu RC, Zhang J, Blum A et al (2004) *HVA1* , a *LEA* gene from barley confers dehydration tolerance in transgenic rice (*Oryza sativa* L .) via cell membrane protection. Plant Sci 166:855–862

Baena-González E, Rolland F, Thevelein JM, Sheen J (2007) A central integrator of transcription networks in plant stress and energy signalling. Nature 448:938–942

Banerjee A, Roychoudhury A (2015) WRKY proteins: signaling and regulation of expression during abiotic stress responses. Sci World J 2015:807560

Banerjee A, Roychoudhury A (2016) Group II late embryogenesis abundant (LEA) proteins: structural and functional aspects in plant abiotic stress. Plant Growth Regul 79:1–17

Banerjee A, Roychoudhury A (2017) Abscisic-acid-dependent basic leucine zipper (bZIP) transcription factors in plant abiotic stress. Protoplasma 254:3–16

Banjara M, Zhu L, Shen G (2012) Expression of an *Arabidopsis* sodium / proton antiporter gene (*AtNHX1*) in peanut to improve salt tolerance. Plant Biotechnol Rep 6:59–67

Bao F, Du D, An Y et al (2017) Overexpression of *Prunus mume* Dehydrin genes in tobacco enhances tolerance to cold and drought. Front Plant Sci 8:151

Battaglia M, Olvera-Carrillo Y, Garciarrubio A et al (2008) The enigmatic LEA proteins and other hydrophilins. Plant Physiol 148:6–24

Biedermannova L, Riley KE, Berka K et al (2008) Another role of proline: stabilization interactions in proteins and protein complexes concerning proline and tryptophane. Phys Chem Chem Phys 10:6350–6359

Bouaziz D, Pirrello J, Charfeddine M et al (2013) Overexpression of *StDREB1* transcription factor increases tolerance to salt in transgenic potato plants. Mol Biotechnol 54:803–817

Cai G, Wang G, Wang L et al (2014) A maize mitogen-activated protein kinase kinase, *ZmMKK1*, positively regulated the salt and drought tolerance in transgenic *Arabidopsis*. J Plant Physiol 171:1003–1016

Cai H, Tian S, Dong H, Guo C (2015) Pleiotropic effects of *TaMYB3R1* on plant development and response to osmotic stress in transgenic *Arabidopsis*. Gene 558:227–234

Cao ZH, Zhang SZ, Wang RK et al (2013) Genome wide analysis of the apple MYB transcription factor family allows the identification of *MdoMYB121* gene confering abiotic stress tolerance in plants. PLoS One 8:e69955

Capell T, Bassie L, Christou P (2004) Modulation of the polyamine biosynthetic pathway in transgenic rice confers tolerance to drought stress. Proc Natl Acad Sci 101:9909–9914

Checker VG, Chhibbar AK, Khurana P (2012) Stress-inducible expression of barley *Hva1* gene in transgenic mulberry displays enhanced tolerance against drought , salinity and cold stress. Transgenic Res 21:939–957

Chen THH, Murata N (2011) Glycinebetaine protects plants against abiotic stress: mechanisms and biotechnological applications. Plant Cell Environ 34:1–20

Chen ZQ, Liu Q, Zhu YS, Li YX (2002) Wheat LEA genes *PMA80* and *PMA 1959*, enhance dehydration tolerance of transgenic rice (*Oryza sativa* L.). Mol Breed 10:71–82

Chen H, An R, Tang JH et al (2007) Over-expression of a vacuolar Na^+/H^+ antiporter gene improves salt tolerance in an upland rice. Mol Breed 19:215–225

Chen J, Xue B, Xia X, Yin W (2013a) A novel calcium-dependent protein kinase gene from *Populus euphratica*, confers both drought and cold stress tolerance. Biochem Biophys Res Commun 441:630–636

Chen JB, Yang JW, Zhang ZY et al (2013b) Two *P5CS* genes from common bean exhibiting different tolerance to salt stress in transgenic *Arabidopsis*. J Genet 92:461–469

Chen L, Wang QQ, Zhou L et al (2013c) *Arabidopsis* CBL-interacting protein kinase (*CIPK6*) is involved in plant response to salt/osmotic stress and ABA. Mol Biol Rep 40:4759–4767

Chen X, Huang Q, Zhang F et al (2014a) *ZmCIPK21*, a maize CBL-interacting kinase, enhances salt stress tolerance in *Arabidopsis thaliana*. Int J Mol Sci 15:14819–14834

Chen X, Wang Y, Lv B et al (2014b) The NAC family transcription factor *OsNAP* confers abiotic stress response through the ABA pathway. Plant Cell Physiol 55:604–619

Chen H, Liu L, Wang L et al (2016) *VrDREB2A*, a DREB-binding transcription factor from *Vigna radiata*, increased drought and high-salt tolerance in transgenic *Arabidopsis thaliana*. J Plant Res 129:263–273

Cheng L, Zou Y, Ding S et al (2009) Polyamine accumulation in transgenic tomato enhances the tolerance to high temperature stress. J Integr Plant Biol 51:489–499

Cheng L, Li S, Hussain J et al (2013) Isolation and functional characterization of a salt responsive transcriptional factor, *LrbZIP* from lotus root (*Nelumbo nucifera* Gaertn). Mol Biol Rep 40:4033–4045

Chu X, Wang C, Chen X et al (2015) The cotton *WRKY* gene *GhWRKY41* positively regulates salt and drought stress tolerance in transgenic *Nicotiana benthamiana*. PLoS One 10:e0143022

Chun J, Li F-S, Ma Y et al (2014) Cloning and characterization of a *SnRK2* gene from *Jatropha curcas* L. Genet Mol Res 13:10958–10975

Coello P, Hey SJ, Halford NG (2011) The sucrose non-fermenting-1-related (SnRK) family of protein kinases: potential for manipulation to improve stress tolerance and increase yield. J Exp Bot 62:883–893

Dalal M, Tayal D, Chinnusamy V, Bansal KC (2009) Abiotic stress and ABA-inducible Group 4 LEA from *Brassica napus* plays a key role in salt and drought tolerance. J Biotechnol 139:137–145

Das K, Roychoudhury A (2014) Reactive oxygen species (ROS) and response of antioxidants as ROS-scavengers during environmental stress in plants. Front Environ Sci 2:53

Dash M, Yordanov YS, Georgieva T et al (2017) Poplar *PtabZIP1-like* enhances lateral root formation and biomass growth under drought stress. Plant J 89:692–705

de Zelicourt A, Colcombet J, Hirt H (2016) The role of MAPK modules and ABA during abiotic stress signaling. Trends Plant Sci 21:677–685

Deng X, Hu W, Wei S et al (2013a) *TaCIPK29*, a CBL-interacting protein kinase gene from wheat, confers salt stress tolerance in transgenic tobacco. PLoS One 8:e69881

Deng X, Zhou S, Hu W et al (2013b) Ectopic expression of wheat *TaCIPK14*, encoding a calcineurin B-like protein-interacting protein kinase, confers salinity and cold tolerance in tobacco. Physiol Plant 149:367–377

Dey A, Samanta MK, Gayen S et al (2016a) Enhanced gene expression rather than natural polymorphism in coding sequence of the *OsbZIP23* determines drought tolerance and yield improvement in rice genotypes. PLoS One 11:e0150763

Dey A, Samanta MK, Gayen S, Maiti MK (2016b) The sucrose non-fermenting 1-related kinase 2 gene *SAPK9* improves drought tolerance and grain yield in rice by modulating cellular osmotic potential, stomatal closure and stress-responsive gene expression. BMC Plant Biol 16:158

Diaz-Vivancos P, Faize M, Barba-Espin G et al (2013) Ectopic expression of cytosolic superoxide dismutase and ascorbate peroxidase leads to salt stress tolerance in transgenic plums. Plant Biotechnol J 11:976–985

Ding Z, Li S, An X et al (2009) Transgenic expression of *MYB15* confers enhanced sensitivity to abscisic acid and improved drought tolerance in *Arabidopsis thaliana*. J Genet Genomics 36:17–29

Dow GJ, Berry JA, Bergmann DC (2014) The physiological importance of developmental mechanisms that enforce proper stomatal spacing in *Arabidopsis thaliana*. New Phytol 201:1205–1217

Dubos C, Stracke R, Grotewold E et al (2010) MYB transcription factors in *Arabidopsis*. Trends Plant Sci 15:573–581

Dubrovina AS, Kiselev KV, Khristenko VS, Aleynova OA (2015) *VaCPK20*, a calcium-dependent protein kinase gene of wild grapevine *Vitis amurensis* Rupr., mediates cold and drought stress tolerance. J Plant Physiol 185:1–12

Fang L, Su L, Sun X et al (2016) Expression of *Vitis amurensis NAC26* in *Arabidopsis* enhances drought tolerance by modulating jasmonic acid synthesis. J Exp Bot 67:2829–2845

Ganguly M, Datta K, Roychoudhury A, Gayen D, Sengupta DN, Datta SK (2012) Overexpression of *Rab16A* gene in indica rice variety for generating enhanced salt tolerance. Plant Signal Behav 7:502–509

Gao SQ, Chen M, Xu ZS et al (2011) The soybean *GmbZIP1* transcription factor enhances multiple abiotic stress tolerances in transgenic plants. Plant Mol Biol 75:537–553

Golldack D, Li C, Mohan H, Probst N (2014) Tolerance to drought and salt stress in plants: unraveling the signaling networks. Front Plant Sci 5:151

Gong X, Zhang J, Hu J et al (2015) *FcWRKY70*, a WRKY protein of *Fortunella crassifolia*, functions in drought tolerance and modulates putrescine synthesis by regulating arginine decarboxylase gene. Plant Cell Environ 38:2248–2262

Gosal SS, Wani SH, Kang MS (2009) Biotechnology and drought tolerance. J Crop Improv 23:19–54

Gu L, Liu Y, Zong X, et al (2010) Overexpression of maize mitogen-activated protein kinase gene, *ZmSIMK1* in *Arabidopsis* increases tolerance to salt stress. Mol Biol Rep 37:4067–4073

Guedes Corrêa LG, Riaño-Pachón DM, Guerra Schrago C, et al. (2008) The role of bZIP transcription factors in green plant evolution: Adaptive features emerging from four founder genes. PLoS One 3:e2944.

Gunapati S, Naresh R, Ranjan S et al (2016) Expression of *GhNAC2* from *G. herbaceum*, improves root growth and imparts tolerance to drought in transgenic cotton and *Arabidopsis*. Sci Rep 6:24978

Guo Y, Halfter U, Ishitani M, Zhu JK (2001) Molecular characterization of functional domains in the protein kinase *SOS2* t7hat is required for plant salt tolerance. Plant Cell 13:1383–1400

Guo Y, Qiu Q-S, Quintero FJ et al (2004) Transgenic evaluation of activated mutant alleles of *SOS2* reveals a critical requirement for its kinase activity and C-terminal regulatory domain for salt tolerance in *Arabidopsis thaliana*. Plant Cell 16:435–449

Halford NG, Hey S, Jhurreea D et al (2003) Metabolic signalling and carbon partitioning: role of Snf1-related (*SnRK1*) protein kinase. J Exp Bot 54:467–475

Hao Y-J, Wei W, Song Q-X et al (2011) Soybean NAC transcription factors promote abiotic stress tolerance and lateral root formation in transgenic plants. Plant J 68:302–313

Hao L, Wen Y, Zhao Y et al (2015) Wheat mitogen-activated protein kinase gene *TaMPK4* improves plant tolerance to multiple stresses through modifying root growth, ROS metabolism, and nutrient acquisitions. Plant Cell Rep 34:2081–2097

Hasthanasombut S, Ntui V, Supaibulwatana K et al (2010) Expression of Indica rice *OsBADH1* gene under salinity stress in transgenic tobacco. Plant Biotechnol Rep 4:75–83

Hasthanasombut S, Supaibulwatana K, Mii M, Nakamura I (2011) Genetic manipulation of Japonica rice using the *OsBADH1* gene from Indica rice to improve salinity tolerance. Plant Cell Tissue Organ Cult 104:79–89

Hayat S, Hayat Q, Alyemeni MN et al (2012) Role of proline under changing environments: a review. Plant Signal Behav 7:1456–1466

He Y, Li W, Lv J et al (2012) Ectopic expression of a wheat MYB transcription factor gene, *TaMYB73*, improves salinity stress tolerance in *Arabidopsis thaliana*. J Exp Bot 63:1511–1522

He G-H, Xu J-Y, Wang Y-X et al (2016) Drought-responsive WRKY transcription factor genes *TaWRKY1* and *TaWRKY33* from wheat confer drought and/or heat resistance in *Arabidopsis*. BMC Plant Biol 16:116

Hirayama T, Shinozaki K (2007) Perception and transduction of abscisic acid signals: keys to the function of the versatile plant hormone ABA. Trends Plant Sci 12:343–351

Hobo T, Kowyama Y, Hattori T (1999) A bZIP factor, *TRAB1*, interacts with *VP1* and mediates abscisic acid-induced transcription. Proc Natl Acad Sci U S A 96:15348–15353

Hong Y, Zhang H, Huang L et al (2016) Overexpression of a stress-responsive NAC transcription factor gene *ONAC022* improves drought and salt tolerance in rice. Front Plant Sci 7:4

Hossain M, Lee Y, Cho J-I et al (2010) The bZIP transcription factor *OsABF1* is an ABA responsive element binding factor that enhances abiotic stress signaling in rice. Plant Mol Biol 72:557–566

Hu DG, Li M, Luo H et al (2012) Molecular cloning and functional characterization of *MdSOS2* reveals its involvement in salt tolerance in apple callus and *Arabidopsis*. Plant Cell Rep 31:713–722

Hu DG, Ma QJ, Sun CH et al (2016) Overexpression of *MdSOS2L1*, a CIPK protein kinase, increases the antioxidant metabolites to enhance salt tolerance in apple and tomato. Physiol Plant 156:201–214

Huang Q, Wang Y (2016) Overexpression of *TaNAC2D* displays opposite responses to abiotic stresses between seedling and mature stage of transgenic *Arabidopsis*. Front Plant Sci 7:1754

Huang X-S, Liu J-H, Chen X-J (2010) Overexpression of *PtrABF* gene, a bZIP transcription factor isolated from *Poncirus trifoliata*, enhances dehydration and drought tolerance in tobacco via scavenging ROS and modulating expression of stress-responsive genes. BMC Plant Biol 10:230

Huang Q, Wang Y, Li B et al (2015) *TaNAC29*, a NAC transcription factor from wheat, enhances salt and drought tolerance in transgenic *Arabidopsis*. BMC Plant Biol 15:268

Huang L, Hong Y, Zhang H et al (2016a) Rice NAC transcription factor *ONAC095* plays opposite roles in drought and cold stress tolerance. BMC Plant Biol 16:203

Huang Z, Zhong X-J, He J et al (2016b) Genome-wide identification, characterization, and stress-responsive expression profiling of genes encoding LEA (late embryogenesis abundant) proteins in Moso Bamboo (*Phyllostachys edulis*). PLoS One 11:e0165953

Hundertmark M, Hincha DK (2008) LEA (late embryogenesis abundant) proteins and their encoding genes in *Arabidopsis thaliana*. BMC Genomics 9:118

Hwang I, Manoharan RK, Kang J et al (2016) Genome-wide identification and characterization of bZIP transcription factors in *Brassica oleracea* under cold stress. Biomed Res Int 2016:4376598

Ibragimova SM, Trifonova EA, Filipenko EA, Shymny VK (2015) Evaluation of salt tolerance of transgenic tobacco plants bearing with *P5CS1* gene of *Arabidopsis thaliana*. Genetika 51:1368–1375

Jakoby M, Weisshaar B, Dröge-Laser W et al (2002) bZIP transcription factors in *Arabidopsis*. Trends Plant Sci 7:106–111

Jeong JS, Kim YS, Baek KH et al (2010) Root-specific expression of *OsNAC10* improves drought tolerance and grain yield in Rice under field drought conditions. Plant Physiol 153:185–197

Jeong JS, Kim YS, Redillas MCFR et al (2013) *OsNAC5* overexpression enlarges root diameter in rice plants leading to enhanced drought tolerance and increased grain yield in the field. Plant Biotechnol J 11:101–114

Jia H, Wang C, Wang F et al (2015) *GhWRKY68* reduces resistance to salt and drought in transgenic *Nicotiana benthamiana*. PLoS One 10:e0120646

Jia H, Hao L, Guo X et al (2016) A Raf-like MAPKKK gene, *GhRaf19*, negatively regulates tolerance to drought and salt and positively regulates resistance to cold stress by modulating reactive oxygen species in cotton. Plant Sci 252:267–281

Jiang S, Zhang D, Wang L et al (2013) A maize calcium-dependent protein kinase gene, *ZmCPK4*, positively regulated abscisic acid signaling and enhanced drought stress tolerance in transgenic *Arabidopsis*. Plant Physiol Biochem 71:112–120

Jing X, Hou P, Lu Y et al (2015) Overexpression of copper/zinc superoxide dismutase from mangrove *Kandelia candel* in tobacco enhances salinity tolerance by the reduction of reactive oxygen species in chloroplast. Front Plant Sci 6:23

Joo J, Choi HJ, Lee YH et al (2013) A transcriptional repressor of the ERF family confers drought tolerance to rice and regulates genes preferentially located on chromosome 11. Planta 238:155–170

Joshi R, Wani SH, Singh B et al (2016) Transcription factors and plants response to drought stress: current understanding and future directions. Front Plant Sci 7:1029

Kang JY, Choi HI, Im MY, Kim SY (2002) *Arabidopsis* basic leucine zipper proteins that mediate stress-responsive abscisic acid signaling. Plant Cell 14:343–357

Kaouthar F, Ameny FK, Yosra K et al (2016) Responses of transgenic *Arabidopsis* plants and recombinant yeast cells expressing a novel durum wheat manganese superoxide dismutase *TdMnSOD* to various abiotic stresses. J Plant Physiol 198:56–68

Kasukabe Y (2004) Overexpression of spermidine synthase enhances tolerance to multiple environmental stresses and up-regulates the expression of various stress-regulated genes in transgenic *Arabidopsis thaliana*. Plant Cell Physiol 45:712–722

Kasukabe Y, He L, Watakabe Y et al (2006) Improvement of environmental stress tolerance of sweet potato by introduction of genes for spermidine synthase. Plant Biotechnol 23:75–83

Ke Q, Wang Z, Ji CY et al (2016a) Transgenic poplar expressing *codA* exhibits enhanced growth and abiotic stress tolerance. Plant Physiol Biochem 100:75–84

Ke YT, Lu CA, Wu SJ, Yeh CH (2016b) Characterization of Rice Group 3 *LEA* genes in developmental stages and under abiotic stress. Plant Mol Biol Report 34:1003–1015

Kerr TCC, Abdel-Mageed H, Aleman L et al (2017) Ectopic expression of two *AREB/ABF* orthologs increase dehydration tolerance in cotton (*Gossypium hirsutum*). Plant Cell Environ. https://doi.org/10.1111/pce.12906

Khan MS, Ahmad D, Khan MA (2015) Utilization of genes encoding osmoprotectants in transgenic plants for enhanced abiotic stress tolerance. Electron J Biotechnol 18:257–266

Kim G-B, Nam Y-W (2013) A novel Δ1-pyrroline-5-carboxylate synthetase gene of *Medicago truncatula* plays a predominant role in stress-induced proline accumulation during symbiotic nitrogen fixation. J Plant Physiol 170:291–302

Kim CY, Vo KTX, Nguyen CD et al (2016) Functional analysis of a cold-responsive rice WRKY gene, *OsWRKY71*. Plant Biotechnol Rep 10:13–23

Kishor PBK, Hong ZL, Miao GH et al (1995) Overexpression of ð-Pyrroline-5-carboxylate synthetase increases proline production and confers osmotolerance in transgenic plants. Plant Physiol 108:1387–1394

Kobayashi Y, Yamamoto S, Minami H et al (2004) Differential activation of the rice sucrose nonfermenting1-related protein kinase2 family by hyperosmotic stress and abscisic acid. Plant Cell 16:1163–1177

Kong X, Pan J, Zhang M et al (2011) *ZmMKK4*, a novel group C mitogen-activated protein kinase kinase in maize (*Zea mays*), confers salt and cold tolerance in transgenic *Arabidopsis*. Plant Cell Environ 34:1291–1303

Kudla J, Batistic O, Hashimoto K (2010) Calcium signals: the lead currency of plant information processing. Plant Cell 22:541–563

Kudo M, Kidokoro S, Yoshida T et al (2017) Double overexpression of *DREB* and *PIF* transcription factors improves drought stress tolerance and cell elongation in transgenic plants. Plant Biotechnol J 15:458–471

Kumar K, Sinha AK (2013) Overexpression of constitutively active mitogen activated protein kinase kinase 6 enhances tolerance to salt stress in rice. Rice 6:25

Kumar V, Shriram V, Kishor PBK et al (2010) Enhanced proline accumulation and salt stress tolerance of transgenic indica rice by over-expressing *P5CSF129A* gene. Plant Biotechnol Rep 4:37–48

Kumar MN, Jane W-N, Verslues PE (2013) Role of the putative osmosensor *Arabidopsis* histidine kinase 1 in dehydration avoidance and low-water-potential response. Plant Physiol 161:942–953

Le DT, Nishiyama R, Watanabe YA et al (2011) Genome-wide survey and expression analysis of the plant-specific NAC transcription factor family in soybean during development and dehydration stress. DNA Res 18:263–276

Lee D-K, Chung PJ, Jeong JS et al (2016) The rice *OsNAC6* transcription factor orchestrates multiple molecular mechanisms involving root structural adaptions and nicotianamine biosynthesis for drought tolerance. Plant Biotechnol J:1–11

Leidi EO, Barragán V, Rubio L et al (2010) The *AtNHX1* exchanger mediates potassium compartmentation in vacuoles of transgenic tomato. Plant J 61:495–506

Li Q, Liu J, Tan D et al (2013) A genome-wide expression profile of salt-responsive genes in the apple rootstock *Malus zumi*. Int J Mol Sci 14:21053–21070

Li C, Chang PP, Ghebremariam KM et al (2014a) Overexpression of tomato *SpMPK3* gene in *Arabidopsis* enhances the osmotic tolerance. Biochem Biophys Res Commun 443:357–362

Li JB, Luan YS, Yin YL (2014b) *SpMYB* overexpression in tobacco plants leads to altered abiotic and biotic stress responses. Gene 547:145–151

Li D, Fu F, Zhang H, Song F (2015a) Genome-wide systematic characterization of the bZIP transcriptional factor family in tomato (*Solanum lycopersicum* L.). BMC Genomics 16:771

Li J b, Luan Y s, Liu Z (2015b) Overexpression of *SpWRKY1* promotes resistance to *Phytophthora nicotianae* and tolerance to salt and drought stress in transgenic tobacco. Physiol Plant 155:248–266

Li X, Feng B, Zhang F et al (2016) Bioinformatic analyses of subgroup-a members of the wheat bZIP transcription factor family and functional identification of *TabZIP174* involved in drought stress response. Front Plant Sci 7:1643

Li K, Xing C, Yao Z, Huang X (2017a) *PbrMYB21*, a novel MYB protein of *Pyrus betulaefolia*, functions in drought tolerance and modulates polyamine levels by regulating arginine decarboxylase gene. Plant Biotechnol J. https://doi.org/10.1111/pbi.12708

Li Y, Cai H, Liu P et al (2017b) *Arabidopsis MAPKKK18* positively regulates drought stress resistance via downstream *MAPKK3*. Biochem Biophys Res Commun 484:292–297

Liang C, Meng Z, Meng Z et al (2016a) *GhABF2*, a bZIP transcription factor, confers drought and salinity tolerance in cotton (*Gossypium hirsutum* L.). Sci Rep 6:35040

Liang Y, Xiong Z, Zheng J et al (2016b) Genome-wide identification, structural analysis and new insights into late embryogenesis abundant (*LEA*) gene family formation pattern in *Brassica napus*. Sci Rep 6:24265

Liao X, Guo X, Wang Q et al (2016) Overexpression of *MsDREB6.2* results in cytokinin-deficient developmental phenotypes and enhances drought tolerance in transgenic apple plants. Plant J 89:510–526

Liu X, Chu Z (2015) Genome-wide evolutionary characterization and analysis of bZIP transcription factors and their expression profiles in response to multiple abiotic stresses in *Brachypodium distachyon*. BMC Genomics 16:227

Liu J, Ishitani M, Halfter U et al (2000) The *Arabidopsis* thaliana *SOS2* gene encodes a protein kinase that is required for salt tolerance. Proc Natl Acad Sci 97:3730–3734

Liu H, Yang W, Liu D et al (2011) Ectopic expression of a grapevine transcription factor *VvWRKY11* contributes to osmotic stress tolerance in *Arabidopsis*. Mol Biol Rep 38:417–427

Liu X, Liu S, Wu J et al (2013) Overexpression of *Arachis hypogaea NAC3* in tobacco enhances dehydration and drought tolerance by increasing superoxide scavenging. Plant Physiol Biochem 70:354–359

Liu C, Mao B, Ou S et al (2014) *OsbZIP71*, a bZIP transcription factor, confers salinity and drought tolerance in rice. Plant Mol Biol 84:19–36

Liu J-H, Wang W, Wu H et al (2015a) Polyamines function in stress tolerance: from synthesis to regulation. Front Plant Sci 6:827

Liu ZB, Zhang WJ, Gong XD et al (2015b) A Cu/Zn superoxide dismutase from *Jatropha curcas* enhances salt tolerance of *Arabidopsis thaliana*. Genet Mol Res 14:2086–2098

Liu L, Zhang Z, Dong J, Wang T (2016a) Overexpression of *MtWRKY76* increases both salt and drought tolerance in *Medicago truncatula*. Environ Exp Bot 123:50–58

Liu X, Song Y, Xing F et al (2016b) *GhWRKY25*, a group I WRKY gene from cotton, confers differential tolerance to abiotic and biotic stresses in transgenic *Nicotiana benthamiana*. Protoplasma 253:1265–1281

Liu Z, Liu P, Qi D et al (2017) Enhancement of cold and salt tolerance of *Arabidopsis* by transgenic expression of the S-adenosylmethionine decarboxylase gene from *Leymus chinensis*. J Plant Physiol 211:90–99

Lovas Á, Bimbó A, Szabó L, Bánfalvi Z (2003) Antisense repression of *StubGAL83* affects root and tuber development in potato. Plant J 33:139–147

Lu Z, Liu D, Liu S (2007) Two rice cytosolic ascorbate peroxidases differentially improve salt tolerance in transgenic *Arabidopsis*. Plant Cell Rep 26:1909–1917

Lu G, Gao C, Zheng X, Han B (2009) Identification of *OsbZIP72* as a positive regulator of ABA response and drought tolerance in rice. Planta 229:605–615

Lu W, Chu X, Li Y et al (2013) Cotton *GhMKK1* induces the tolerance of salt and drought stress, and mediates defence responses to pathogen infection in transgenic *Nicotiana benthamiana*. PLoS One 8:e68503

Luo X, Bai X, Sun X et al (2013) Expression of wild soybean *WRKY20* in *Arabidopsis* enhances drought tolerance. J Exp Bot 64:2155–2169

Lv F, Zhang H, Xia X, Yin W (2014) Expression profiling and functional characterization of a CBL-interacting protein kinase gene from *Populus euphratica*. Plant Cell Rep 33:807–818

Lv Z, Wang S, Zhang F et al (2016) Overexpression of a novel NAC domain-containing transcription factor (*AaNAC1*) enhances the content of artemisinin and increases tolerance to drought and *Botrytis cinerea* in *Artemisia annua*. Plant Cell Physiol 57:1961–1971

Ma DM, Xu WR, Li HW et al (2014) Co-expression of the *Arabidopsis SOS* genes enhances salt tolerance in transgenic tall fescue (*Festuca arundinacea* Schreb.). Protoplasma 251:219–231

Mallikarjuna G, Mallikarjuna K, Reddy MK, Kaul T (2011) Expression of *OsDREB2A* transcription factor confers enhanced dehydration and salt stress tolerance in rice (*Oryza sativa* L.). Biotechnol Lett 33:1689–1697

Mao X, Zhang H, Tian S et al (2010) *TaSnRK2.4*, an SNF1-type serine/threonine protein kinase of wheat (*Triticum aestivum* L.), confers enhanced multistress tolerance in *Arabidopsis*. J Exp Bot 61:683–696

Mao X, Chen S, Li A et al (2014) Novel NAC transcription factor *TaNAC67* confers enhanced multi-abiotic stress tolerances in *Arabidopsis*. PLoS One 9:e84359

Meng X, Wang J-R, Wang G-D et al (2015) An R2R3-MYB gene, *LeAN2*, positively regulated the thermo-tolerance in transgenic tomato. J Plant Physiol 175:1–8

Mohanty A, Kathuria H, Ferjani A et al (2002) Transgenics of an elite indica rice variety Pusa Basmati 1 harbouring the *codA* gene are highly tolerant to salt stress. Theor Appl Genet 106:51–57

Monshausen GB, Gilroy S (2009) Feeling green: mechanosensing in plants. Trends Cell Biol 19:228–235

Nakashima K, Shinwari ZK, Sakuma Y et al (2000) Organization and expression of two *Arabidopsis DREB2* genes encoding DRE-binding proteins involved in dehydration- and high-salinity-responsive gene expression. Plant Mol Biol 42:657–665

Negi NP, Shrivastava DC, Sharma V, Sarin NB (2015) Overexpression of *CuZnSOD* from *Arachis hypogaea* alleviates salinity and drought stress in tobacco. Plant Cell Rep 34:1109–1126

Nguyen TX, Sticklen M (2013) Barley *HVA1* gene confers drought and salt tolerance in transgenic maize (*Zea mays* L.). Adv crop. Sci Technol 1:105

Nijhawan A, Jain M, Tyagi AK, Khurana JP (2008) Genomic survey and gene expression analysis of the basic leucine zipper transcription factor family in rice. Plant Physiol 146:333–350

Ning J, Li X, Hicks LM, Xiong L (2010) A Raf-like MAPKKK gene *DSM1* mediates drought resistance through reactive oxygen species scavenging in rice. Plant Physiol 152:876–890

Niu X, Zheng W, Lu B-R et al (2007) An unusual posttranscriptional processing in two betaine aldehyde dehydrogenase loci of cereal crops directed by short, direct repeats in response to stress conditions. Plant Physiol 143:1929–1942

Niu CF, Wei W, Zhou QY et al (2012) Wheat WRKY genes *TaWRKY2* and *TaWRKY19* regulate abiotic stress tolerance in transgenic *Arabidopsis* plants. Plant Cell Environ 35:1156–1170

Nuruzzaman M, Manimekalai R, Sharoni AM et al (2010) Genome-wide analysis of NAC transcription factor family in rice. Gene 465:30–44

Oraby HF, Ransom CB, Kravchenko AN, Sticklen MB (2005) Barley *HVA1* gene confers salt tolerance in R3 transgenic oat. Crop Sci 45:2218–2227

Osakabe Y, Yamaguchi-Shinozaki K, Shinozaki K, Tran LS (2014) ABA control of plant macroelement membrane transport systems in response to water deficit and high salinity. New Phytol 202:35–49

Pasquali G, Biricolti S, Locatelli F et al (2008) *Osmyb4* expression improves adaptive responses to drought and cold stress in transgenic apples. Plant Cell Rep 27:1677–1686

Pathak MR, Teixeira JA, Wani SH (2014) Polyamines in response to abiotic stress tolerance through transgenic approaches. GM Crop Food 5:87–96

Paz-Ares J, Ghosal D, Wienand U et al (1987) The regulatory c1 locus of *Zea mays* encodes a protein with homology to myb proto-oncogene products and with structural similarities to transcriptional activators. EMBO J 6:3553–3558

Pehlivan N, Sun L, Jarrett P et al (2016) Co-overexpressing a plasma membrane and a vacuolar membrane sodium/proton antiporter significantly improves salt tolerance in transgenic *Arabidopsis* plants. Plant Cell Physiol 57:1069–1084

Peng X, Zhang L, Zhang L et al (2013) The transcriptional factor *LcDREB2* cooperates with *LcSAMDC2* to contribute to salt tolerance in *Leymus chinensis*. Plant Cell Tissue Organ Cult 113:245–256

Puranik S, Sahu PP, Srivastava PS, Prasad M (2012) NAC proteins: regulation and role in stress tolerance. Trends Plant Sci 17:369–381

Puranik S, Sahu PP, Mandal SN et al (2013) Comprehensive genome-wide survey, genomic constitution and expression profiling of the NAC transcription factor family in foxtail millet (*Setaria italica* L.). PLoS One 8:e64594

Qi L, Yang J, Yuan Y et al (2015) Overexpression of two R2R3-MYB genes from *Scutellaria baicalensis* induces phenylpropanoid accumulation and enhances oxidative stress resistance in transgenic tobacco. Plant Physiol Biochem 94:235–243

Qin Y, Wang M, Tian Y et al (2012) Over-expression of *TaMYB33* encoding a novel wheat MYB transcription factor increases salt and drought tolerance in *Arabidopsis*. Mol Biol Rep 39:7183–7192

Qin Y, Tian Y, Liu X (2015) A wheat salinity-induced WRKY transcription factor *TaWRKY93* confers multiple abiotic stress tolerance in *Arabidopsis thaliana*. Biochem Biophys Res Commun 464:428–433

Qiu Y, Yu D (2009) Over-expression of the stress-induced *OsWRKY45* enhances disease resistance and drought tolerance in *Arabidopsis*. Environ Exp Bot 65:35–47

Qiu Q-S, Guo Y, M a D et al (2002) Regulation of SOS1, a plasma membrane Na$^+$/H$^+$ exchanger in *Arabidopsis thaliana*, by SOS2 and SOS3. Proc Natl Acad Sci U S A 99:8436–8441

Rahman H, Ramanathan V, Nallathambi J et al (2016) Over-expression of a *NAC 67* transcription factor from finger millet (*Eleusine coracana* L.) confers tolerance against salinity and drought stress in rice. BMC Biotechnol 16:35

Ravikumar G, Manimaran P, Voleti SR et al (2014) Stress-inducible expression of *AtDREB1A* transcription factor greatly improves drought stress tolerance in transgenic indica rice. Transgenic Res 23:421–439

Redillas MCFR, Jeong JS, Kim YS et al (2012) The overexpression of *OsNAC9* alters the root architecture of rice plants enhancing drought resistance and grain yield under field conditions. Plant Biotechnol J 10:792–805

Riechmann JL, Meyerowitz EM (1998) The AP2/EREBP family of plant transcription factors. Biol Chem 379:633–646

Rong W, Qi L, Wang A et al (2014) The ERF transcription factor *TaERF3* promotes tolerance to salt and drought stresses in wheat. Plant Biotechnol J 12:468–479

Roychoudhury A, Banerjee A (2016) Endogenous glycine betaine accumulation mediates abiotic stress tolerance in plants. Tropical Plant Res 3:105–111

Roychoudhury A, Chakraborty M (2013) Biochemical and molecular basis of varietal difference in plant salt tolerance. Annu Rev Res Biol 3:422–454

Roychoudhury A, Das K (2014) Functional role of polyamines and polyamine-metabolizing enzymes during salinity, drought and cold stresses. In: Anjum N, Gill S, Gill R (eds) Plant adaptation to environmental change. CAB International, London, UK, pp 141–156

RoyChoudhury A, Roy C, Sengupta DN (2007) Transgenic tobacco plants overexpressing the heterologous lea gene *Rab16A* from rice during high salt and water deficit display enhanced tolerance to salinity stress. Plant Cell Rep 26:1839–1859

Roychoudhury A, Paul S, Basu S (2013) Cross-talk between abscisic acid-dependent and abscisic acid-independent pathways during abiotic stress. Plant Cell Rep 32:985–1006

Roychoudhury A, Banerjee A, Lahiri V (2015) Metabolic and molecular-genetic regulation of proline signaling and its crosstalk with major effectors mediates abiotic stress tolerance in plants. Turk J Bot 39:887–910

Rushton PJ, Somssich IE, Ringler P, Shen QJ (2010) WRKY transcription factors. Trends Plant Sci 15:247–258

Saad AS, Li X, Li H-P et al (2013) A rice stress-responsive NAC gene enhances tolerance of transgenic wheat to drought and salt stresses. Plant Sci 203–204:33–40

Sah SK, Kaur G, Wani SH (2016) Metabolic engineering of compatible solute trehalose for abiotic stress tolerance in plants. In: Osmolytes and plants acclimation to changing environment: emerging omics technologies. Springer, New Delhi, pp 83–96

Saha J, Brauer EK, Sengupta A et al (2015) Polyamines as redox homeostasis regulators during salt stress in plants. Front Environ Sci 3:21

Sakamoto A, Alia MN (1998) Metabolic engineering of rice leading to biosynthesis of glycinebetaine and tolerance to salt and cold. Plant Mol Biol 38:1011–1019

Sakuma Y, Liu Q, Dubouzet JG et al (2002) DNA-binding specificity of the ERF/AP2 domain of *Arabidopsis DREBs*, transcription factors involved in dehydration- and cold-inducible gene expression. Biochem Biophys Res Commun 290:998–1009

Sakuma Y, Maruyama K, Osakabe Y et al (2006) Functional analysis of an *Arabidopsis* transcription factor, *DREB2A*, involved in drought-responsive gene expression. Plant Cell 18:1292–1309

Shafi A, Chauhan R, Gill T et al (2015) Expression of *SOD* and *APX* genes positively regulates secondary cell wall biosynthesis and promotes plant growth and yield in *Arabidopsis* under salt stress. Plant Mol Biol 87:615–631

Sharma A, Kumar D, Kumar S et al (2016) Ectopic expression of an atypical hydrophobic group 5 LEA protein from wild peanut, *Arachis diogoi* confers abiotic stress tolerance in tobacco. PLoS One 11:e0150609

Sharoni AM, Nuruzzaman M, Satoh K et al (2011) Gene structures, classification and expression models of the AP2/EREBP transcription factor family in Rice. Plant Cell Physiol 52:344–360

Shen H, Liu C, Zhang Y et al (2012) *OsWRKY30* is activated by MAP kinases to confer drought tolerance in rice. Plant Mol Biol 80:241–253

Shi H, Lee B, Wu S-J, Zhu J-K (2003) Overexpression of a plasma membrane Na$^+$/H$^+$ antiporter gene improves salt tolerance in *Arabidopsis thaliana*. Nat Biotechnol 21:81–85

Shin D, Moon S-J, Han S et al (2011) Expression of *StMYB1R-1*, a novel potato single MYB-like domain transcription factor, increases drought tolerance. Plant Physiol 155:421–432

Shriram V, Kumar V, Devarumath RM et al (2016) MicroRNAs as potential targets for abiotic stress tolerance in plants. Front Plant Sci 7:817

Shukla PS, Gupta K, Agarwal P et al (2015) Overexpression of a novel *SbMYB15* from *Salicornia brachiata* confers salinity and dehydration tolerance by reduced oxidative damage and improved photosynthesis in transgenic tobacco. Planta 242:1291–1308

Song Y, Jing SJ, Yu DQ (2010) Overexpression of the stress-induced *OsWRKY08* improves osmotic stress tolerance in *Arabidopsis*. Chinese Sci Bull 54:4671–4678

Song X, Yu X, Hori C et al (2016) Heterologous overexpression of poplar *SnRK2* genes enhanced salt stress tolerance in *Arabidopsis thaliana*. Front Plant Sci 7:612

Su J, Hirji R, Zhang L et al (2006) Evaluation of the stress-inducible production of choline oxidase in transgenic rice as a strategy for producing the stress-protectant glycine betaine. J Exp Bot 57:1129–1135

Su LT, Li JW, Liu DQ et al (2014) A novel MYB transcription factor, *GmMYBJ1*, from soybean confers drought and cold tolerance in *Arabidopsis thaliana*. Gene 538:46–55

Sun P, Zhu X, Huang X, Liu JH (2014a) Overexpression of a stress-responsive MYB transcription factor of *Poncirus trifoliata* confers enhanced dehydration tolerance and increases polyamine biosynthesis. Plant Physiol Biochem 78:71–79

Sun Z-M, Zhou M-L, Xiao X-G et al (2014b) Genome-wide analysis of AP2/ERF family genes from *Lotus corniculatus* shows *LcERF054* enhances salt tolerance. Funct Integr Genomics 14:453–466

Sun J, Hu W, Zhou R et al (2015) The *Brachypodium distachyon BdWRKY36* gene confers tolerance to drought stress in transgenic tobacco plants. Plant Cell Rep 34:23–35

Szabados L, Savouré A (2010) Proline: a multifunctional amino acid. Trends Plant Sci 15:89–97

Taj G, Agarwal P, Grant M, Kumar A (2014) MAPK machinery in plants. Plant Signal Behav 5:1370–1378

Tak H, Negi S, Ganapathi TR (2016) Banana NAC transcription factor *MusaNAC042* is positively associated with drought and salinity tolerance. Protoplasma 254:803–816

Tang L, Kwon S-Y, Kim S-H et al (2006) Enhanced tolerance of transgenic potato plants expressing both superoxide dismutase and ascorbate peroxidase in chloroplasts against oxidative stress and high temperature. Plant Cell Rep 25:1380–1386

Tang N, Zhang H, Li X et al (2012a) Constitutive activation of transcription factor *OsbZIP46* improves drought tolerance in rice. Plant Physiol 158:1755–1768

Tang Y, Liu M, Gao S et al (2012b) Molecular characterization of novel *TaNAC* genes in wheat and overexpression of *TaNAC2a* confers drought tolerance in tobacco. Physiol Plant 144:210–224

Tran L-SP, Urao T, Qin F et al (2007) Functional analysis of *AHK1/ATHK1* and cytokinin receptor histidine kinases in response to abscisic acid, drought, and salt stress in *Arabidopsis*. Proc Natl Acad Sci U S A 104:20623–20628

Tripathi V, Parasuraman B, Laxmi A, Chattopadhyay D (2009) *CIPK6*, a CBL-interacting protein kinase is required for development and salt tolerance in plants. Plant J 58:778–790

Tseng MJ, Liu CW, Yiu JC (2007) Enhanced tolerance to sulfur dioxide and salt stress of transgenic Chinese cabbage plants expressing both superoxide dismutase and catalase in chloroplasts. Plant Physiol Biochem 45:822–833

Tu M, Wang X, Feng T et al (2016) Expression of a grape (*Vitis vinifera*) bZIP transcription factor, *VlbZIP36*, in *Arabidopsis thaliana* confers tolerance of drought stress during seed germination and seedling establishment. Plant Sci 252:311–323

Tuteja N, Mahajan S (2007) Calcium signaling network in plants: an overview. Plant Signal Behav 2:79–85

Umezawa T, Yoshida R, Maruyama K et al (2004) *SRK2C*, a SNF1-related protein kinase 2, improves drought tolerance by controlling stress-responsive gene expression in *Arabidopsis thaliana*. Proc Natl Acad Sci 101:17306–17311

Urao T, Yakubov B, Satoh R et al (1999) A transmembrane hybrid-type histidine kinase in *Arabidopsis* functions as an Osmosensor. Plant Cell 11:1743–1754

Virk N, Li D, Tian L et al (2015) Arabidopsis Raf-like mitogen-activated protein kinase kinase kinase gene *Raf43* is required for tolerance to multiple abiotic stresses. PLoS One 10:e0133975

Vivek PJ, Tuteja N, Soniya EV (2013) *CDPK1* from ginger promotes salinity and drought stress tolerance without yield penalty by improving growth and photosynthesis in *Nicotiana tabacum*. PLoS One 8:e76392

Wang X-S, Zhu H-B, Jin G-L et al (2007) Genome-scale identification and analysis of LEA genes in rice (*Oryza sativa* L.). Plant Sci 172:414–420

Wang BQ, Zhang QF, Liu JH, Li GH (2011a) Overexpression of *PtADC* confers enhanced dehydration and drought tolerance in transgenic tobacco and tomato: effect on ROS elimination. Biochem Biophys Res Commun 413:10–16

Wang J, Sun PP, Chen CL et al (2011b) An arginine decarboxylase gene *PtADC* from *Poncirus trifoliata* confers abiotic stress tolerance and promotes primary root growth in *Arabidopsis*. J Exp Bot 62:2899–2914

Wang R-K, Li L-L, Cao Z-H et al (2012) Molecular cloning and functional characterization of a novel cotton CBL-interacting protein kinase gene (*GhCIPK6*) reveals its involvement in multiple abiotic stress tolerance in transgenic plants. Plant Mol Biol 79:123–135

Wang C, Deng P, Chen L et al (2013) A wheat *WRKY* transcription factor *TaWRKY10* confers tolerance to multiple abiotic stresses in transgenic tobacco. PLoS One 8:e65120

Wang L, Liu Y, Cai G et al (2014a) Ectopic expression of *ZmSIMK1* leads to improved drought tolerance and activation of systematic acquired resistance in transgenic tobacco. J Biotechnol 172:18–29

Wang RK, Cao ZH, Hao YJ (2014b) Overexpression of a R2R3 MYB gene *MdSIMYB1* increases tolerance to multiple stresses in transgenic tobacco and apples. Physiol Plant 150:76–87

Wang T, Tohge T, Ivakov A et al (2015a) Salt-related MYB1 coordinates abscisic acid biosynthesis and signaling during salt stress in *Arabidopsis*. Plant Physiol 169:1027–1041

Wang X, Zeng J, Li Y et al (2015b) Expression of *TaWRKY44*, a wheat WRKY gene, in transgenic tobacco confers multiple abiotic stress tolerances. Front Plant Sci 6:615

Wang C, Lu W, He X et al (2016a) The cotton mitogen-activated protein kinase kinase 3 functions in drought tolerance by regulating stomatal responses and root growth. Plant Cell Physiol 57:1629–1642

Wang FW, Wang ML, Guo C et al (2016b) Cloning and characterization of a novel betaine aldehyde dehydrogenase gene from *Suaeda corniculata*. Genet Mol Res 15:gmr7848

Wang H, Wang H, Shao H, Tang X (2016c) Recent advances in utilizing transcription factors to improve plant abiotic stress tolerance by transgenic technology. Front Plant Sci 7:67

Wang J, Li Q, Mao X et al (2016d) Wheat transcription factor *TaAREB3* participates in drought and freezing tolerances in *Arabidopsis*. Int J Biol Sci 12:257–269

Wang Y, Sun T, Li T et al (2016e) A CBL-interacting protein kinase *TaCIPK2* confers drought tolerance in transgenic tobacco plants through regulating the stomatal movement. PLoS One 11:e0167962

Wang Z, Su G, Li M et al (2016f) Overexpressing *Arabidopsis ABF3* increases tolerance to multiple abiotic stresses and reduces leaf size in alfalfa. Plant Physiol Biochem 109:199–208

Wang H, Wu Y, Yang X et al (2017) *SmLEA*, a gene for late embryogenesis abundant protein isolated from *Salvia miltiorrhiza,* confers tolerance to drought and salt stress in *Escherichia coli* and *S. miltiorrhiza*. Protoplasma 254:685–696

Wani SH, Singh NB, Haribhushan A, Mir JI (2013) Compatible solute engineering in plants for abiotic stress tolerance - role of glycine betaine. Curr Genomics 14:157–165

Wani SH, Kumar V, Shriram V, Sah SK (2016) Phytohormones and their metabolic engineering for abiotic stress tolerance in crop plants. Crop J 4:1–15

Wei KA, Chen JUAN, Wang YA et al (2012) Genome-wide analysis of bZIP-encoding genes in maize. DNA Res 19:463–476

Wei S, Hu W, Deng X et al (2014) A rice calcium-dependent protein kinase *OsCPK9* positively regulates drought stress tolerance and spikelet fertility. BMC Plant Biol 14:133

Wei T, Deng K, Liu D et al (2016) Ectopic expression of DREB transcription factor, *AtDREB1A*, confers tolerance to drought in transgenic *Salvia miltiorrhiza*. Plant Cell Physiol 57:1593–1609

Wen XP, Pang XM, Matsuda N et al (2008) Over-expression of the apple spermidine synthase gene in pear confers multiple abiotic stress tolerance by altering polyamine titers. Transgenic Res 17:251–263

Wen X-P, Ban Y, Inoue H et al (2010) Spermidine levels are implicated in heavy metal tolerance in a spermidine synthase overexpressing transgenic European pear by exerting antioxidant activities. Transgenic Res 19:91–103

Wi SJ, Kim WT, Park KY (2006) Overexpression of carnation S-adenosylmethionine decarboxylase gene generates a broad-spectrum tolerance to abiotic stresses in transgenic tobacco plants. Plant Cell Rep 25:1111–1121

Xia Z, Wei T, Jia L, Yongsheng L (2014) Co-expression of rice *OsP5CS1* and *OsP5CS2* genes in transgenic tobacco resulted in elevated proline biosynthesis and enhanced abiotic stress tolerance. Chinese J Appl Environ Biol 20:717–722

Xiang Y, Tang N, Du H et al (2008) Characterization of *OsbZIP23* as a key player of the basic leucine zipper transcription factor family for conferring abscisic acid sensitivity and salinity and drought tolerance in rice. Plant Physiol 148:1938–1952

Xiao B, Huang Y, Tang N, Xiong L (2007) Over-expression of a LEA gene in rice improves drought resistance under the field conditions. Theor Appl Genet 115:35–46

Xing Y, Chen W-H, Jia W, Zhang J (2015) Mitogen-activated protein kinase kinase 5 (*MKK5*)-mediated signalling cascade regulates expression of iron superoxide dismutase gene in *Arabidopsis* under salinity stress. J Exp Bot 66:5971–5981

Xiong H, Li J, Liu P et al (2014) Overexpression of *OsMYB48-1*, a novel MYB-related transcription factor, enhances drought and salinity tolerance in rice. PLoS One 9:e92913

Xu D, Duan X, Wang B et al (1996) Expression of a late embryogenesis abundant protein gene, *HVA1*, from barley confers tolerance to water deficit and salt stress in transgenic rice. Plant Physiol 110:249–257

Xu Z-S, Liu L, Ni Z-Y et al (2009) *W55a* encodes a novel protein kinase that is involved in multiple stress responses. J Integr Plant Biol 51:58–66

Xu H, Li K, Yang F et al (2010a) Overexpression of *CsNMAPK* in tobacco enhanced seed germination under salt and osmotic stresses. Mol Biol Rep 37:3157–3163

Xu J, Tian Y-S, Peng R-H et al (2010b) *AtCPK6*, a functionally redundant and positive regulator involved in salt/drought stress tolerance in *Arabidopsis*. Planta 231:1251–1260

Xu J, Duan X, Yang J et al (2013) Coupled expression of Cu/Zn-superoxide dismutase and catalase in cassava improves tolerance against cold and drought stresses. Plant Signal Behav 8:e24525

Yamaguchi-shinozaki K, Shinozaki K (2006) Transcriptional regulatory networks in cellular responses and tolerance to dehydration and cold stresses. Annu Rev Plant Biol 57:781–803

Yan H, Li Q, Park SC et al (2016) Overexpression of *CuZnSOD* and *APX* enhance salt stress tolerance in sweet potato. Plant Physiol Biochem 109:20–27

Yang W, Kong Z, Omo-Ikerodah E et al (2008) Calcineurin B-like interacting protein kinase *OsCIPK23* functions in pollination and drought stress responses in rice (*Oryza sativa* L.). J Genet Genomics 35:531–543

Yang Q, Chen ZZ, Zhou XF et al (2009) Overexpression of *SOS* (salt overly sensitive) genes increases salt tolerance in transgenic *Arabidopsis*. Mol Plant 2:22–31

Yang A, Dai X, Zhang W-H (2012a) A R2R3-type MYB gene, *OsMYB2*, is involved in salt, cold, and dehydration tolerance in rice. J Exp Bot 63:2541–2556

Yang L, Ji W, Gao P et al (2012c) *GsAPK*, an ABA-activated and calcium-independent SnRK2-type kinase from *G. soja*, mediates the regulation of plant tolerance to salinity and ABA stress. PLoS One 7:e33838

Yang X, Wang X, Ji L et al (2015) Overexpression of a *Miscanthus lutarioriparius* NAC gene *MlNAC5* confers enhanced drought and cold tolerance in *Arabidopsis*. Plant Cell Rep 34:943–958

Yang G, Yu L, Zhang K et al (2017) A *ThDREB* gene from *Tamarix hispida* improved the salt and drought tolerance of transgenic tobacco and *T. hispida*. Plant Physiol Biochem 113:187–197

Yao L, Jiang Y, Lu X et al (2016) A R2R3-MYB transcription factor from *Lablab purpureus* induced by drought increases tolerance to abiotic stress in *Arabidopsis*. Mol Biol Rep 43:1089–1100

Yin X-Y, Yang A-F, Zhang K-W, Zhang J-R (2004) Production and analysis of transgenic maize with improved salt tolerance by the introduction of *AtNHX1* gene. Acta Bot Sin 7:854–861

Ying S, Zhang D-F, Fu J et al (2012) Cloning and characterization of a maize bZIP transcription factor, *ZmbZIP72*, confers drought and salt tolerance in transgenic *Arabidopsis*. Planta 235:253–266

Yokotani N, Ichikawa T, Kondou Y et al (2009) Tolerance to various environmental stresses conferred by the salt-responsive rice gene *ONAC063* in transgenic *Arabidopsis*. Planta 229:1065–1075

Yu J, Lai Y, Wu X et al (2016) Overexpression of *OsEm1* encoding a group I LEA protein confers enhanced drought tolerance in rice. Biochem Biophys Res Commun 478:703–709

Yuan F, Yang H, Xue Y et al (2014) *OSCA1* mediates osmotic-stress-evoked Ca^{2+} increases vital for osmosensing in *Arabidopsis*. Nature 514:367–371

Yue Y, Zhang M, Zhang J et al (2012) *SOS1* gene overexpression increased salt tolerance in transgenic tobacco by maintaining a higher K $^+$/Na $^+$ ratio. J Plant Physiol 169:255–261

Zhai Y, Wang Y, Li Y et al (2013) Isolation and molecular characterization of *GmERF7*, a soybean ethylene-response factor that increases salt stress tolerance in tobacco. Gene 513:174–183

Zhang H-X, Hodson JN, Williams JP, Blumwald E (2001) Engineering salt-tolerant *Brassica* plants: characterization of yield and seed oil quality in transgenic plants with increased vacuolar sodium accumulation. Proc Natl Acad Sci 98:12832–12836

Zhang H, Liu W, Wan L et al (2010a) Functional analyses of ethylene response factor *JERF3* with the aim of improving tolerance to drought and osmotic stress in transgenic rice. Transgenic Res 19:809–818

Zhang H, Mao X, Wang C, Jing R (2010b) Overexpression of a common wheat gene *Tasnrk2.8* enhances tolerance to drought, salt and low temperature in *Arabidopsis*. PLoS One 5:e16041

Zhang H, Mao X, Jing R et al (2011a) Characterization of a common wheat (*Triticum aestivum* L.) *TaSnRK2.7* gene involved in abiotic stress responses. J Exp Bot 62:975–988

Zhang L, Xi D, Li S et al (2011b) A cotton group C MAP kinase gene, *GhMPK2*, positively regulates salt and drought tolerance in tobacco. Plant Mol Biol 77:17–31

Zhang X, Wang L, Meng H et al (2011c) Maize *ABP9* enhances tolerance to multiple stresses in transgenic *Arabidopsis* by modulating ABA signaling and cellular levels of reactive oxygen species. Plant Mol Biol 75:365–378

Zhang D, Jiang S, Pan J et al (2014a) The overexpression of a maize mitogen-activated protein kinase gene (*ZmMPK5*) confers salt stress tolerance and induces defence responses in tobacco. Plant Biol 16:558–570

Zhang J, Zou D, Li Y et al (2014b) *GhMPK17*, a cotton mitogen-activated protein kinase, is involved in plant response to high salinity and osmotic stresses and ABA signaling. PLoS One 9:e95642

Zhang L, Liu G, Zhao G et al (2014c) Characterization of a wheat R2R3-MYB transcription factor gene, *TaMYB19*, involved in enhanced abiotic stresses in *Arabidopsis*. Plant Cell Physiol 55:1802–1812

Zhang L, Zhang L, Xia C et al (2015a) The novel wheat transcription factor *TaNAC47* enhances multiple abiotic stress tolerances in transgenic plants. Front Plant Sci 6:1174

Zhang L, Zhang L, Xia C et al (2015b) A novel wheat bZIP transcription factor, *TabZIP60*, confers multiple abiotic stress tolerances in transgenic *Arabidopsis*. Physiol Plant 153:538–554

Zhang X, Liu X, Wu L et al (2015c) The *SsDREB* transcription factor from the succulent halophyte *Suaeda salsa* enhances abiotic stress tolerance in transgenic tobacco. Int J Genomics 2015:875497

Zhao J, Sun Z, Zheng J et al (2009) Cloning and characterization of a novel CBL-interacting protein kinase from maize. Plant Mol Biol 69:661–674

Zhao X, Yang X, Pei S et al (2016) The *Miscanthus* NAC transcription factor *MlNAC9* enhances abiotic stress tolerance in transgenic *Arabidopsis*. Gene 586:158–169

Zheng L, Liu G, Meng X et al (2013) A WRKY gene from *Tamarix hispida*, *ThWRKY4*, mediates abiotic stress responses by modulating reactive oxygen species and expression of stress-responsive genes. Plant Mol Biol 82:303–320

Zhou S, Zhang Z, Tang Q et al (2011) Enhanced V-ATPase activity contributes to the improved salt tolerance of transgenic tobacco plants overexpressing vacuolar Na$^+$/H$^+$ antiporter *AtNHX1*. Biotechnol Lett 33:375–380

Zhou X, Jin Y, Yoo C, Lin X (2013) CYCLIN H; 1 regulates drought stress responses and blue light-induced stomatal opening by inhibiting reactive oxygen species accumulation in *Arabidopsis*. Plant Physiol 162:1030–1041

Zhou L, Wang NN, Gong SY et al (2015) Overexpression of a cotton (*Gossypium hirsutum*) WRKY gene, *GhWRKY34*, in *Arabidopsis* enhances salt-tolerance of the transgenic plants. Plant Physiol Biochem 96:311–320

Zhu N, Cheng S, Liu X et al (2015) The R2R3-type MYB gene *OsMYB91* has a function in coordinating plant growth and salt stress tolerance in rice. Plant Sci 236:146–156

Zhu Y, Yan J, Liu W et al (2016) Phosphorylation of a NAC transcription factor by *ZmCCaMK* regulates abscisic acid-induced antioxidant defense in maize. Plant Physiol 171:1651–1664

Zou J-J, Wei F-J, Wang C et al (2010) *Arabidopsis* calcium-dependent protein kinase *CPK10* functions in abscisic acid- and Ca^{2+}-mediated stomatal regulation in response to drought stress. Plant Physiol 154:1232–1243

Zou J-J, Li X-D, Ratnasekera D et al (2015) *Arabidopsis* CALCIUM-DEPENDENT PROTEIN KINASE8 and CATALASE3 function in abscisic acid-mediated signaling and H_2O_2 homeostasis in stomatal guard cells under drought stress. Plant Cell 27:1445–1460

Chapter 8
Engineering Disease Resistance in Rice

K. K. Kumar, E. Kokiladevi, L. Arul, S. Varanavasiappan, and D. Sudhakar

Abstract Rice diseases cause substantial yield loss in rice. Through conventional breeding, resistance genes (R-gene) were transferred into elite rice genotypes particularly against the fungal blast and bacterial blight diseases. Main drawback of this approach is that, in the long term, breakdown of resistance occurs due to evolution of new virulent pathogen strains. In the current scenario, developing rice with durable broad-spectrum resistance through genetic transformation is gaining importance. In this direction, genetic transformation of rice was being carried out for the past two decades via expressing pathogenesis-related (PR) proteins, antimicrobial peptide, and genes governing signaling pathways as well as elicitor proteins. In spite of several reports, the expression of PR proteins and antimicrobial peptides did not yield desirable disease control in rice. Better understanding of disease resistance mechanism in plants helped in identifying critical transcription factors (TFs) involved in disease resistance. Overexpression of *NPR1* encoding non-expressor of pathogenesis-related protein 1 and *OsWRKY45* transcription factors in rice showed strong disease resistance to multiple pathogens and at the same time resulted in fitness cost. Recently, transgenic rice with high level of resistance to important rice diseases was achieved by expressing *NPR1* and *WRKY45* under tissue-specific/pathogen-responsive promoter; thereby agronomic traits are not altered. Rice transformants expressing the pathogen-derived elicitor proteins particularly from rice blast pathogen, *Magnaporthe oryzae* is a promising approach for imparting broad-spectrum disease resistance without yield penalty. Host-delivered RNAi technology is the latest of the approaches toward enhancing disease resistance against sheath blight and viral disease of rice. Recently, genome-editing tools are being deployed in rice to enhance resistance against diseases of rice.

Keywords Rice transgenic · Disease resistance · Signalling pathway · Elicitor protein · RNAi · Genome editing

K. K. Kumar · E. Kokiladevi · L. Arul · S. Varanavasiappan · D. Sudhakar (✉)
Department of Plant Biotechnology, Centre for Plant Molecular Biology & Biotechnology,
Tamil Nadu Agricultural University, Coimbatore, India

© Springer International Publishing AG, part of Springer Nature 2018
S. S. Gosal, S. H. Wani (eds.), *Biotechnologies of Crop Improvement, Volume 2*,
https://doi.org/10.1007/978-3-319-90650-8_8

8.1 Introduction

Rice (*Oryza sativa* L.), being one of the important cereal crops of the world, is affected by more than 70 diseases. The most important rice diseases are the blast caused by fungus *Magnaporthe oryzae* and the bacterial blight (BB) caused by *Xanthomonas oryzae* pv. *oryzae* (*Xoo*). Sheath blight (ShB) caused by fungus *Rhizoctonia solani* is also one of the important diseases of rice along with a few viral diseases. These diseases are responsible for causing annual yield losses of up to 50% of rice productivity (Datta et al. 2002). Rice is known to possess many disease resistance genes (R-gene) associated with blast and bacterial blight diseases. More than 40 genes conferring BB resistance (Sundaram et al. 2014), 101 blast-resistant genes (Rajashekara et al. 2014), and 350 quantitative trait loci (QTLs) have been identified (Sharma et al. 2012). Deployment of disease resistance (R) genes and quantitative trait loci through backcross breeding method has contributed greatly to increasing rice resistance against diverse pathogens (Kou and Wang 2010). However, such effort is hampered by the resistance breakdown due to the variability in pathogen population or development of new strains due to mutation (Jones and Dangl 2006; Dangl et al. 2013). Unlike the BB or blast disease, no major resistant gene is known in rice germplasm for sheath blight disease. Breeding for sheath blight resistance has not been successful as the resistance is controlled by multiple loci, and there is no reliable source of rice germplasm with complete resistance to the disease (Liu et al. 2009; Zuo et al. 2014). In the absence of suitable genetic resistance for ShB, chemical method is the only option for its control. Therefore, breeding for varieties with durable and broad-spectrum disease resistance is critical to sustainable agricultural development.

More than 25 viruses are known to infect rice. Important viral diseases of rice include rice dwarf virus (RDV), rice black-streaked dwarf virus (RBSDV), rice stripe virus (RSV), rice tungro bacilliform virus (RTBV), and rice tungro spherical virus (RTSV). In Southern Vietnam during 2006–2007, more than 485,000 hectares of paddy fields were severely affected by rice grassy stunt virus (RGSV) or co-infection by RGSV and rice ragged stunt virus (RRSV), resulting in heavy loss and directly affecting millions of rice farmers (Cabauatan et al. 2009). In China, epidemic outbreaks of the rice black-streaked dwarf disease resulted in a grain yield decrease of 10–40%, resulting in a total loss of grain production in the rice planting areas of southern China (Li et al. 1999). Rice tungro is one of the important viral diseases of rice, which is caused by the joint infection of two unrelated viruses *Rice tungro bacilliform virus* (RTBV), a double-stranded DNA-containing virus, and *Rice tungro spherical virus* (RTSV), a single-stranded RNA virus. The most conspicuous symptoms of tungro are the stunting of plants and yellow-orange discoloration of leaves. In the recent times, genetic engineering has been sought as the method of choice for achieving disease resistance in rice.

8.2 Engineering Disease Resistance in Rice by Overexpressing Antimicrobial Proteins

Earlier generation of transgenic rice for disease resistance focused on expressing the antimicrobial proteins belonging to pathogenesis-related (PR) proteins or antimicrobial peptide to engineer rice disease resistance. Enhanced disease resistance was observed via expressing the PR proteins or antimicrobial peptide, but the level of resistance conferred was not sufficient enough for commercial cultivation.

8.2.1 Overexpression of Pathogenesis-Related (PR) Proteins

Pathogenesis-related (PR) proteins are a group of plant proteins that express during pathogen infection as a defense mechanism. Several classes of PR proteins are known in plants. Plant chitinase (PR-3) and β-1-3-glucanase (PR-2) are two hydrolytic enzymes produced by the plants to break down the chitin (N-acetyl-D-Glucosamine) and glucan (laminarin) polymer, respectively, which constitute the major components of the fungal cell wall.

8.2.1.1 Overexpression of Chitinase

First attempt to engineer disease resistance in rice was done by Lin et al. (1995) by overexpressing rice chitinase gene, *chi11*, using constitutive maize ubiquitin promoter and showed enhanced resistance to *R. solani*. Subsequently, rice *chi11* gene was used to transform different genotypes of rice (Nishizawa et al. 1999; Datta et al. 2000, 2001; Kumar et al. 2003; Sridevi et al. 2003; Kalpana et al. 2006; Maruthasalam et al. 2007). Recently, a high expressing novel chitinase gene was isolated from the sheath blight-resistant QTL region (qSBR11-1 on chromosome 11) of resistant *indica* rice variety Tetep (Richa et al. 2016). Transformation of susceptible *japonica* rice line Taipei 309 (TP309) with the novel rice chitinase gene provided enhanced resistance against sheath blight pathogen, *R. Solani* (Richa et al. 2017). Li et al. (2009) transformed rice overexpressing *Momordica charantia* class I chitinase gene (McCHIT1) and showed an enhanced resistance to *R. solani* and *M. oryzae*. Compared to chitinases of plant origin, chitinases from biocontrol agents exhibit higher antifungal activity. Shah et al. (2009) transformed rice cv. PB1 with an endo-chitinase gene (*cht42*) from a fungus *Trichoderma virens* and recorded 62% reduction in sheath blight disease index.

Reports on co-expression of rice chitinase gene along with other PR protein showed synergistic effect for disease control in rice. The co-transformants expressing both *tlp* and *chi11* in rice showed an elevated resistance against *R. solani*

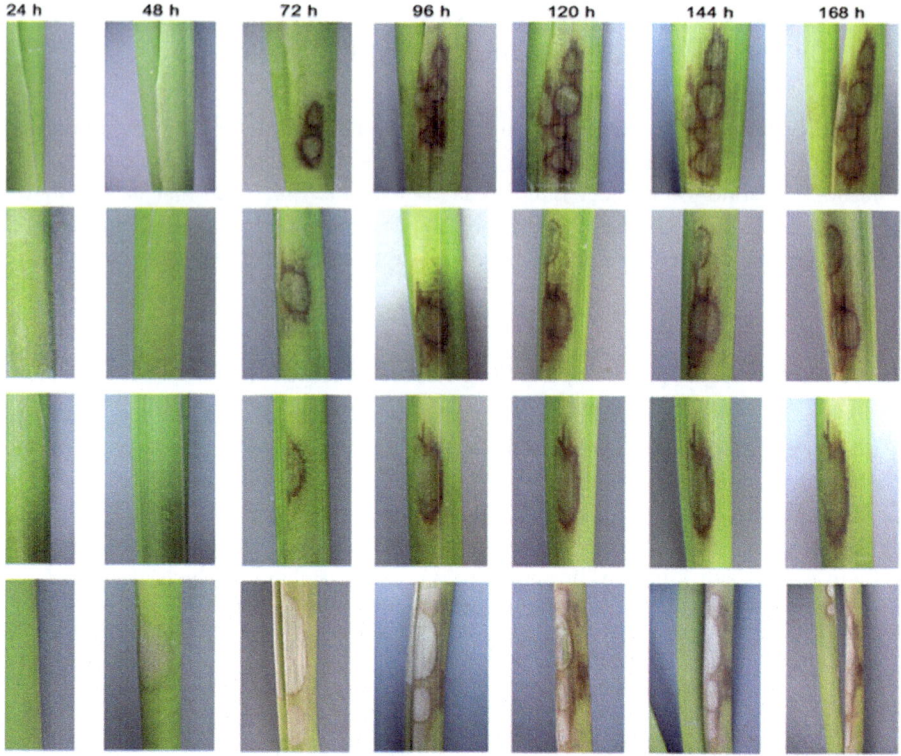

Fig. 8.1 Pyramiding of PR proteins (*chi11* and *tlp*) in transgenic rice *cv*. Pusa Basmati1 (PB1) demonstrated enhanced level of resistance to sheath blight disease. Bioassay was done in intact leaf sheaths of non-transgenic and transgenic PB1 lines using sheath blight pathogen. Reaction of SM-PB1-9 (*chi11*) (**a**) SM-PB1-5 (*tlp*) (**b**) SM-PB1-1 (*Chi11 + tlp + Xa21*) (**c**) and untransformed PB1 (**d**) to sheath blight pathogen infection at 24, 48, 72, 96, 120, 144, and 168 h after infection (HAI). (Source: Maruthasalam et al. 2007)

(Fig. 8.1) and *Sarocladium oryzae* than plants expressing either *tlp* or *Chi11* (Kalpana et al. 2006; Maruthasalam et al. 2007). Transgenic rice plants transformed with *Chi11*, *tlp*, and *Xa21* and displayed resistance to both sheath blight and bacterial leaf blight (Fig. 8.2) (Maruthasalam et al. 2007). Transgenic rice constitutively co-expressing *tlp-D34* (thaumatin-like protein) gene and *chi11* showed enhancement of sheath blight resistance with disease index reduced to 39% (Shah et al. 2013). Co-expression of a rice basic chitinase gene and a ribosome-inactivating protein in rice caused a significant reduction in sheath blight development (Kim et al. 2003). Co-expression of *OsChi11* and *Osoxo4* genes in a green tissue-specific manner provided 63% resistance against sheath blight without affecting agronomically important traits (Karmakar et al. 2016). Maize phosphoenolpyruvate carboxylase (PEPC) promoter was used for *OsChi11* expression, and rice $P_{D540-544}$ promoter was used for *Osoxo4* gene expression.

Fig. 8.2 Transgenic rice plants pyramided with *chi11+ tlp + Xa21* showed resistance to sheath blight and bacterial leaf blight. Reaction of SM-PB1-1 (**a**) and untransformed PB1 (**b**) to ShB infection at 168 HAI. Reaction of SM-PB1-1 (**c**) and non-transgenic PB1 (**d**) to *Xoo* infection at 14 days after inoculation. (Source: Maruthasalam et al. 2007)

8.2.1.2 Overexpression of other PR Proteins in Rice

Oxalic acid (OA) is a nonhost-specific toxin secreted by certain plant pathogens during infection (Dutton and Evans 1996). Plant oxalate oxidase (OxO) enzyme degrades OA into CO_2 and hydrogen peroxide (H_2O_2). OxO-generated H_2O_2 may function as a secondary messenger in the activation of phytoalexin biosynthetic pathways, hypersensitive response (HR), systemic acquired resistance (SAR), and PR gene expression in plants. Genome analysis of rice showed four tandemly duplicated *oxo* genes (*Osoxo1–Osoxo4*) in chromosome 3, with *Osoxo4* playing a role in disease resistance (Carrillo et al. 2009). Transgenic rice overexpressing the rice oxalate oxidase 4 (*Osoxo4*) gene under a green tissue-specific promoter (rice PD540–544) exhibited 50% protection against *R. solani* without any agronomic imbalance (Molla et al. 2013).

Germin-like protein (GLP) gene family is one of the important defense gene families that have been considered to play an important role in several aspects of plant development or stress tolerance (Knecht et al. 2010). One of the rice GLP genes, *OsGLP2-1*, was significantly induced by blast fungus (Liu et al. 2016). Overexpression of *OsGLP2-1* quantitatively enhanced resistance to leaf blast, panicle blast, and bacterial blight (Liu et al. 2016). *OsGLP2-1*-mediated resistance to blast and bacterial blight was involved in the activation of jasmonic acid (JA)-dependent pathway instead of salicylic acid (SA)-dependent pathway.

Osmotin and osmotin-like proteins (OLP) belong to thaumatin-like proteins (TLP) of the PR-5 family because they all contain a typical thaumatin domain. It is involved in plant permeability stress and defense responses because of its antibacterial properties in vivo against a broad range of plant pathogens (Narasimhan et al. 2005). Xue et al. (2016) found that *OsOSM1* expression is strongly induced by *R. solani* in ShB-resistant rice variety YSBR1. Overexpression of *OsOSM1* (OsOSM1ox) in susceptible variety Xudao 3 significantly increases resistance to SB in transgenic rice (Xue et al. 2016). They found that JA-responsive marker genes are induced in OsOSM1ox lines and suggest that the activation of JA signalling pathway may account for the increased resistance in transgenic OsOSM1ox lines.

8.3 Engineering Disease Resistance in Rice by Overexpressing Antimicrobial Peptides

Antimicrobial peptides (AMPs) are used by both plant and animal systems to destroy microorganism, including bacteria, fungi, mycoplasma, and viruses. AMPs are characterized to possess high anti microbial activity and be very quick in killing microbes and at the same time are nontoxic to eukaryotic cells. Defensins are small antifungal peptides (~5 KDa) of eukaryotic origin present in plant, animal, and insects. Plant defensins (PR-12) are low molecular weight cysteine-rich peptide thought to affect cell membrane of microbes and prevent its ion uptake. Transgenic rice expressing *Dahlia merckii* defensin (*DM-AMP1*) gene gave better level of protein (up to 80%) to the two important rice fungal pathogens *M. oryzae* and *R. solani* (Jha et al. 2009). The Dm-AMP1 signal peptide had successfully targeted the Dm-AMP1 to apoplast in transgenic rice. Transgenic rice expressing the antimicrobial peptide from onion (*Ace*-AMP1) improved their resistances to blast, sheath blight, and bacterial blight by 86%, 67%, and 82%, respectively (Patkar and Chattoo 2006). Rice overexpressing *Rs-AFP2* defensin gene from *Raphanus sativus* suppressed the growth of *M. oryzae* and *R. solani* by 77% and 45%, respectively (Jha and Chattoo 2010). Transgenic expression of Rs-AFP2 was not accompanied by an induction of PR gene expression, suggesting that the expression of Rs-AFP2 directly inhibits the pathogens. The antimicrobial peptide of humans, LL-37, is a 37-residue-long peptide which possesses broad-spectrum antibacterial activity and was used for rice transformation. Transgenic rice expressing the LL-37 peptide in the intercellular space showed enhanced disease resistance against bacterial leaf blight and blast (Lee et al. 2017). To avoid degradation by the plant proteases, the fusion of vicilin signal peptide at the N-terminal of LL-37 directed it to intercellular space. The *pGD1* (phosphogluconate dehydrogenase) promoter from rice was used to induce stable expression of SP-LL-37 in transgenic rice. Giant silk moth (*Hyalophora cecropia*) encodes the antimicrobial protein cecropin A and cecropin B. Transgenic rice plant expressing plant codon-optimized *cecropin A* gene exhibited resistance to rice blast without an induction of PR gene expression (Coca et al. 2006). Similarly, transgenic rice plants expressing *cecropin B* exhibited reduction in lesion due to BB pathogen infection (Sharma et al. 2000; Coca et al. 2004). Overall, the overexpression of antimicrobial protein in transgenic rice confers enhanced level of resistance to all important diseases of rice.

8.4 Engineering Broad-Spectrum Disease Resistance in Rice

Plants defend against microbial pathogen attack by activating a variety of defense systems that are mediated through multiple signalling pathways. Plant defense signalling is mainly mediated through the plant hormones salicylic acid (SA), jasmonic acid (JA), and ethylene (ET). In general, plants upon exposure to pathogen induce two well-known forms of immune responses: SA-mediated systemic acquired resistance (SAR) and JA/ET-mediated inducible systemic resistance (ISR). The induced immune response often confers durable, broad-spectrum, and systemic resistance against different pathogens at distal tissue from the infection or treatment site. Many transcription factors are successfully used for engineering disease resistance in rice (Table 8.1). Transcription factors NPR1 and WRKY45 act as key positive regulator of SA-mediated pathway in plants. The SA pathway in rice appears to branch into OsNPR1/NH1 and WRKY45-mediated sub-pathway. Plant-inducible immune response can also be triggered by exogenous application of a number of elicitors or elicitor transgene expression.

8.4.1 Rice Transgenic Plants Expressing NPR1 Gene

8.4.1.1 AtNPR1

Arabidopsis thaliana NPR1 (AtNPR1) is a key positive regulator, which acts downstream of the signal molecule SA in regulating gene expression of SAR pathway (Cao et al. 1994). Transgenic rice constitutively expressing the *AtNPR1* gene results in disease resistance to bacterial pathogen *Xoo* but had a negative impact on growth and agronomic traits due to triggering lesion-mimic/cell death (LMD) phenotype (Fitzgerald et al. 2004). In another study, transgenic rice plants constitutively expressing *At*NPR1 have been reported to exhibit negative physiological consequences in the form of growth retardation, height reduction, and decreased seed production (Quilis et al. 2008). Green tissue-specific expression of *AtNPR1* using the $P_{D540-544}$ promoter in rice confers resistance to the sheath blight pathogen, with no concomitant abnormalities in plant growth and yield parameters (Molla et al. 2016). They demonstrated that an increase in the *AtNPR1* transcript levels in the transgenic rice plants resulted in the activation of many defense-related PR genes, and the elevated induction of PR genes appeared to translate into enhanced resistance

Table 8.1 Genetic transformation of rice with transcription factor associated with disease resistance

Sl. No.	Gene	Promoter	Type of expression	Transgenic indica/japonica variety	Disease resistance	Alteration in phenotype	Reference
1.	*OsWRKY6* NCBI GenBank: BK005009.1	Constitutive; CaMV35S	Overexpression	*Japonica* cv Nipponbare	Strong resistance to BLB	Growth retardation	Choi et al. (2015)
2.	*OsWRKY11* RGAP[a]: LOC_Os01g43650	Constitutive; CaMV35S	Overexpression	*Japonica* cv. Nipponbare	Enhanced resistance to BLB	Enhanced tolerance to drought stress, reduction in height	Lee et al. (2018)
3.	*OsWRKY13* NCBI GenBank: EF143611	Constitutive; maize *ubiquitin*	Overexpression	*Japonica* cv. Mudanjiang 8	Resistance to BLB and blast	Decreases tolerance to cold and salt stresses as well as to retarded growth and development	Qiu et al. (2007, 2008)
4.	*OsWRKY30* NCB GenBank: DQ298180	Constitutive; maize *ubiquitin*	Overexpression	*Japonica* cv. Xiushui 11	Enhanced resistance to rice sheath blight and blast	Increased drought stress, exhibited no obvious morphological changes	Peng et al. (2012)
5.	*OsWRKY45-1* NCBI GenBank: AK066255 RAP-DB[b]: Os05g0322900	Constitutive; maize *ubiquitin*	Overexpression	*Japonica* cv. Nipponbare	Strong resistance to blast but reduces resistance to BLB	Transgenic rice plants varied with growth conditions, adverse effects on agronomic traits	Shimono et al. (2007)
6.	*OsWRKY45-2* indica cv. Minghui 63 *NCBI GenBank:* GQ331927 *OsWRKY45-1* japonica cv. Nipponbare NCBI GenBank: GQ331932	Constitutive; maize *ubiquitin*	Downregulation of OsWRKY45-1 and overexpression of OsWRKY45-2	*Indica* cv. Minghui 63 or *Japonica* cv. Mudanjiang 8	Enhanced resistance to Xoo and Xoc (*Xanthomonas oryzae pv oryzicola*). However, overexpression of both allele of WRKY confers enhanced resistance to blast	Reduces adaptation to salt, cold, and drought stresses	Tao et al. (2009)

7.	OsWRKY45 NCBI GenBank: AK066255 RAP-DBᵇ: Os05g0322900	Constitutive; OsUbi7	Overexpression	Japonica cv. Nipponbare	Resistance to BLB and blast	Minor negative effects on agronomic traits	Goto et al. (2015)
8.	OsWRKY45 NCBI GenBank: AK066255 RAP-DBᵇ: Os05g0322900	Pathogen inducible; PR1b with ADH 5'-UTR	Inducible expression	Japonica cv. Nipponbare	Resistance to blast and blast	Agronomic traits comparable to control untransformed rice	Goto et al. (2016)
9.	OsWRKY62 NCBI GenBank: AK067834 RAP-DBᵇ: Os09g0417800	Constitutive; maize ubiquitin	Downregulation (RNAi)	Japonica cv. Nipponbare	Susceptible to blast and BLB, thus act as positive regulator	Plays a negative role in pathogen defense under hypoxia stress	Fukushima et al. (2016)
10.	OsWRKY67 RAP-DBᵇ: Os05g0183100	Constitutive; CaMV35S	Overexpression	Japonica cv. Dongjin (DJ)	Enhanced resistance to BLB and blast	Plant growth retardation	Vo et al. (2018)
11.	OsWRKY76 NCBI GenBank: AK068337	Constitutive; maize ubiquitin	Overexpression	Japonica cv. Nipponbare	Enhances susceptibility to M. oryzae and Xoo	Increases cold tolerance	Yokotani et al. (2013)
12.	OsEREBP1 RGAPᵃ: LOC_Os02g54160	Constitutive; maize ubiquitin	Overexpression	Japonica cv. Kitaake	Resistance to BLB	Confers drought and submergence tolerance	Jisha et al. (2015)
13.	OsMYC2 RGAPᵃ: Os10g42430 or RAP-DBᵇ: Os10g0575000	Constitutive; maize ubiquitin	Downregulation (RNAi)	Japonica cv. TP309	Showed enhanced resistance against BLB	Suppressed seedling growth compared to wild-type plants under blue light and showed little effect under white light	Giri et al. (2017)

(continued)

Table 8.1 (continued)

Sl. No.	Gene	Promoter	Type of expression	Transgenic indica/japonica variety	Disease resistance	Alteration in phenotype	Reference
14.	*OsDC* gene from *Bacillus subtilis* named as *Bacisubin* NCBI GenBank: HQ452341	Constitutive; CaMV35S	Overexpression	*Japonica* cv. Nipponbare	Resistance to rice blast and sheath blight	Normal growth, development, and grain production in rice	Qi et al. (2017)
15.	*OsXrn4* gene NCBI GenBank: AK070933	Constitutive; CaMV 35S	Overexpression	*Japonica* cv. Nipponbare	Conferred resistance to rice stripe virus	–	Jiang et al. (2018)
16.	MAPK kinase 10.2 RGAP: Loc_Os03g12390	Constitutive; maize *ubiquitin*	Overexpression	*Japonica* cv. Nipponbare	Enhances resistance to *Xanthomonas oryzae* pv. *oryzicola* (*Xoc*)	Increases rice tolerance to drought stress	Ma et al. (2017)
17.	*OsCIPK30* (calcineurin B-like proteins) interaction protein kinase protein. RAP-DB[b]: Os01g0759200	Constitutive; maize *ubiquitin*	Overexpression	*Japonica* cv. Nipponbare	Delay the RSV symptoms and show milder RSV symptoms	Transgenic plants showed normal growth	Liu et al. (2017)
18.	*OsMADS26* RAP-DB[b]: Os08g0112700	Constitutive; maize *ubiquitin*	RNAi lines	*Japonica* cv. Nipponbare	Resistance to blast and BLB	Displayed enhanced tolerance to water deficit. Moderate impact on plant development	Khong et al. (2015)

No.	Gene	Promoter	Expression	Cultivar	Trait	Effects	Reference
19.	OsACS2 NCBI GenBank:AK064250	Pathogen-inducible; *PBZ1* promoter	Inducible expression	*Japonica* cv. Kitaake	Broad-spectrum disease resistance to necrotrophic and hemibiotrophic fungal. Resistance to a field isolate of *R. solani*, as well as different races of *M. oryzae*	Enhance resistance to fungal pathogens without negatively impacting crop productivity	Helliwell et al. (2013)
20.	*AtNPR1* *NCBI GenBank:* ATU76707	Constitutive; maize *ubiquitin*	Overexpression	*Japonica* cv. Taipei 309 (TP309)	Resistance to the bacterial pathogen *Xoo*	Detrimental side effects under controlled environment to develop lesion-mimic/cell death (LMD) phenotype	Fitzgerald et al. (2004)
21.	*AtNPR1* *NCBI GenBank:* ATU76707	Constitutive; maize *ubiquitin*	Overexpression	*Japonica* cv. Senia	Confers resistance against blast, bakanae diseases (caused by *Fusarium verticillioides*) and foot root disease (caused by *Erwinia chrysanthemi*) of rice. Higher susceptibility to infection by the *Rice yellow mottle virus (RYMV)*	Growth retardation and reduced height, observed spontaneous lesions when growing under suboptimal conditions (growth chamber). Susceptible to salt and drought stress	Quilis et al. (2008)

(continued)

Table 8.1 (continued)

Sl. No.	Gene	Promoter	Type of expression	Transgenic indica/japonica variety	Disease resistance	Alteration in phenotype	Reference
22.	*AtNPR1* NCBI GenBank:NM_105102	Green tissue-specific expression; $P_{D540-544}$ promoter NCBI GenBank: KJ857554	Tissue-specific expression	*Indica* cv. Pusa SugandhII II (PSII)	Confers resistance to the sheath blight pathogen	Transgenic plants are phenotypically similar to the non-transgenic wild type under greenhouse conditions	Molla et al. (2016)
23.	*AtNPR1*	*Pathogen inducible*; *TFB1* promoter plus 5′ leader sequence TAIR[c]: AT4G36988	Inducible expression	*Japonica* cv. Zhonghua 11 (*ZH11*)	Broad-spectrum disease resistance. Resistance to BLB, fungal blast, and bacterial leaf streak	No compromise on rice plant fitness	Xu et al. (2017b)
24.	*BjNPR1* *NCBI GenBank*: AY667498	Constitutive; CaMV35S	Overexpression	*Indica* cv. Chaitanya and Samba Mahsuri	Enhanced resistance to rice blast, sheath blight, and bacterial leaf blight diseases	Improvement in certain agronomic traits such as increases in plant height, flag-leaf length, panicle length, number of seeds/panicle, and seed yield/plant	Sadumpati et al. (2013)

No.	Gene	Expression/promoter	Method	Variety	Resistance	Main agronomic traits	Reference
25.	*OsERF922* RAP-DB[b]: *Os01g0752500*	Expression of sgRNA under rice U6 promoter	Gene knockout using CRISPR/Cas9	*Japonica* cv. Kuiku131	Enhanced resistance to blast	Main agronomic traits are not altered	Wang et al. (2016a)
26.	*Bsr-d1* RGAP[a]: *LOC_Os03g32230*	Constitutive; maize *ubiquitin for RNAi*	Downregulation (RNAi) and gene knockout using CRISPR/Cas9	*Japonica* cv. TP309 rice	Broad-spectrum blast disease resistance	Agronomic character not altered	Li et al. (2017)

[a]RGAP Rice Genome Annotation Project
[b]RAP-DB Rice Annotation Project Database
[c]TAIR the Arabidopsis Information Resource

of transgenic rice to *R. solani*. Expression of *AtNPR1* under pathogen-inducible promoter can overcome fitness cost in rice. Earlier plant defense gene expression was thought to be regulated at transcriptional level by the pathogen-inducible promoter. However recently, Xu et al. (2017a) did global transcriptome profiling on *Arabidopsis* plant exposed to elf18 elicitor and discovered that fundamental layer of regulation also occurs at translation level during defense response. In this study, they identified a pathogen-inducible TFB1 gene in *Arabidopsis* that is rapidly and transiently induced upon pathogen challenge. *TFB1* promoter with 5′ leader sequence (before the start codon for *TFB1*) contains two untranslated ORF (uORFs) in it. Translation of TBF1 is normally suppressed by these two uORFs within the 5′ leader sequence (Pajerowska-Mukhtar et al. 2012). Xu et al. (2017b) transformed rice with a construct that expresses *AtNPR1* under *TFB1* promoter cassette (*TFB1* promoter plus 5′ leader sequence with two pathogen-responsive upstream open reading frames, uORFs$_{TFB1}$). Thus, they engineered broad-spectrum disease resistance in rice without compromising on rice plant fitness. The rice plants displayed resistance to BLB, fungal blast, and bacterial leaf streak. Thus using *TFB1* cassette, it is possible to develop transgenic plants with enhanced broad-spectrum disease resistance with minimal adverse effects on growth and development. In an another study, rice co-expressing *AtNPR1* and *OsCHI11* under green tissue-specific promoter showed enhanced sheath blight tolerance as compared to single-gene transformants (Karmakar et al. 2017).

8.4.1.2 OsNPR1/NH1

Overexpression of *OsNPR1/NH1* was shown to confer strong resistance to both *Xoo* and *M. oryzae* (Chern et al. 2005; Sugano et al. 2010). Overexpression of *OsNPR1/* NH1 in rice induced constitutive activation of *PR* gene expression, accompanied by lesion-mimic symptoms and light hypersensitivity (Chern et al. 2005). Overexpression of *OsNPR1* conferred disease resistance to bacterial blight but also enhanced herbivore susceptibility in transgenic plants (Yuan et al. 2007). Sugano et al. (2010) conducted experiments to determine the function of *OsNPR1* and found that overexpression of *OsNPR1* led to increased activity in defense mechanisms against pathogens but reduced cellular activity with regard to photosynthesis and protein synthesis that leave the plant more vulnerable to herbivore predation.

8.4.1.3 BjNPR1

Transgenic *indica* rice expressing *Brassica juncea NPR1* (*BjNPR1*) exhibits enhanced resistance to rice blast, sheath blight, and bacterial leaf blight diseases (Sadumpati et al. 2013). Rice transformants with higher levels of *BjNPR1* revealed improvement in certain agronomic traits such as increases in plant height, panicle length, flag-leaf length, number of seeds/panicle, and seed yield/plant as compared to the untransformed plants.

8.4.2   Rice Transgenic Plants Expressing OsWRKY45

Although discovered recently, WRKY transcription factors are becoming one of the best characterized classes of plant transcription factors and are at the forefront of research on plant defense responses. More recent studies have provided direct evidence for the involvement of specific WRKY proteins in plant defense responses. Interaction between rice and *Xoo* is a classical example of host-pathogen interaction and serves as an ideal model system for investigation. WRKYs, a class of plant-specific transcription factors, act as a key regulator of plant immune response (Ulker and Somssich 2004). The "WRKY" domain of ~60 amino acids binds to the cognate *cis*-acting "W" box motif (C/T)TGAC(C/T) in the promoter of several downstream target genes. Rice overexpressing *OsWRKY45* under strong constitutive promoter (maize ubiquitin promoter, *PZmUbi*) showed extremely strong disease resistance to both rice blast and leaf blight but at significant costs on rice growth and yield (Shimono et al. 2007; Shimono et al. 2012). The WRKY45-OX rice plants cultivated in a growth chamber showed restricted growth, and those cultivated in a greenhouse showed only minor growth retardation (Shimono et al. 2007).To reduce the negative effect of *WRKY45* overexpression in rice, Goto et al. (2015) optimized expression of *WRKY45* gene in rice using a moderate-strength constitutive rice ubiquitin promoter (POsUbi7). Transgenic rice plants expressing WRKY gene at moderate level showed strong resistance to both blast and BLB diseases in a greenhouse, although the degree of resistance was a little weaker than that of the representative PZmUbi line (Goto et al. 2015). At the same time, adverse effects of environmental factors on WRKY45-ox lines are alleviated in POsUbi7 lines, whereas most of the PZmUbi plants died after the low-temperature treatment, indicating that a high level of WRKY45 expression rendered rice plants cold sensitive.

Blast pathogen, *M. oryzae* hyphae, invades rice cells within 24 h post-inoculation. However, WRKY45 is induced after the *M. oryzae* invasion in rice (Shimono et al. 2007). Due to the time lag in WRKY45 protein induction, it is unable to exert its full defense potential against blast pathogen (Shimono et al. 2007, 2012). To address this issue, Goto et al. (2016) developed rice lines in which WRKY45 induction occurs soon after pathogen challenge using an early pathogen-responsive promoter. Goto et al. (2016) developed transgenic rice with strong disease resistance to blast and BB by expressing WRKY45 under the control of pathogen-responsive promoters in combination with a translational enhancer derived from a 5'-untranslated region (UTR) of rice alcohol dehydrogenase (ADH). Although pathogen-responsive promoters alone failed to confer effective disease resistance, the use of the ADH 5'-UTR in combination with them, in particular the PR1b and GST promoters, enhanced disease resistance. The 2-kb upstream sequence of PR1b showed a very early pathogen response with high level of WRKY45 expression confined to infection site. This early and strong local induction of WRKY45 may be critical for the strong disease resistance in WRKY45-expressing lines. Field trials showed that overall PR1b promoter-driven (with ADH 5'-UTR) lines performed the best without any negative effects on agronomic traits, which is comparable to control untrans-formed rice.

Recently, *OsWRKY67* was found to be upregulated against pathogen challenges. Activation of *OsWRKY67* by T-DNA tagging significantly improved the resistance against two rice pathogens, blast and BB. Subsequently, overexpression of *OsWRKY67* in rice confirmed enhanced disease resistance but led to a restricted plant growth of the transgenic plants with high levels of *OsWRKY67* protein. OsWRKY67 RNAi lines significantly reduced resistance to *M. oryzae* and *Xoo* isolates tested and abolished XA21-mediated resistance, implying the possibility of broad-spectrum resistance from *OsWRKY67* (Vo et al. 2018). On the other side, *OsWRKY62* was reported to act as negative regulator of innate and *Xa21*-mediated resistance against bacterial blight (Peng et al. 2008). Further in the study, transgenic rice lines overex-pressing *OsWRKY62* challenged with *Xoo* were found to show significantly longer lesions than the wild-type controls. Thus, the negative role played by *OsWRKY62* was evident and suggests suppression of such a kind of negative players could be employed towards enhancing the innate defense system in rice. Similarly, overex-pression of *OsWRKY72* was found to be negatively influencing BB resistance in rice (Seo et al. 2011).

8.4.3 Engineering Disease Resistance in Rice by Enhancing Ethylene Biosynthesis

Recent evidence indicates that ethylene (ET) pathway also plays a major role in mediating plant disease resistance. Six rice *ACS* genes (*OsACS1-6*) are reported to exist in the rice genome. During the rice-*M. oryzae* interaction, endogenous ET levels increased within 48 h after inoculation with a significantly higher production of ET in the incompatible Pii R-gene-mediated interaction (Iwai et al. 2006). OsACS1 and OsACS2 were significantly induced upon *M. oryzae* infection, along with the induction of an ACC oxidase (ACO) gene, *OsACO7*. Silencing of *OsACS2* and *OsACO7* by RNA interference (RNAi) resulted in increased susceptibility to rice blast (Seo et al. 2011), suggesting that OsACS2 and ET production play a posi-tive role in rice resistance to *M. oryzae* infection. Helliwell et al. (2013) genetically manipulated the endogenous ET level in transgenic rice by expressing *OsACS2* (1-aminocyclopropane-1-carboxylic acid synthase) transgene under control of a strong rice pathogen-inducible promoter *PBZ1*. Rice plants generated exhibited increased resistance to *R. solani* and different races of *M. oryzae*. These results sug-gest that pathogen-inducible production of ET in transgenic rice can enhance broad-spectrum disease resistance to necrotrophic and hemibiotrophic fungal pathogens without negatively impacting crop productivity.

8.4.4 Engineering Rice Disease Resistance by Expressing Pathogen Protein Elicitor Gene

One promising approach to the achievement of broad-spectrum resistance is to incorporate genes that elicit general defense responses in plants. Several microbial protein elicitors have been shown to induce systemic acquired resistance in plants by activation of both SA- and ET/JA- mediated signalling pathway. The bacterial harpin and flagellin protein acts as elicitor in plants. Shao et al. (2008) introduced a harpin-encoding gene, *hrf1*, derived from *Xoo* into rice. Transgenic rice expressing *Xoo* harpin gene was highly resistant to all major races of *M. oryzae*. Bacterial flagellin expression induces disease resistance in transgenic rice (Takakura et al. 2008). Expression of the *PemG1* gene from *M. oryzae* in transgenic rice results in enhanced resistance to the rice blast fungus (Qiu et al. 2009). By characterizing the protein in the culture filtrate of rice blast fungus, two novel protein elicitors, MoHrip1 and MoHrip2, were identified, and subsequently their gene was isolated from the *M. oryzae* (Chen et al. 2012, 2014). The MoHrip1- and MoHrip2-expressing transgenic rice plants displayed higher resistance to rice blast and stronger tolerance to drought stress than wild-type rice (Wang et al. 2017). The MoHrip1 and MoHrip2 transgenic rice also exhibited enhanced agronomic traits such as increased plant height, tiller number, thousand-kernel weight, and ear number. Rice transformants overexpressing *MoSM1* protein elicitor gene from *M. oryzae* confers broad-spectrum resistance to both *Xoo and BLB* but at the same time had no effect on drought, salinity, or grain yield (Hong et al. 2017). The MoSM1-OE plants contained elevated levels of salicylic acid (SA) and jasmonic acid (JA) and constitutively activated the expression of SA and JA signaling-related regulatory and defense genes. However, no alteration in resistance to sheath blight disease was observed in MoSM1-OE lines.

8.5 RNAi-Mediated Gene Silencing in Rice to Engineer Disease Resistance

RNA interference, an evolutionarily conserved process that is active in a wide variety of eukaryotic organisms, is a sequence-specific gene-silencing mechanism that is induced by dsRNA (Baulcombe 2004). The dsRNA is diced into small interfering RNAs (siRNAs) of 21–24 nucleotides by an endonuclease called dicer. These siRNAs are then incorporated into the RNA- induced silencing complex to guide degradation or translational repression in a sequence-specific manner. Host-delivered RNAi (HD-RNAi) is a method which involves the production of double-stranded RNA (dsRNA) molecules targeting pathogen genes in the host plant, which will be processed further into small interfering RNA molecules (siRNAs). HD-RNAi has been successful in engineering resistance against plant virus (Duan et al. 2012), insects (Huvenne and Smagghe, 2010), nematode (Fairbairn et al. 2007), and fungi (Nunes and Dean, 2012). Recently, Tiwari et al. (2017) demonstrated that host-delivered RNAi method can be used for the control of sheath blight in rice. They transformed rice with the hairpin RNAi construct containing fusion of two pathogenicity Map Kinase 1 (*PKM1*) genes, *RPMK1-1* and *RPMK1-2* of *R. solani*. Due to host-delivered siRNA-mediated silencing of the target genes, the expression level of *RPMK1-1* and *RPMK1-2* was significantly lower in *R. solani* infecting transgenic rice, thereby enhancing sheath blight resistance in rice.

Ding et al. (2006) has developed a Brome Mosaic Virus (BMV)-based VIGS (virus-induced gene silencing) system to produce the siRNA of the target gene in rice. BMV-based system was employed to target the three predicted pathogenicity genes, *MoABC1*, *MoMAC1*, and *MoPMK1*. Zhu et al. (2017) studied the effectiveness of BMV-mediated HIGS (host-induced gene silencing) in silencing three predicted pathogenicity genes of *M. oryzae*. Inoculation of BMV viral vectors in rice resulted in systemically generating fungal gene-specific small interfering RNA(siRNA) molecules, which inhibited disease development and reduced the transcription of targeted fungal genes after subsequent *M. oryzae* inoculation (Zhu et al. 2017).

Virus resistance mediated by natural resistance genes and RNA silencing-mediated virus resistance are currently two major research focuses (Sasaya et al. 2014). Plant uses RNA silencing as a natural defense mechanism against plant viruses. Thus RNA silencing has been successfully exploited for engineering virus resistance in plants including rice. So far no natural resistance gene discovered for RBSDV in rice germplasm (Nicaise 2014). Rice black-streaked dwarf virus (RBSDV) is a dsRNA virus that causes severe yield loss in rice grown in Asia. Wang et al. (2016b) transformed rice with hairpin RNAi (hpRNAi) construct targeting four RBSDV genes, *S1*, *S2*, *S6*, and *S10*, encoding the RNA-dependent RNA polymerase, the putative core protein, the RNA silencing suppressor, and the outer capsid protein, respectively. Transgenic rice plants expressing the RBSDV hpRNA showed strong virus resistance in both the field and artificial assay system. Wang et al. (2016b) showed that long hpRNA targeting multiple viral genes can be used to generate stable and durable virus resistance in rice. They did small RNA deep sequencing on the RBSDV-resistant transgenic lines and detected siRNAs from all four viral gene sequences in the hpRNA transgene, indicating that the whole chimeric fusion sequence can be efficiently processed by dicer into siRNAs. Earlier to this report, transgenic rice plants containing an hpRNA transgene targeting the *P9-1*-encoding gene were almost immune to RBSDV infection (Shimizu et al. 2011).

8.6 Genome Engineering in Rice for Disease Resistance

Genome-editing technologies offer possibility of genome modification in a site-directed manner. Three popular genome-editing methods are zinc finger nucleases (ZNFs), transcription activator-like effector nucleases (TALENs), and CRISPER/

Cas9 system. Among the three methods, CRISPER/Cas9 system is an effective system for introducing mutation in the gene of interest in crop plants. Gene editing was successfully used for engineering disease resistance in rice. The rice bacterial blight susceptibility gene *Os11N3* (also called *OsSweet14*) was disrupted using TALEN genome-editing tool to provide *Xoo* resistance (Li et al. 2012). The *SWEET* gene encodes sucrose efflux transporter family and is hijacked by *Xoo*, using its endogenous TAL effectors AvrXa7 or PthXo3, to activate the gene and thus divert sugars from the plant cell so as to satisfy the pathogen's nutritional needs and enhance its persistence. Recently, Wang et al. (2016a) mutated (loss-of-function) the *OsERF922* gene by CRISPR/Cas9 method. Mutated rice lines thus created showed enhanced rice blast resistance without affecting the main agronomic traits. A natural allele of a C2H2-domain transcription factor gene, *bsr-d1*, confers broad-spectrum resistance to rice blast in Digu rice variety (Li et al. 2017). CRISPR/Cas9-mediated knockout of Bsr-d1 enhanced blast resistance without alteration in agronomic character (Li et al. 2017).

References

Baulcombe D (2004) RNA silencing in plants. Nature 431:356–363

Cao H, Bowling SA, Gordon AS, Dong X (1994) Characterization of an Arabidopsis mutant that is nonresponsive to inducers of systemic acquired resistance. Plant Cell 6:1583–1592

Cabauatan PQ, Cabunagan RC, Choi IR (2009) Rice viruses transmitted by the brown plant hopper Nilaparvata lugens Stål. In: Heong KL, Hardy B (eds) Planthoppers: New Threats to the Sustainability of Intensive Rice Production Systems in Asia. IRRI, Los Baños, pp 357–368

Carrillo MGC, Goodwin PH, Leach JE et al (2009) Phylogenomic relationships of Rice Oxalate Oxidases to the Cupin Superfamily and their association with disease resistance QTL. Rice 2:67–79. https://doi.org/10.1007/s12284-009-9024-0

Chen M, Zeng H, Qiu D, Guo L, Yang X, Shi H et al (2012) Purification and characterization of a novel hypersensitive response-inducing elicitor from Magnaporthe oryzae that triggers defense response in Rice. PLoS One 7:e37654

Chen M, Zhang C, Zi Q, Qiu D, Lie W, Zeng H (2014) A novel elicitor identified from Magnaporthe oryzae triggers defense responses in tobacco and rice. Plant Cell Rep 33:1865–1879

Chern M, Fitzgerald HA, Canlas PE, Navarre DA, Ronald PC (2005) Overexpression of a rice NPR1 homolog leads to constitutive activation of defense response and hypersensitivity to light. Mol Plant-Microbe Interact 18:511–520. https://doi.org/10.1094/MPMI-18-0511

Choi C, Hwang SH, Fang IR, Kwon SI, Park SR, Ahn I, Kim JB, Hwang DJ (2015) Molecular characterization of Oryza sativa WRKY6, which binds to W-box-like element 1 of the Oryza sativa pathogenesis-related (PR) 10a promoter and confers reduced susceptibility to pathogens. New Phytol 208:846–859. https://doi.org/10.1111/nph.13516

Coca M, Bortolotti C, Rufat M, Peñas G, Eritja R, Tharreau D, del Pozo AM et al (2004) Transgenic rice plants expressing the antifungal AFP protein from Aspergillus giganteus show enhanced resistance to the rice blast fungus Magnaporthe grisea. Plant Mol Biol 54:245–259

Coca M, Penas G, Gomez J, Campo S, Bortolotti C, Messeguer J, Segundo BS (2006) Enhanced resistance to the rice blast fungus Magnaporthe grisea conferred by expression of a cecropin A gene in transgenic rice. Planta 223:392–406. https://doi.org/10.1007/s00425-005-0069-z

Dangl JL, Horvath DM, Staskawicz BJ (2013) Pivoting the plant immune system from dissection to deployment. Science 341:746–751. https://doi.org/10.1126/science.1236011

Datta K, Baisakh N, Thet KM, Tu J, Datta SK (2002) Pyramiding transgenes for multiple resistance in rice against bacterial blight, yellow stem borer and sheath blight. Theor Appl Genet 106:1–8

Datta K, Koukolikova NZ, Baisakh N, Oliva N, Datta SK (2000) *Agrobacterium*-mediated engineering for sheath blight resistance of *indica* rice cultivars from different ecosystems. Theor Appl Genet 100:832–839

Datta K, Tu J, Oliva N, Ona I, Velazhahan R, Mew TW, Muthukrishnan S, Datta SK (2001) Enhanced resistance to sheath blight by constitutive expression of infection-related rice chitinase in transgenic elite *indica* rice cultivars. Plant Sci 60:405–414

Ding XS, Schneider WL, Chaluvadi SR, Mian MA, Nelson RS (2006) Characterization of a Brome mosaic virus strain and its use as a vector for gene silencing in monocotyledonous hosts. Mol Plant-Microbe Interact 19:1229–1239

Duan CG, Chun-Han W, Hui-Shan G (2012) Application of RNA silencing to plant disease resistance. Silence 3:5. silencejournal.com/content/3/1/5

Dutton MV, Evans CS (1996) Oxalate production by fungi : its role in pathogenicity and ecology in the soil environment. Can J Microbiol 42:881–895

Fitzgerald HA, Chern MS, Navarre R, Ronald PC (2004) Overexpression of (At)NPR1 in rice leads to a BTH-and environment-induced lesion-mimic/cell death phenotype. Mol Plant-Microbe Interact 17:140–151

Fairbairn DJ et al (2007) Host-delivered RNAi: an effective strategy to silence genes in plant parasitic nematodes. Planta 226:1525–1533

Fukushima S, Mori M, Sugano S, Takatsuji H (2016) Transcription factor WRKY62 plays a role in pathogen defense and hypoxia-responsive gene expression in rice. Plant Cell Physiol 57:2541–2551. https://doi.org/10.1093/pcp/pcw185

Giri MK, Gautam JK, Babu Rajendra Prasad V, Chattopadhyay S, Nandi AK (2017) Rice MYC2 (OsMYC2) modulates light-dependent seedling phenotype, disease defence but not ABA signalling. J Biosci 42:501–508. https://doi.org/10.1007/s12038-017-9703-8

Goto S, Sasakura-Shimoda F, Suetsugu M, Selvaraj MG, Hayashi N, Yamazaki M, Ishitani M, Shimono M, Sugano S, Matsushita A, Tanabata T, Takatsuji H (2015) Development of disease-resistant rice by optimized expression of WRKY45. Plant Biotechnol J 13:753–765. https://doi.org/10.1111/pbi.12303

Goto S, Sasakura-Shimoda F, Yamazaki M, Hayashi N, Suetsugu M, Ochiai H, Takatsuji H (2016) Development of disease-resistant rice by pathogen-responsive expression of WRKY45. Plant Biotechnol J 14:1127–1138

Helliwell EE, Wang Q, Yang YN (2013) Transgenic rice with inducible ethylene production exhibits broad-spectrum disease resistance to the fungal pathogens *Magnaporthe oryzae* and *Rhizoctonia solani*. Plant Biotechnol J 11:33–42

Hong Y, Yang Y, Zhang H, Huang L, Li D, Song F (2017) Overexpression of MoSM1, encoding for an immunity-inducing protein from *Magnaporthe oryzae*, in rice confers broad-spectrum resistance against fungal and bacterial diseases. Sci Rep 7:4103. https://doi.org/10.1038/srep41037

Huvenne H, Smagghe G (2010) Mechanisms of dsRNA uptake in insects and potential of RNAi for pest control: a review. J Insect Physiol 56:227–235

Iwai T, Miyasaka A, Seo S, Ohashi Y (2006) Contribution of ethylene biosynthesis for resistance to blast fungus infection in young rice plants. Plant Physiol 142:1202–1215

Jha S, Chattoo B (2010) Expression of a plant defensin in rice confers resistance to fungal phytopathogens. Transgenic Res 19:373–384. https://doi.org/10.1007/s11248-009-9315-7

Jha S, Tank HG, Prasad BD, Chattoo BB (2009) Expression of Dm-AMP1 in rice confers resistance to *Magnaporthe oryzae* and *Rhizoctonia solani*. Transgenic Res 18:59–69

Jiang S, Jiang L, Yang J, Peng J, Lu Y, Zheng H, Lin L, Chen J, Yan F (2018) Over-expression of *Oryza sativa* Xrn4 confers plant resistance to virus infection. Gene 639:44–51. https://doi.org/10.1016/j.gene.2017.10.004

Jisha V, Dampanaboina L, Vadassery J, Mithöfer A, Kappara S, Ramanan R (2015) Overexpression of an AP2/ERF type transcription factor OsEREBP1 confers biotic and abiotic stress tolerance in rice. PLoS One 10:e0127831. https://doi.org/10.1371/journal.pone.0127831

Jones JDG, Dangl JL (2006) The plant immune system. Nature 444:323–329

Kalpana K, Maruthasalam S, Rajesh T, Poovannan K, Kumar KK, Kokiladevi E, Raja JA, Sudhakar D, Velazhahan R, Samiyappan R (2006) Engineering sheath blight resistance in elite *indica* rice cultivars using genes encoding defense proteins. Plant Sci 170:203–215

Karmakar S, Molla KA, Chanda PK, Sarkar SN, Datta SK, Datta K (2016) Green tissue-specific co-expression of chitinase and oxalate oxidase 4 genes in rice for enhanced resistance against sheath blight. Planta 243:115–130

Karmakar S, Molla KA, Das K, Sarkar SN, Datta SK, Datta K (2017) Dual gene expression cassette is superior than single gene cassette for enhancing sheath blight tolerance in transgenic rice. Sci Rep 7:7900. https://doi.org/10.1038/s41598-017-08180-x

Khong GN, Pati PK, Richaud F, Parizot B, Bidzinski P et al (2015) OsMADS26 negatively regulates resistance to pathogens and drought tolerance in rice. Plant Physiol 169:2935–2949. https://doi.org/10.1104/pp.15.01192

Kim JK, Jang IC, Wu R, Zuo WN, Boston RS, Lee YH, Ahn IP, Nahm BH (2003) Co-expression of a modified maize ribosome-inactivating protein and a rice basic chitinase gene in transgenic rice plants confers enhanced resistance to sheath blight. Transgenic Res 12:475–484

Kou Y, Wang S (2010) Broad-spectrum and durability: understanding of quantitative disease resistance. Curr Opin Plant Biol 13:181–185

Knecht K, Seyffarth M, Desel C, Thurau T, Sherameti I, Lou B, Oelmüller R, Cai D (2010) Expression of BvGLP-1 encoding a germin-like protein from sugar beet in *Arabidopsis thaliana* leads to resistance against phytopathogenic fungi. Mol Plant-Microbe Interact 23:446–457. https://doi.org/10.1094/MPMI-23-4-0446

Kumar KK, Poovannan K, Nandakumar R, Thamilarasi K, Geetha C, Jayashree N, Kokiladevi E, Raja JAJ, Samiyappan R, Sudhakar D, Balasubramanian P (2003) A high throughput functional expression assay system for a defence gene conferring transgenic resistance on rice against the sheath blight pathogen, *Rhizoctonia solani*. Plant Sci 165:969–976

Lee H, Cha J, Choi C, Choi N, Ji HS, Park SR, Lee S, Hwang DJ (2018) Rice WRKY11 plays a role in pathogen defense and drought tolerance. Rice (N Y) 11:5. https://doi.org/10.1186/s12284-018-0199-0

Lee IH, Jung YJ, Cho YG, Nou IS, Huq MA et al (2017) SP-LL-37, human antimicrobial peptide, enhances disease resistance in transgenic rice. PLoS One 12:e0172936. https://doi.org/10.1371/journal.pone.0172936

Li C, Song J, Jiang L (1999) Research progress on the maize rough dwarf virus disease. Plant Prot 25:34–37

Li T, Liu B, Spalding MH, Weeks DP, Yang B (2012) High-efficiency TALEN-based gene editing produces disease resistant rice. Nat Biotechnol 30:390–392

Li P, Pei Y, Sang X, Ling Y, Yang Z, He G (2009) Transgenic indica rice expressing a bitter melon (*Momordica charantia*) class I chitinase gene (McCHIT1) confers enhanced resistance to *Magnaporthe grisea* and *Rhizoctonia solani*. Eur J Plant Pathol 125:533–543

Li W, Zhu Z, Chern M, Yin J, Yang C, Ran L et al (2017) A natural allele of a transcription factor in rice confers broad-Spectrum blast resistance. Cell 170:114–126.e15. https://doi.org/10.1016/j.cell.2017.06.008

Lin W, Anuratha C, Datta K, Potrykus I, Muthukrishnan S, Datta SK (1995) Genetic engineering of rice for resistance to sheath blight. Nat Biotechnol 13:686–691

Liu G, Jia Y, McClung KM, Datta A, Correll JC (2009) Mapping quantitative trait loci responsible for resistance to sheath blight in rice. Phytopathology 99:1078–1084

Liu Q, Yang J, Yan S, Zhang S, Zhao J, Wang W et al (2016) The germin-like protein OsGLP2-1 enhances resistance to fungal blast and bacterial blight in rice. Plant Mol Biol 92(4–5):411–423

Liu Z, Li X, Sun F, Zhou T, Zhou Y (2017) Over expression of OsCIPK30 enhances plant tolerance to rice stripe virus. Front Microbiol 8:2322. https://doi.org/10.3389/fmicb.2017.02322

Ma H, Chen J, Zhang Z, Ma L, Yang Z, Zhang Q, Li X, Xiao J, Wang S (2017) MAPK kinase 10.2 promotes disease resistance and drought tolerance by activating different MAPKs in rice. Plant J 92:557–570. https://doi.org/10.1111/tpj.13674

Maruthasalam S, Kalpana K, Kumar KK, Loganathan M, Poovannan K, Raja JAJ (2007) Pyramiding transgenic resistance in elite *indica* rice cultivars against the sheath blight and bacterial blight. Plant Cell Rep 26:791–804

Molla KA, Karmakar S, Chanda PK, Ghosh S, Sarkar SN, Datta SK, Datta K (2013) Rice oxalate oxidase gene driven by green tissue specific promoter increases tolerance to sheath blight pathogen (*Rhizoctonia solani*) in transgenic rice. Mol Plant Pathol 14:910–922

Molla KA, Karmakar S, Chanda PK, Sarkar SN, Datta SK, Datta K (2016) Tissue-specific expression of Arabidopsis NPR1 gene in rice for sheath blight resistance without compromising phenotypic cost. Plant Sci 250:105–114. https://doi.org/10.1016/j.plantsci.2016.06.005

Narasimhan ML, Coca MA, Jin J, Yamauchi T, Ito Y, Kadowaki T et al (2005) Osmotin is a homolog of mammalian adiponectin and controls apoptosis in yeast through a homolog of mammalian adiponectin receptor. Mol Cell 17:171–180

Nicaise V (2014) Crop immunity against viruses: outcomes and future challenges. Front Plant Sci 5:660. https://doi.org/10.3389/fpls.2014.00660

Nishizawa Y, Nishio Z, Nakazono K, Soma M, Nakajima E, Ugaki M (1999) Enhanced resistance to blast (*Magnaporthe grisea*) in transgenic japonica rice by constitutive expression of rice chitinase. Theor Appl Genet 99:383–390

Nunes CC, Dean RA (2012) Host-induced gene silencing: a tool for understanding fungal host interaction and for developing novel disease control strategies. Mol Plant Pathol 13:519–529

Pajerowska-Mukhtar K, Wang W, Tada Y, Dong X (2012) The HSF-like transcription factor TBF1 is a major molecular switch for plant growth-to-defense transition. Curr Biol 22:103–112. https://doi.org/10.1016/j.cub.2011.12.015

Patkar RN, Chattoo BB (2006) Transgenic *indica* rice expressing ns-LTP-like protein shows enhanced resistance to both fungal and bacterial pathogens. Mol Breed 17:159–171

Peng X, Hu Y, Tang X, Zhou P, Deng X, Wang H, Guo Z (2012) Constitutive expression of rice WRKY30 gene increases the endogenous jasmonic acid accumulation, PR gene expression and resistance to fungal pathogens in rice. Planta 236:1485–1498. https://doi.org/10.1007/s00425-012-1698-7

Peng Y, Le B, Chern X, Dardick C, Chern M, Ruan R, Canlas PE, Ronald PC (2008) OsWRKY62 is a negative regulator of basal and Xa21-mediated defence against *Xanthomonas oryzae* pv. *oryzae* in rice. Mol Plant 1:446–458

Qi Z, Yu J, Shen L, Yu Z, Yu M et al (2017) Enhanced resistance to rice blast and sheath blight in rice (*Oryza sativa* L.) by expressing the oxalate decarboxylase protein Bacisubin from Bacillus subtilis. Plant Sci 265:51–60. https://doi.org/10.1016/j.plantsci.2017.09.014

Qiu D, Xiao J, Ding X, Xiong M, Cai M, Cao Y, Li X, Xu C, Wang S (2007) OsWRKY13 mediates rice disease resistance by regulating defense-related genes in salicylate- and jasmonate-dependent signaling. Mol Plant-Microbe Interact 20:492–449

Qiu D, Xiao J, Xie W, Liu H, Li X, Xiong L, Wang S (2008) Rice gene network inferred from expression profiling of plants overexpressing OsWRKY13, a positive regulator of disease resistance. Mol Plant 1:538–551. https://doi.org/10.1093/mp/ssn012

Qiu D, Mao J, Yang X, Zeng H (2009) Expression of an elicitor-encoding gene from *Magnaporthe grisea* enhances resistance against blast disease in transgenic rice. Plant Cell Rep 28:925–933. https://doi.org/10.1007/s00299-009-0698-y

Quilis J, Peñas G, Messeguer J, Brugidou C, Segundo BS (2008) The Arabidopsis AtNPR1 inversely modulates defense responses against fungal, bacterial, or viral pathogens while conferring hypersensitivity to abiotic stresses in transgenic rice. Mol Plant-Microbe Interact 21:1215–1231

Rajashekara H, Ellur RK, Khanna A, Nagarajan M, Gopala Krishnan S et al (2014) Singh inheritance of blast resistance and its allelic relationship with five major R genes in a rice landrace 'Vanasurya'. Ind Phytopathol 67:365–369

Richa K, Tiwari IM, Devanna BN, Botella JR, Sharma V, Sharma TR (2017) Novel Chitinase gene LOC_Os11g47510 from Indica Rice Tetep provides enhanced resistance against sheath blight pathogen *Rhizoctonia solani* in Rice. Front Plant Sci 8:596. https://doi.org/10.3389/fpls.2017.00596

Richa K, Tiwari IM, Kumari M, Devanna BN, Sonah H, Kumari A et al (2016) Functional characterization of novel chitinase genes present in the sheath blight resistance QTL: qSBR11-1 in rice line tetep. Front Plant Sci 7:244. https://doi.org/10.3389/fpls.2016.00244

Sadumpati V, Kalambur M, Vudem DR, Kirti PB, Khareedu VR (2013) Transgenic indica rice lines expressing *Brassica juncea* Nonexpressor of pathogenesis-related genes 1 (BjNPR1), exhibit enhanced resistance to major pathogens. J Biotechnol 166:114–121

Sasaya T, Nakazono-Nagaoka E, Saika H, Aoki H, Hiraguri A, Netsu O et al (2014) Transgenic strategies to confer resistance against viruses in rice plants. Front Microbiol 4:409. https://doi.org/10.3389/fmicb.2013.00409

Seo S, Mitsuhara I, Feng J, Iwai T, Hasegawa M, Ohashi Y (2011) Cyanide, a coproduct of plant hormone ethylene biosynthesis, contributes to the resistance of rice to blast fungus. Plant Physiol 155:502–514

Shah JM, Raghupathy V, Veluthambi K (2009) Enhanced sheath blight resistance in transgenic rice expressing an endochitinase gene from *Trichoderma virens*. Biotechnol Lett 31:239–244

Shah JM, Singh R, Veluthambi K (2013) Transgenic rice lines constitutively co-expressing tlp-D34 and chi11 display enhancement of sheath blight resistance. Biol Plant 57:351–358

Shao M, Wang J, Dean RA, Lin Y, Gao X, Hu S (2008) Expression of a harpin-encoding gene in rice confers durable nonspecific resistance to *Magnaporthe grisea*. Plant Biotechnol J 6:73–81. https://doi.org/10.1111/j1467-7652200700304x. PMID: 18005094

Sharma TR, Rai AK, Gupta SK, Vijayan J, Devanna BN, Ray S (2012) Rice blast management through host-plant resistance: retrospect and prospects. Agric Res 1:37–52

Sharma A, Sharma R, Imamura M, Yamakawa M, Machii H (2000) Transgenic expression of cecropin B, an antibacterial peptide from Bombyx mori, confers enhanced resistance to bacterial leaf blight in rice. FEBS Lett 484:7–11

Shimizu T, Nakazono-Nagaoka E, Akita F, Uehara-Ichiki T, Omura T, Sasaya T (2011) Immunity to rice black streaked dwarf virus, a plant reovirus, can be achieved in rice plants by RNA silencing against the gene for the viroplasm component protein. Virus Res 160:400–403

Shimono M, Sugano S, Nakayama A, Jiang CJ, Ono K, Toki S et al (2007) Rice WRKY45 plays a crucial role in benzothiadiazole-inducible blast resistance. Plant Cell 19:2064–2076. https://doi.org/10.1105/tpc.106.046250

Shimono M, Koga H, Akagi A, Hayashi N, Goto S, Sawada M et al (2012) Rice WRKY45 plays important roles in fungal and bacterial disease resistance. Mol Plant Pathol 13:83–94. https://doi.org/10.1111/j.1364-3703.2011.00732.x

Sridevi G, Sabapathi N, Meena P, Nandakumar R, Samiyappan R, Muthukrishnan S, Veluthambi K (2003) Transgenic *indica* rice variety Pusa basmati 1 constitutively expressing a rice chitinase gene exhibits enhanced resistance to *Rhizoctonia solani*. J Plant Biochem Biotechnol 12:93–101

Sugano S, Jiang C-J, Miyazawa S-I, Masumoto C, Yazawa K, Hayashi N et al (2010) Role of OsNPR1 in rice defense program as revealed by genome-wide expression analysis. Plant Mol Biol 74:549–562. https://doi.org/10.1007/s11103-010-9695-3

Sundaram RM, Chatterjee S, Oliva R, Laha GS, Cruz LJE, Sonti RV (2014) Update on bacterial blight of rice: fourth international conference on bacterial blight. Rice (N Y) 7:12. https://doi.org/10.1186/s12284-014-0012-7

Takakura Y, Che FS, Ishida Y, Tsutsumi F, Kurotani K, Usami S et al (2008) Expression of a bacterial flagellin gene triggers plant immune responses and confers disease resistance in transgenic rice plants. Mol Plant Pathol 9:525–529. https://doi.org/10.1111/j1364-3703200800477x. PMID: 18705865

Tao Z, Liu H, Qiu D, Zhou Y, Li X, Xu C, Wang S (2009) A pair of allelic WRKY genes play opposite roles in rice–bacteria interactions. Plant Physiol 151:936–948

Tiwari IM, Jesuraj A, Kamboj R, Devanna BN, Botella JR (2017) Sharma TR (2017) host delivered RNAi, an efficient approach to increase rice resistance to sheath blight pathogen (*Rhizoctonia solani*). Sci Rep 7:7521. https://doi.org/10.1038/s41598-017-07749-w

Ulker B, Somssich IE (2004) WRKY transcription factors: from DNA binding towards biological function. Curr Opin Plant Biol 7:491–498

Vo KTX, Kim CY, Hoang TV, Lee SK, Shirsekar G, Seo YS et al (2018) OsWRKY67 plays a positive role in basal and XA21-mediated resistance in rice. Front Plant Sci 8:2220. https://doi.org/10.3389/fpls.2017.02220

Wang F, Wang C, Liu P, Lei C, Hao W, Gao Y, Liu YG, Zhao K (2016a) Enhanced rice blast resistance by CRISPR/Cas9-targeted mutagenesis of the ERF transcription factor gene OsERF922. PLoS One 11:e0154027. https://doi.org/10.1371/journal.pone.0154027. eCollection 2016

Wang F, Li W, Zhu J, Fan F, Wang J, Zhong W, Wang MB, Liu Q et al (2016b) Hairpin RNA targeting multiple viral genes confers strong resistance to Rice Black-Streaked Dwarf Virus. Int J Mol Sci 17(5). pii:E705). https://doi.org/10.3390/ijms17050705

Wang Z, Han Q, Zi Q, Lv S, Qiu D, Zeng H (2017) Enhanced disease resistance and drought tolerance in transgenic rice plants overexpressing protein elicitors from *Magnaporthe oryzae*. PLoS One 12:e0175734. https://doi.org/10.1371/journal.pone.0175734

Xu G, Greene GH, Yoo H, Liu L, Marqués J, Motley J, Dong X (2017a) Global translational reprogramming is a fundamental layer of immune regulation in plants. Nature 545:487–490. https://doi.org/10.1038/nature22371

Xu G, Yuan M, Ai C, Liu L, Zhuang E, Karapetyan S, Wang S, Dong X (2017b) uORF-mediated translation allows engineered plant disease resistance without fitness costs. Nature 545:491–494. https://doi.org/10.1038/nature22372

Xue X, Cao ZX, Zhang XT, Wang Y, Zhang YF, Chen ZX, Pan XB, Zuo SM (2016) Overexpression of OsOSM1 enhances resistance to rice sheath blight. Plant Dis 100:1634–1642

Yokotani N, Sato Y, Tanabe S, Chujo T, Shimizu T, Okada K, Yamane H, Shimono M, Sugano S, Takatsuji H, Kaku H, Minami E, Nishizawa Y (2013) OsWRKY76 is a rice transcriptional repressor playing opposite roles in blast disease resistance and cold stress tolerance. J Exp Bot 64:5085–5097. https://doi.org/10.1093/jxb/ert298

Yuan Y, Zhong S, Li Q, Zhu Z, Lou Y, Wang L, Wang J, Wang M, Li Q, Yang D (2007) Functional analysis of rice NPR1-like genes reveals that OsNPR1/NH1 is the rice orthologue conferring disease resistance with enhanced herbivore susceptibility. Plant Biotechnol J 5: 313–324

Zuo S, Zhang Y, Chen Z, Jiang W, Feng M, Pan X (2014) Improvement of rice resistance to sheath blight by pyramiding QTLs conditioning disease resistance and tiller angle. Rice Sci 21:318–326

Zhu L, Zhu J, Liu Z, Wang Z, Zhou C, Wang H (2017) Host-Induced Gene Silencing of Rice Blast Fungus *Magnaporthe oryzae* Pathogenicity Genes Mediated by the Brome Mosaic Virus. Genes (Basel) 8(10.) pii: E241). https://doi.org/10.3390/genes8100241

Chapter 9
Genetic Transformation of Sugarcane and Field Performance of Transgenic Sugarcane

Gauri Nerkar, Avinash Thorat, Suman Sheelavantmath, Harinath Babu Kassa, and Rachayya Devarumath

Abstract Sugarcane is an important industrial cash crop contributing more than 70% of the sugar and 40% of biofuel production globally. The complex polyploid-aneuploid type of genome of sugarcane makes it difficult to generate hybrids through conventional breeding programs. Thus, genetic improvement of sugarcane through transgenic approaches has fascinated the attention of most biotechnologists around the world. Moreover, plant biotechnology has the potential to improve economically important traits in sugarcane as well as diversify sugarcane beyond traditional applications such as sucrose production. Although being a recalcitrant species for transformation, several advances have been made in the area of sugarcane transformation. Traits such as disease resistance, improved tolerance to salt and drought, and increased sucrose content through metabolic engineering and expression of recombinant proteins (biopharming) have been some of the areas which appear promising as far as the application of transgenic sugarcane is concerned. Stability of the transgene expression is another major bottleneck when transforming a polyploid crop like sugarcane. This chapter will help to focus on the efficient molecular tools and improved transgenic methodologies used during sugarcane transformation in addition to the field performance of transgenic sugarcane.

Keywords *Agrobacterium* · Biolistic · Field performance · Minimal gene cassettes · Promoters · Sugarcane · Transgenic

G. Nerkar · A. Thorat · H. B. Kassa · R. Devarumath (✉)
Molecular Biology and Genetic Engineering Laboratory, Vasantdada Sugar Institute, 412307, Pune, Maharashtra, India

S. Sheelavantmath
Molecular Biology and Genetic Engineering Laboratory, Vasantdada Sugar Institute, 412307, Pune, Maharashtra, India

Department of Biotechnology, Sinhgad College of Science, Ambegaon, 412307, Pune, Maharashtra, India

© Springer International Publishing AG, part of Springer Nature 2018
S. S. Gosal, S. H. Wani (eds.), *Biotechnologies of Crop Improvement, Volume 2*,
https://doi.org/10.1007/978-3-319-90650-8_9

9.1 Introduction

Sugarcane is an important food and bioenergy source and a significant component of the economy in many countries in the tropics and subtropics. The *Saccharum* genus is a group of crop species particularly challenging for improvement. Cultivars are interspecific aneuploid hybrids. The crossing of large genomes (with multiple recent duplications that allow chromosome pairing and recombination) makes each progeny genotype a unique genome. To improve yield and other traits of interest for the development of an energy cane, research must unravel the complexities of the sugarcane genome, develop statistical genetics for highly polyploid genomes, and identify genes associated with sucrose content, drought resistance, biomass, and cell wall recalcitrance (Waclawovsky et al. 2010). Despite its enormous potential as a bioenergy crop, breeding efforts for improving the narrow genetic base of modern cultivars of sugarcane are often constrained by the high ploidy level and interspecific origins (Lakshmanan 2005; Meng et al. 2006; D'Hont et al. 2008). In the past three decades, sugarcane has benefited from biotechnology through engineering genes for the improvement of traits of interest (Lakshmanan 2005; Meng et al. 2006; Brumbley et al. 2008; D'Hont et al. 2008; Altpeter and Oraby 2010) following the first report of transgenic sugarcane plants (Bower and Birch 1992). Sugarcane transformation (Fig. 9.1) has come a far way during the past two decades. The availability of reliable and efficient methods for transformation has prompted transformation of a large number of agronomic traits into sugarcane. Indonesia has become the first country to commercialize transgenic sugarcane cultivation (Parisi

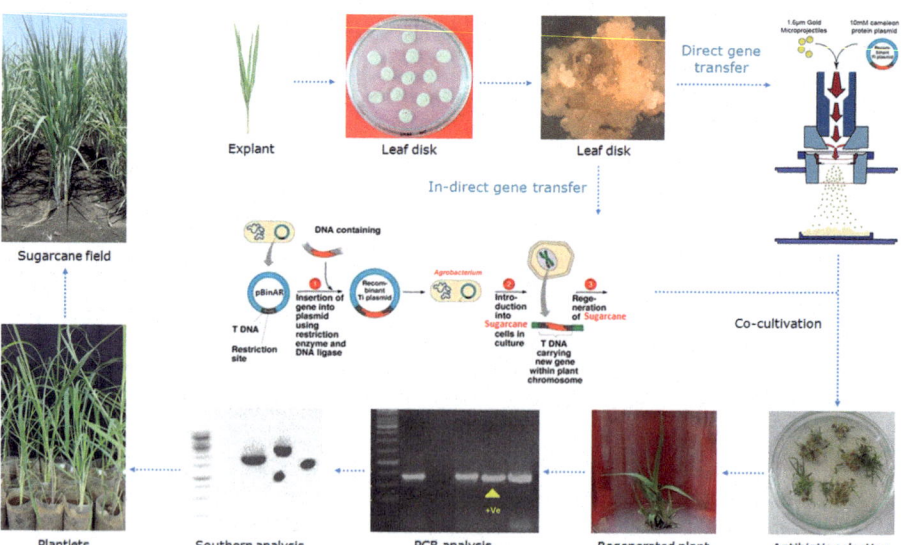

Fig. 9.1 Schematic representation of biolistic and *Agrobacterium*-mediated transformation methods for sugarcane

et al. 2016) and is likely to be commercialized in other countries like Brazil, India, and Australia. Thus, sugarcane is a potential candidate as a far as the release of transgenic crops is concerned in the future. Hence, it is important to understand the important factors that contribute to the success of transgenic sugarcane. This chapter discusses the advances in the field of sugarcane transformation by reviewing the molecular tools, the transformation methods, the selection system, the agronomic traits transformed into sugarcane, the field performance of transgenic sugarcane, and finally the biosafety consideration for transgenic sugarcane.

9.2   Molecular Tools for Sugarcane Transformation

9.2.1   Promoters Used for Enhanced Transgene Expression

Transgene expression mainly depends on the use of potent promoters that regulate the temporal and spatial expression patterns of transgenes. Hence the development of genomic tools or the identification of potent promoters is the major priority for the genetic improvement of major crops especially for the production of useful products that can be expressed to levels suitable for commercialization. The use of potent promoters becomes even more important in crops like sugarcane where transgene silencing is the major bottleneck in the production of transgenic lines that have high transgene expression. The currently used monocot promoters are relatively few and mostly derived from highly expressed constitutive plant genes, such as the ubiquitin (Ubi) promoters, maize Ubi1 (Christensen and Quail 1996), sugarcane ub4 and ub9 (Wei et al. 2003), rice RUBQ2 (Liu et al. 2003), *Porteresia coarctata* Ubi2.3 (Philip et al. 2013), and *Erianthus arundinaceus* Eriubi D7 (Chakravarthi et al. 2015). Recently, tissue-specific monocot promoters have also been developed that target gene expression in stems. Promoters functional in the sugarcane culm include sugarcane dirigent and O-methyltransferase from putative defense and fiber biosynthesis-related genes (Damaj et al. 2010), maize phosphoenolpyruvate carboxylase (Harrison et al. 2011; Kinkema and Miles 2013), and sugarcane Loading Stem Gene (Moyle and Birch 2013). Kinkema et al. (2014) showed that the CmYLCV promoter combined with maize polyubiquitin-1 intron (CMP-Zm-iUbi1) represents one of the strongest promoters described to date for driving transgene expression in sugarcane leaf and stem. Gene modification and a dual transcriptional enhancer sequence were shown to be valuable tools in sugarcane for substantially improving transgene expression levels from different promoters (Kinkema et al. 2014). Petrasovits et al. (2012) tested plant and viral promoters, viz., the maize and rice polyubiquitin promoters, the maize chlorophyll A-/B-binding protein promoter, and a Cavendish banana streak badnavirus promoter, in combination with multigene or single-gene constructs to increase PHB levels. At the seedling stage, the highest levels of polymer were produced in sugarcane plants when the Cavendish banana streak badnavirus promoter was used. However, in all cases, this promoter underwent silencing as the plants matured. The rice Ubi promoter enabled the production of PHB at levels similar to the maize Ubi promoter. The maize chlorophyll A-/B-binding protein promoter enabled the production of PHB to levels as high as 4.8% of the leaf dry weight, which is approximately 2.5 times higher than previously reported levels in sugarcane (Petrasovits et al. 2012). However, till date, the maize ubiquitin promoter has been the benchmark for the transgene expression in sugarcane. Most of the sugarcane transgenic plants assessed for their field performance are developed using this promoter.

9.2.2   Use of Minimal Gene Cassettes for Efficient Sugarcane Transformation

Use of minimal gene cassette comprising of the transgene of interest, promoter, and terminator has been tested in sugarcane (Beyene et al. 2011; Taparia et al. 2012a; Taparia et al. 2012b; Jackson et al. 2013; Wu et al. 2015) and may have great potential for the generation of commercial transgenic sugarcane events mainly because it avoids using the whole plasmid vectors that contain undesirable bacterial selection genes and vector backbone sequence. Reducing the amount of MC to 10 ng per shot led to simple transgene integration and stable transgene expression in sugarcane (Taparia et al. 2012a) due to the absence of prokaryotic backbone sequences which can contribute to DNA recombination (Kohli et al. 1999) leading to posttranscriptional gene silencing (Plasterk and Ketting 2000; Hammond et al. 2001) and the absence of bacterial sequences that may induce methylation at the transgene loci, causing transcriptional gene silencing (Clark et al. 1997; Jakowitsch et al. 1999). In addition, prokaryotic vector backbone sequences in transgenic events may negatively impact regulatory approval (Zhao et al. 2007).

9.3   Comparison of Biolistic- and *Agrobacterium*-Mediated Methods for Sugarcane Transformation

Biolistic- and *Agrobacterium*-mediated transformation methods in sugarcane were first reported by Bower and Birch (1992) and Arencibia et al. (1998), respectively. Most published reports to date assessing transgene expression in sugarcane are based on biolistic transformation (Braithwaite et al. 2004; Hamerli and Birch 2011; Harrison et al. 2011; Petrasovits et al. 2012; Moyle and Birch 2013; Mudge et al. 2013) due to its applicability to a wide range of genotypes (Altpeter and Oraby 2010). *Agrobacterium* tumefaciens-mediated gene transfer is typically limited to few genotypes (Jackson et al. 2013; Joyce et al. 2014) but is preferred mostly due to its ability to generate single-copy insertions (Dong et al. 2014; Joyce et al. 2014). Both methods result in variable transgene integration complexities with subsequent consequences for the transgene performance. Transgenes which are inserted in multiple copies are more likely to be silenced (Meyer 1995; Schubert et al. 2004; Meng et al. 2006). But even single-copy transgenic events can undergo silencing depending on where in the genome they are inserted (Stoger et al. 1998; Kohli et al. 1999). Single-copy integration of transgenes into the plant genome also facilitates structural characterization (Que et al. 2014).

Only a few reports exist which compare the efficiency of biolistic- and *Agrobacterium*-mediated methods of transformation in sugarcane (Jackson et al. 2013; Joyce et al. 2014; Wu et al. 2015). A direct comparison of *Agrobacterium*-mediated transformation (AMT) and particle bombardment using whole plasmid (WP) and excised minimal cassettes (MC), for transformation efficiency, transgene integration complexity, and transgene expression, was made in sugarcane (Jackson et al. 2013) using a selectable marker gene and a luciferase reporter. They concluded that assuming a preference for low-copy transformants with high expression, the preferred methods would be bombardment using MC at low DNA concentration (6.6 ng per shot in our example) or AMT (in an amenable cultivar). Both of these methods yielded several likely single-copy transformants with LUC activity above 108 units per microgram of cellular protein, within easily produced population sizes of transgenic sugarcane plant lines.

Joyce et al. (2014) assessed the field performance of transgenic sugarcane plants for 3 years. Their results showed an overall reduction in growth and cane yield in biolistic, *Agrobacterium*-mediated method as well as untransformed tissue culture (TC) events, compared with the parent clone (PC) Q117 (no transformation or tissue culture) in the plant, first-ratoon and second-ratoon crops. However, when individual events were analyzed separately, yields of some transgenic events from both *Agrobacterium*-mediated and biolistic methods were comparable to PC, suggesting that either transformation method can produce commercially suitable clones. Interestingly, a greater percentage of biolistic transformants were similar to PC for growth and yield than *Agrobacterium*-mediated transformed clones. Crop ratoonability and sugar yield components [Brix%, Pol%, and commercial cane sugar (CCS)] were unaffected by transformation or tissue culture. Transgene expression remained stable over different crop cycles and increased with plant maturity. Transgene copy number did not influence transgene expression, and both transformation methods produced low transgene copy number events. No consistent pattern of genetic changes was detected in the test population using three DNA fingerprinting techniques.

The transformation efficiency, transgene integration pattern, expression level, and expression stability were compared in the commercially important sugarcane cultivar CP88-1762 (Wu et al. 2015). There were no significant differences between the two transformation systems for transformation efficiency, the frequency of single-copy integration, or the level and stability of transgene expression when carried out with the same expression cassette, tissue culture, and selection procedure in 12 independent experiments.

From all the three reports discussed above, it is evident that both biolistic transformations using minimal gene cassettes and *Agrobacterium*-mediated transformation (in amenable genotypes) provide suitable platforms for generation of elite sugarcane events.

9.4 Selectable Markers Used for Sugarcane Transformation

The most crucial step in plant transformation is the selection of transformed cells that are regenerable. Therefore employing a selection system that efficiently kills the non-transformed cells as well as allows the proliferation of the transformed cells is required for the development of a reliable transformation protocol (Wilmink and Dons 1993). Antibiotics or herbicides have been most commonly used for sugarcane transformation, so far, which include (i) npt II (neomycin phosphotransferase)/G418 (Geneticin), (ii) bar (phosphinothricin acetyltransferase)/PPT (phosphinothricin), and (iii) hpt (hygromycin phosphotransferase)/hygromycin. Neomycin phosphotransferase (npt-II) selectable marker gene in combination with Geneticin selection is the most frequently used selection system. Transgenic lines grew rapidly and produced roots on medium containing 0.04 mM (25 mg/L) Geneticin which eliminated all escapes (Bower and Birch 1992). Falco et al. (2000) reported only 3% of the total transgenic sugarcane plant population escaping Geneticin selection, whereas the escape rate was 42% in case of plants regenerated in the presence of bialaphos for resistance to the bar gene in immature leaf and immature inflorescence-derived calli (Gallo-Meagher and Irvine 1996). However, for axillary bud transformation, a comparison of kanamycin, Geneticin, and phosphinothricin (PPT) selection showed that PPT was the most effective selection agent (Manickavasagam et al. 2004).

Consumer concern over the presence of antibiotic and herbicide resistance genes in genetically modified crops has led to the development of alternative selection procedures. A "positive" selection regimen essentially incorporates a physiologically inert metabolite as the selection agent and a corresponding selectable marker gene that confers a metabolic advantage, thus alleviating the growth inhibitory effects of selection for the transformed cells (Altpeter and Oraby 2010). The utility of the phosphomannose isomerase (PMI)/mannose-based selection system has been established for sugarcane by Jain et al. (2007) with an escape rate of 44%. Zhang et al. (2014) compared the pmi/mannose selection system with bar/PPT and nptII/G-418 and reported that pmi/mannose system had a transformation frequency of 4.2%, which were higher than the frequencies obtained in the widely used negative selection systems bar/PPT (0.47%), nptII/G418 (1.38%), and hpt/hygromycin (0.63%).

In the recent years, plant genes containing amino acid substitutions leading to decreased herbicide binding have been used for selection in sugarcane (van der Vyver et al. 2013). This report uses a selectable marker gene of plant origin – acetolactate synthase (ALS) – which is targeted by a sulfonylurea herbicide. However, the ALS selection system was less effective in sugarcane when compared to other well-established systems, such as the nptII gene and use of Geneticin. When the

Geneticin/nptII system was applied for selecting transgenic sugarcane calli, 1–5% of the total bombarded calli would survive with almost no escapes (Bower and Birch 1992; Snyman et al. 1996). In chlorsulfuron/als system, however, up to 20% of the bombarded calli survived after 8 weeks of in vitro selection with insufficient new growth to easily differentiate between transgenic and non-transgenic calli. Also, from the surviving calli, 30 putative transgenic plants regenerated, of which 6 contained the als gene, an escape rate of 20% (van der Vyver et al. 2013).

Since the existing selection systems for use in sugarcane transformation are predominantly based on antibiotic resistance, there is an urgent need for the development of alternative selection systems that might be more acceptable to producers and consumers. Developing additional, alternative selection systems for specific crops will also allow for repeated transformations where more than one selection system is needed for multiple gene transfers in gene stacking approaches into a particular plant species.

9.5   Sugarcane Transformation for Agronomic Traits

Several genes of agronomic importance like disease/pest resistance, salt and drought tolerance, and sugar accumulation have been engineered in sugarcane (Altpeter and Oraby 2010; Srikanth et al. 2011; Hotta et al. 2011).

9.5.1   Disease/Pest Resistance

Genes from *Bacillus thuringiensis* (Bt) have been used for development of borer resistance sugarcane using various cry genes such as Cry1A (Arencibia et al. 1999), Cry1Ab (Braga et al. 2003), Cry1Ac (Jing-Sheng et al. 2008), and Cry1Aa3 (Kalunke et al. 2009). Transgenic sugarcane for borer resistance was also reported using Cry1Aa3 gene. The efficacy of native Cry1Aa, Cry1Ab, and Cry1Ac against *C. infuscatellus* was tested using in vitro bioassays through diet surface contamination method and Cry1Ab as the most toxic among the three compounds (Arvinth et al. 2010). Christy et al. (2009) transferred aprotinin genes to sugarcane cultivars and showed in the in vivo bioassay studies that larvae of top borer *Scirpophaga excerptalis* Walker (Lepidoptera: Pyralidae) fed on transgenics showed a significant reduction in weight and impairment of larval development. Zhu et al. (2011) and Gilbert et al. (2005) produced transgenic sugarcane with an untranslatable coat protein gene of SCYLV; Joyce et al. (1998) developed sugarcane mosaic virus-resistant sugarcane using coat protein gene SCMV with the aim of improving resistance to this virus. On the other hand, transgenic sugarcane resistant to leaf scald disease (Zhang et al. 1998) and Fiji leaf gall disease (McQualter et al. 2001) was developed by using albD and FDVS9 genes, respectively. A recent report based on red rot-resistant transgenic sugarcane through the expression of β-1,3-glucanase gene from

Trichoderma spp. described the stable transfer of the second clonal generation raised from resistant transgenic plants and showed the transgene expression was upregulated (up to 2.0-fold in leaves and 5.0-fold in roots) after infection, as compared to before infection in resistant plants (Nayyar et al. 2017).

9.5.2   Herbicide Resistance

Owing to the lack of herbicide-resistant genes in gene pool, just like other crop species, sugarcane is sensitive to the herbicide, and thus it needs exogenous genes to improve its herbicide resistance. Genetically modified sugarcane resistant to herbicide, usually with transformation of the genes such as bar and epsps into sugarcane genome, has been reported (Gallo-Meagher and Irvine 1996; Falco et al. 2000). Transformation of bar gene conferring glufosinate ammonium tolerance has also been reported in sugarcane (Enriquez-Obregon et al. 2000; Gallo-Meagher and Irvine 1996; Falco et al. 2000; Leibbrandt and Snyman 2001). Recently, it is demonstrated that the mutant sorghum acetolactate synthase (mALS) gene from sorghum supports the production of intragenic, herbicide-resistant sugarcane (Dermawan et al. 2016).

9.5.3   Abiotic Stress Tolerance

Trehalose synthase gene (TSase) was transferred into sugarcane using *Agrobacterium*- mediated method to improve sugarcane drought tolerance (Zhang et al. 2000; Wang et al. 2005). Kumar et al. (2014) reported the transformation of sugarcane cultivar CP-77-400 with AVP1 (*Arabidopsis* vacuolar pyrophosphatase) gene using apical meristem as the target tissue. This report suggested that transgenic plants expressing higher levels of AVP1 transcripts in sugarcane are able to withstand salt and drought stress regimes probably due to profuse root system development in these plants.

In sugarcane, there are reports on proline accumulation in calli, plantlets, and whole plants in field trials when exposed to salt stress (Gandonou et al. 2006; Wahid and Ghazanfar 2006; Patade et al. 2008) and water deficit (Errabii et al. 2006). Several works report that higher proline accumulation in P5CS-transgenic plants confers increased tolerance to abiotic stress (Hong et al. 2000; Molinari et al. 2007; Kumar et al. 2010). Guerzoni et al. (2014) evaluated the response to salt stress of sugarcane plants transformed with the *Vigna aconitifolia* P5CS gene, which encodes Δ1- pyrroline-5-carboxylate synthetase, under the control of a stress-induced promoter AIPC (ABA-inducible promoter complex) and found that the transgenic lines accumulated up to 25% higher amounts of proline when compared with non-transformed control plants.

To improve the drought and salinity tolerance of sugarcane, a DEAD-box helicase gene isolated from pea (pea DNA helicase 45, PDH45) with a constitutive promoter, Port Ubi2.3, was transformed into the commercial sugarcane variety Co 86032 through *Agrobacterium*-mediated transformation, and the transgenics were screened for tolerance to soil moisture stress and salinity (Augustine et al. 2015). The V_1 transgenic events exhibited significantly higher cell membrane thermostability, transgene expression, relative water content, gas exchange parameters, chlorophyll content, and photosynthetic efficiency under soil moisture stress compared to the wild type (WT). The overexpression of PDH45 transgenic sugarcane also led to the upregulation of DREB2-induced downstream stress-related genes. The transgenic events demonstrated higher germination ability and better chlorophyll retention than WT under salinity stress. These results suggest the possibility for development of increased abiotic stress-tolerant sugarcane cultivars through overexpression of PDH45 gene. The expression of the gene encoding the enzyme isopentenyl transferase (ipt) under control of the cold-inducible gene promoter AtCOR15a has been reported (Belintani et al. 2012) which has also been shown not to affect plant growth while providing a greater tolerance to cold stress. Recently, Ramiro et al. (2016) have shown that transgenic expression of a highly conserved cell death suppressor, Bax Inhibitor-1 from *Arabidopsis thaliana* (AtBI-1), can confer increased tolerance of sugarcane plants to long-term (>20 days) water stress conditions which especially in the roots can be correlated to induction of endoplasmic reticulum (ER) stress by the protein glycosylation inhibitor tunicamycin. These findings suggest that suppression of ER stress in C4 grasses, which include important crops such as sorghum and maize, can be an effective means of conferring improved tolerance to long-term water deficit. This result could potentially lead to improved resilience and yield of major crops in the world.

9.6 Field Performance of Transgenic Sugarcane

Field evaluation of transgenic sugarcane is the first step toward the commercial release of transgenic cultivars with improved traits of agronomic importance. Stable and uniform expression of an introduced trait(s) and their comparable agronomic performance to elite commercial cultivars are critical factors for the commercial release of any transgenic event (Joyce et al. 2014). Transgenic sugarcane requires extensive field assessment of a large population of independent transgenic events because of the large genetic variability inherent in transgenic sugarcane populations (Gilbert et al. 2005; Joyce et al. 1998; Vickers et al. 2005; Pribil et al. 2007). Field evaluations of transgenic sugarcane, however, have been limited to a few studies (Table 9.1), with most of them focusing on a few selected events. Transgenic sugarcane evaluated for field trials has been mainly produced using biolistic method except for one report on electroporation (Arencibia et al. 1999) and one report on *Agrobacterium*-mediated transformation (Joyce et al. 2014). Arencibia et al. (1999) concluded that field trials of the five selected clones confirmed the expression of the

Table 9.1 Field evaluation of transgenic sugarcane

Sr. No.	Sugarcane genotype	Gene	Method of transformation	Explant	Parameters tested during field evaluation	Selectable marker/agent	Author and year
1	Ja60-5	Truncated *cryIA(b)* gene from *Bacillus thuringiensis*	Electroporation	Cell suspensions from embryogenic calli	Infection with artificial borer (*Diatraea saccharalis* Fab.)	*nptII* (Geneticin)	Arencibia et al. (1999)
2	NCo310	*pat* gene (herbicide resistance)	Biolistic	Embryogenic callus	Effect of expression of transgene on agronomic performance	*nptII* (Geneticin)	Leibbrandt and Snyman (2003)
3	Q117	PPO (polyphenol oxidase), SPS (sucrose phosphate synthase)	Biolistic	Embryogenic callus	Yield, sucrose content, CCS%	*nptII* (Geneticin)	Vickers et al. (2005)
4	CP 92-1666	SCYLV (sugarcane yellow leaf virus)	Biolistic	Embryogenic callus	Agronomic performance, field expression of transformed gene and selectable marker, genetic differences using SSR genotyping	*nptII* (Geneticin)	Gilbert et al. (2009)
5	Q117	Trehalulose synthase	Biolistic	Embryogenic callus	Growth and trehalulose production in field	*nptII* (Geneticin)	Hamerli and Birch (2011)
6	ROC16, YT79-177	Truncated *cryIAc*	Biolistic	Embryogenic callus	Transgenic plants infested with Procerasvenosatum Walker (sugarcane stalk borer) evaluated for agronomic traits, viz., stalk height, stalk diameter, stalk weight, percentage of damaged stalks and attacked internodes, Brix, and Pol	*nptII* (Geneticin)	Weng et al. (2011)

7	Q127, Q135, Q138, Q174, Q183, Tellus	Isomaltulose synthase	Biolistic	Embryogenic callus	Isomaltulose (IM) in basal internode, IM in whole-cane juice, Brix in whole-cane juice, stalk height and stalk diameter, cane yield	*nptII* (Geneticin)	Basnayake et al. (2012)
8	Q117, Q138, KQ228	Vacuole-targeted NTPP signal peptide fused to *isomaltulose synthase* (IMS) and *trehalulose synthase* (THS)	Biolistic	Embryogenic callus	IMS transgene expression and stability of transgene of the vegetative generations VG1 and VG2	*nptII* (Geneticin)	Mudge et al. (2013)
9	Q117, KQ228, L97-128, CP84-1198, CP89-2143, and SP70-1143	*Neomycin phosphotransferase II* (*nptII*) and *phosphomannose isomerase* (*pmi*)	*Agrobacterium*-mediated	Leaf roll disks, embryogenic callus	Stability of transgene expression across multiple growing seasons in V_1	*nptII* (Geneticin)*pmi* (sucrose and mannose)	Dong et al. (2014)
10	Q117	*Malus domestica* sorbitol-6-phosphate dehydrogenase gene (*mds6pdh*)	Biolistic and *Agrobacterium*-mediated	Embryogenic callus	Stalk height, stalk diameter, stalk weight, and stalk-expressed juice parameters, tons cane per hectare, Brix, Pol, ratoonability, transgene expression, and genotypic analysis using AFLP to determine methylation patterns in transgenic events	*nptII* (Geneticin)	Joyce et al. (2014)

(continued)

Table 9.1 (continued)

Sr. No.	Sugarcane genotype	Gene	Method of transformation	Explant	Parameters tested during field evaluation	Selectable marker/ agent	Author and year
11	Q117	CBHI (cellobiohydrolase I), CBHII (cellobiohydrolases II), EG (endoglucanase)	Biolistic	Embryogenic callus	Effect of subcellular targeting on cellulase accumulation studied in T_0R_1 and stability of transgene expression studied in T_0 generations	*nptII* (Geneticin)	Harrison et al. (2011)
12	RA87-3	Glyphosate tolerance	Biolistic	Embryogenic callus	Glyphosate tolerance and phenotypic growth characteristics	*nptII* (Geneticin)	Noguera et al. (2015)
13	Badila	Sugarcane mosaic virus coat protein (SCMV-CP)	Biolistic	Embryogenic callus	Agronomic traits (stalk diameter, stalk height, stalk weight, millable stalk numbers, length of internode and tons cane per hectare), sucrose concentrations, cane fiber content, Brix and Pol; field virus resistance and stability of transgene expression	*nptII* (kanamycin)	Yao et al. (2017)
14	FN15	*cry1Ac*	Biolistic	Embryogenic callus	Yield estimates (height, stem diameter, brix, number of millable stalks, cane yield per unit area, sucrose content, and sucrose yield); resistance of transgenic sugarcane to sugarcane borer	*Bar* (phosphinothricin)	Gao et al. (2016)
15	L97-128, CP84-1198, SP70-1143	AmCYAN1 (*Anemonia majano cyan* fluorescent protein)	Agrobacterium-mediated	Leaf roll discs, embryogenic callus	T-DNA insert stability and AmCYAN1 protein accumulation over three ratoon generations	PMI (*phosphomannose isomerase*)	Caffall et al. (2016)

insect-resistance trait under natural borer infection and showed that the agronomic traits were similar, but not identical, to those of the original genotype. Leibbrandt and Snyman (2003) and Gilbert et al. (2005) performed field trials over 3 years and compared one and four transgenic events, respectively, to parent clones. Both reports concluded that stable expression of the transgene and agronomic performance equivalent to parent clones were achievable. In contrast, Vickers et al. (2005) and Gilbert et al. (2009) showed that most transgenic events suffered a significant yield reduction (tons of cane/ha) in comparison with the parent clone. Although the reports on field performance of transgenic sugarcane for yield and agronomic characteristics are contradictory, the sugar content, % Brix, % Pol, and % purity measurements have consistently been reported as unaffected by the tissue culture and transformation process. The first reported field trials showed substantial TH accumulation in a few tested sugarcane lines engineered to produce a vacuole-targeted TH synthase (ths gene) (Hamerli and Birch 2011). Sucrose isomerase activity was low in these transgenic lines, and the results indicate the strong potential to develop sugarcane for commercial-scale production of IM if the higher activity can be engineered in appropriate developmental patterns (Basnayake et al. 2012). Developmentally controlled expression of a silencing-resistant gene for vacuolar SI activity allows high IM yields across multiple generations in field-grown plants (Mudge et al. 2013). Transgene expression of these five transgenic events in three different varieties was stable through multiple growing seasons in the field (Dong et al. 2014).

Joyce et al. (2014) investigated the field performance of tissue-cultured and transgenic sugarcane events produced using four different strains of *Agrobacterium tumefaciens* and biolistics (using linearized minimal DNA cassettes or whole plasmid DNA vectors) for transgene expression stability, agronomic performance, and yield characteristics over several years, with the objective of determining the most appropriate transformation technology for commercial transgenic sugarcane development.

Transgenic plants expressing the SCMV-CP gene generated from Badila had enhanced cane yield (TCH) and sugar content (TSH) as well as lower SCMV disease incidence (Yao et al. 2017). Although Dale and McPartlan (1992) found that GUS expression had a significant negative effect in potato, nptII expression did not affect agronomic and yield performance, indicating that nptII does not affect primary metabolism of the plants and thus transgenic plants can be compared to wild-type counterparts under similar growth conditions (Yao et al. 2017).

Extensive field trials have paralleled the analysis of genomic changes in transgenic plants. The most relevant conclusion is that, although the level of genomic changes was low, nevertheless selected transgenic plants have undergone minor but clear morphological, physiological, and phytopathological variations. In more general terms, the results suggest that both biological activity of the foreign gene and somaclonal variations should be evaluated in population studies and that the extent of somaclonal variation should be adequately determined in transgenic populations in order to allow the appropriate management of field trials.

Gao et al. (2016) developed sugarcane transgenic plants with improved resistance to the sugarcane borer, *Diatraea saccharalis* (F). Their results suggest that a medium-copy number of cry1Ac gene in transgenic sugarcane may be more desirable than a too high or too low number since this appears to compromise gene expression. Higher Cry1Ac protein expression may not be optimum because high protein expression consumes more plant energy and negatively affects agronomic traits. All transgenic lines with medium-copy number expressing Cry1Ac protein had relatively equivalent or lower theoretical sucrose yield, compared to controls, and showed significantly improved sugarcane borer resistance. These lines can be potentially used for commercial purposes.

9.7 Biosafety Considerations for Transgenic Sugarcane

The transfer of stress tolerance genes into a crop generates concern that the transgenic plant will become a weed, that the gene will be transferred to wild relatives increasing their weediness, or that intraspecific gene transfer by pollen may prevent effective segregation of transgenic and non-transgenic products. Vegetative propagation in sugarcane prevents segregation of multiple transgenes needed for trait stacking or pathway engineering as well as offers high transgene containment (Altpeter and Oraby 2010). Most importantly, compared with the other flowering crops, such as rice or sorghum, GM sugarcane belongs to one of the lowest risk plant species when considering food and environmental safety because of its flowering mechanisms and vegetative propagation characteristics and the fact that sugar were derived from high-temperature boiling process at 107 °C (Altpeter and Oraby 2010; Zhou et al. 2016). These attributes contribute to the biosafety of sugarcane for the production of value-added products.

There have been several biosafety reports on transgenic sugarcane lines (Gilbert et al. 2005; Ruan et al. 2007). Gilbert et al. (2005) evaluated the variability in agronomic characteristics and field disease resistance of transgenic sugarcane transformed for resistance to sugarcane mosaic virus (SCMV) strain E. Ruan et al. (2007) investigated the effects on enzyme activities and microbe communities in rhizosphere soil of sugarcane mosaic virus coat protein (ScMV-CP) transgenic sugarcane and found that there was no change in the soil bacterial diversity and no apparent effect on soil enzyme activities or the population number of soil microbes in the rhizosphere soil. Zhou et al. (2016) found that no horizontal gene flows from cry1Ac sugarcane to the soil. No significant difference in the population of culturable microorganisms between the non-GM and cry1Ac transgenic sugarcane was observed, and there were no significant interactions between the sugarcane lines and the growth stages. Thus, they concluded that the cry1Ac sugarcane lines may not affect the microbial community structure and functional diversity of the rhizosphere soil and have a few negative effects on soil enzymes (Zhou et al. 2016).

9.8   Conclusion

The sugarcane transformation has come a long way from being one of the recalcitrant species for transformation to be successfully transformed and also being evaluated at the field level. Transgenic sugarcane with improved agronomic traits like improved resistance to diseases and pests and herbicide and tolerance to abiotic stresses has been developed during the last two decades. Although there exist only a few reports on extensive field assessment of a large population of independent transgenic events, these studies have thrown a light on important aspects like effect of transformation or tissue culture procedures on crop ratoonability and sugar yield components, stability of transgene expression over different crop cycles and during maturity, effect of transgene copy number on transgene expression, and choice of transformation method for producing low-copy and stable transgene expressing lines. The commercial success of transgenic sugarcane will depend on consistently efficient commercial sugarcane transformation platform across different sugarcane varieties and achievement of predictable and stable transgene expression through vegetative propagation and over multiple sugarcane growing seasons under field conditions.

References

Altpeter F, Oraby H (2010) Sugarcane. In: Kempken F, Jung C (eds) Genetic modification of plants, biotechnology in agriculture and forestry, vol 64. Springer-Verlag, Berlin/Heidelberg. https://doi.org/10.1007/978-3-642-02391-0_23

Arencibia AD, Carmona ER, Cornide MT, Castiglione S, O'Relly J, Chinea A, Oramas P, Sala F (1999) Somaclonal variation in insect-resistant transgenic sugarcane (Saccharum hybrid) plants produced by cell electroporation. Transgenic Res 8:349–360

Arencibia AD, Carmona ER, Tellez P, Chan M, Yu S, Trujillo LE, Oramas P (1998) An efficient protocol for sugarcane (Saccharum spp. L.) transformation mediated by *Agrobacterium* tumefaciens. Transgenic Res 7:213–222

Arvinth S, Arun S, Selvakesavan RK, Srikanth J, Mukunthan N, Ananda Kumar P, Premachandran MN, Subramonian N (2010) Genetic transformation and pyramiding of aprotinin-expressing sugarcane with cry1Ab for shoot borer (Chilo infuscatellus) resistance. Plant Cell Rep 29(4):383–395

Augustine SM, Ashwin Narayan J, Syamaladevi DP, Appunu C, Chakravarthi M, Ravichandran V, Tuteja N, Subramonian N (2015) Introduction of pea DNA helicase 45 into sugarcane (Saccharum spp. hybrid) enhances cell membrane thermostability and upregulation of stress-responsive genes leads to abiotic stress tolerance. Mol Biotechnol 57:475–488. https://doi.org/10.1007/s12033-015-9841-x

Basnayake SWV, Morgan TC, Wu LG, Birch RG (2012) Field performance of transgenic sugarcane expressing isomaltulose synthase. Plant Biotechnol J 10:217–225

Belintani NG, Guerzoni JTS, Moreira RMP, Vieira LGE (2012) Improving low-temperature tolerance in sugarcane by expressing the ipt gene under a cold inducible promoter. Biol Plant 56(1):71–77

Beyene G, Buenrostro-Nava MT, Damaj MB, Gao SJ, Molina J, Mirkov TE (2011) Unprecedented enhancement of transient gene expression from minimal cassettes using a double terminator. Plant Cell Rep 30:13–25

Bower R, Birch RG (1992) Transgenic sugarcane plants via microprojectile bombardment. Plant J 2:409–416

Braga DPV, Arrigoni EDB, Silva-Filho MC, Ulian EC (2003) Expression of the Cry1Ab protein in genetically modified sugarcane for the control of Diatraea saccharalis (Lepidoptera: Crambidae). J New Seeds 5:209–222

Braithwaite KS, Geijskes RJ, Smith GR (2004) A variable region of the sugarcane bacilliform virus (SCBV) genome can be used to generate promoters for transgene expression in sugarcane. Plant Cell Rep 23:319–326

Brumbley SM, Snyman SJ, Gnanasambandam A, Joyce P, Hermann SR, da Silva JAG, McQualter RB, Wang ML, Egan BT, Patterson AH, Albert HH, Moore PH (2008) Sugarcane. In: Kole C, Hall TC (eds) Compendium of transgenic crop plants: transgenic sugar, tuber and fiber crops. Wiley-Blackwell, Oxford, pp 1–58

Caffall HK, He C, Smith-Jones M, Mayo K, Mai P, Dong S, Ke J, Dunder E, Yarnall M, Whinna R, DeMaio J, Weining G, Sheldon J, Allen M, Costello T, Setliff K, Jain R, Snyder A, Lovelady C, Jepson I (2016) Long-term T-DNA insert stability and transgene expression consistency in field propagated sugarcane. Plant Mol Biol 93:451–463. https://doi.org/10.1007/s11103-016-0572-6

Chakravarthi M, Philip A, Subramonian N (2015) Truncated ubiquitin 5′ regulatory region from Erianthus arundinaceus drives enhanced transgene expression in heterologous systems. Mol Biotechnol 57:820–835

Christensen AH, Quail PH (1996) Ubiquitin promoter-based vectors for high level expression of selectable and/or screenable marker genes in monocotyledonous plants. Transgenic Res 5:213–218

Christy LA, Aravinth S, Saravanakumar M, Kanchana M, Mukunthan N, Srikanth J, Thomas G, Subramonian N (2009) Engineering sugarcane cultivars with bovine pancreatic trypsin inhibitor (aprotinin) gene for protection against top borer (Scirpophaga excerptalis Walker). Plant Cell Rep 28:175–184

Clark AJ, Harold G, Yull FE (1997) Mammalian cDNA and prokaryotic reporter sequences silence adjacent transgenes in transgenic mice. Nucleic Acids Res 25:1009–1014

D'Hont A, Souza GM, Menossi M, Vincentz M, van-Sluys MA, Glaszmann JC, Ulian E (2008) Sugarcane: a major source of sweetness, alcohol, and bio-energy. In: Moore PH, Ming R (eds) Plant genetics and genomics: crops and models, vol 1. Springer, New York, pp 483–513

Dale P, McPartlan H (1992) Field performance of transgenic potato plants compared with controls regenerated from tuber discs and shoot cuttings. Theor Appl Genet 84:585–591

Damaj MB, Kumpatla SP, Emani C, Beremand PD, Reddy AS, Rathore KS, Buenrostro-Nava MT, Curtis IS, Thomas TL, Mirkov TE (2010) Sugarcane DIRIGENT and O-methyl transferase promoters confer stem-regulated gene expression in diverse monocots. Planta 231:1439–1458

Dermawan H, Karan R, Jung JH, Zhao Y, Parajuli S, Sanahuja G, Altpeter F (2016) Development of an intragenic gene transfer and selection protocol for sugarcane resulting in resistance to acetolactate synthase-inhibiting herbicide. Plant Cell Tissue Organ Cult 126:459. https://doi.org/10.1007/s11240-016-1014-5

Dong S, Delucca P, Geijskes RJ, Ke J, Mayo K, Mai P, Sainz M, Caffall K, Moser T, Yarnall M, Setliff K, Jain R, Rawls E, Smith-Jones M, Dunder E (2014) Advances in *Agrobacterium*-mediated sugarcane transformation and stable transgene expression. Sugar Tech 16(4):366–371

Enriquez GA, Trujillo LA, Menndez C, Vazquez RI, Tiel K, Dafhnis F, Arrieta J, Selman G, Hernandez L (2000) Sugarcane (Saccharum hybrid) genetic transformation mediated by *Agrobacterium* tumefaciens: production of transgenic plants expressing proteins with agronomic and industrial value. Dev Plant Genet Breed 5:76–81

Errabii T, Gandonou C, Essalmani H, Abrini J, Idaomar M, Senhaji N (2006) Growth, proline and ion accumulation in sugarcane callus cultures under drought-induced osmotic stress and its subsequent relief. Afr J Biotechnol 4:1250–1255

Falco MC, Tulmann Neto A, Ulian EC (2000) Transformation and expression of a gene for herbicide resistance in Brazilian sugarcane. Plant Cell Rep 19:1188–1194

Gallo-Meagher M, Irvine JE (1996) Herbicide resistant sugarcane containing the bar gene. Crop Sci 36:1367–1374

Gandonou CB, Errabii T, Abrini J, Idaomar M, Senhaji NS (2006) Selection of callus cultures of sugarcane (Saccharum sp.) tolerant to NaCl and their response to salt stress. Plant Cell Tissue Organ Cult 87:9–16

Gao S, Yang Y, Wang C, Guo J, Zhou D, Wu Q et al (2016) Transgenic sugarcane with a cry1ac gene exhibited better phenotypic traits and enhanced resistance against sugarcane borer. PLoS One 11(4):e0153929. https://doi.org/10.1371/journal.pone.0153929

Gilbert RA, Gallo-Meagher M, Comstock JC, Miller JD, Jain M, Abouzid A (2005) Agronomic evaluation of sugarcane lines transformed for resistance to sugarcane mosaic virus strain E. Crop Sci 45:2060–2067

Gilbert RA, Glynn NC, Comstock JC, Davis MJ (2009) Agronomic performance and genetic characterization of sugarcane transformed for resistance to sugarcane yellow leaf virus. Field Crop Res 111:39–46

Guerzoni JTS, Belintani NG, Moreira RMP, Hoshino AA, Domingues DS, Filho JCB, Vieira LGE (2014) Stress-induced D1-pyrroline-5-carboxylate synthetase (P5CS) gene confers tolerance to salt stress in transgenic sugarcane. Acta Physiol Plant 36:2309–2319

Hamerli D, Birch RG (2011) Transgenic expression of trehalulose synthase results in high concentrations of the sucrose isomer trehalulose in mature stems of field-grown sugarcane. Plant Biotechnol J 9:32–37

Hammond SM, Caudy AA, Hannon GJ (2001) Post-transcriptional gene silencing by double-stranded RNA. Nat Rev Genet 2:110–119

Harrison MD, Geijskes J, Coleman HD, Shand K, Kinkema M, Palupe A, Hassall R, Sainz M, Lloyd R, Miles S, Dale JL (2011) Accumulation of recombinant cellobiohydrolase and endoglucanase in the leaves of mature transgenic sugar cane. Plant Biotechnol J 9:884–896

Hong Z, Lakkineni K, Zhang Z, Verma DPS (2000) Removal of feedback inhibition of D1-pyrroline-5-carboxylate synthetase results in increased proline accumulation and protection of plants from osmotic stress. Plant Physiol 122:1129–1136. https://doi.org/10.1104/122.4.1129

Hotta CT, Lembke CG, Domingues DS, Ochoa EA, Cruz GMQ, Melotto-Passarin DM, Marconi TG, Santos MO, Mollinari M, Margarido GRA, Crivellari AC, dos Santos WD, de Souza AP, Hoshino AA, Carrer H, Souza AP, Garcia AAF, Buckeridge MS, Menossi M, Marie-Anne VS, Souza GM (2011) The biotechnology roadmap for sugar cane improvement. Trop Plant Biol 3:75–87

Jackson MA, Anderson DJ, Birch RG (2013) Comparison of Agrobacterium and particle bombardment using whole plasmid or minimal cassette for production of high-expressing, low-copy transgenic plants. Transgenic Res 22:143–151

Jain M, Chengalrayan K, Abouzid A, Gallo M (2007) Prospecting the utility of a PMI/mannose selection system for the recovery of transgenic sugarcane (Saccharum spp. hybrid) plants. Plant Cell Rep 26:581–590

Jakowitsch J, Papp I, Moscone EA, van Der Winden J, Matzke M, Matzke Antonius JM (1999) Molecular and cytogenetic characterization of a transgene locus that induces silencing and methylation of homologous promoters intrans. Plant J 17:131–140

Jing-Sheng X, Shiwu G, Liping X, Rukai C (2008) Construction of expression vector of CryIA(c) gene and its transformation in sugarcane. Sugar Tech 10(3):269–273

Joyce P, Hermann S, O'Connell A, Dinh Q, Shumbe L, Lakshmanan P (2014) Field performance of transgenic sugarcane produced using Agrobacterium and biolistics methods. Plant Biotechnol J 12:411–424

Joyce PA, McQualter RB, Bernard MJ, Smith GR (1998) Engineering for resistance to SCMV in sugarcane. Acta Hortic 461:385–392

Kalunke RM, Kolge AM, Babu KH, Prasad DT (2009) Agrobacterium-mediated transformation of sugarcane for borer resistance using cry 1Aa3 gene and one-step regeneration of transgenic plants. Sugar Tech 11(4):355–359

Kinkema M, Geijskes J, de Lucca P, Palupe A, Shand K, Coleman HD, Brinin A, Brett Williams B, Sainz M, Dale JL (2014) Improved molecular tools for sugar cane biotechnology. Plant Mol Biol 84:497–508. https://doi.org/10.1007/s11103-013-0147-8

Kinkema M, Miles S (2013) Compositions and methods for increased expression in sugar cane. Patent application WO/2013/090137

Kohli A, Griffiths S, Palacios N, Twyman R, Vain P, Laurie D, Christou P (1999) Molecular characterization of transforming plasmid rearrangements in transgenic rice reveals a recombination hotspot in the CaMV 35S promoter and confirms the predominance of microhomology mediated recombination. Plant J 17:591–601

Kumar V, Shriram V, Kavi-Kishor PB, Jawali N, Shitole MG (2010) Enhanced proline accumulation and salt stress tolerance of transgenic indica rice by over-expressing P5CSF129A gene. Plant Biotechnol Rep 4:37–48. https://doi.org/10.1007/s11816-009-0118-3

Kumar T, Uzma, Khan MR, Abbas Z, Ali GM (2014) Genetic Improvement of Sugarcane for Drought and Salinity Stress Tolerance Using Arabidopsis Vacuolar Pyrophosphatase (AVP1) Gene. Mol Biotechnol 56(3):199–209. https://doi.org/10.1007/s12033-013-9695-z

Lakshmanan P (2005) Somatic embryogenesis in sugarcane- an addendum: sugarcane biotechnology: challenges and opportunities. In Vitro Cell Dev Biol-Plant 42:202–205

Leibbrandt N, Snyman S (2001) Initial field testing of transgenic glufosinate ammonium-resistant sugarcane. In: Proceedings of the South African Sugar Technologists' Association pp 108–111

Leibbrandt NB, Snyman SJ (2003) Stability of gene expression and agronomic performance of a transgenic herbicide-resistant sugarcane line in South Africa. Crop Sci 43(2):671–677

Liu AW, Oard SV, Oard JH (2003) High transgene expression levels in sugarcane (Saccharum officinarum L.) driven by the rice ubiquitin promoter RUBQ2. Plant Sci 165:743–750

Manickavasagam M, Ganapathi A, Anbazhagan VR, Sudhakar B, Selvaraj N, Vasudevan A, Kasthurirengan S (2004) *Agrobacterium*-mediated genetic transformation and development of herbicide-resistant sugarcane (Saccharum species hybrids) using axillary buds. Plant Cell Rep 23:134–143

McQualter RB, Harding RM, Dale JL, Smith GR (2001) Virus derived transgenes confer resistance to Fiji disease in transgenic sugarcane plants. Proc Int Soc Sugar Cane Technol 24(2):584–585

Meng L, Ziv M, Lemaux PG (2006) Nature of stress and transgene locus influences transgene expression stability in barley. Plant Mol Biol 62:15–28

Meyer P (1995) Understanding and controlling transgene expression. Trends Biotechnol 13:332–337

Molinari HBC, Marur CJ, Daros E, Campos MKF, Carvalho JFRP, Bespalhok JCF, Pereira LFP, Vieira LGE (2007) Evaluation of the stress-inducible production of proline in transgenic sugarcane (Saccharum spp.): osmotic adjustment, chlorophyll fluorescence and oxidative stress. Physiol Plant 130:218–229. https://doi.org/10.1111/1399-3054.2007.00909

Moyle RL, Birch RG (2013) Sugarcane loading stem gene promoters drive transgene expression preferentially in the stem. Plant Mol Biol 82:51–58

Mudge SR, Basnayake SWV, Moyle RL, Osabe K, Graham MW, Morgan TE, Birch RG (2013) Mature-stem expression of a silencing-resistant sucrose isomerase gene drives isomaltulose accumulation to high levels in sugarcane. Plant Biotechnol J 11:502–509

Nayyar S, Sharma BK, Kaur A, Kalia A, Sanghera GS, Thind KS et al (2017) Red rot resistant transgenic sugarcane developed through expression of β-1,3-glucanase gene. PLoS One 12(6):e0179723. https://doi.org/10.1371/journal.pone.0179723

Noguera A, Enrique R, Perera MF, Ostengo S Racedo J, Costilla D, Zossi S, Cuenya MI, Filippone MP, Welin B, Castagnaro AP (2015) Genetic characterization and field evaluation to recover parental phenotype in transgenic sugarcane: a step toward commercial release. Mol Breed 35:115. https://doi.org/10.1007/s11032-015-0300-y

Parisi C, Tillie P, Rodríguez-Cerezo E (2016) The global pipeline of GM crops out to 2020. Nat Biotechnol 34:32–36

Patade VY, Suprasanna P, Bapat VA (2008) Gamma irradiation of embryogenic callus cultures and in vitro selection for salt tolerance in sugarcane (Saccharum officinarum L.). Agric Sci China 7:101–105

Petrasovits LA, Zhao L, McQualter RB, Snell KD, Somleva MN, Patterson NA, Nielsen LK, Brumbley SM (2012) Enhanced polyhydroxybutyrate production in transgenic sugarcane. Plant Biotechnol J 10:569–578. https://doi.org/10.1111/j.1467-7652.2012.00686.x

Philip A, Syamaladevi DP, Chakravarthi M, Gopinath K, Subramonian N (2013) 5′ regulatory region of ubiquitin 2 gene from Porteresia coarctata makes efficient promoters for transgene expression in monocots and dicots. Plant Cell Rep 32:1199–1210

Plasterk RHA, Ketting RF (2000) The silence of the genes. Curr Opin Genet Dev 10:562–567

Pribil M, Hermann SR, Dun GD, Karno XX, Ngo C, O'Neill S, Wang L, Bonnett GD, Chandler PM, Beveridge CA, Lakshmanan P (2007) Altering sugarcane shoot architecture through genetic engineering: prospects for increasing cane and sugar yield. In: Proceedings of the 2007 conference of the Australian Society of Sugar Cane Technologists held at Cairns, Queensland, Australia, 8–11 May 2007, pp 251–257. Australian Society of Sugar Cane Technologists

Que Q, Elumalai S, Li X, Zhong H, Nalapalli S, Schweiner M, Fei X, Nuccio M, Kelliher T, GuW CZ, Chilton M-DM (2014) Maize transformation technology development for commercial event generation. Front Plant Sci 5:379. https://doi.org/10.3389/fpls.2014.00379

Ramiro DA, Melotto-Passarin DM, Barbosa MDA, Santos FD, Gomez SGP, Junior NSM, Lam E, Carrer H (2016) Expression of Arabidopsis Bax Inhibitor-1 in transgenic sugarcane confers drought tolerance. Plant Biotechnol J 14:1826–1837

Ruan M, Yan X, Yao Z, Chuanyu Y, Bixia Z, Ying G et al (2007) Effects on enzyme activities and microbe in rhizosphere soil of ScMV-CP transgenic sugarcane. Chin Agric Sci Bull 23:381–386. https://doi.org/10.3969/j.issn.1000-6850.2007.04.088

Schubert D, Lechtenberg B, Forsbach A, Gils M, Bahadur S, Schmidt R (2004) Silencing in Arabidopsis T-DNA transformants: the predominant role of a gene-specific RNA sensing mechanism versus position effects. Plant Cell 16:2561–2572

Snyman SJ, Meyer GM, Carson D, Botha FC (1996) Establishment of embryogenic callus and transient gene expression in selected sugarcane varieties. S Afr J Bot 62:151–154

Srikanth J, Subramonian N, Premachandran MN (2011) Advances in transgenic research for insect resistance in sugarcane. Trop Plant Biol 4:52–61

Stoger E, Williams S, Keen D, Christou P (1998) Molecular characteristics of transgenic wheat and the effect on transgene expression. Transgenic Res 7:463–471

Taparia Y, Fouad WM, Gallo M, Altpeter F (2012a) Rapid production of transgenic sugarcane with the introduction of simple loci following biolistic transfer of a minimal expression cassette and direct embryogenesis. In Vitro Cell Dev Biol Plant 48:15

Taparia Y, Gallo M, Altpeter F (2012b) Comparison of direct and indirect embryogenesis protocols, biolistic gene transfer and selection parameters for efficient genetic transformation of sugarcane. Plant Cell Tissue Organ Cult 111:131–141

van der Vyver C, Conradie T, Kossmann J, Lloyd J (2013) In vitro selection of transgenic sugarcane callus utilizing a plant gene encoding a mutant form of acetolactate synthase. In Vitro Cell Dev Biol-Plant 49:198–206

Vickers JE, Grof CPL, Bonnett GD, Jackson PA, Morgan TE (2005) Effects of tissue culture, biolistic transformation, and introduction of PPO and SPS gene constructs on performance of sugarcane clones in the field. Aust J Agric Res 56:57–68

Waclawovsky AJ, Paloma MS, Lembke CG, Moore PH, Souza GM (2010) Sugarcane for bioenergy production: an assessment of yield and regulation of sucrose content. Plant Biotechnol J 8:263–276

Wahid A, Ghazanfar A (2006) Possible involvement of some secondary metabolites in salt tolerance of sugarcane. J Plant Physiol 163:723–730. https://doi.org/10.1016/005.07.007

Wang Z-Z, Zhang S-Z, Yang B-P, Li Y-R (2005) Trehalose synthase gene transfer mediated by Agrobacterium tumefaciens enhances resistance to osmotic stress in sugarcane. Sugar Tech 7(1):49–54

Wei H, Wang ML, Moore PH, Albert HH (2003) Comparative expression analysis of two sugarcane polyubiquitin promoters and flanking sequences in transgenic plants. J Plant Physiol 160:1241–1251

Weng LX, Deng HH, Xu JL, Li Q, Zhang YQ, Jiang ZD, Li QW, Chen JW, Zhang LH (2011) Transgenic sugarcane plants expressing high levels of modified cry1Ac provide effective control against stem borers in field trials. Transgenic Res 20(4):759–772

Wilmink A, Dons JJM (1993) Selective agents and marker genes for use in transformation of monocotyledonous plants. Plant Mol Biol Rep 11:165–185

Wu H, Awan FS, Vilarinho A, Zeng Q, Kannan B, Phipps T, McCuiston J, Wang W, Caffall K, Altpeter F (2015) Transgene integration complexity and expression stability following biolistic or *Agrobacterium*-mediated transformation of sugarcane. In Vitro Cell Dev Biol Plant 51:603

Yao W, Ruan M, Qin L, Yang C, Chen R, Chen B, Zhang M (2017) Field performance of transgenic sugarcane lines resistant to sugarcane mosaic virus. Front Plant Sci 8:104. https://doi.org/10.3389/fpls.2017.00104

Zhang L, Xu J, Birch RG (1998) Engineered detoxification confers resistance against a pathogenic bacterium. Nat Biotechnol 17:1021–1024

Zhang M, Zhuo X, Wang J, Wu Y, Yao W, Chen R (2014) Effective selection and regeneration of transgenic sugarcane plants using positive selection system. In Vitro Cell Dev Biol-Plant. https://doi.org/10.1007/s11627-014-9644-y

Zhang SZ, Zheng XQ, Lin JF, Guo LQ, Zan LM (2000) Cloning of trehalose synthase gene and transformation into sugarcane. J Agric Biotechnol 8(4):385–388

Zhao Y, Qian Q, Wang H, Huang D (2007) Hereditary behavior of bar gene cassette is complex in rice mediated by particle bombardment. J Genetics Genomics 34:824–835

Zhou D, Xu L, Gao S, Guo J, Luo J, You Q, Que Y (2016) Cry1Ac Transgenic Sugarcane Does Not Affect the Diversity of Microbial Communities and Has No Significant Effect on Enzyme Activities in Rhizosphere Soil within One Crop Season. Front Plant Sci 7:265

Zhu YZ, McCafferty H, Osterman G, Lim S, Agbayani R, Lehrer A, Schenck S, Komor E (2011) Genetic transformation with untranslatable coat protein gene of sugarcane yellow leaf virus reduces virus titers in sugarcane. Transgenic Res 20:503–512. https://doi.org/10.1007/s11248-010-9432-3

Chapter 10
Insect Smart Pulses for Sustainable Agriculture

Meenal Rathore, Alok Das, Neetu S. Kushwah, and Narendra Pratap Singh

Abstract The development of high-yielding insect-tolerant cultivars using conventional methods has been slow due to a number of reasons. With the advent of recombinant tools and genetic transformation systems, it has been possible to harness gene pool(s) by crossing the species barrier and utilize them for desired trait. Insect pest resistance has largely been introgressed in many crops, including pulses, by using the *cry* genes from *Bacillus thuringiensis*. However, many plant genes like lectins, protein inhibitors, etc. are also available that impart tolerance to insect pests and can be used for developing insect-tolerant plants. In comparison with other crops, relatively less work is available in this context in pulses because of their recalcitrant nature and biosafety issues related to candidate gene(s). In the regime of climate change, plant-pest dynamics has also witnessed change, and the need to develop transgenic tolerant to both pest and diseases is desirable. In context of sustainable pulse production, it is essential to develop and use insect-tolerant transgenics that have been developed by following the biosafety regulations, are high yielding, fit into popular cropping systems, and are expected to be remunerative to the stakeholders.

Keywords Insect resistance · Transgenics · Chickpea · Pigeon pea · Pea · Cowpea

10.1 Introduction

Pulses are a major source of vegetarian protein that contain and add 15 essential minerals and vitamins in the cereal-based diet of Indian subcontinent. They are an integral component for sustainable agriculture as they enhance soil quality through atmospheric nitrogen fixation and by adding organic matter to it. Though recognized as the principal producer and consumer of pulses in the world and witness

M. Rathore (✉) · A. Das · N. S. Kushwah · N. P. Singh
Division of Plant Biotechnology, ICAR-Indian Institute of Pulses Research, Kanpur 208024, India

© Springer International Publishing AG, part of Springer Nature 2018   227
S. S. Gosal, S. H. Wani (eds.), *Biotechnologies of Crop Improvement, Volume 2*, https://doi.org/10.1007/978-3-319-90650-8_10

to an increase in average pulse production over the years, India still needs to import pulses to fulfill its domestic needs. Production of pulses reached a total of about 22.95 million tons recently that was far more than the produce of about 17 million tons in 2015–2016 (Third Advance Estimates of Production of Commercial Crops for 2016–2017, Agricultural Statistics Division Directorate of Economics & Statistics). This was possible not only due to better varieties and favorable climate but also due to improved market and government policies. Effective area under pulses cultivation in the country is estimated at about 25–26 million hectares, while the realized productivity is less than 1 ton per hectare. Shortfall in pulses has been attributed to a number of factors, the major being ever-increasing population, dependency on climate, geographical shift in the area, complex disease-pest syndrome, post-harvest losses, socioeconomic conditions of the farmers, and many more.

10.2   Insect Pests

The major constraints that limit the realization of potential yield of pulses include biotic and abiotic stresses prevalent in the pulse-growing areas besides socioeconomic factors. Among biotic stresses, fungal and viral diseases like *Fusarium* wilt, root rot, *Cercospora* leaf spot, *Phytopthora* blight, powdery mildew, yellow mosaic disease, etc. are a few that cause economic loss to pulse crops. Weeds also cause substantial loss to pulses. Among key insect pests, gram pod borer (*Helicoverpa armigera*) in chickpea (*Cicer arietinum* L.) and pigeon pea (*Cajanus cajan* L. Millsp); pod fly in pigeon pea; whitefly, jassids, and thrips in mung bean (*Vigna radiata* L. Wilczek); and urd bean (*Vigna mungo* L. Hepper) cause severe damage to the respective crops. Bruchids are the most serious pest of stored pulse grains and require topmost priority in management. Recently, nematodes have emerged as potential threat to the successful cultivation of pulses in many areas.

An average loss of 30% in pulses in India is caused by insects that mount to an economic loss of ca. $815 million (Dhaliwal and Arora 1994). This loss can even reach 100% during incidences of heavy infestation. Dependence on insecticides and their incessant use for managing the insect pests has had adverse effect on beneficial organisms and quality of produce and has also led to resistance in pest(s). Resistance against insecticides in insects has already been reported in tobacco caterpillar (*Spodoptera litura*), whitefly (*Bemisia tabaci*), potato aphid (*Myzus persicae*), cotton aphid (*Aphis gossypii*), and also in gram pod borer (*Helicoverpa armigera*) (Armes et al. 1996; Kranthi et al. 2002). This clearly indicates that a sole solution cannot manage these insect pests. The integrated pest management (IPM) system needs to be in place as a viable and feasible solution and improved varieties that are tolerant or resistant to such insect pests would be a boon to this system.

10.3 Insect-Tolerant Cultivars

While conventionally bred varieties that are tolerant/resistant to certain pests are already in use, there are many pests for which natural resistance in crop plants are not available, and hence varieties/lines with natural tolerance to these pests are not available, e.g., gram pod borer, bruchids, whitefly, etc. It is here that genetic engineering can aid in developing better pulse(s) that are tolerant to desired insect pests using novel genes. Genetic transformation of a legume, as such, is relatively difficult because they are recalcitrant to transformation, i.e., have poor regeneration ability and also because *in vitro* regeneration is genotype dependent. However, considerable progress has been made in establishing *in vitro* regeneration systems in several legumes like chickpea (Batra et al. 2002; Sharma et al. 2006a), pigeon pea (Sharma et al. 2006c; Krishna et al. 2011), cowpea (Obembe 2008), green gram (Yadav et al. 2010), black gram (Acharjee et al. 2012; Adlinge et al. 2014), and many others. With an established regeneration system, genetic transformation can be worked upon to develop transgenic legumes tolerant to insect pests.

10.4 Cry Genes

The preferred gene(s) for imparting insect pest tolerance in many crops till date have been the *cry* genes obtained from the bacterium *Bacillus thuringiensis* (Bt). This Bt is a Gram-positive, aerobic spore forming, soil bacterium with several dozens subspecies. The cry genes encode crystal proteins that accumulate in crystalline inclusion bodies produced by the bacterium on sporulation and bear insecticidal properties. Genes encoding different *Bt* toxins (Cry1Aa, Cry1Ab, Cry1Ac, Cry1Ba, Cry1Ca, Cry1H, Cry2Aa, Cry3A, Cry6A, and Cry9C) have been expressed in different crop plants for resistance to different groups of insect pests (Schuler et al. 1998). The use of these genes for imparting resistance against different insect pests in different pulses is also underway.

10.5 Other Potential Genes

Though cry genes have been the preferred choice for many years, there are many other potential non-Bt genes like lectins, amylase inhibitors, chitinases, ribosome-inactivating proteins, protease inhibitors, vegetative insecticidal proteins, etc. that also bear insecticidal properties. Several Bt strains produce a protein during their vegetative growth that possess toxicity of the same intensity as that of $Bt\delta$-endotoxins against susceptible insects. This protein, known as vegetative insecticidal proteins

(VIPs), induces gut paralysis that is followed by complete lysis of the gut epithelium cells, causing larval mortality (Yu et al. 1997). Estruch et al. (1997) stated that VIPs can be used to target $Bt\delta$-endotoxins-resistant insect pests. Different alpha-amylase inhibitors have different modes of inhibition and different specificity profiles against diverse alpha-amylases. The sugar-binding proteins, the plant lectins, have a protective function against a range of organisms. They are particularly more efficient toward the sap-sucking insects (Czapla and Lang 1990; Hilder et al. 1995). Snowdrop, garlic, and chickpea lectins are known to have adverse effects on the survival, growth, and development of gram pod borer (Shukla et al. 2005; Sharma et al. 2005). But these lectins are also known for their toxicity to mammals and humans. Neurotoxin isolated from spider has also been used to develop transgenic tobacco that showed resistance to gram pod borer (Jiang et al. 1996); however, possible toxicity to mammals was a serious issue here.

10.6 Transgenic Pulses

The first insect-pest-tolerant transgenic plant developed was in tobacco using the cowpea trypsin inhibitor gene that expressed to levels of up to 1% of the soluble protein and gave protection against the lepidopteran pest *Heliothis virescens* (Hilder et al. 1987). A number of transformation attempts have been made to introgress genes having insecticidal property into legumes, but only few have attained success. The status of transgenics in pulses resistant to insect pests and their efficacy is briefed below. Table 10.1 enlists major insects pests of pulses in India.

Table 10.1 The major insect pests of pulses in India

Insect pest	Common name	Pulse crop
H. armigera	Gram pod borer	Chickpea, pigeon pea
Maruca vitrata (Geyer)	Spotted pod borer	Cowpea, pigeon pea
Melanagromyza obtusa Malloch	Pod fly	Pigeon pea
Tanaostigmodes cajaninae	Pod wasp	Pigeon pea
Liriomyza cicerina (Rondani)	Leaf miner	Chickpea
Etiella zinckenella Triet	Spiny pod borer	Pigeon pea, field pea, lentil
Aphis craccivora Koch	Aphid	Cowpea, field pea, faba bean, and *Phaseolus* beans
Acyrthocyphon pisum (Harris)	Pea aphid	Field pea
Bemisia tabaci	Whitefly	Phaseolus sp., green gram, black gram
Spilosoma oblique Walker	Bihar hairy caterpillar	Green gram, black gram
Empoasca spp.	Leafhoppers	Black gram, green gram, and *Phaseolus* beans
Callosobruchus chinensis L. and *C. maculatus*	Bruchids	All legumes
Bruchus pisorum L.	Pea weevil	Field pea

10.6.1   Chickpea

During the last two decades, successful attempts of genetic transformation of chickpea through *Agrobacterium* and particle gun delivery have been reported. Chickpea has been shown amenable to both *A. tumefaciens-* and *A. rhizogenes*-mediated infection, leading to the formation of crown gall and hairy roots, respectively. The first report of genetic transformation of chickpea began with the transformation of calli (Srinivasan et al. 1991). Subsequently, several reports were made of chickpea transformation using the embryonic axis or parts thereof. Much of the transgenic research in chickpea has been devoted toward the development of insect resistance, predominantly against gram pod borer utilizing synthetic Bt genes, including cry1Ac, cry1Ab, cry2Aa, cry1Aa/Ab, and cry1Aabc. A high level of *Helicoverpa* larval mortality (98.0–100.0%) was reported with synthetic Bt genes either singly or in combination (Sanyal et al. 2005; Acharjee et al. 2010; Mehrotra et al. 2011; Ganguly et al. 2014; Das et al. 2017). Strong developmental inhibition of stored grain pests (*Callosobruchus chinensis* and *Callosobruchus maculatus*: bruchids) using bean alpha-amylase inhibitor (αAI) was also reported (Sarmah et al. 2004; Ignacimuthu and Prakash 2006). However, expression of αAI in pea resulted in altered structure and immunological cross-priming in mouse (Prescott et al. 2005). This calls for a thorough biosafety assessment before the incorporation of such lines in breeding program. Survival and fecundity were reduced in a sap-sucking pest, *Aphis craccivora, in planta* bioassay of transgenic chickpea harboring *Allium sativum* leaf agglutinin (ASAL; Chakraborti et al. 2009). The antibiotic selection marker, *nptII* encoding neomycin phosphotransferase II conferring resistance to the antibiotic kanamycin monosulfate, has been extensively used in chickpe a, as chickpea is fairly sensitive to kanamycin selection. The status of chickpea transgenics resistant to various insect pests are listed in Table 10.2.

10.6.2   Pigeon Pea

Successful transformation was achieved by common methods such as particle bombardment and *Agrobacterium tumefaciens* (including in planta)-mediated genetic transformation. However, a wide variation in transformation efficiency (0.2–8.0%) was reported in pigeon pea. Synthetic Bt genes like cry1Ab, cry1A (b), cry1 E–C, cry1Ac + cry2Aa, cry1Aabc, and cowpea protease inhibitor (CPI) were used for engineering insect (gram pod borer) resistance (Lawrence and Koundal 2001; Surekha et al. 2005; Verma and Chand 2005; Sharma et al. 2006b; Ramu et al. 2011; Das et al. 2017; Ghosh et al. 2017). In genetic transformation, antibiotic selection (*nptII* and *hptII*) and reporter (*gus* and *gfp*) genes were used for selection and tracking. A novel method of identifying transgenic progeny was reported based on lateral

Table 10.2 Insect-resistant transgenic chickpea

Sl. no.	Cultivar	Method of transformation	Explant	Gene/trait	Transformation frequency	Stage	Efficacy	References
1	ICCV 1, ICCV 6, *desi* genotypes	*Agrobacterium tumefaciens*	Embryonic axis	*cry1Ac*	45.8%	T1	Developmental abnormalities of *H. armigera* larvae	Kar et al. (1997)
2	Semsen	*Agrobacterium tumefaciens*	Cotyledon with half embryonic axis	*α-ai*	0.72%	T2/ T3	Inhibited development of *Callosobruchus chinensis* and *C. maculatus*	Sarmah et al. (2004)
3	C-236, BG-256, Pusa 362, Pusa 372	*Agrobacterium tumefaciens*	Embryonic axis	*cry1Ac*	1.12%	T0/ T1	>80% larval mortality	Sanyal et al. (2005)
4	K850	*Agrobacterium tumefaciens*	Embryo axis	*αAI*	6.7%	T0/ T1	66–80% *C. maculatus* larval mortality	Ignacimuthu and Prakash (2006)
5	Chaffa, PG12, 1CCC37,1CCC32	*Particle gun bombardment*	Epicotyl and embryonic axis	*cry1Ac*	18%	T0	*Mortality up to 60% in H. armigera and Spodoptera litura*	Indurker et al. (2007)
6	ICCV 89314	*Agrobacterium tumefaciens*	Dissected cotyledon with half embryo	*ASAL*	–	T1/ T2	Survival and fecundity of *Aphis craccivora* decreased to 11–26% and 22–42%, respectively	Chakraborti et al. (2009)
7	ICCV 89314, Semsen	*Agrobacterium tumefaciens*	Cotyledon with half embryonic axis	*cry2Aa*	Co-transformation frequency 87%	T1	Up to 100%	Acharjee et al. (2010)
8	P-362	*Agrobacterium tumefaciens*	Cotyledonary node	*cry1Ab,cry1Ac*	1.6–2.7%	T1/ T2	>90–95%	Mehrotra et al. (2011)
9	DCP92-3	*Agrobacterium tumefaciens*	Embryonic axis	Fused*cry1Ab/Ac*	–	T1/ T2	Up to 100%	Ganguly et al. (2014)
10	DCP92-3	*Agrobacterium tumefaciens*	Cotyledon with embryonic axis	*Cry1Aabc*	0.076%	T4/ T5	80–100% larval mortality	Das et al. (2017)

αAI alpha-amylase inhibitor, *ASAL Allium sativum* leaf agglutinin, *cry1Ab* crystal 1Ab (Bt), *cry1Ac* crystal 1Ac (Bt), *cry2Aa* crystal 2Aa (Bt), *Cry1Aabc* Crystal1Aabc (Bt)

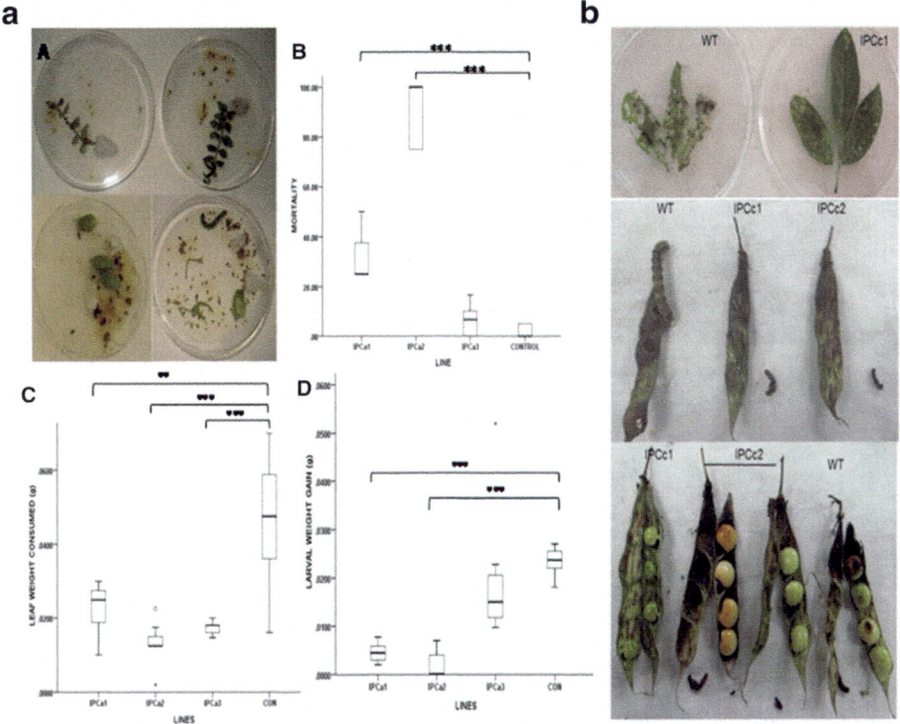

Fig. 10.1 (a, b) Insect bioassay of transgenic chickpea and pigeon pea lines harboring Bt cry-1Aabc gene (a) Detached leaf and pod assay, larval mortality, leaf weight consumed and larval weight gain (b) Detached leaf and pod bioassay of transgenic pigeon pea lines

root inhibition (Das et al. 2016). Codon-optimized, truncated *Bt* gene (*cry1Ac*) and domain shuffled (*cry1Aabc*) were used to generate transgenic chickpea (*cv.* DCP92-3) and pigeon pea (*cv.* Asha) lines, utilizing *Agrobacterium tumefaciens*. Transgenic chickpea (ca. 500 lines) and pigeon pea lines (ca. 500) harboring *Bt* gene(s) were generated, and few efficacious lines were characterized in details. Molecular analyses indicate integration, transmission, and expression of transgene in subsequent generations. Insect bioassays (detached leaf, pod, and whole-plant bioassay) of the generated chickpea and pigeon pea lines using larvae of the pod borer indicate higher mortality (90–100%) in few of the lines. Five characterized events each in chickpea and pigeon pea exhibiting higher insect mortality were permitted by the Review Committee on Genetic Manipulation (RCGM), Department of Biotechnology, Ministry of Science and Technology and Genetic Engineering Appraisal Committee (GEAC), and Ministry of Environment, Forest, and Climate Change, Government of India, for event selection trial at ICAR -Indian Institute of Pulses Research, Kanpur (Fig. 10.1a, b). The status of pigeonpea transgenics resistant to various insects pests are listed in Table 10.3.

Table 10.3 Insect-resistant transgenic pigeon pea

Sl. no.	Cultivar	Method of transformation	Explant	Gene/trait	Transformation frequency	Stage	Efficacy	Reference
1	Pusa 855	*Agrobacterium tumefaciens*	Embryonic axis	CPI	30–59%	T1	–	Lawrence and Koundal (2001)
2	ICPL87	*Agrobacterium tumefaciens*	Embryonal segment	*cry1E-C*	15%	T1/T2	–	Surekha et al. (2005)
3	Bahar, UPAS 120	*Agrobacterium tumefaciens*	Decapitated embryo axis	*cry1Ab*	0.2–0.3%	T1	–	Verma and Chand (2005)
4	ICPL 87	*Agrobacterium tumefaciens*	Axillary meristem	Cry1Ab	54–70%	T3	–	Sharma et al. (2006b)
5	TTB 7	*Agrobacterium tumefaciens* (*In planta*)	Seedling axis	Cry1Ac-F	–	T3	80–100% gram pod borer larval mortality	Ramu et al. (2011)
6	UPAS 120	*Agrobacterium tumefaciens*	Embryonic axis	Cry1Ac + cry2Aa	–	T2	80–100% gram pod borer larval mortality	Ghosh et al. (2017)
7	Asha (ICPL 87119)	*Agrobacterium tumefaciens*	Cotyledon with embryonic axis	Cry1Aabc	0.06%	T4/T5	80–100% gram pod borer larval mortality	Das et al. (2017)

CPI cowpea protease inhibitor, *cry1E-C* crystal 1E-C (Bt), *cry1Ab* crystal 1Ab (Bt), *cry1Ac-F* crystal 1Ac-F (Bt), *cry1Ac* crystal 1Ac (Bt), *cry2Aa* crystal 2Aa (Bt), *Cry1Aabc* Crystal1Aabc (Bt)

10.6.3 Pea

In pea, 10–70% yield losses can occur due to insect pests (Biddle and Cattlin 2001; Clement et al. 2002; Korth 2008). They are susceptible to insect pests of stored grains. *Bruchus pisorum*, commonly known as pea weevil, attacks pea during seed development in a growing crop and can cause losses up to 40% in seed yield (Smith 1990). Schroeder et al. (1995) developed transgenics in pea cv. Green feast harboring the alpha-amylase inhibitor-1 (αAI-1) regulated for expression in cotyledon and embryonic axis of developing seed (Shade et al. 1994). Of the many transgenics developed, the F10 line, homozygous for the desired gene, was selected for high expression of αAI-1 protein, and its progeny (T4) was subjected to pea weevil infestation assays along with non-transgenic control plants. It was observed that about 70% of T5 seeds and also control seeds were invaded with newly hatched larvae. In control, 87% of infested seeds had adult weevil emergence in a mean time of 85.3 days, while in T5 seeds, even after 140 days, no adult weevil emergence was observed. Further investigation revealed that only dead larvae in first or second instar were present in the T5 seeds. With no variability in morphological parameters between control and transgenic pea, these transgenics were protected from adult weevil damage as weevil development was blocked at early larval stage(s). Field trials of this transgenic line in Wagga Wagga, New South Wales, in 1996 revealed that only 7% of larvae developed into adults in transgenic seeds 75 days post-harvest, whereas in control, 98% of the larvae had developed into adults (Morton et al. 2000).

The pea cultivar Laura was also transformed separately with αAI-1 and αAI-2 using the method described by researchers (Schroeder et al. 1993; Morton et al. 2000). T3 seed of two lines of Laura having αAI-1 and T2 seed of one line of Laura having αAI-2 were sown in field trials at WaggaWagga, New South Wales; Horsham, Victoria; and Katanning, Western Australia, in 1997 as a randomized block design. Analysis revealed that both the αAI-1 lines provided complete protection against adult pea weevil damage at all the three sites. Also, the level of protection was higher in Laura transgenics in comparison with Greenfeast lines probably due to higher expression level of the αAI-1 protein, as documented during the study. No yield penalty was reported in transgenic pea. However, Laura having αAI-2 recorded a lesser percent adult emergence in comparison with control Laura, yet it was relatively higher than that for Laura having αAI-1. Also, the larvae collected from the αAI-2 Laura seeds were still alive, and they developed into adults by 52% 110 days post-harvest. This clearly indicates that αAI-2 significantly i.e. delays the emergence of adults by approximately a month but is unable to cause mortality in them. Molecular and biochemical studies suggest that the documented difference in efficacy between the two inhibitors is probably due to their different chemical properties. Environmental stresses play a major role in affecting crop production, productivity, as well as harvest quality (Xiong et al.

1999), and pea is not an exception when exposed to abiotic stress and related biotic stress like pea weevil infestations. As performance of transgenic plants is dependent on the continued expression of introgressed gene under stress conditions (Sachs et al. 1998; Traore et al. 2000), protection given against pea weevil in transgenic pea is subject to continuous and stable expression of αAI and its activity. And these properties tend to vary during the exposure of plants to stress. Hence researchers studied the response of transgenic pea cv. Greenfeast having αAI-1 to water deficit and high temperature and its effect on pea weevil survival (Sousa-Majer et al. 2004). The status of transgenic pea resistant to various insect pests are listed in Table 10.4.

A study was conducted to see the effect of water deficit on transgenic plants relative to control plants by exposing them to simulated temperature and relative humidity conditions in a glasshouse until maturity. A week post first pod set, irrigation was completely withheld in a set of pots bearing the transgenic and control plants for the water-deficit experiment. Control condition pots were watered daily to pot capacity (Jones et al. 1980). While water deficit conditions revealed similar changes in leaf relative water content measurements, significant decrease in number of seeds per pod, and pod wall weight among control and transgenic plants, no significant effect on the level of αAI-1 expression per unit protein of transgenic pea seed was detected (Sousa-Majer et al. 2004).

To assess the effect of high temperature, transgenic and control plants were exposed to simulated normal weather conditions until flowering. Thereafter, a set of transgenic and control pots were exposed to 27/22 °C day/night for 2 weeks and a continuous vapor pressure deficit (VPD) of 1.3 kPa in a growth cabinet and then to 32/27 °C day/night for next 2 weeks at same VPD. At maturity, pods were harvested and studied. Pots were watered regularly to be maintained at pot capacity. High temperature significantly reduced the number of seeds per pod and pod wall weight in both control and transgenic plants. The weight per seed was lower for transgenic in comparison with control under normal- and high-temperature conditions. No difference was recorded for total protein per seed in transgenic plants under both temperatures; however, a reduction in expression level of αAI-1was recorded. Percent adult emergence of weevils increased from 92% to 94% in control plants when exposed from 27/22 °C day/night to 32/27 °C day/night temperature regime. However, a significant increase from 1.2% to 39% was observed in transgenic pea under similar conditions. This clearly suggests that high temperatures render the transgenic pea cv. Laura more susceptible to pea weevil attack (Sousa-Majer et al. 2004, 2007).

Heliothis virescens is a polyphagous insect pest which can feed and incur damage to many plant species including legumes. Being a Lepidopteran insect, the cry1 Ac gene was expected to manage the insect, and in this effort, Negawo et al. (2016) developed transgenic pea lines and confirmed them for integration and expression of *cry1Ac*. The lines were taken to advanced generations and also tested for their efficacy against *H. virescens*. Progeny of developed lines, in different generations (T0–T4), were subjected to insect bioassay. The total larval mortality recorded

Table 10.4 Insect-resistant transgenic pea

S. no.	Cultivar	Method of transformation	Gene/trait	Stage	Efficacy	References
1	Greenfeast	*Agrobacterium-* mediated transformation	αAI-1/ protection against pea weevil (*Bruchus pisorum*)	T5	Provides complete protection against the pea weevil. αAI-1 inhibits pea bruchid α-amylase by 80% over a broad pH range (pH 4.5–6.5) Reduces pea weevil survival by 93–98%. Larval mortality occurs at an early instar	Schroeder et al. (1995) Morton et al. (2000) Sousa-Majer et al. (2007)
	Greenfeast	*Agrobacterium-* mediated transformation	αAI-1/ protection against pea weevil (*Bruchus pisorum*)	–	High temperature reduces protective capacity of αAI-1 against pea weevil	Sousa-Majer et al. (2004).
2.	Laura	*Agrobacterium-* mediated transformation	αAI-1/ protection against pea weevil (*Bruchus pisorum*)	T3	Provides complete protection against the pea weevil. αAI-1 inhibits pea bruchid α-amylase by 80% over a broad pH range (pH 4.5–6.5) Reduces pea weevil survival by 93–98%. Larval mortality occurs at an early instar	Morton et al. (2000) Sousa-Majer et al. (2007)
3.	Laura	*Agrobacterium-* mediated transformation	αAI-2/ protection against pea weevil (*Bruchus pisorum*)	T3	αAI-2 provides partial protection as it is a much less effective inhibitor of pea bruchid α-amylase and inhibits the enzyme by only 40% and only in the pH 4.0–4.5 range Does not affect pea weevil survival but delays larval development based on head capsule size	Morton et al. (2000) Sousa-Majer et al. (2007)
4.	Sponsor	*Agrobacterium-* mediated transformation	*Cry1Ac*	T2–T7	100% larval mortality of tobacco budworm (*Heliothis virescens*)	Negawo et al. (2016) Negawo et al. (2013)

α-ai 1, alpha-amylase inhibitor 1; *α-ai* 2, alpha-amylase inhibitor 2; *cry*1Ac, crystal 1Ac (Bt) revealed four lines to have about 73–92% larval mortality, while three lines showed less than 20% larval mortality (Negawo et al. 2013, 2016).

10.6.4 Cowpea

Cowpea production is constrained by insect pests including sucking bugs, flower bud thrips, weevils, and pod borers (Singh and van Embden 1979) and cause significant loss (Jackai et al. 1995). Storage pests cause severe damage to the cowpea seeds during storage. Transgenics in cowpea cv. Pusa Komal were developed using the common bean α-amylase inhibitor-1 (αAI-1) gene under the bean phytohemagglutinin promoter resulting in accumulation of αAI-1 in transgenic seeds. The use of thiol compounds during co-cultivation and a geneticin-based selection system gave an increase in stable transformation frequency in cowpea. Transgenic cowpeas expressing the αAI strongly inhibited development of *C. maculatus* and *C. chinensis* in insect bioassays (Solleti et al. 2008).

Legume pod borer (*Maruca vitrata*) also known as Maruca pod borer (MPB) causes colossal loss in cowpea yield (Jackai et al. 1995; Tamo et al. 2003). Bett et al. (2017) identified, cloned, and expressed a vegetative insecticidal protein (VIP) gene that is effective against MPB and developed a gene cassette for transforming cowpea for resistance against the pest. They redesigned the gene-coding region so as to increase the GC content, deleted the polyadenylation signals and mRNA stabilizing motifs, and used plant-preferred codons. The vip3Ba gene was under the control of *Arabidopsis* small subunit 1A promoter and tobacco small subunit3′end derived from pSF12 (Tabe et al. 1995).

Genetic transformation of cowpea was achieved by *Agrobacterium*-mediated method that included a sonication step (Popelka et al. 2006). Molecular analysis confirmed integration and expression of the transgene. Leaves from the vip3Ba-expressing cowpea lines (having the protein in range of 155–895 ng/ml TSP) were subjected to insect bioassay using 6-day-old *M. vitrata* larvae with both negative control, a non-transgenic parent line, and a positive control, Bt cowpea line 709A, having *cry1Ab* (260 ng//ml TSP) (Higgins et al. 2012). Ten days later, no larvae died in control leaves and had developed to their final instar stage, while 100% mortality was seen in positive control as well as *vip3Ba*-bearing transgenic cowpea lines. Leaf damage was negligible in transgenic leaf samples while negative control leaf revealed voracious feeding. Bioassay also revealed the efficacy of *VIP3Ba* at low doses as 155 ng/mg TSP was sufficient for MPB larval death (Bett et al. 2017).

The Commonwealth Scientific and Industrial Research Organization laboratory, Canberra-Australia, developed two transgenic lines in cowpea in IT86D-1010 that was resistant to *M. vitrata* (Higgins 2007). These were used as male parents and crossed with three other cowpea genotypes, namely, IT97K-499-35, IT86D-1010, and IT93K-693-2 (female parents) to generate F1s that were screen tested under field conditions, along with their parents, in a completely randomized block design during 2012 at CFT site of institute for Agricultural Research of Ahmadu Bello University Samaru-Zaria. Six infestations were carried out at an interval of 4 days each by releasing the first instar larvae of MPB on flowers of plants at increasing levels of pressure. Infestations were initiated on the 45th day after planting, and data was recorded on the 77th day after planting. The transgenic parents had a minimum number of pods damaged per plant, while other non-transgenic genotypes had damage in the range of 5.0–11.07 pods per plant. The F1 fared well and had relatively no pod damage (Mohammed et al. 2014).

Transgenics in cv. Tvu201 were developed using *cry1Ab* insecticidal gene by the method of nodal electroporation. The gene was under control of constitutive promoter CaMV35S. Transgenic leaves were subjected to insect bioassay using third instar MPB larvae. The number of larvae surviving, feeding deterrence score (FDS), and larval weight was measured. Insect bioassay was done through three progenies and in T3 progeny wherein 50% inhibition in feeding was observed, and percent larval survival was low as compared to control. In transgenic lines where mortality was not seen, the larval weight was found to be highly reduced (Adesoye et al. 2008). Bakshi and coworkers developed an improved method of *Agrobacterium*-mediated genetic transformation wherein sonication and vacuum infiltration were also put to use and an enhanced transformation efficiency of 3.09% was obtained. The cultivar Pusa Komal was introgressed with Bt *cry1Ac* driven by CaMV35 along with *nptII* and *gus* genes. The regenerated tranformants were screened for stable integration and expression of *cry1Ac* and taken forward to T1 generation. Segregation analysis reveals single dominant gene inheritance (Bakshi et al. 2011). The status of cowpea transgenics resistant to various insects pests is listed in Table 10.5.

10.6.5   Green Gram and Black Gram

Sonia and coworkers reported the *Agrobacterium*-mediated transformation protocol using cotyledonary node explant. Various factors that influence Agrobacterium-mediated transformation like the age of explant, removal of both the cotyledon from explant, preculture and wounding of explants, addition of acetosyringone, PPT-based selection of transformants, and pH of co-cultivation medium were optimized to improve transformation efficiency in green gram. Using this protocol, bruchid resistant transgenic green gram (cv. Pusa 105) was developed by transforming cotyledonary node explant with the *Agrobacterium tumefaciens* strain EHA105 harboring a binary vector pKSB carrying bialaphos resistance (*bar*) gene and *Phaseolus vulgaris* α-amylase inhibitor-1 (αAI-1) gene. Phosphinothricin (PPT)-selected shoots were

Table 10.5 Insect-resistant transgenic cowpea

S. no.	Cultivar	Method of transformation	Gene/trait	Transformation frequency	Stage	Efficacy	References
1	Pusa Komal	Agrobacterium-mediated transformation	(Phaseolus vulgaris) aai-1/ resistance against bruchids	–	–	Inhibits the development of C. maculatus and C. chinensis	Solleti et al. (2008)
2.	–	Agrobacterium-mediated transformation	vip3Ba resistance against Maruca vitrata	–	–	100% mortality of M. vitrata larvae	Bett et al. (2017)
3	TCL 709 genotype	Agrobacterium-mediated transformation	Cry1Ab resistance against M. vitrata	–	Transformation event derived from IT86D-1010	–	Mohammed et al. (2014)
4	TCL 711 genotype	Agrobacterium-mediated transformation	Cry1Ab resistance against M. vitrata	–	Transformation event derived from IT86D-1010	–	Mohammed et al. (2014)
5	TVu 201	Electroporation-mediated genetic transformation	Cry1Ab resistance against M. vitrata	–	T3 progeny	Reduced larval survival and larval weight and inhibited M. vitrata feeding on cowpea leaves	Adesoye et al. (2008)
6	Pusa Komal	Improved Agrobacterium-mediated transformation	Cry1Ac	3.09	T1	–	Bakshi et al. (2011)

α-ai 1, alpha-amylase inhibitor 1; cry1Ac, crystal 1Ac (Bt); cry1Ab, Crystal1Ab (Bt); vip3Ba, vegetative insecticidal protein 3Ba

Table 10.6 Insect-resistant transgenic green gram and black gram

S. no.	Cultivar	Method of transformation	Gene/trait	Transformation frequency	Stage	Efficacy	References
Vigna radiata							
1	Pusa 105	Direct using cotyledonary node explant	αai-1 resistance against bruchid	1.51%	T1	–	Sonia et al. (2007)
Vigna mungo							
1	Co5	Direct using cotyledonary node explant	*Cry1Ac*	0.25%	T1	–	Acharjee et al. (2004)

α-ai, alpha-amylase inhibitor; *cry 1Ac*, crystal 1Ac (Bt); *cry1Aabc*, Crystal1Aabc (Bt)

regenerated into complete plantlets. The protocol is rapid and takes around 8–10 weeks for recovery of transgenic plants with the transformation efficiency of 1.51%. Transgenic plants showed stable integration and expression of transgene as evidenced by Southern hybridization and PPT leaf paint assay, respectively. Most of the T_1 progeny inherited both the transgenes. However, insect bioassay to test the efficacy of gene construct has not been studied (Sonia et al. 2007).

Direct shoot organogenesis using cotyledonary node as explant for genetic transformation of black gram using *Agrobacterium* harboring the *Bacillus thuringiensis cry1Ac* gene and nptII as a selectable marker gene was carried out, and a transformation frequency of 0.25% was obtained. Expression and integration of transgene were confirmed by dot blot; Western blot analyses in T0 and T1 plants and transmission of transgene were confirmed by PCR. However, insect bioassay to test the efficacy of gene construct for *Helicoverpa armigera* and *Maruca testulalis* has not been reported (Acharjee et al. 2004). The status of green gram and blackgram transgenics resistant to various insects pests is listed in Table10.6.

10.7   Potential Benefits of Insect-Resistant Transgenics

Use of insecticidal proteins in transgenic will take care of the invading pests, and hence, the benefit will be in terms of reduced use of chemical sprays, thus reducing the costs of application. Also, tissue-specific expression of these insecticidal proteins will ensure reduced collateral damage that is often associated with the use of broad-spectrum insecticides. As a result, the quality of produce will be relatively better, and residue issues in harvest will be minimized and will attract remunerative prices. In short, use of insect smart transgenics will benefit growers by giving them higher yields through less input costs and easier agronomic management practices. Another advantage is reduced exposure of farmer to insecticide and also to the environment.

10.8 Biosafety Issues for Insect-Tolerant Transgenics

A major issue of concern is the possibility of allergenicity and/or toxicity in developed transgenics for both mammals and off-target organisms. In India, the strict biosafety regulations imposed by the Government of India ensure that developed transgenics are safe for humans, organisms, and the environment before they are released for public use. As such, no specific receptors for *Bt* protein have been identified in the gastrointestinal tract of mammals, including man (Kuiper and Noteborn 1994). Novel genes expressing insecticidal proteins in transgenics have to undergo several rigorous tests that ensure safety of the protein to mammals and other living organisms. The usefulness of insect-tolerant transgenics has already been documented in crops like Bt cotton in the USA and India, but having been in the fields for more than a decade, issues like resistance to insecticide in pod borer in fields have also been emerging. (Tabashnik and Carrière 2010). Efforts to minimize such incidences may include crop rotation leading to diversity in pest incidences and agronomic management practices.

10.9 Future Prospects

The development of transgenics using gene cassettes that takes care of issues like antibiotic-resistant genes as selectable marker, continuous expression of desired gene/protein leading to increased load of expression on plant, expression in all tissue types, etc. will enhance the efficiency of developed transgenics. It is in this context that use of clean gene technology or the two-vector system for development of marker-free transgenics is helpful. Use of inducible and tissue-specific promoters will enhance temporal and spatial regulation of desired transgene. The development and use of multigene constructs having more than one gene for desired traits are also a feasible option for better and efficient transgenics.

10.10 Conclusion

Genetically engineered insect-tolerant food legumes will definitely be an advantage when incorporated into the cropping system. Though most transgenics developed in food legumes have used either the Bt gene or the alpha-amylase gene, more of the available genes need to be tested for their bio-efficacy and safety in the context of mammalian toxicity and/or off-target toxicity. Additional advantage may be taken if the trait for insect resistance is combined with other beneficial traits like disease resistance, and the transgenics are developed based on the location/area of specific problems. Use of multigene constructs, inducible promoter-driven gene expression, and biotic and abiotic stress tolerance in combination are some aspects that may be worked upon for developing better transgenics more amenable to the regime of climate change. In short, tailoring of crops for desired traits by genetic engineering is an option that needs to be strengthened and supported for developing better cultivars for sustainable production of food legumes.

References

Acharjee S, Das P, Sen S, Bordoloi S, Kumar PA, Sarmah BK (2004) *Agrobacterium*-mediated genetic transformation of black gram for resistance against pod borers. Proc. Indian Society of Agricultural Biochemists. 12–14 Nov, pp 188–192

Acharjee S, Sarmah BK, Kumar PA, Olsen K, Mohan R, Moar WJ, Moore A, Higgins TJV (2010) Transgenic chickpeas expressing a sequence-modified cry2Aa gene. Plant Sci 178:333–339

Acharjee S, Handique PJ, Sarmah BK (2012) Effect of thidiazuron (TDZ) on *in vitro* regeneration of blackgram (*Vigna mungo* L.) embryonic axes. J Crop Sci Biotech 15(4):311–331. https://doi.org/10.1007/s12892-011-0122-3

Adlinge PM, Samal KC, Kumara Swamy RV, Rout GR (2014) Rapid in vitro plant regeneration of black gram (*Vigna mungo* L. Hepper) var. sarala, an important legume crop. Proc Natl Acad Sci, India, Sect B: Biol Sci 84:823–827

Armes NJ, Jadhav DR, De Souza KR (1996) A survey of insecticide resistance in *Helicoverpa armigera* in the Indian sub-continent. Bull Entomol Res 86:499–514

Adesoye A, Machuka J, Togun A (2008) *CRY 1AB* transgenic cowpea obtained by nodal electroporation. Afr J Biotechnol 7(18):3200–32108

Bakshi A, Mishra SS, Sahoo L (2011) Improved *Agrobacterium*-mediated transformation of cowpea via sonication and vacuum infiltration. Plant Cell Rep. https://doi.org/10.1007/s00299-011-1133-8

Batra P, Yadav NR, Sindhu A, Yadav RC, Chowdhury VK, Chowdhury JB (2002) Efficient protocol for *in vitro* direct plant regeneration in chickpea *Cicer arietinum* L. Indian J Exp Biol 40(5):600–602

Bett B, Gollasch S, Moore A, James W, Armstrong J, Walsh T, Harding R, Higgins TJV (2017) Transgenic cowpeas (*Vigna unguiculata*L. Walp) expressing *Bacillus thuringiensis*Vip3Ba protein are protected against the *Maruca* pod borer (*Maruca vitrata*). Plant Cell Tiss Org Cult 131:335–345

Biddle AJ, Cattlin ND (2001) Pests and diseases of peas and beans – a colour handbook. Manson Publishing Ltd, London

Chakraborti D, Sarkar A, Mondal HA, Das S (2009) Tissue specific expression of potent insecticidal, *Allilum sativum* leaf agglutinin (ASAL) in important pulse crop, chickpea (*Cicer arietinium* L.) to resist the phloem feeding *Aphis craccivora*. Transgenic Res 18:529–544

Clement SL, Hardie DC, Elberson LR (2002) Variation among accessions of *Pisum fulvum* for resistance to pea weevil. Crop Sci 42:2167–2173

Czapla TH, Lang BA (1990) Effects of plant lectins on the larval development of European corn borer (Lepidoptera: Pyralidae) and Southern corn rootworm (Coleoptera: Chysomelidae). J Econ Entomol 83:2480–2485

Das A, Datta S, Sujayanand GK, Kumar M, Singh AK, Arpan SA, Ansari J, Kumar M, Faruqui L, Thakur S, PA K, Singh NP (2016) Expression of chimeric Bt gene, Cry1Aabc in transgenic pigeonpea (cv. Asha) confers resistance to gram pod borer (*Helico verpa armigera* Hubner.). Plant Cell Tiss Org Cult 127:705–715

Das A, Datta S, Thakur S, Shukla A, Ansari J, Sujayanand GK, Kumar PA, Chaturvedi SK, Singh NP (2017) Expression of a chimeric gene encoding insecticidal crystal protein *Cry1Aabc* of *Bacillus thuringiensis* in chickpea (*Cicer arietinum* L.) confers resistance to gram pod borer (*Helicoverpa armigera* Hubner.). Front Plant Sci 8:1423

Dhaliwal GS, Arora R (1994) Trends in agricultural insect pest management. Commonwealth Publishers, New Delhi

Estruch JJ, Carozzi NB, Desai N, Duck NB, Warren GW, Koziel M (1997) Transgenic plants: an emerging approach to pest control. Nat Biotechnol 15:137–141

Ganguly M, Molla KA, Karmakar S, Datta K, Datta SK (2014) Development of pod borer-resistant transgenic chickpea using a pod-specific and a constitutive promoter-driven fused cry1Ab/Ac gene. Theor Appl Genet 127(12):2555–2565

Ghosh G, Ganguly S, Purohit A, Chowdhury RK, Das A, Chakroborti D (2017) Transgenic pigeon-pea events expressing cry1Ac and cry2Aa exhibit resistance to *Helicoverpa armigera*. Plant Cell Rep 36(7):1037–1051

Higgins TJ (2007) Bt cowpea with protection against pod borer for transfer to Africa. http://www.Publications.csiro.au/rpr/download?pid=csiro:EP124059&dsid=DS1

Higgins TJV, Gollasch S, Molvig L et al (2012) Insect-protected cowpeas using gene technology. In: Boukar O, Coulibaly O, Fatokun CA et al (eds) Innovative research along the cowpea value chain. Proceedings of the Fifth World Cowpea Conference on improving livelihoods in the cowpea value chain through advancement in science. Saly, Senegal 27 September–1 October 2010. International Institute of Tropical Agriculture, Ibadan, pp 131–137

Hilder VA, Gatehouse AM, Sheerman SE, Barker RF, Boulter D (1987) A novel mechanism of insect resistance engineered into tobacco. Nature 300:160–163

Hilder VA, Powell KS, Gatehouse AMR, Gatehouse JA, Gatehouse LN, Shi Y, Hamilton WDO, Merryweather A, Newell CA, Timans JC, Peumans WJ, Van Damme E, Boulter D (1995) Expression of snowdrop lectin in transgenic tobacco plants results in added protection against aphids. Trans Res 4:18–25

Ignacimuthu S, Prakash S (2006) *Agrobacterium*-mediated transformation of chickpea with α-amylase inhibitor gene for insect resistance. J Biosci 31:339–345

Indurker S, Misra HS, Eapen S (2007) Genetic transformation of chickpea (*Cicer arietinum* L.) with insecticidal crystal protein gene using particle gun bombardment. Plant Cell Rep 26:755–763

Jackai LEN (1995) The legume pod borer *Maruca testulalis*, and its principal host plant, *Vigna unguiculata* (L.) Walp.- use of selective insecticide sprays as an aid in the identification of useful levels of resistance. Crop Prot, 14:299–306

Jiang H, Zhu YX, Chen ZL (1996) Insect resistance of transformed tobacco plants with a gene of a spider insecticidal peptide. Acta Bot Sin 38:95–99

Jones MM, Osmond CB, Turner NC (1980) Accumulation of solutes in leaves of sorghum and sunflower in response to water defcits. Aust J Plant Physiol 7:193–203

Kar S, Basu D, Das S, Ramakrishnan NA, Mukherjee P, Sen SK (1997) Expression of cry1Ac gene of *Bacillus thuringiensis* in transgenic chickpea plants inhibits development of pod borer (*Heliothis armigera*) larvae. Transgenic Res 6:177–185

Korth KL (2008) Genes and traits of interest for transgenic plants. In: Stewart CN (ed) Plant biotechnology and genetics: principles, techniques, and applications. John Wiley & Sons, Inc., Hoboken

Kranthi KR, Jadhav DR, Kranthi S, Wanjari RR, Ali S, Russell DA (2002) Insecticide resistance in five major insect pests of cotton in India. Crop Prot 21:449–460

Krishna G, Reddy PS, Ramteke PW, Rambabu P, Sohrab SS, Rana D, Bhattacharya P (2011) In vitro regeneration through organogenesis and somatic embryogenesis in pigeon pea [*Cajanus cajan* (L.) Millsp.] cv. JKR105. Physiol Mol Biol Pl 17(4):375–385

Kuiper HA, Noteborn HJM (1994) Food safety assessment of transgenic insect-resistant Bt tomatoes. Food safety evaluation. In: Proceedings of an OECD-sponsored Workshop, 12–15 September 1994, Oxford, UK. Organisation for Economic Cooperation and Development (OECD), Paris, France, pp 50–57

Lawrence PK, Koundal KR (2001) *Agrobacterium tumefaciens* mediated transformation of pigeonpea (Cajanus cajan L. Millsp.) and molecular analysis of regenerated plants. Curr Sci 80:1428–1432

Mehrotra M, Singh AK, Sanyal I, Altosaar I, Amla DV (2011) Pyramiding of modified cry1Ab and cry1Ac genes of *Bacillus thuringiensis* in transgenic chickpea for improved resistance to pod borer insect Helicoverpa armigera. Euphytica 182:87–102

Mohammed BS, Ishikayu MF, Abdullahi US, Katung MD (2014) Response of transgenic Bt cowpea lines and their hybrids under field conditions. J Plant Breed Crop Sci 6(8):91–96

Morton RL, Schroeder HE, Bateman KS, Chrispeels MJ, Armstrong E, Higgins TJV (2000) Bean alpha-amylase inhibitor 1 in transgenic peas (*Pisum sativum*) provides complete protection from pea weevil (*Bruchus pisorum*) under field conditions. PNAS 97(8):3820–3825

Negawo AT, Aftabi M, Jacobsen HJ, Altosaar I, Hassan FS (2013) Insect resistant transgenic pea expressing cry1Ac gene product from *Bacillus thuringiensis*. Biol Control 67:293–300

Negawo AT, Baraneka L, Jacobsena HJ, Hassan F (2016) Molecular and functional characterization of cry1Ac transgenic pea lines. GM Crops & Food 7:159–174

Obembe OO (2008) Exciting times for cowpea genetic transformation research. Life Sci J 5(2):50–52

Popelka JC, Gollasch S, Moore A, Molvig L, Higgins TJV (2006) Genetic transformation of cowpea (*Vignaunguiculata*L.) and stable transmission of the transgenes to progeny. Plant Cell Rep 25:304–312

Prescott VE, Campbell PM, Moore A, Mattes J, Rothenberg ME, Foster PS, Higgins TJ, Hogan SP (2005) Transgenic expression of bean alpha-amylase inhibitor in peas results in altered structure and immunogenicity. J Agric Food Chem 53(23):9023–9030

Ramu SV, Rohini S, Keshavareddy G, Neelima MG, Shanmugham NB, Kumar ARV, Sarangi SK, Kumar PA, Udayakumar M (2011) Expression of a synthetic cry1AcF gene in transgenic Pigeon pea confers resistance to *Helicoverpa armigera*. J Appl Entomol, 136:675–687.

Sachs ES, Benedict JH, Stelly DM, Taylor JF, Altaman DW, Berberich SA, Davis SK (1998) Expression and segregation of genes encoding CryIA insecticidal proteins in cotton. Crop Sci 38:1–11

Sanyal I, Singh AK, Kaushik M, Amla DV (2005) *Agrobacterium* mediated transformation of chickpea (*Cicer arietinum* L.) with *Bacillus thuringiensis* cry1Ac gene for resistance against pod borer insect *Helicoverpa armigera*. Plant Sci 168:1135–1146

Sarmah BK, Moore A, Tate W, Molvig L, Morton RL, Rees DP, Chiaiese P, Chrispeels MJ, Tabe LM, Higgins TJV (2004) Transgenic chickpea seeds expressing high levels of a bean amylase inhibitor. Mol Breed 14:73–82

Schroeder HE, Schotz AH, Wardley-Richardson T, Spencer D, Higgins TJV (1993) Transformation and regeneration of two cultivars of pea (*Pisum sativum* L.). Plant Physiol 101:751–757

Schroeder HE, Gollasch S, Moore A, Tabe LM, Craig S, Hardie DC, Chrispeels MJ, Spencer D, Higgins TJV (1995) Bean a-amylase inhibitor confers resistance to the pea weevil (*Bruchus pisorum*) in transgenic peas (*Pisum sativum* L.). Plant Physiol 107(1233-1):239

Schuler TH, Poppy GM, Kerry BR, Denholm I (1998) Insect resistant transgenic plants. TIBTECH 16:168–175

Singh SR, Van Emden HF (1979) Insect pests of grain legumes. Annu Rev Entomol 24:255–278

Shade RE, Schroeder HE, Pueyo JJ, Tabe LM, Murdock LL, Higgins TJV, Chrispeels MJ (1994) Transgenic pea seeds expressing the alpha-amylase inhibitor of the common bean are resistant to bruchid beetles. BioTechnology 12:793–796

Sharma KK, Ananda Kumar P, Singh NP, Sharma HC (2005) Insecticidal genes and their potential in developing transgenic crops for resistance to *Heliothis/Helicoverpa*. In: Sharma HC (ed) *Heliothis/ Helicoverpa* management: emerging trends and strategies for future research. Oxford & IBH Publishing Co. Pvt. Ltd., New Delhi, pp 255–274

Sharma KK, Bhatnagar-Mathur P, Jayanand B (2006a) Chickpea (*Cicer arietinum* L.). In: Wang K (ed) *Agrobacterium* protocols: methods in molecular biology, vol 44. Humana Press Inc., Totowa, pp 313–323

Sharma KK, Lavanya K, Anjalah A (2006b) *Agrobacterium tumefaciens*- mediated production of transgenic pigeon pea (*Cajanus cajan* [L.]Millsp.) expressing the synthetic BT CRY1AB gene. In Vitro Cell Dev Biol 42:165–173

Sharma KK, Sreelatha G, Dayal S (2006c) Pigeonpea [*Cajanus cajan* L. (Millsp.)]. In: Wang K (ed) *Agrobacterium* protocols: methods in molecular biology, vol 44. Humana Press Inc., Totowa, pp 359–367

Shukla S, Arora R, Sharma HC (2005) Biological activity of soybean trypsin inhibitor and plant lectins against cotton bollworm/legume pod borer, *Helicoverpa armigera*. Plant Biotechnol 22:1–6

Smith AM (1990) Pea weevil (*Bruchus pisorum* L.) and crop loss implications for management. In: Fujii K, Gatehouse AMR, Johnson CD, Mitchell R, Yoshida T (eds) Bruchids and legumes: economics, ecology and coevolution. Kluwer Academic Publishers, Dordrecht, pp 105–114

Solleti SK, Bakshi S, Purkayastha J, Panda SK, Sahoo L (2008) Transgenic cowpea (*Vigna unguiculata*) seeds expressing a bean alpha-amylase inhibitor 1 confer resistance to storage pests, bruchid beetles. Plant Cell Rep 27(12):1841–1850.

Sonia SR, Singh RP, Jaiwal PK (2007) *Agrobacterium tumefaciens* mediated transfer of *Phaseolus vulgaris* alpha-amylase inhibitor-1 gene into mung bean *Vigna radiata* (L.) Wilczek using bar as selectable marker. Plant Cell Rep 26:187–198

Sousa-Majer MJ, Turner NC, Hardie DC, Morton RL, Lamont B, Higgins TJV (2004) Response to water deficit and high temperature of transgenic peas (*Pisum sativum* L.) containing a seed specific a-amylase inhibitor and the subsequent effects on pea weevil (*Bruchus pisorum* L.) survival. J Exp Bot 55(396):497–505

Sousa-Majer MJ, Hardie DC, Turner NC, Higgins TJV (2007) Bean α-amylase inhibitors in transgenic peas inhibit development of pea weevil larvae. J Econ Entomol 100(4):1416–1422

Srinivasan MT, Sharma RP (1991) Agrobacterium-mediated genetic transformation of chickpea (*Cicer arietinum*). Indian J Exp Biol, 29:758–761

Surekha C, Beena MR, Arundhati A, Singh PK, Tuli R, Dutta-Gupts A, Kirti PB (2005) Agrobacterium mediated genetic transformation of pigeonpea (*Cajanus cajan* L.) using embryonal segments and development of transgenic plants for resistance against *Spodoptera*. Plant Sci 169:1074–1080

Tabashnik BE, Carrière Y (2010) Field-evolved resistance to Bt cotton: bollworm in the U.S. and pink bollworm in India. Southwestern Entomologist 35(3):417–424

Tabe LM, Wardley-Richardson T, Ceriotti A, Aryan A, McNabb W, Moore A, Higgins TJV (1995) A biotechnological approach to improving the nutritive value of alfalfa. J Anim Sci 73:2752–2759

Tamo M, Ekesi S, Maniania NK, Cherry A (2003) Biological control, a non-obvious component of IPM for cowpea. In: Neuenschwander P, Borgemeister C, Langewald J (eds) Biological control in IPM systems in Africa. CABI Publishing, Wallingford, pp 295–309

Third Advance Estimates of Production of Commercial Crops for 2016–17, Agricultural Statistics Division Directorate of Economics & Statistics

Traore SB, Carlson RE, Pilcher CD, Rice ME (2000) Bt and non-Bt maize growth and development as affected by temperature and drought stress. Agron J 92:1027–1035. www.gktoday. in/blog/pulses-production-consumption-and-international-trade-in-india/#Consumption_and_ Import_dependency

Verma AK and Chand L (2005) Agrobacterium-mediated transformation of pigeonpea (Cajanus cajan L.) with uidA and CryIA(b) genes. Physiol Mol Biol Plant. 11:99–109

Xiong L, Ishitani M, Zhu JK (1999) Interaction of osmotic stress, temperature and abscisic acid in the regulation of gene expression in *Arabidopsis*. Plant Physiol 119:205–212

Yadav SK, Sreenu P, Maheshwari M, Vanaja M, Venkateswarlu B (2010) Efficient shoot regeneration from double cotyledonary node explants of green gram (*Vigna radiata* (L.) Wilczek.). Indian J Biotechnol 9:403–407

Yu CG, Mullins MA, Warren GW, Koziel MG, Estruch JJ (1997) The *Bacillus thuringiensis* vegetative insecticidal protein Vip3A lyses midgut epithelium cells of susceptible insects. Appl Enzviron Microbiol 63:532–536

Chapter 11
Genetic Transformation of Millets: The Way Ahead

Sweta Dosad and H. S. Chawla

Abstract Millets are a group of small-seeded cereals and forage grasses grown in arid and semiarid regions of Asia and Africa, where majority of cereals cannot be relied upon to provide sustainable yield. While major cereals such as wheat, rice, and maize provide only food security, millets provide multiple securities, viz., food, fodder, health, nutrition, livelihood, and ecological. In the present chapter, recent advances in genetic transformation studies conducted in millets to date have been summarized. Millets have been transformed primarily by particle bombardment, whereas, *Agrobacterium*-mediated transformation is still lagging behind. Efforts need to be made to genetically improve millets by incorporating certain agronomically important traits, such as resistance to biotic and abiotic stresses, resistance to lodging, increased seed size, and palatability along with softness of grain to make these crops more desirable for consumer.

Keywords *Agrobacterium* · Biolistic · Millets · Transformation

11.1 Introduction

Millets comprise of at least 10 genera and 14 species belonging to the Poaceae (Gramineae) family of the monocotyledonous group that includes pearl millet (*Pennisetum glaucum*), finger millet (*Eleusine coracana*), Kodo millet (*Paspalum scrobiculatum*), barnyard millet (*Echinochloa frumentacea*), foxtail millet (*Setaria italica*), little millet (*Panicum sumatrense*), elephant grass (*Pennisetum purpureum*), guinea grass (*Panicum maximum*), etc. Millets are considered to be the first cultivated cereal in the world (O'Kennedy et al. 2004). They constitute major source of food for poor hence called "poor man's cereal." Salient features of millets are their adaptability to adverse environmental conditions, requirement of minimal inputs, and good nutritional properties making them the crops of agricultural security of poor farmers that inhabit arid, infertile, and marginal lands. Millets are usually

S. Dosad · H. S. Chawla (✉)
Genetics & Plant Breeding Department, G.B. Pant University of Agriculture & Technology, Pantnagar, India

© Springer International Publishing AG, part of Springer Nature 2018
S. S. Gosal, S. H. Wani (eds.), *Biotechnologies of Crop Improvement, Volume 2*,
https://doi.org/10.1007/978-3-319-90650-8_11

tolerant to drought, because of their rapid growth, short life cycle, high temperature tolerance, and deep root system (O'Kennedy et al. 2011). Each of the millets possess nutraceutical properties, but despite all the nutritional benefits, millets face several production constraints such as their low yield since they are mostly cultivated in marginal areas with poor moisture and fertility conditions (Plaza-wuthrich and Tadele 2012). Inherent characteristics, such as susceptibility to fungal blast disease and lodging, further cause a significantly high seed yield loss (50%) (Latha et al. 2005).

Conventional plant breeding along with molecular breeding and improved agricultural practices has resulted in dramatic crop improvements over the past 50 years (Sharma and Ortiz 2000). However, there is intense pressure for further improvements in crop quality and quantity as a result of population growth, public and industrial demands, health requirements, and environmental constraints. Improvement of millets through traditional methods has limited applications due to the narrow gene pool; in addition these are time-consuming methods which are often accompanied with linkage drag. Thus, conventional breeding alone cannot solve the problem of food insecurity and benefit smallholder farmers. The best possible way to combat this problem is by adopting genetic transformation tools which has made it feasible to transfer agronomically important genes into crop plants from organisms that are outside the range of conventional breeding techniques such as microorganisms, unrelated plants, and animals in a more precise, reliable, and speedy manner. It holds great potential to assuage some of the major constraints which affect productivity of these crops, by developing plants resistant to biotic and abiotic stress conditions prevailing in semiarid tropics. According to James (2014), if 377 million tons of additional food, feed and fiber produced by biotech crops during the period 1996–2012 had been grown conventionally, it is estimated that an additional 123 million hectares of conventional crops would have been required to produce the same tonnage.

Furthermore, transformation procedures are simple and cost-effective techniques for functional genomics studies and have recently been used to investigate function of novel genes in millets by inducing overexpression (Liu et al. 2009; Wang et al. 2014), co-suppression (Qin et al. 2008), antisense plants (Liu et al. 2009), and RNA interference (Liu et al. 2009; Ceasar et al. 2017) through introduction of dsRNA or DNA construct. Transient gene expression of genes by *Agrobacterium* and microprojectile are the most favored techniques in many plant species. In addition to that, transient gene expression in a model system provides insight into many protein functions such as subcellular localization, trafficking, protein-protein interaction, protein activity, stability, and degradation (Ueki et al. 2009). Development of *in vitro* regeneration system will be essential for the genetic modification of these important millets as successful development of transgenic plants can only be achieved by developing an efficient, reliable, genotype-independent regeneration system which will enable the introduction of many agronomically important genes into millets in the near future. Despite numerous reports available on millet tissue culture, the difficulties encountered in obtaining the desired response are numerous since the morphogenic potential of *in vitro* culture depends on several factors such as physiological status of the explant material, medium composition, culture conditions, growth regulators, and the genotype of the donor plants. Thus, a prerequisite for efficient use of plant tissue culture techniques requires a thorough knowledge of these factors. Detailed information regarding *in vitro* regeneration studies in millets can be obtained from our review (Dosad and Chawla 2016). In the present chapter, focus is on progress made in genetic transformation of millets via *Agrobacterium* and biolistic gene delivery methods, factors that critically affect transformation efficiency and future prospects.

11.2 Explants for Transformation

Successful *Agrobacterium*-mediated genetic transformation and regeneration depend on the choice of explant, its developmental stage, and source. Varying reports on regeneration and transformation response of explants, even from the same genotype, have been obtained due to the variation in endogenous hormone levels in different explants. Explants collected from the same inflorescence behave differently in culture, depending on size and location on the inflorescence. Response of explants from well-nourished plants is different from those of nutrient-deficient plants. Explants collected from plants grown in different seasons exhibited different regeneration response. Even in a single experiment, with similar explants, culture response is not 100% in most of the times (Bhaskaran and Smith 1990). Genetic transformation requires cytoplasmically rich, actively dividing cells (Meyer et al. 1985) because such cells are better able to overcome the stress induced by transformation and provide high frequency transgene integration. Various explants have been used in millets to initiate embryogenic tissue to facilitate stable transformation which have been given below.

11.2.1 Shoot Apical Meristem

Shoot apices have been found to be highly regenerative in long-term cultures (Vikrant and Rashid 2003). Moreover, shoot meristem based multiplication systems are in general genotype independent, and thus transformation can be extended to different genotypes (Sai et al. 2006). They have been successfully used to transform pearl millet (Devi and Sticklen 2002; Jha et al. 2011; Ignacimuthu and Kannan 2013), finger millet (Lakkakula et al. 2017), and foxtail millet (Ceasar et al. 2017). Alternatively, shoot tip-derived calli have also been used as explant for co-cultivation with *Agrobacterium*, e.g., pearl millet (Latha et al. 2006) and finger millet (Latha et al. 2005; Mahalakshmi et al. 2006; Ceasar and Ignacimuthu 2011).

11.2.2 *Immature Embryonic Tissues*

Freshly isolated immature embryos are considered as one of the best explants in cereals as well as in millets for regeneration (Bhaskaran and Smith 1990). They have been used as explant for transformation in pearl millet (Girgi et al. 2002, 2006; O'Kennedy et al. 2004, 2011). In addition, immature inflorescence-derived callus has also been used as explant in pearl millet (Jalaja et al. 2016) and foxtail millet (Qin et al. 2008; Liu et al. 2009; Wang et al. 2014; Pan et al. 2015). Although calli derived from immature inflorescence and immature zygotic embryos have been considered as an important source of totipotent cultures in many cereals, millets, and grasses (Vasil 1982), a serious drawback associated with them is seasonal availability which require planting at different time intervals for continuous supply of the immature embryos and inflorescence followed by collection of large number of immature tissues and laborious routine tissue culture practices within a limited time period in a year. Quick loss of regeneration potential in calli and problems associated with isolation and sterilization of immature embryos are also serious limitations for the use of these explants in regeneration (Sai et al. 2006).

11.2.3 *Seed-Derived Calli*

Among all the explants used for conducting regeneration and transformation studies, seed-derived embryogenic calli are the most preferred explants because of their several distinct advantages over other explants as starting material: (1) They are available more easily around the year in bulk quantities, (2) can be stored and easily handled, and (3) calli induced from scutellar tissue of mature seeds are suitable for gene delivery and genetic transformation, actively dividing and capable of regenerating into fertile plants. Successful transformation using seed-derived embryogenic calli have been reported in pearl millet (Lambe et al. 1995; Ramadevi et al. 2014; Ramineni et al. 2014), finger millet (Gupta et al. 2001; Sharma et al. 2011; Jagga-Chugh et al. 2012; Ramegowda et al. 2013; Bayer et al. 2014; Hema et al. 2014), foxtail millet (Martins et al. 2015b), barnyard millet (Gupta et al. 2001) and Bahia grass (Grando et al. 2002; Smith et al. 2002; Altpeter and James 2005; Gondo et al. 2005; Luciani et al. 2007; James et al. 2008; Sandhu and Altpeter 2008; Mancini et al. 2014).

11.3 Antinecrotic Compound

Monocots are generally recalcitrant to *in vitro* regeneration, which is a major obstacle faced during transformation of monocot species. This problem is further exacerbated by hypersensitivity of several millet genotypes to *Agrobacterium*- and bombardment-induced stress which causes tissue browning and necrosis following co-cultivation and bombardment, respectively. This nuisance can be alleviated by reducing the oxidative burst with the help of antioxidants such as polyvinylpyrrolidone (PVPP), dithiothreitol (DTT), cysteine, glutamine, proline, ascorbic acid, citric acid, and silver nitrate. The explants are either pretreated with antinecrotic mix or the coculture, and regeneration media is fortified with it. These antinecrotic compounds, along with reducing necrosis, also improve regeneration response of explants.

11.4 Effect of Vitamins, Amino Acids, and Inorganic Nutrients on Regeneration and Transformation

11.4.1 Casein, Glutamine and Proline

To maximize regeneration and transformation efficiency, investigators have included complex substances such as casein hydrolysate, glutamine, and proline into the media. A number of workers have evaluated the effects of these compounds on embryogenesis and shoot regeneration in millets. Addition of casein hydrolysate and glutamine has been proved to improve the initiation of embryogenic cultures in finger millet (Mohanty et al. 1985; Eapen and George 1990). Mohanty et al. (1985) reported 500 mgl^{-1} casein hydrolysate was effective in initiating callus induction in finger millet. Eapen and George (1990) employed 100 mgl^{-1} proline, 800 mgl^{-1} glutamine, and 250 mgl^{-1} casein to improve regeneration response. Yemets et al. (2013) demonstrated that addition of 100 mgl^{-1} glutamine, 200 mgl^{-1} casein hydrolysate, and 500 mgl^{-1} proline improved callus quality and regeneration response. In another study, Lakkakula et al. (2015) studied the effect of various concentrations of proline and casein hydrolysate on finger millet and claimed that 500 mgl^{-1} casein hydrolysate induced the highest frequency of embryogenic callus induction and shoot regeneration. In foxtail millet, Wang et al. (2011), Pan et al. (2015), and Lakkakula et al. (2016) supplemented the culture media with 800 mgl^{-1} casein hydrolysate; Liu et al. (2009) supplemented with 750–1000 mgl^{-1} proline to facilitate efficient transformation and regeneration. While in pearl millet, addition of 500 mgl^{-1} proline, 300mgl^{-1} casein to the tissue culture induction medium resulted in a highly efficient regeneration and transformation procedure (O'Kennedy et al. 2004; Jha et al. 2011).

11.4.2 Inorganic Nutrients

Biotechnological research still lags behind in millets, and one of the main reasons is poor plant regeneration and transformation efficiencies of most of the genotypes (Kothari et al. 2005). One of the important factors that influence the regeneration and transformation potential of plants is the presence of nutrients in tissue culture media, and also, they are necessary components of various enzymes (Maksymiec 1997).

They act as secondary messengers and help in regulating and controlling plant tissue growth (Niedz and Evens 2007). Inorganic nutrients are major components of MS medium and hence can be manipulated to study their effect on the morphogenetic potential of the plants. Several workers have conducted experimental studies to investigate the effect of manipulated levels of minerals and have demonstrated that every species has its own requirements for particular minerals. Since inorganic nutrient levels used in most plant tissue culture media were initially standardized for tobacco tissue culture (Murashige and Skoog 1962), these nutrient levels which were optimum for tobacco tissue culture may not necessarily be ideal for other plant species like graminaceous monocots (Dahleen 1995).

11.4.2.1 Copper Sulfate (CuSO₄)

Of all the micronutrients, $CuSO_4$ has gained utmost importance. It is known to play an important role in several metabolic activities like protein and carbohydrate metabolism (Kothari-Chajer et al. 2008); it is also directly involved in the photosynthetic electron transport chain as a constituent of plastocyanin, a copper-containing protein. Thus, copper might be affecting regeneration process by affecting photosynthesis process (Sahrawat and Chand 1999). The stimulatory effect of copper may be attributed to the fact that numerous copper-containing proteins naturally occur in plants, and it is a part of several enzymes and hence might play a key role on callus induction and regeneration. In conclusion, copper can be incorporated into MS basal medium to improve morphogenic response of explant.

In finger millet, many times an increase in regeneration response in the presence of elevated levels of copper was observed by Kothari et al. (2004), Kothari-Chajer et al. (2008), and Sharma et al. (2011). Kothari et al. (2004) studied the effect of micronutrient manipulation on callus induction and plant regeneration. $CuSO_4$ increased to five times the normal concentration in the media resulted in a fourfold increase in number of regenerated shoots. Kothari-Chajer et al. (2008) found a marked improvement in quality, fresh weight, and percent response with increasing concentration of copper up to a threshold limit beyond which there is a decline in regeneration efficiency. Best regeneration response was observed at ten times Cu concentration; on the other hand, callus induced on medium devoid of Cu absolutely failed to regenerate. Similar results were reported by Sharma et al. (2011) where the presence of ten times $CuSO_4$ in the media was found effective for enhancing the number of responding calli as well as for recovery of plantlets.

11.4.2.2 Silver Nitrate (AgNO₃)

In addition to *Agrobacterium*- and microprojectile bombardment-induced necrosis and tissue browning, another limiting factor in regeneration and transformation is ethylene produced by cultured plant cells and perhaps by the gelling agent such as agar in culture medium. This effect is further aggravated by 2,4-D (widely used for callus induction) which strongly enhances ethylene production. *In vitro* studies have indicated that ethylene inhibits somatic embryogenesis and shoot regeneration (Wu et al. 2006). Supplementation of coculture or regeneration media with silver nitrate inhibit *Agrobacterium* growth without affecting T-DNA delivery and integration and also inhibit ethylene action, hence greatly improving the callus induction frequency, friable embryogenic callus frequency, somatic embryogenesis, shoot formation, efficient root formation, and transformation efficiency (Benson 2000). Regeneration medium can easily be fortified with $AgNO_3$ since it is soluble in water and lack of phytotoxicity at effective concentrations.

Oldach et al. (2001) noticed 5 mgl^{-1} $AgNO_3$ significantly increased regeneration rate by 14–87% depending upon genotype of Sorghum. Vikrant and Rashid (2002) and O'Kennedy et al. (2004) reported that addition of 10 mgl^{-1} $AgNO_3$ into the media significantly increased frequency of embryogenesis in Kodo and pearl millet, respectively. Kothari-Chajer et al. (2008) have reported that inclusion of $AgNO_3$ favored growth and development of callus in Kodo as well as finger millet. They further observed approximately three and two times increase in plant regeneration after incorporation of 0.17 mgl^{-1} and 1.0 mgl^{-1} $AgNO_3$ in regeneration medium of Kodo and finger millet, respectively. Sharma et al. (2011) and Wang et al. (2011) have also reported improvement in callus development, growth, and regeneration with the inclusion of 5–10 mgl^{-1} $AgNO_3$ in finger millet and foxtail millet, respectively.

11.5 Promoters for Gene of Interest

Selection of promoter is one of the key factors in developing successful procedure for genetic transformation. Three major classes of promoters are currently used in plant biotechnology: (1) those able to drive expression constitutively, (2) those driving expression specifically in certain tissues and at certain developmental stages, and (3) those able to drive expression under inductive conditions.

11.5.1 Constitutive Promoters

Constitutive promoters drive high level of gene expression in all the cells throughout the entire growth and developmental period. One of the first promoters to be used in cereal transformation was the *CaMV35S* derived from the 35S RNA transcript of tobacco cauliflower mosaic virus (Franck et al. 1980). This promoter has been used extensively to drive gene expression at high levels in a broad range of dicots; however, in monocots the *CaMV35S* promoter exhibits lower activity (Christensen et al. 1992). Nevertheless, *CaMV35S* promoter has been used extensively for driving transgene expression in finger millet by Latha et al. (2005), Ceasar and Ignacimuthu (2011), Hema et al. (2014), and Jayasudha et al. (2014); pearl millet by Latha et al. (2006), Ramadevi et al. (2014), and Ramineni et al. (2014); and in foxtail millet by Liu et al. (2009).

The different performance of promoters in dicotyledonous or monocotyledonous cells might be due to differences in the respective transcription factors; it appears that dicot promoters are better recognized in dicots than in monocots and vice versa (Barcelo et al. 2001). Hence, it can be concluded that monocot promoters are the best choice for monocot transformation when a high level of expression is needed. Therefore, several other strong constitutive promoter sequences have been isolated from monocots to achieve high level of transgene expression in monocotyledonous plants. These include the maize alcohol dehydrogenase (*Adh1*), rice actin (*Act1*), maize ubiquitin (*Ubi1*) (Christensen et al. 1992), and the modified maize alcohol dehydrogenase 1 promoter (*Emu*) (Last et al. 1991) which was constructed by adding a set of enhancer elements to the 5′ end of a truncated *adh1* promoter and first intron. Lambe et al. (2000) and Tiecoura et al. (2015) in pearl millet have observed better efficiency of *Emu* than *CaMV35S* in regulating *gus* expression. However, in stable transformation it is much less effective (Barcelo et al. 2001).

Comparison of transcriptional efficiencies of different promoters or the identification and isolation of novel constitutive promoters might further enhance the expression of selected gene. In 1997, Li and coworkers revealed the order of promoter strength (from strong to weak) in rice was *Ubi1* > *Act1* > *Emu* > *CaMV35S*. However, in millets evaluation of promoter efficiency was first performed in 2001 by Gupta and coworkers. They studied five gene promoters (*CaMV35*, *Act1*, *Ubi1*, rice ribulose 1, 5-biphosphate carboxylase small subunit (*RbcS*) gene promoter, and *Flaveria trinervia* (*Ft*) gene promoter) for the expression of *gus* reporter gene in finger millet and barnyard millet. Among them, *Ubi1* was found to be the most efficient promoter for both the millets in terms of bringing about maximum *gus* activity. Similar results were reported in foxtail millet by Saha and Blumwald (2016); a comparison of expression patterns among transgenic lines revealed that reporter genes driven by the *Ubi1* promoter had high levels of expression as compared with the *CaMV35S* promoter. *Ubi1* has been successfully employed in driving transgene expression in millets, e.g., pearl millet (Girgi et al. 2002; Goldman et al. 2003; O'Kennedy et al. 2004, 2011; Ignacimuthu and Kannan 2013), finger millet (Ignacimuthu and Ceasar 2012; Bayer et al. 2014), Bahia grass (Grando et al. 2002; Altpeter and James 2005; Luciani et al. 2007), and foxtail millet (Wang et al. 2014; Girgi et al. 2006).

11.5.2 Improvement of Promoters

The activity of promoters in monocots can be enhanced by the inclusion of an intron between promoter and coding region; this phenomenon is known as intron-mediated enhancement (Vain et al. 1996). For example, low activity of *CaMV35S* in monocots was improved by fusing a monocot intron sequence such as maize alcohol dehydrogenase *adh1* intron 1 (Mancini et al. 2014) or *Act1*intron 1 or Maize *Shrunken-1* intron or *hsp70* intron (Sandhu and Altpeter 2008, Luciani et al. 2007 and Zhang et al. 2007) between promoter sequence and the gene which resulted in sufficient level of gene expression. The same is true for promoters isolated from monocot genes. *Adh1*, *Ubi1*, *Act1*, or *Emu* in combination with a monocot intron between promoter and gene greatly enhance expression of the gene and can be used to improve transformation efficiency in monocots (Li et al. 1997). Lambe et al. (2000) reported highest transient expression of *gus* gene by *Emu* promotor with *Adh1* intron in pearl millet. Enhanced gene expression in the presence of maize ubiquitin promoter (*Ubi1*) with its first intron was also reported in pearl millet (Girgi et al. 2002; Goldman et al. 2003; O'Kennedy et al. 2011; Ignacimuthu and Kannan 2013) and Bahia grass (Altpeter and James 2005; Luciani et al. 2007; James et al. 2008; Sandhu and Altpeter 2008).

11.5.3 Organ- or Tissue-Specific Promoter

Regulated promoters are generally preferred over constitutive promoters because of practical and biosafety advantages. Since a gene regulated by constitutive promoter is expressed in all the plant tissues, it may lead to undesirable pleiotropic effects in transgenic plants and could expose herbivores, pollinating insects and microbes in the rhizosphere to the recombinant protein. Restriction of protein accumulation to seeds helps in reducing these risks (Stoger et al. 2000; Commandeur et al. 2003). Thus, organ- or tissue-specific promoter can be used to limit recombinant protein expression in a specific organ as it drives the transgene expression at a specific site or time. However, there are only few reports available on tissue-specific promoters, e.g., maize pollen-specific promoter *Zm13* was used by Qin et al. (2008) in foxtail millet to study the function of the gene *Si401* in anther development. Ramegowda et al. (2013) have shown that wheat endosperm-specific promoter *Bx17* performs better than *CaMV35S* in driving *OsZIP1* gene (responsible for improving seed Zn concentration) in finger millet seeds. A novel promoter, *F128* from foxtail millet, was used to drive *gus* expression specifically in transgenic seeds of foxtail millet, maize, and *Arabidopsis* with higher activity than the constitutive *CaMV35S* promoter and the maize seed-specific promoter *19Z* by Pan et al. (2015).

11.5.4 Inducible Promoter

These promoters drive gene expression only under specific conditions such as pathogens or wounding, abiotic stresses, or by the action of chemicals. An abiotic stress-inducible promoter, *Hva1s* from wild barley, was used to drive stress-inducible expression of a DREB1A transcription factor ortholog in Bahia grass (James et al. 2008). O'Kennedy et al. (2011) employed a pathogen/wound-induced promoter, the *pin2* (potato proteinase inhibitor IIK wound-inducible promoter), to drive expression of transgene *gluc78*, encoding β-1, 3-glucanase in pearl millet during wounding and/or pathogen attack only. Recently, Sen and Dutta (2016) isolated and characterized abiotic stress-inducible bidirectional promoter *EcBDP* from finger millet which drove expression of *gus* and *gfp* reporter genes only when induced either by abscisic acid or cold treatment.

11.5.5 Databases for Plant Promoter Sequences

Efficient expression of a gene is dependent on the selection of an appropriate promoter but limited number of promoters are available in vectors and provide only little variation in gene expression patterns. Thus, with an increasing number of sequenced plant genomes, it has become necessary to develop a robust computational method for detecting novel plant promoters (Anami et al. 2013). To date, a wide variety of programs for predicting promoters are available, e.g., PlantPAN (Chang et al. 2008), GRASSIUS (Yilmaz et al. 2009), PlantCARE (Lescot et al. 2002), and TransGene Promoters (TGP) database (Smirnova et al. 2012).

11.6 Selectable Marker Genes

Only a small fraction of the cells, exposed to the transformation process, accept and integrate foreign DNA; rest of the cells remains untransformed. In order to detect transformed cells, marker genes are transferred into the recipient cells along with the gene of interest. Marker genes are characterized into two types.

11.6.1 Selectable Marker

Selectable marker genes allow the transformed cells to proliferate on media containing toxic levels of selection agent, while non-transformed cells die facilitating efficient selection of transformed cells expressing the chosen marker gene. A large number of such selectable markers are available, but the most popular selectable marker genes used in plant transformation vectors include constructs providing resistance to antibiotics such as kanamycin and hygromycin or herbicides such as phosphinothricin, glyphosate, bialaphos, etc.

11.6.2 Antibiotic Resistance Marker

One of the most commonly used selectable marker genes is *nptII* gene (neomycin phosphotransferase II), which imparts resistance to kanamycin or G418 (Geneticin). It is being used successfully for genetic transformation of dicot species. However, it cannot be used efficiently for transformation of monocots, especially Gramineae because they show high resistance to this antibiotic resulting in higher frequencies of escapes. Therefore, transformation of monocots mostly involves the use of *hpt* gene (hygromycin phosphotransferase) conferring resistance to hygromycin for which monocot cells exhibit comparatively higher sensitivity. Hygromycin is more toxic than kanamycin and kills sensitive cells faster leading to relatively low escapes. It has been employed successfully as selectable marker in finger millet (Ceasar and Ignacimuthu 2011; Jagga-Chugh et al. 2012; Ignacimuthu and Ceasar 2012;

Ramegowda et al. 2013; Hema et al. 2014; Jayasudha et al. 2014; Lakkakula et al. 2017), pearl millet (Lambe et al. 1995; Jha et al. 2011), and foxtail millet (Wang et al. 2011; Martins et al. 2015a, b; Pan et al. 2015; Saha and Blumwald 2016).

11.6.3 Herbicide-Resistant Markers

Another alternative is to use herbicides as selective agent. In some cases, the herbicide resistance traits also offer an opportunity to improve crop agronomy. Herbicide resistance genes work either by coding for modified target proteins insensitive to the herbicide or for an enzyme that degrades or deactivates the herbicide in the plant. The most commonly used herbicide resistance gene is *bar* gene encoding phosphinothricin acetyltransferase (PAT) enzyme which converts PPT/bialaphos into a non-herbicidal acetylated form and confers resistance to the cell and has been used successfully in detection of transformed finger millet (Latha et al. 2005; Bayer et al. 2014), pearl millet (Lambe et al. 2000; Girgi et al. 2002; Goldman et al. 2003; Latha et al. 2006; Ramadevi et al. 2014), and Bahia grass (Smith et al. 2002; Gondo et al. 2005).

Antibiotic and herbicide selection markers are negative selection markers. During negative selection, the transgenic cells survive by converting the selective agent to detoxified compound that may still exert a negative influence on plant cells. In addition, the release of toxic substances by dying adjacent cells adversely affects the growth and proliferation of transformed cells by interfering with their metabolic activities. Further, dying cells may act as a barrier between the medium and the transgenic cells, thereby preventing or slowing the uptake of essential nutrients. Selection systems based on herbicide or antibiotic resistance either allow the regeneration of escapes, even at a high selection pressure, or adversely affect the regeneration process.

Unlike antibiotic and herbicide markers, which kill the untransformed cells, there are positive selection markers also like mannose-positive selection system where untransformed cells are starved; hence they cease to grow but do not die whereas transformed cells proliferate and regenerate faster (O'Kennedy et al. 2004, 2011). The *man*A gene encodes phosphomannose isomerase (PMI), and mannose acts as selective agent. Transformed cells with *man*A gene can convert the mannose selective agent to easily metabolizable compound, fructose-6-phosphate. It overcomes some of the limitations encountered by the negative selection regarding interference with growth and regeneration by toxic agents, and it is also considered safe for animals, humans, and environment. O'Kennedy et al. (2004, 2011) have used *man*A gene to produce transgenic pearl millet with increased transformation efficiency.

11.7 Reporter Genes

A number of reporter genes are available which show immediate expression in transformed cells. Reporter genes are included in transformation vectors for the given reasons: (i) to enable easy identification of potential transformants during the development of a transformation protocol, (ii) as a means of quantitative analysis of cell- and tissue-specific expression and quantifying activity of promoters, and (iii) reporter system is useful in the analysis of plant gene expression and standardization of parameters for successful gene transfer in a particular technique. The most commonly used reporter genes in plant transformation are the *uid*A gene and the green fluorescent protein (GFP) (Barampuram and Zhang 2011). *Uid*A gene encodes an enzyme β-glucuronidase (GUS) which in turn breaks down glucuronide substrate to give blue color, so that its presence can be detected in situ. GUS is easily visualized, and the enzyme is relatively stable.

The green fluorescent protein (GFP) system is an ideal reporter and selectable marker for gene expression analysis and plant transformation. GFP was discovered by Shimomura et al. (1962), and subsequently, Prasher et al. (1992) identified and cloned the *gfp* gene from jellyfish, *Aequorea victoria*. It is a very stable protein that emits green fluorescent light in blue to ultraviolet range which can easily be detected under fluorescent microscope without an external substrate. Niedz et al. (1995) were the first to show that wild-type *Aequorea* GFP could be visualized in plant protoplasts. Although wild-type GFP was used successfully in plant cell and tissue expression studies, it had some disadvantages, such as aberrant splicing in plants and formation of cytotoxic and nonfunctional aggregates. Effective expression in whole plants was achieved upon modification of the GFP-coding sequence (Haseloff et al. 1997) that improved fluorescence intensity and thermostability. Another advantage of GFP is its molecular size facilitates the construction of fusion proteins for subcellular protein localization or protein-protein interactions.

11.8 Promoters for Marker Genes

The choice of promoters that drive the selectable marker or reporter gene affects the efficiency of transformation. Selectable markers and reporters need to be expressed constitutively in various tissues and organs, calli, or in cells. Popular constitutive promoters for monocotyledonous plants include the *CaMV35S*, *Ubi1*, and *Act1*. In some of the vectors, the promoter for the selectable marker, reporter gene, and gene of interest is the same, but ideally, avoidance of duplication of the same components is recommended (Komari et al. 2006).

11.9 Improvements in Marker Genes

Since *CaMV35S* is a prokaryotic promoter, it could result in leaky expression of plant selectable marker genes in *Agrobacterium* which will lead to selection of "false-positives" in plant transformation experiments. The same is true for the nopaline synthase promoter (*nos*) which was used in many early vector constructs. Furthermore, being prokaryotic in origin, *CaMV35S* promoter works efficiently in other prokaryotic microorganisms that inhabit plants (Maas et al. 1997). So, there is a concern that expression in microorganisms may not only interfere with the study of the early events in transformation (Vancanneyt et al. 1990) but also result in horizontal transfer of the antibiotic resistant gene (Libiakova et al. 2001). Therefore, a strategy of inserting introns in marker genes was adopted which is based on the fact that RNA is processed in plants but not in *Agrobacterium*; hence the expression of marker gene is derived only from plant cells not from the residual *A. tumefaciens* cells. Vancanneyt et al. (1990) placed the intron 2 of the potato STLS1 into the coding sequence of the *gus* marker gene. In another study, Ohta et al. (1990) placed the castor bean catalase gene intron within the N-terminal part of the coding sequence of *gus*; the same strategy was practiced in pearl millet by Jha et al. (2011) in foxtail millet (Martins et al. 2015b; Saha and Blumwald 2016) and in finger millet (Ceasar and Ignacimuthu 2011; Ignacimuthu and Ceasar 2012; Lakkakula et al. 2017) to ensure that expression of β-glucuronidase activity was derived from plant cells only not from residual *A. tumefaciens* cells. In all the reports, GUS activity was limited only to transformed tissues and remain unexpressed in *Agrobacterium*. Similarly, Maas et al. (1997) placed the intron 2 of the potato *STLS 1* gene in the N-terminal part of the *nptII* which completely eliminated leakiness in kanamycin selection. *Shrunken-1* intron 1 also blocks transcriptional activity of *CaMV35S* in *E. coli* as well as in *A. tumefaciens* cells. However, genes containing insertion of the *Shrunken-1* intron 1 are only useful for expression and transformation of monocotyledonous plants; genes containing the potato *STLS 1* intron 2 can be universally used in monocotyledonous as well as in dicotyledonous plants (Maas et al. 1997).

11.10 Strain and Vector System

11.10.1 Vir Helper Strain

To date, only a limited number of vir helper strains have been used in most of the transformation studies. One of the most commonly used vir helper strain is LBA4404 that harbors the disarmed *Ach5* Ti plasmid and a binary vector such as *pBin19* (Bevan 1984). In addition, GV3101 strain is one of the commonly used laboratory strains useful for several dicot transformations (Koncz and Schell 1986). However, a major limitation of *Agrobacterium* strains is the narrow host range. This problem was resolved when it was found that some *Agrobacterium* strains harbor broad host range Ti plasmid. One of them was L,L-succinamopine A281 strain hypervirulent on several solanaceous plants (Hood et al. 1986). Detailed study of this strain by Hood et al. (1986) revealed that the sequences present outside the T-DNA on its Ti plasmid (pTiBo542) are responsible for the hypervirulence of the *A. tumefaciens* strain, A281. Later, the superior transformation ability of A281 was exploited to enhance efficiency of other strains by introducing the supervirulent *vir* genes (VirB, VirG, and VirC) of pTiBo542 (the hypervirulent Ti plasmid of *A. tumefaciens* A281), and supervirulent strains were designed, e.g., EHA101, EHA105, AGL0, and AGL1. The bacterial kanamycin resistance gene in EHA101 (Hood et al. 1986) was deleted to develop the *vir* helper strain EHA105 (Hood et al. 1993). Presence of "supervirulent" *vir* genes endows AGL1, EHA101, and EHA105 the property of broader host range and higher transformation efficiency. They contain succinamopine type Ti plasmid with C58 origin (Hamilton and Fall 1971) and have been shown to be suitable for monocot transformation (Hiei and Komari 2008). Sharma et al. (2011) and Saha and Blumwald (2016) measured GUS activity in transformed finger millet and foxtail millet, respectively, and concluded that the efficiency of EHA105 is higher than LBA4404. Similar findings were also reported in rice by Cho et al. (2014).

11.10.2 Binary and Superbinary Vectors

For *Agrobacterium*-mediated transformation to occur, the T-DNA and the *vir* region must be present in the bacterium. One of the first vectors developed for transformation of plants involved the removal of wild-type T-DNA or oncogenes, to create a disarmed strain (Hoekama et al.1983). The introduction of engineered T-DNA into *A. tumefaciens* involved the insertion of genes into an *E. coli* vector, such as pBR322, that could be integrated into the disarmed Ti plasmid to create a cointegrative vector (Fraley et al. 1986). Although the system was successful, the resulting vector of ~150 kb was difficult to handle in the laboratory because of its large size and instability. To obviate the problems associated with cointegrate vector, a binary vector system was designed by Hoekema et al. (1983), based on the fact that the T-DNA and the *vir* region could operate on separate plasmids. When these replicons are within the same *Agrobacterium* cell, the products of *vir* genes operate in trans to transfer T-DNA into the plant cell. The plasmid harboring the T-region is called the binary vector, whereas, the plasmid containing the *vir* genes is called as the *vir* helper.

T-DNA binary vectors revolutionized the use of *Agrobacterium* to introduce genes into plants. These plasmids are small and easy to manipulate in both *E. coli* and *Agrobacterium* and contain borders of T-DNA, multiple cloning site, markers for selection and maintenance in both *E. coli* and *Agrobacterium*, plant selectable marker gene between the right and left borders of T-DNA, and origin of replication (*ori*) for replication in *E. coli* and *Agrobacterium*. In the binary Ti vectors, the plant selectable marker genes are placed near the left border (LB), while the gene of interest is placed near the right border (RB). Since during T-DNA transfer, the RB precedes the LB, therefore, placing the gene of interest closer to the RB ensures that it will be transferred before the selectable marker gene.

11.10.3 Examples of Binary Vectors Used for Millet Transformation

1. One of the first binary vectors, pBIN19 developed in 1984 by Bevan, is one of the most widely used binary vectors for the *Agrobacterium*-mediated transformation of plants. It contains kanamycin resistance gene for selection in

bacteria and plants. pBIN19 is the progenitor of several binary vectors developed later. Its improved version, pBIN20, contains many additional single restriction sites in the MCS 26. Another derivative of pBIN19 is pBI101 a promoterless binary vector. This vector allows evaluation of promoters which can easily be ligated upstream of the *gus* gene and then transferred to the host. pBI121 is an example of widely used binary expression vector; it was also derived from pBIN19. It contains two complete expression cassettes, one for *nptII* and other for *gus* gene under *CaMV35S*. *gus* can be replaced by gene of interest to check its level of expression under *CaMV35S* promoter.

2. pGreen is one of a large family of plant transformation vectors much smaller than pBIN19 and its derivatives.

3. pCAMBIA is a series of binary vectors each having different characteristics. The pPZP vector backbone was used to construct the pCAMBIA series of vectors with *nptII*, *hpt*, or *bar* as selection markers and *gus* or *gfp* as reporters. pCAMBIA series of vectors are represented by a four-digit number:

   (i) The first digit indicates plant selection: 0 for absence, 1 for hygromycin resistance, and 2 for kanamycin resistance.

   (ii) The second digit indicated bacterial selection: 1 for spectinomycin/ streptomycin resistance, 2 for chloramphenicol resistance, 3 for kanamycin resistance, and 4 for both streptinomycin/spectinomycin and kanamycin resistance.

   (iii) The third digit indicates the polylinker: 0 for puc18 polylinker, 8 for puc8 polylinker, and 9 for puc9 polylinker.;

   (iv) The fourth digit indicates reporter gene: 0 for no reporter, 1 for *E. coli gus*, 2 for mgfp5, 3 for gusA:mgfp 5 fusion, 4 for mgfp5:*gus*A fusion and 5 for *GUS*Plus. pCAMBIA 1201, pCAMBIA 1301, pCAMBIA 1380, pCAMBIA 1381, and pCAMBIA 2300 have been used successfully for millet transformation.

In spite of all these, until the early 1990s *Agrobacterium*-mediated transformation was mainly confined to dicots, whereas, monocots were considered recalcitrant for transformation. However, scientists came up with an idea of further improving transformation efficiency of binary vectors and expanding the host range to recalcitrant monocot genotypes by increasing the level of expression of the *vir* genes (Gelvin 2003). Thus, superbinary vectors were constructed by introducing the 14.8 kb KpnI fragment that contains the *vir*B, *vir*G, and *vir*C genes from pTiBo542 (Ti plasmid of hypervirulent strain A281) into small T-DNA-carrying plasmid (Komari 1990). The development of these superbinary vectors made transformation of various recalcitrant monocots easy which was earlier considered intractable. The superbinary vector has been highly efficient in transforming various plants, particularly recalcitrant species, such as important cereal crops (Barampuram and Zhang 2011).

In general, four systems are used for transformation using supervirulent strains (EHA101, EHA105, AGL1) carrying binary or superbinary vectors as well as commonly used "ordinary" strains with binary or superbinary vectors. However, the capability of a superbinary vector is most evident when it is combined with strain LBA4404, whereas the performance of a superbinary vector is not very good when it is carried by strains derived from A281, such as EHA101, EHA105, or AGL1 (Komari et al. 2006).

11.11 *Agrobacterium*-Mediated Transformation

Various methods have been employed to genetically transform plants, but the simplest and one of the most preferred approaches for introducing the gene of interest into target crops is *Agrobacterium*-mediated genetic transformation (Gelvin 2003) as it offers several advantages over others: (1) a relative high transformation efficiency, (2) the ability to transfer large DNA segments, (3) facilitates precise introgression of single/low copy number of transgene with fewer arrangements thus minimizing the incidence of gene silencing in subsequent generations, and (4) stable transformation. A breakthrough in plant transformation came in the late 1970s with the discovery of crown gall formation by *Agrobacterium tumefaciens* (Larebeke et al. 1974). The first transgenic tobacco plant expressing recombinant genes was generated by Estrella et al. (1983). It was shown that wounded tobacco cells synthesize phenolic compounds such as acetosyringone and a hydroxyacetosyringone to induce expression of *vir* genes that are responsible for the transfer of T-DNA from *A. tumefaciens* to the wounded host cell (Stachel et al. 1985). Monocotyledons secrete little or no phenolic compounds thus cannot activate the *vir* genes of *Agrobacterium* (Usami et al. 1987). Therefore, for successful T-DNA transfer in monocotyledons, these inducing compounds are added into the *A. tumefaciens* suspension culture to activate the *vir* genes prior to inoculation with explant.

The very first attempt to develop transgenic monocot was made by Bytebier in 1987 via *Agrobacterium*-mediated transformation of *Asparagus officinalis*. Evidence for T-DNA integration and the expression of reporter and selectable marker genes in transformed cereals was reported first for rice (Raineri et al. 1990). However, the efforts made by Hiei et al. (1994) for transformation of several japonica rice cultivars with up to 30% transformation rate proved the utility *of Agrobacterium tumefaciens* for genetic modification of cereals.

11.11.1 *Factors Influencing* Agrobacterium-*Mediated Transformation*

Numerous factors such as genotype, types and stages of the explant, *Agrobacterium* strain, vector system, selectable marker system, inoculation and coculture medium, osmotic treatment, antinecrotic solution, *Agrobacterium* density and infection time, surfactants, and tissue culture are critical factors for stable transformation. Out of all the mentioned factors, the genotype dependence is a major hurdle. There are only a limited number of genotypes that respond toward tissue culture and transformation which restrict introduction of desired gene into elite cultivars of agronomically important cereals. Despite these difficulties significant improvements in plant tissue culture, construction of supervirulent strains of *Agrobacterium*, and the engineering of novel superbinary vectors have enabled this natural vector system to be used for the production of transgenic plants in a wide range of millets (Table 11.1).

11.11.2   Progress Made in Agrobacterium-*Mediated Transformation of Millets*

11.11.2.1   Pearl Millet

The first report for pearl millet transformation by *Agrobacterium* was provided by Jha et al. (2011). They optimized optical density, inoculation duration, co-cultivation time, acetosyringone concentration in co-cultivation medium, and vacuum infiltration assisted inoculation by using supervirulent *Agrobacterium* strain EHA105 harboring the binary vector pCAMBIA 1301 containing the hygromycin phosphotransferase (*hpt*) and β-glucuronidase (*gus*) genes to investigate and optimize T-DNA delivery into shoot apices of pearl millet. The highest transformation frequency of 5.79% was obtained when the shoot apex explants were infected for 30 min with *Agrobacterium*. One of the major constraints that can be alleviated by genetic transformation is fungal diseases. Pearl millet is susceptible to several fungal diseases, but the most economically damaging are downy mildew, caused by the oomycete *Sclerospora graminicola* and rust disease caused by the basidiomycete *Puccinia substriata* and *Puccinia penniseti*. Transgenic lines showing high level of resistance to rust pathogen *Puccinia penniseti* were developed by Ignacimuthu and Kannan (2013) using shoot apex derived embryogenic callus as explants for transformation with EHA105 strain harboring rice chitinase gene (chi11) under the control of maize ubiquitin (*Ubi1*) promoter. Ramineni et al. (2014) introduced *Brassica juncea* nonexpressor of pathogenesis-related genes 1 (*BjNPR1*) into pearl millet male fertility restorer line ICMP451 by *A. tumefaciens* strain LBA4404 harboring Ti plasmid pSB111. The transgenic plants showed resistance to multiple strains of *S. graminicola* which seems promising for the development of durable and broad spectrum downy mildew resistant hybrids. In the same year, Ramadevi et al. (2014) transferred a synthetic gene encoding the antimicrobial peptide magainin regulated by the *CaMV35S* promoter and the nopaline synthase (*nos*) terminator. The plant expression cassette was introduced into the vector pSB11-bar, and the recombinant plasmid was mobilized into *A. tumefaciens* strain LBA4404 for the generation of a superbinary vector pSB111-bar-mag which was used to infect pearl millet calli resulting in a transformation frequency of 2.73%. However, expression of the antimicrobial peptide magainin failed to provide resistance to the downy mildew. In another effort, *A. tumefaciens* strains, GV2600 possessing the binary vector pCAMBIA2300 with osmotin gene from brassica and rice class I

Table 11.1 *Agrobacterium*-mediated transformation of millets

Crop	Explant	Promoter/marker	Promoter/gene	Strain	Vector	Phenotype	Reference
Finger millet	Callus	*35S/nptII*		EHA 105	pCNL 56		Sharma et al. (2011)
		35S/gus		LBA 4404	Pbi 121		
	Shoot apex	*35S/hpt*	*35S/gus*	LBA4404	pCAMBIA 1301		Ceaser and Ignacimuthu (2011)
	Shoot apex	*35S/hpt and gus*	*Ubi1/Chi11*	LBA 4404	pCAMBIA 1301	Leaf blast resistance	Ignacimuthu and Ceaser (2012)
	Callus	*35S/hpt*	*35S and Bx 17/OsZIP1*	EHA 105	p Green 0179	Zn accumulation	Ramegowda et al. (2013)
	Callus		*35S/TuAm1 and HvTUB1*	EHA105	pBITUBA8	Dinitroaniline resistance	Bayer et al. (2014)
	Callus	*35S/hpt*	*35S/mtlD*	EHA 105	pCAMBIA 1380	Multiple stress tolerance	Hema et al. (2014)
		35S/gus			pCAMBIA 1301		
	Callus	*hpt*	*35S/PgNHX1 and AVP1*	EHA 105	pCAMBIA 1301	Salinity tolerance	Jayasudha et al. (2014)
	Shoot apex	*35S/hpt*	*35S/gus*	EHA 105	pCAMBIA 1301		Lakkakula et al. (2017)
	Seedling leaves	*Nos/nptII*	*EcBPD/gus and gfp*	LBA4404	pBI101		Sen and Dutta (2016)
Pearl millet	Shoot apex	*35S/hpt*	*35S/gus*	EHA 105	pCAMBIA 1301		Jha et al. (2011)
	Callus	*35S/gus*	*Ubi1/chit11*	LBA 4404	pSB1	Rust resistance	Ignacimuthu and Kannan (2013)
	Callus	*35S/bar*	*35S/mag*	LBA 4404	pSB11	Downy mildew resistance	Ramadevi et al. (2014)
	Callus	*35S/bar*	*35S/BjNPR1*	LBA 4404	pSB111	Downy mildew resistance	Ramineni et al. (2014)
			35S/osmotin and chitinase	GV2600	pCAMBIA 2300	Downy mildew resistance	Jalaja et al. (2016)
		35S/bar			pCAMBIA 1300		

Foxtail	Callus	35S/hpt	Zm13/Si401	LBA4404	pBin19	Male sterility	Qin et al. (2008)
	Immature inflorescence	35S/hpt	PF128/SBgLR	LBA4404	pSB-SBgLR		Wang et al. (2011)
		35S/hpt	Ubi1/SiLEA4	pCOU		Drought and salt resistance	Wang et al. (2014)
	Inflorescence	hpt	RFP	AGL1	pANIC 6A		Martins et al. (2015a)
	Callus	hpt	gfp and gus	EHA105	p6mD		Martins et al. (2015b)
	Callus		F128/gus	LBA4404	pCAMBIA 2300		Pan et al. (2015)
	Spike	Hpt and bar	35S/gus; 35S/GUSPlus, Ubi/GUSPlus, 35S/GFP, Ubi/DsRed	AGL1 / EHA105	pCAMBIA 1201		Saha and Blumwald(2016)
	Shoot apex		SiPHT	LBA4404	pFGC		Ceasar et al. (2017)
Tef	Callus	Nos/hpt	35S/PcGA2ox1	LBA4404	pGPTV	Semi-dwarfism	Gebre et al. (2013)

35S, cauliflower mosaic virus 35S promoter, *Uq1* ubiquitin gene promoter from maize, *Bx17* endosperm-specific promoter from foxtail millet, *EcBDP* bidirectional promoter from *Eleusine coracana*, *nos* nopaline synthase, *gus* or *uidA* β-glucuronidase gene, *bar* phosphinothricin acetyltransferase gene, *hpt* hygromycin phosphotransferase gene, *nptII* neomycin phosphotransferase gene, *TUAm1* α1-tubulin gene from *Eleusine indica*, *HvTUB1* β1-tubulin gene from barley, *OsZIP1* rice Zn transporter, *mtlD* mannitol-1-phosphate dehydrogenase, *NHX1* Na$^+$/H$^+$ antiporter from *Pennisetum glaucum*, *AVP1* H$^+$-PPase from *Arabidopsis thaliana*, *chi11* rice chitinase gene, *mag* antimicrobial peptide magainin, *BjNPR1 Brassica juncea* nonexpressor of pathogenesis-related genes 1, *SBgLR* lysine-rich protein-encoding gene from potato, *RFP* red fluorescent protein, *SiPHT* phosphate transporter 1 from *Setaria italic*, *PcGA2ox1 Phaseolus coccineus* GA2 oxidase endochitinase gene, and pCAMBIA 1300 containing the plant selection marker *bar gene* (that confers phosphinothricin resistance) in pPur vector under the control of *CaMV35S* promoter was co-cultivated with immature inflorescence-derived embryogenic calli. The resulting transformants were highly resistant to downy mildew pathogen, *Sclerospora graminicola* (Jalaja et al. 2016*)*.

11.11.2.2 Finger Millet

In 2011, Sharma et al. and Ceaser and Ignacimuthu made efforts to standardize *Agrobacterium*-mediated transformation protocol taking into account the influence of antinecrotics, incubation temperature, pH, and nutrients. Once the transformation protocol is standardized, it is possible to transfer genes into crop plants from unrelated plants, microbes, and animals. Sharma et al. (2011) found highest frequency of transformation when green nodular calli were infected with *Agrobacterium* strain EHA 105 possessing plasmid pCNL-56 of 0.8 OD600 for 25 min at 22 °C. High concentrations of Tween-20 (0.1%) and 200 μM acetosyringone in infection medium, pretreatment of target tissue with antioxidant mix containing ascorbic acid (14.20 μM), cysteine (82.64 μM), and silver nitrate (29.41 μM) and incorporation of ten times $CuSO_4$ and three times NH_4NO_3 led to high frequency of transformed plants. Ceaser and Ignacimuthu (2011) optimized transformation conditions by using *Agrobacterium* strain LBA4404 harboring binary vector pCAMBIA 1301, which contained *hpt* as selectable marker gene and *uidA* as reporter gene. Exposure of explants for 30 min to agrobacterial inoculum and 3 days of co-cultivation on filter paper placed on medium supplemented with 100 μM acetosyringone (AS) was found to be optimum. Addition of 100 μM L-cysteine in the selection medium enhanced the frequency of transgenic plant recovery. Few years later, Hema et al. (2014) developed transgenic finger millet plants expressing mannitol-1-phosphate dehydrogenase (*mtlD*) gene from bacteria through *Agrobacterium*-mediated genetic transformation. The binary vectors pCAMBIA 1301 having *uidA* and pCAMBIA 1380 having *mtlD*, both driven by *CaMV35S* promoter, were used with 6% transformation efficiency. In the same year, Bayer and colleagues transformed finger millet by bioballistic and *Agrobacterium*-mediated transformation with a mutant α-tubulin gene (*TUAml*) for imparting resistance against dinitroaniline. Jayasudha et al. (2014) transformed finger millet for salt tolerance through *in vitro Agrobacterium*-mediated transformation. They transferred a double gene construct of *PgNHX1* (from *Pennisetum glaucum*) and *AVP1* (from *Arabidopsis thaliana*) using the plant binary expression vector pCAMBIA 1301. Transgenic plants thus developed were found to exhibit high salt tolerance of 300 mM compared to wild-type plants. Recently, Lakkakula et al. (2017) have developed an improved *Agrobacterium*-mediated transformation and rapid regeneration system (45 days) using optimized transformation and direct plant regeneration conditions. The shoot apical meristems were cocultured with *Agrobacterium* strain EHA105 carrying binary vector pCAMBIA 1301 with *hpt* as a selectable marker gene and *gusA* as a reporter gene both driven by separate *CaMV35S* promoters.

11.11.2.3 Foxtail Millet

Foxtail millet (*Setaria italica*) has small diploid genome (2n = 18, approximately 510 Mb) (Bennetzen et al. 2012), short stature, simple growth requirements, prolific seed production, short life cycle, and C4 photosynthesis, and the most important feature is its ability to be grown in controlled environmental conditions, under relatively low light intensity, and is amenable to regeneration and transformation owing to these properties where it offers several research advantages and has emerged as model plant system to dissect gene functions in millets (Doust et al. 2009). *Agrobacterium*-mediated transformation system for foxtail millet was first reported by Liu et al. (2005). They obtained 6.6% transformation frequency through this approach. *Agrobacterium*-mediated transformation has been considered as a simple and cost-effective tool for the development of transgenic plants and studying functional genomics. Transgene-mediated RNA interference or co-suppression is a simplest way to investigate gene function. Co-suppression, one kind of RNA silencing, was first reported in petunia. A pigment-producing gene under the control of a powerful promoter was introduced in order to deepen the purple color of the flowers. Instead of expected enhancement of purple color, many of the flowers appeared variegated or even white. This phenomenon was named "co-suppression" afterward, since both the expression of exogenous and endogenous homologous genes were suppressed (Napoli et al. 1990). This phenomenon was adopted by Qin et al. (2008) to study the function of gene *Si401* in pollen development by introducing it into calli induced from 0.5 to 1 cm length panicle. *Agrobacterium* strain LBA4404 carrying vector pBin19 with *Si401* gene under the control of maize pollen-specific promoter *Zm13* and *nptII* gene as selection marker was used for transformation which resulted in vacant aborted pollen grains. Their experiment demonstrated that *Si401* plays an essential role in anther development in foxtail millet and would be a good candidate gene that can be useful for generation of male-sterile plants in millets as well as in other crops. Similarly, RNA silencing triggered by transgene has a broad application for studying gene functions in plant. In order to assess the function of phosphate transporters, Ceasar et al. (2017) downregulated expression of three members of the PHT1 phosphate transporter family of *Setaria italica*, *SiPHT1;2*, *SiPHT1;3*, and *SiPHT1;4*, by RNA interference (RNAi) after treating shoot apex explant strain LBA4404 carrying three RNAi vectors (*pFGC-SiPHT1;2*, *pFGC-SiPHT1;3*, or *pFGC-SiPHT1;4*) to understand the function of PHT1 transporters on Pi transport with a view to improve Pi uptake and efficient utilization under low Pi stress conditions. Transformants were recovered with 10% efficiency and showed reduction in the total and inorganic P contents in shoot and root tissues confirming the role of these transporters in both uptake of Pi from the soil and transport within the plant. Another way to assess the function of a novel gene is by overexpressing the gene and then screening the transgenics for altered trait. Wang et al. (2014) transformed foxtail millet by overexpressing *SiLEA14*. The transgenic foxtail millet showed higher tolerance and improved growth under salt and drought stresses compared with the wild type indicating the important role of novel gene *SiLEA14* in abiotic stress resistance.

A novel seed-specific promoter, *F128*, was first used by Wang et al. (2011); transformation efficiency of 5.5% was obtained with strain LBA4404 harboring superbinary vector pSB130 carrying the lysine-rich protein-encoding gene *SBgLR* driven *F128* and hygromycin resistance gene *hpt* under the control of the *CaMV35S* promoter. The efficiency of the same promoter, *F128*, was evaluated by Pan et al. (2015) by fusing it to the *gus* reporter gene of pCAMBIA2300. The construct was introduced into foxtail millet calli induced from immature inflorescences mediated by *A. tumefaciens* strain LBA4404. GUS analysis revealed that p*F128* drove GUS expression efficiently and specifically in the seeds with significantly higher activity than the constitutive *CaMV35S* promoter and the maize seed-specific 19 Zein (19Z) promoter. In another study, a highly efficient regeneration and transformation protocol was developed by Martins et al. (2015b) which allowed rapid regeneration of plants in 15 weeks with transformation efficiency up to 29%. They used seed-derived callus as explants and co-cultivated them with *Agrobacterium* strain EHA105 harboring p6Md vector containing the *hpt* marker gene.

However, monocots are generally recalcitrant for *in vitro* tissue culture procedure; hence plant regeneration becomes time-consuming and labor intensive and requires specialized equipment and highly qualified personnel. In addition, tissue culture is genotype dependent and frequently results in somaclonal variations due to epigenetic changes or chromosomal rearrangements (Janice et al. 2009); regenerated plants are often chimeric with morphological anomalies and reduced fertility. It is therefore highly desirable to develop an *Agrobacterium*-mediated transformation procedure for monocots that obviates the need of tissue culture. Therefore, a new strategy, *in planta* transformation system, has been developed in which DNA is directly introduced into intact plant, and there is no need to regenerate plants under *in vitro* conditions.

11.12   *In Planta* Transformation Method

In planta gene transfer method offers several advantages such as integration of large segments of transfer DNA (T-DNA) into the host genome, low copy number of transgene, stable inheritance, and fewer rearrangements of T-DNA with less transgene silencing in subsequent generations (Birch 1997). The first *in planta* transformation procedure was performed on *Arabidopsis* seeds by Feldman and Marks (1987). *Agrobacterium*-mediated transformation by vacuum infiltration of *Arabidopsis* inflorescences was first reported by Bechtold et al. (1993) and further modified as floral dip method by Clough and Bent (1998). Floral dip method has been used in genetic engineering strategies as it directly produces genetically modified seeds bypassing the laborious tissue culturing procedures. Martins et al. (2015a) were the first to transform foxtail millet using floral dip method. They infiltrated the spikes of 1-month-old plants at boot stage with bacterial suspension (AGL1 strain) carrying the pANIC 6 A for 10 min which resulted in 0.6% transformation efficiency. Soon after, Saha and Blumwald (2016) optimized the conditions for stable transformation via spike-dip method using AGL1 harboring the *35S::GUS* binary vector; a range of transformation efficiency of 0.5–0.7% was obtained with 20–40 min of dipping. They evaluated the efficiency of four strains (AGL1, EHA105, GV3101, and LBA4404) by measuring *gus* activity; the order of efficiency was EHA105 = AGL1 > LBA4404 = GV3101. After that they transformed foxtail with EHA105 harboring four reporters, namely, *35S::GUS*Plus, *Ubi::GUS*Plus, 35S::GFP, and *Ubi*::DsRed and revealed that reporter genes driven by the *Ubi1* promoter had high levels of expression as compared with the *CaMV35S* promoter. In a recent study, a novel bidirectional promoter EcBDP was isolated and characterized in finger millet by Sen and Dutta (2016). *Agrobacterium* strain LBA4404 containing pBI101-mGFP5-EcBDP vector was used for *in planta Agrobacterium* infiltration of young finger millet leaves. Simultaneous expression of GUS and GFP under *EcBDP* established it as a potent natural bidirectional promoter from monocot origin which is induced by abiotic stresses. Even though, *in planta* transformation method is very useful for plant species that are recalcitrant to tissue culture, it is not a widely adopted method perhaps due to less transformation rate and requirement of more expertise.

11.13   Direct DNA Transfer

A number of alternative plant transformation methods were developed to facilitate gene transfer in recalcitrant genotypes or species. These methods can be divided into two categories: (1) physical gene transfer methods which are based on the direct delivery of DNA into the plant cells. It can be achieved by various methods such as particle bombardment, macroinjections, microinjections, liposome-mediated transformation, silicon carbide fiber-mediated transformation, ultrasound-mediated transformation, etc. and (2) chemical gene transfer which involves the use of plasma membrane destabilizing or precipitating agents such as polyethylene glycol, calcium phosphate coprecipitation, polycation DMSO, and DEAE (diethyl amino ethyl). However, out of all these techniques, only particle bombardment has gained widespread use.

11.14   Microprojectile Bombardment or Biolistic or Particle Bombardment

Gene transfer by high-velocity microprojectiles is widely used as it is genotype and plant tissue independent (O'Kennedy et al. 2011). The principle involved in gene gun is that metal particles coated with naked plasmid DNA containing the gene of interest are accelerated to high velocity penetrating the plant cells allowing foreign DNA to be released and integrate into the target cell genome. The first particle gun (Sanford et al. 1987) used an explosive charge to accelerate tungsten, but this was superseded by a helium-driven gun (Kikkert 1993). Another simple and inexpensive particle bombardment device for delivery of DNA into plant cells is the Particle Inflow Gun (PIG) (Finer et al. 1992), in which the DNA-coated microprojectiles are accelerated directly in a pressurized stream of helium rather than being supported by a macrocarrier. Microprojectile bombardment is the preferred method for introduction of two or more genes into a single genotype because it eliminates the need for multiple *Agrobacterium* strains. Therefore, the production of transgenic millets by biolistic gene delivery method has now become the method of choice compared to *Agrobacterium*-mediated transformation method (Table 11.2).

11.14.1 Factors Influencing Microprojectile Bombardment

Several factors are considered critical for successful gene transfer using particle bombardment technology. These factors which include the target tissue, vector, and chemical and physical properties of the metal are important because they affect the depth of penetration as well as the extent of damage to the target cells, helium pressures, the level of vacuum generated, and target distance. All of these parameters are under the experimenter's control and can be optimized according to one's need (Barampuram and Zhang 2011).

11.14.2 Progress in Millet Transformation

11.14.2.1 Pearl Millet

Among all the millets, pearl millet (*Pennisetum glaucum*) has been given the highest priority in genetic transformation studies. In most of the available reports, pearl millet was transformed by biolistic method of gene delivery. The first attempt to transform millets was carried out by Hauptmann et al. (1987). They electroporated the protoplasts of *Pennisetum purpureum* and *Panicum maximum* with plasmids carrying a chloramphenicol acetyltransferase gene. In 1991, Taylor et al. bombarded immature embryos with a plasmid pMON8678 containing *gus* gene under the control of maize alcohol dehydrogenase gene (*Adh1*) promoter. Later in 1993, they transformed pearl millet with plasmids pBARGUS and pAHC25 and observed better expression of the *uid*A gene in plasmid pAHC25 as compared to pBARGUS. Lambe et al. (1995) transformed pearl millet by the biolistic method with two plasmids p35SGUS and pROB5 containing the *GUS* gene and *hpt*, respectively, under *CaMV35S* promoter. However, they could not regenerate plants. In 2000, they tested several vectors having *gus, hpt, bar*, or *nptII* regions in various combinations of *CaMV35S, Adh1*, and *Emu* promoters. They reported highest transient expression of *gus* gene with recombinant *Emu* promoter. In an effort to further optimize pearl millet transformation, Girgi et al. (2002) delivered genes to

Table 11.2 Biolistic transformation of millets

Crop	Explant	Promoter/ marker	Promoter/gene	vector	Phenotype	Reference
Finger millet	Callus	35S/bar	35S/pin	pPin	Leaf blast resistance	Latha et al. (2005)
	Callus		Act1/PcSrp	TG0063 of pCAMBIA series	Salt tolerance	Mahalakshmi et al. (2006)
	Callus	35S/hpt	35S/gusA	pCAMBIA 1381		Jagga-Chugh et al. (2012)
	callus		CaMV35, Act1, Ubi1, RbcS, Flt/gus			Gupta et al. (2001)
	Callus	Ubi/bar	Ubi/α1-tubulinTUAm1 Ubi/β1-tubulinHvTUB1	pAHTU-Am1 pAHTUB1	Dinitroaniline resistance	Bayer et al. (2014)
Pearl millet	Embryogenic callus	35S/hpt	35S/gus			Lambe et al. (1995)
	Shoot-tip clumps		Act1/gus	pAct1-F		Devi and Sticklen (2002)
	IZE	Ubi1/bar	Ubi1/ gus	pAHC25	Herbicide resistance (BASTA)	Girgi et al. (2002)
	IE, spikelet, apical meristem	Ubi1/bar and gus	Ubi1/bar and gus 35S/GFP	pAHC25 p524EGFP.1	Herbicide resistance	Goldman et al. (2003)
	IZE	Ubi1/manA	Ubi1/manA	PNOV3604		Kennedy et al. (2004)
	IZE	35S/bar	Ubi1/afp		Fungal resistance	Girgi et al. (2006)
	Callus	35S/bar	35S/pin	pPur	Resistance to downy mildew	Latha et al. (2006)
	IZE	Ubi1/manA	Ubi1 or pin2/gluc78	pAHC25	Resistance to downy mildew	Kennedy et al. (2011)
	Callus	Emu or 35S/ gus	35S/als	pULGU1 pEmuGN	Herbicide resistance (Chlorsulfuron)	Tiecoura et al. (2015)

(continued)

Table 11.2 (continued)

Crop	Explant	Promoter/marker	Promoter/gene	vector	Phenotype	Reference
Foxtail	Callus	35S/hpt	35S/SiPf40	pROKf40	Extra tillering	Liu et al. (2009)
Bahia grass	Callus		35S/bar	pAHC20	Phosphinothricin resistance	Smith et al.(2002)
			Ubi1/gus	pAHC25		Grando et al. (2002)
		35S/bar	Act1/gus	PDB1		Gondo et al. (2005)
		Ubi1/nptII		pPZP111		Altpeter and James (2005)
		35S/nptII	Ubi1/Cry1FA	PHZUbi-oxi		Luciani et al. (2007)
		35S/nptII	Ubi1/bar	pJFbar		Sandhu and Altpeter(2008)
		Ubi1/nptII	HVA1s/DREB1A		Salt and drought resistance	James et al. (2008)
			Act1/egfp Ubi/bar			Mancini et al. (2014)

35S cauliflower mosaic virus 35S promoter, Uq1 ubiquitin gene promoter from maize, Act1 rice actin gene promoter, RbcS rice ribulose 1, 5-biphosphate carboxylase small subunit gene promoter, FT Flaveria trinervia gene promoter, pin2 wound-inducible potato proteinase inhibitor IIK gene promoter, pin an antifungal protein (PIN)-encoding gene from prawn, PcSrp gene-encoding Porteresia coarctata serine-rich-protein, AFP antifungal protein from the Aspergillus giganteus, gluc78 β-1,3-glucanase, als acetolactate synthase gene, SiLEA late embryogenesis abundant proteins from Setaria italica, hpt hygromycin phosphotransferase gene, nptII neomycin phosphotransferase gene scutellar tissue by two biolistic delivery systems (particle gun 1000/He (PDS) or particle inflow gun (PIG)). For PIG transformation calli were bombarded with plasmid pAHC25 harboring the *gus* reporter gene and *bar* (phosphinothricin acetyltransferase) marker gene; both the genes were driven by *Ubi1* promoter. For PDS plasmid calli were transformed with plasmid p35SAcS containing *bar* selection marker gene under the control of the *CaMV35S* promoter and *gus* gene regulated by the maize *Ubi1* promoter. However, PIG and PDS yielded only 0.02% and 0.18% transformants, respectively. In the same year, Devi and Sticklen (2002) developed a genotype-independent pearl millet transformation system by using shoot apical meristems as target tissue for microprojectile bombardment. Plasmid pAct1-F containing *uid*A gene driven by rice *Act1* promoter was used for transformation; they used tungsten particles of two different sizes (0.9 and 1.2 μm). The important parameters influencing the transformation like the size of the tungsten particles, density of the particles, pressure of the helium gas, distance from the target, and osmoticum of the culture medium were optimized.

The transformation frequency of pearl millet was greatly increased by Goldman et al. (2003) using three different explants (embryogenic tissue, inflorescences, and apical meristems) from a diploid hybrid HGM100 and a partial inbred tetraploid IA4X with transformation frequency ranged from 5 to 85%. They used plasmids pAHC25 and p524EGFP; plasmid pAHC25 contains the selectable bar gene, encoding the enzyme PAT, and the reporter gene (*uidA*) encoding *GUS*, both under the control of separate maize ubiquitin promoters (*Ubi1*). Plasmid p524EGFP.1 carrying an enhanced green fluorescent protein-encoding (*gfp*) gene driven by *CaMV35S*. Inflorescence of the tetraploid genotype was found to be most capable of generating transgenics. Subsequently, the transformation protocol was optimized by O'Kennedy et al. (2004) using a phosphomannose isomerase (*man*A) as positive selectable marker driven by *Ubi1* promoter placed in the plasmid pNOV3604ubi. The use of *man*A selection limited the number of escapes to less than 10%, whereas, using the *bar* gene and selecting with 3–5 mg l⁻¹bialaphos resulted in more than 90% nontransformed escapes (Girgi et al. 2002). Tiecoura et al. (2015) obtained transgenic plants resistant to chlorsulfuron herbicide, by co-bombardment of embryogenic calli with the plasmids pULGU1 and p35S*GUS* or p*Emu*GN. The plasmid pULGU1 contained the mutant acetolactate synthase (*ALS*) gene responsible for resistance to the chlorsulfuron herbicide, driven by *CaMV35S* promoter and terminator. The plasmid p35S*GUS* harbored the reporter *gus* gene regulated by the *CaMV35S* promoter and octopine synthase (*ocs*) terminator. The plasmid p*Emu*GN carried the reporter *gus* gene controlled by the *Emu* promoter and the nopaline synthase (*nos*) terminator. They observed that *Emu* was more efficient than *CaMV35S* in deriving *gus* gene expression.

Downy mildew, caused by *Sclerospora graminicola*, is the most significant biotic constraint of pearl millet production leading to devastating annual crop losses of 20–40% (Thakur 2008). The pathogen is highly virulent and several pathotypes have been identified. The first transgenic pearl millet expressing functionally active foreign gene conferring resistance to downy mildew and rust was produced by Girgi et al. (2006) with 0.13–0.15% transformation rates. The antimicrobial protein gene *afp* from the mold *Aspergillus giganteus* was put under the control of *Ubi1* promoter; a *pat* gene under the control of *CaMV35S* was used as selection marker. *In vitro* infection of detached leaves and *in vivo* inoculation of whole plants with the basidiomycete *P. substriata*, the causal agent of rust disease, and the oomycete *S. graminicola*, causal agent of downy mildew, transgenic plants, showed up to 90% enhanced disease resistance for *P. substrata* and *S. graminicola*, causal agent of rust and downy mildew, respectively. Shortly after that, a chemically synthesized, prawn antifungal protein encoding gene (*pin*) was used to develop transgenic pearl millet resistant to downy mildew by Latha et al. (2006). Shoot tip-derived embryogenic calli were co-bombarded with pPin35S and pBar35S constructs containing *pin* and *bar* genes, respectively, both driven by *CaMV35S* promoter. The transformed plants showed high level of resistance to fungal pathogen when compared to untransformed control plants. O'Kennedy et al. (2011) cloned a gene (*gluc78*) encoding an antifungal hydrolytic enzyme, and 1,3-β-glucosidase was cloned from strain P1 of a biocontrol fun*gus Trichoderma atroviride* and introduced into immature zygotic embryos with constructs containing gluc78 driven by either the constitutive ubiquitin promoter or the wound-inducible potato proteinase inhibitor IIK gene promoter (*pin2*). The positive selectable marker gene, *man*A, under the control of the *Ubi1* promoter, was used for cotransformation. Unfortunately, transgenic pearl millet plants expressing *gluc78* were more susceptible to downy mildew and rust and displayed severe disease symptoms relative to wild-type plants. Recently, Jalaja et al. (2016) developed downy mildew-resistant plants by transferring *osmotin* and *chitinase* genes using pCAMBIA2300 vector.

11.14.2.2 Finger Millet

Only a limited number of studies are available on finger millet transformation by biolistic method. Preliminary work on finger millet transformation was performed by Gupta et al. (2001) via biolistic means. They evaluated the efficiency of five gene promoters (*CaMV35*, rice *Act1*, maize *ubiquitin 1*, rice ribulose 1, 5-biphosphate carboxylase small subunit (*RbcS*) gene promoter, and *Flaveria trinervia* (*Ft* gene promoter)) driving the *GUS* gene in leaf lamina, leaf sheath, and regenerating seed callus and found that promoters *ubiquitin 1* and *actin* gave the best response for *GUS* gene expression; *RbcS* and *CaMV35S* promoters produced medium response while *Ft* promoter was found to be ineffective for GUS expression. Jagga-Chugh et al. (2012) standardized microprojectile bombardment-mediated genetic transformation parameters from seed-derived callus of *Eleusine coracana* using pCAMBIA 1381 *hpt* as selectable marker gene and *gus*A as reporter gene. They found that 1100 psi rupture disk pressure with 3 cm distance from rupture disk to macrocarrier, 12 cm microprojectile travel distance, osmotic treatment of callus with 0.4 M sorbitol, and double bombardment with gold particles of 1.0 μm size provided maximum transient GUS expression and transformation efficiency. Transgenic finger millet resistant to leaf blast disease was successfully developed by Latha et al. (2005). They chemically synthesized an antifungal protein (PIN) from prawn and cloned into plasmid pPin35S along with a bar reporter gene; both the genes were under the control of *CaMV35S* promoter. The transformed plants were selected on phosphinothricin-supplemented medium. By using the same protocol, Mahalakshmi et al. (2006) developed salt-tolerant plant by bombarding seed-derived calli with vector TG0063 harboring *PcSrp* gene driven by rice actin 1 promoter. Later, in 2012 Ignacimuthu and Ceasar developed resistance for the same disease by introducing rice chitinase gene (*chi11*) under the control of maize ubiquitin promoter. Transformed plants were selected and regenerated on hygromycin-supplemented medium. Ramegowda et al. (2013) successfully overexpressed *OsZIP1* gene in finger millet under the transcriptional control of constitutive *CaMV35S* and endosperm-specific *Bx17* promoters to improve Zn accumulation. They reported stronger activity of endosperm-specific promoter (*Bx17*) over *CaMV35S* promoter.

11.14.2.3   Foxtail Millet

Liu et al. (2009) identified a novel gene *SiPf40* from an immature seed cDNA library of foxtail millet (*Setaria italica*). To investigate the role of *SiPf40* gene in millet, embryogenic calli were transformed with plasmids pROKf40s, pROKf40an, and pROKf40i containing *SiPf40* gene in sense, antisense, and a fragment of *SiPf40* in sense or antisense orientation, respectively, under the control of *CaMV35S* promoter. The resultant plants overexpressing *SiPf40* displayed extra tillering associated with vessel enlarging and xylary fibers increasing, whereas, the tiller number decreases in *SiPf40* gene silenced plants.

11.14.2.4   Barnyard Millet

So far there is only one report available on barnyard millet transformation. Gupta et al. (2001) evaluated the efficiency of five gene promoters (*CaMV35, Act1, Ubil, rice ribulose 1,* 5-biphosphate carboxylase small subunit *(RbcS)* gene promoter, and *Flaveria trinervia (Ft)* gene promoter) by measuring expression of *gus* reporter gene. Their study demonstrated the efficiency of only *Ubi1* promoter in driving the expression of the transgene, while other promoters turned out to be inefficient.

11.14.2.5   Bahia Grass

The warm-season grass *Paspalum notatum* (Bahia grass) is an important forage crop in tropical and subtropical regions around the world. Its mode of reproduction is either diploid sexual or obligate apomictic in tetraploid genotypes (Burton and Forbes 1960). The transformation system for this important forage grass was first established by Smith et al. (2002) using seed-derived embryogenic as explant for bombardment with a vector containing the *bar* selectable marker/reporter gene under the control of *CaMV35S*. The overall average rate of transgenic plants was 4.7%. Grando et al. (2002) published their work on gene transfer using the same explant with *uid*A reporter driven by maize *Ubi1* promoter placed in the plasmid pAHC25. Further improvements in transformation protocol were made by Gondo et al. and Altpter and James independently in 2005. Gondo et al. bombarded seed-derived embryogenic calli with plasmids containing *gus* reporter gene regulated by *Act1* promoter and the bialaphos resistance gene (*bar*) under control of the *CaMV35S* promoter. While Saltpeter and James used selectable *nptII* gene under control of the maize *Ubi1* for the transformation experiments. Altpeter's group utilized this protocol to produce transgenic plants displaying reduced levels of endogenous gibberellins (Agharkar et al. 2007) and studied *bar* transgene expression in apomictic progeny of Bahia grass (Sandhu et al. 2007) and also developed transgenics with enhanced turf quality by overexpressing the Arabidopsis ATHB16 transcription factor regulated by *CaMV35S* promoter which led to significant changes in plant architecture. Overexpression of Arabidopsis ATHB16 transcription factor in Bahia grass produced significantly more vegetative and fewer reproductive tiller (Zhang et al. 2007). Later the same laboratory investigated co-transfer and expression of two unlinked gene (Sindhu and Altpeter 2008) and also developed abiotic stress tolerant plants by introducing genes for DREB1A and WRKY38 transcription factors from barley and wheat, respectively (James et al. 2008; Xiong et al. 2009). DREB1A ortholog from *Hordeum spontaneum* was driven by abiotic stress-inducible barley HVA1s promoter, and selection marker gene *nptII* was under the control of the maize ubiquitin 1 promoter (James et al. 2008). DREB1A gene encodes transcription factors which enhance expression of stress-protective target genes. Hence, transgenic plants expressing DREB1A showed high tolerance for drought and salt stresses. Co-transfer of two unlinked, minimal, linear transgene expression cassettes lacking vector backbone (MCs) was carried out by Sindhu and Altpeter (2008) to evaluate co-integration and quantify co-expression. The expression cassettes contain the *nptII* gene under the transcriptional regulation of the *CaMV35S* promoter and the bar gene under the control of the constitutive maize ubiquitin promoter. Co-expression of unlinked *nptII* and bar genes occurred with 95% frequency. It will allow transfer of multiple genes into a single genotype simultaneously.

One of the major problems of Bahia grass is high susceptibility to insect fall armyworm (*Spodoptera frugiperda*) which causes significant seasonal economic loss in turf grass. To build resistance against fall armyworm (*Spodoptera frugiperda*), Luciani et al. (2007) delivered *cry1Fa* gene encoding a δ-endotoxin from *Bacillus thuringiensis* under constitutive *Ubi1* promoter and *nptII* as selection marker under transcriptional control of the *CaMV35S* promoter into cultivar Tifton 9. An average mortality rate of 83% was observed when neonate larvae of fall armyworm were fed on transgenic leaves of *cry1Fa* expressing line. Mancini et al. (2014) developed a transformation protocol based on previous reports by Smith et al. (2002), Altpeter and James (2005), and Gondo et al. (2005) with modifications aimed at increasing transformation rates. Transformation experiments were carried out with two different constructs containing the reporter *gfp* gene cloned under the rice *Act1* promoter and the selector *bar* gene driven by the maize *Ubi1* promoter obtaining 40.7% cotransformation frequency. Although the method of introducing DNA into cells by microprojectile bombardment has revolutionized the field of genetic transformation of crop plants, few major drawbacks associated with this system are (1) considerable variation seen in stability, integration, and expression of the introduced transgene; (2) results in high transgene copy number and a high frequency of transgene rearrangement, which may lead to transgene silencing or co-suppression; (3) only DNA fragments of less than 10 kb in size can be transferred by the biolistic technology because large fragments get destroyed during the bombardment or adhere poorly to the metal particles; and (4) this technique is expensive and requires specialized instruments such as biolistic/DNA gun for bombardment.

11.15 Future Prospects and Concluding Remarks

A major concern of present day agricultural scientists is to increase crop yield by overcoming drastic reduction in crop productivity caused by various biotic and abiotic stresses. In order to achieve this objective, it is necessary to develop novel crop genotypes that are able to tolerate various stresses without having a significant negative effect on their growth and productivity. Millets, which are called miracle grains due to their compliant nature toward changing climate, can survive and flourish in harsh environmental conditions. However, these crops have not been sufficiently studied, and their potential remains largely unexploited. Both conventional and modern improvement techniques have not been used appropriately for their improvement. Moreover, majority of labs in developed as well as in developing countries focus on major cereals such as wheat, rice, and maize, undermining the hidden potential of millets. Although *in vitro* plant regeneration studies started in the mid-1980s, till now very little efforts have been made to genetically transform millets, which could be due to less responsiveness of millets to transformation protocols. There are no model cultivars, which can be transformed at an efficient rate for any of the millet species. Genetic transformation system is highly cultivar dependent; hence a large-scale screening of genotypes is needed to be done to identify highly responsive genotypes for transformation protocols. Moreover, most of the available reports on millet transformation are restricted to the analysis of marker or reporter gene expression only. Efforts need to be made for production of transgenic millets expressing agronomically important genes to improve millet production by conferring resistance to biotic and abiotic stresses. Furthermore, genes for resistance to lodging, seed softness and boldness, improved milling, processing qualities and palatability, male sterility and increased tillering should be targeted to make these crops widely accepted by consumers. Moreover, a new branch of genetics, i.e., reverse genetics, has emerged as a powerful tool to investigate the gene function. In reverse genetics a gene is disrupted, and the effect of its abolition is observed on an organism. The most widely used approach to abolish gene function is RNA-induced gene silencing or co-suppression by delivering dsRNA or antisense RNA in an organism that effectively silences target gene. The most common method to develop knockout mutants is bombardment with gold or tungsten particles that have been coated with DNA. Hence transgenic will not only enhance quality and quantity of agronomically important crops but will also provide an insight into function of novel genes and promoters from millets.

References

Agharkar M, Lomba P, Altpeter F, Zhang H, Kenworthy K, Lange T (2007) Stable expression of AtGA2ox1 in a low-input turfgrass (*Paspalum notatum* Flugge) reduces bioactive gibberellin levels and improves turf quality under field conditions. Plant Biotechnol J 5(6):791–801

Altpeter F, James VA (2005) Genetic transformation of turf type bahiagrass *(Paspalum notatum Flugge)* by biolistic gene transfer. Int Turfgrass Soc Res J 10:1–5

Anami S, Njuguna E, Coussens G, Aesaert S, Van Lijsebettens M (2013) Higher plant transformation: principles and molecular tools. Int J Dev Biol 57(6-7-8):483–494

Barampuram S, Zhang ZJ (2011) Recent advances in plant transformation. Methods Mol Biol 701:1–35

Barcelo P, Rasco-Gaunt S, Thorpe C, Lazzeri PA (2001) Transformation and gene expression. Adv Bot Res 34:59–126

Bayer GY, Yemets AI, Blume YB (2014) Obtaining the transgenic lines of finger millet *Eleusine coracana* (L.). with dinitroaniline resistance. Cytol Genet 48(3):139–144

Bevan M (1984) Binary *Agrobacterium* vectors for plant transformation. Nucleic Acids Res 12:8711–8721

Bechtold N, Ellis J, Pelletier G (1993) In planta *Agrobacterium* mediated gene transfer by infiltration of adult Arabidopsis thaliana plants. Comptes rendus de l'Académie des sciences. Série 3, Sciences de la vie 316(10):1194–1199

Bennetzen JL, Schmutz J, Wang H, Percifield R, Hawkins J, Pontaroli AC, Estep M, Feng L, Vaughn JN, Grimwood J, Jenkins J (2012) Reference genome sequence of the model plant Setaria. Nat Biotechnol 30(6):555–561

Benson EE (2000) Sepecial symposium: in vitro plant recalcitrance in vitro plant recalcitrance: an introduction. In Vitro Cell Dev Biol Plant 36(3):141–148

Bhaskaran S, Smith RH (1990) Regeneration in cereal tissue culture: a review. Crop Sci 30:1328–1336

Birch RG (1997) Plant transformation: problems and strategies for practical application. Annu Rev Plant Physiol Plant Mol Biol 48:297–326

Burton GW, Forbes I (1960) The genetics and manipulation of obligate apomixis in common bahiagrass (*Paspalum notatum* Flu¨gge). In: Proceedings of the VIII International Grassland Congress. Alden Press, Oxford, pp 66–71

Bytebier B, Deboek F, De Greve H, van Montagu M, Hemalsteens JP (1987) T-DNA organisation in tumour cultures and transgenic plants of the monocotyledon *Asparagus officinalis*. Proc Natl Acad Sci U S A 84:5345–5349

Ceasar SA, Ignacimuthu S (2011) *Agrobacterium* mediated transformation of Finger millet (*Eleusine coracana* (L.) Gaertn.) using shoot apex explants. Plant Cell Rep 30:1759–1770

Ceasar SA, Baker A, Ignacimuthu S (2017) Functional characterization of the PHT1 family transporters of foxtail millet with development of a novel *Agrobacterium*-mediated transformation procedure. Sci Rep. https://doi.org/10.1038/s41598-017-14447-0

Chang WC, Lee TY, Huang HD, Huang HY, Pan RL (2008) PlantPAN: plant promoter analysis navigator, for identifying combinatorial cis-regulatory elements with distance constraint in plant gene groups. BMC Genomics 9(1):561

Cho MJ, Wu E, Kwan J, Yu M, Banh J, Linn W, Anand A, Li Z, TeRonde S, Register JC, Jones TJ (2014) *Agrobacterium*-mediated high-frequency transformation of an elite commercial maize (*Zea mays* L.) inbred line. Plant Cell Rep 33:1767–1777

Christensen AH, Sharrock RA, Quail P (1992) Maize polyubiquitin genes: structure, thermal perturbation of expression and transcript splicing, and promoter activity following transfer to protoplasts by electroporation. Plant Mol Biol 18:675–689

Clough SJ, Bent AF (1998) Floral dip: a simplified method for *Agrobacterium* – mediated transformation of Arabidopsis thaliana. Plant J 16(6):735–743

Commandeur U, Twyman RM, Fischer R (2003) The biosafety of molecular farming in plants. AgBiotechNet 5(110):1–9

Dahleen LS (1995) Improved plant regeneration from barley callus cultures by increased copper levels. Plant Cell Tiss Org Cult 43:267–269

Devi P, Sticklen M (2002) Culturing shoot-tip clumps of pearl millet [*Pennisetum glaucum* (L.) R. Br.] and optimal microprojectile bombardment parameters for transient expression. Euphytica 125:45–50

Dosad S, Chawla HS (2016) In vitro plant regeneration and transformation studies in millets: current status and future prospects. Indian J Plant Physiol 21(3):239–254

Doust AN, Kellogg EA, Devos KM, Bennetzen JL (2009) Foxtail millet: a sequence-driven grass modelsystem. Plant Physiol 149:137–141. https://doi.org/10.1104/pp.108.129627

Eapen S, George L (1990) Influence of phytohormones, carbohydrates, amino acids, growth supplements and antibiotics on somatic embryogenesis and plant differentiation in finger millet. Plant Cell Tiss Org Cult 22:87–93

Estrella LH, Depicker A, Montagu MV, Schell J (1983) Expression of chimaeric genes transferred into plant cells using a Ti-plasmid-derived vector. Nature 303:209–213

Feldmann KA, Marks MD (1987) *Agrobacterium* – mediated transformation of germinating seeds of Arabidopsis thaliana: a non-tissue culture approach. Mol Gen Genet MGG 208(1–2):1–9

Finer JJ, Vain P, Jones MW, McMullen MD (1992) Development of the particle in flow gun for DNA delivery to plant cells. Plant Cell Rep 11:323–328

Franck A, Guilley H, Jonard G, Richards K, Hirth L (1980) Nucleotide sequence of cauliflower mosaic virus DNA. Cell 21:285–294

Fraley RT, Rogers SG, Horsch RB, Gelvin SB (1986) Genetic transformation in higher plants. Crit Rev Plant Sci 4(1):1–46

Gebre E, Gugsa L, Schlüter U, Kunert K (2013) Transformation of tef (*Eragrostis tef*) by Agrobacterium through immature embryo regeneration system for inducing semi-dwarfism. S Af J Bot 87:9–17

Girgi M, O'Kennedy MM, Morgenstern A, Smith G, Lorz H, Oldach KH (2002) Transgenic and herbicide resistant pearl millet (*Pennisetum glaucum* L.) R.Br. via microprojectile bombardment of scutellar tissue. Mol Breed 10:243–252

Girgi M, Breese WA, Lörz H, Oldach KH (2006) Rust and downy mildew resistance in pearl millet (*Pennisetum glaucum*) mediated by heterologous expression of the afp gene from Aspergillus giganteus. Transgenic Res 15(3):313–324

Goldman JJ, Hanna WW, Fleming G, Ozias-Akins P (2003) Fertile transgenic pearl millet [*Pennisetum glaucum* (L.) R. Br.] plants recovered through microprojectile bombardment and phosphinothricin selection of apical meristem-, inflorescence-, and immature embryo-derived embryogenic tissues. Plant Cell Rep 21:999–1009

Gondo T, Shin-ichi T, Ryo A, Osamu K, Franz H (2005) Green, herbicide-resistant plants by particle inflow gun mediated gene transfer to diploid bahiagrass (*Paspalum notatum*). J Plant Physiol 16:1367–1375

Grando MF, Franklin CI, Shatters JRG (2002) Optimizing embryogenic callus production and plant regeneration from 'Tifton 9' bahiagrass seed explants for genetic manipulation. Plant Cell Tiss Org Cult 71:213–222

Gupta P, Raghuvanshi S, Tyagi AK (2001) Assessment of the efficiency of various gene promoters via biolistics in leaf and regenerating seed callus of millets, *Eleusine coracana* and Echinochloa crusgalli. Plant Biotechnol 18:275–282

Gelvin SB (2003) *Agrobacterium* -mediated plant transformation: the biology behind the "Gene-Jockeying" tool. Microbiol Mol Biol Rev 67(1):16–37

Hamilton RH, Fall MZ (1971) The loss of tumor-initiating ability in *Agrobacterium* tumefaciens by incubation at high temperature. Cell Mol Life Sci 27(2):229–230

Haseloff J, Siemering KR, Prasher DC, Hodge S (1997) Removal of a cryptic intron and subcellular localization of green fluorescent protein are required to mark transgenic Arabidopsis plants brightly. Proc Natl Acad Sci U S A 94:2122–2127

Hauptmann RM, Ozias-Akins P, Vasil V, Tabaeizadeh Z, Rogers SG, Horsch RB, Vasil I, Fraley RT (1987) Transient expression of electroporated DNA in monocotyledonous and dicotyledonous species. Plant Cell Rep 6:265–270

Hema R, Vemanna RS, Sreeramulu S, Reddy CP, Kumar MS, Udayakumar M (2014) Stable expression of mtlD gene imparts multiple stress tolerance in Finger millet. PLoS One. https://doi.org/10.1371/journal.pone.0099110

Hiei Y, Komari T (2008) Agrobacterium -mediated transformation of rice using immature embryos or calli induced from mature seed. Nat Protoc 3:824–834

Hiei Y, Ohta S, Komari T, Kumashiro T (1994) Efficient transformation of rice (Oryza sativa L.) mediated by Agrobacterium and sequence analysis of the boundaries of the DNA. Plant J 6:271–282

Hoekema A, Hirsch PR, Hooykaas PJJ, Schilperoort RA (1983) A binary plant vector strategy based on separation of vir-and T-region of the Agrobacterium tumefaciens Ti-plasmid. Nature 303(5913):179–180

Hood EE, Helmet GL, Fraley RT, Chilton M-D (1986) The hypervirulence of Agrobacterium tumefaciens A281 is encoded in a region of pTiBo542 outside of T-DNA. J Bacteriol 168:1291–1301

Hood EE, Gelvin SB, Melchers LS, Hoekema A (1993) New Agrobacterium helper plasmids for gene transfer to plants. Transgenic Res 2(4):208–218

Ignacimuthu S, Ceasar SA (2012) Development of transgenic finger millet (Eleusine coracana (L.) Gaertn.) resistant to leaf blast disease. J Biosci 37:135–147

Ignacimuthu S, Kannan P (2013) Agrobacterium – mediated transformation of pearl millet (Pennisetum typhoides (L.) R. Br.) for fungal rust. Asian J Plant Sci 112:97–108

Jagga-Chugh S, Kachhwaha S, Sharma M, Kothari-Chajer A, Kothari SL (2012) Optimization of factors influencing microprojectile bombardment-mediated genetic transformation of seed-derived callus and regeneration of transgenic plants in Eleusine coracana (L.) Gaertn. Plant cell. Tissue Organ Cult (PCTOC) 109(3):401–410

Jalaja N, Maheshwari P, Naidu KR, Kavi Kishor PB (2016) In vitro regeneration and optimization of conditions for transformation methods in Pearl millet, Pennisetum glaucum (L.). Int J Clin Biol Sci 1:34–52

James C (2014) Executive summary. In: Global status of commercialized biotech/GM crops. ISAAA brief no. International Service for the Acquisition of Agri-Biotech Applications, Ithaca, p 2013

James VA, Neibaur JI, Altpeter F (2008) Stress inducible expression of the DREB1A transcription factor from xeric, Hordeum spontaneum L. in turf and forage grass (Paspalum notatum Flugge) enhances abiotic stress tolerance. Transgenic Res 17:93–104

Janice M, Zale S, Agarwal S, Loar CMS (2009) Evidence for stable transformation of wheat by floral dip in Agrobacterium tumefaciens. Plant Cell Rep 28(6):903–913

Jayasudha BG, Sushma AM, Prashantkumar HS, Sashidhar VR (2014) An efficient in vitro Agrobacterium–mediated transformation protocol for raising salinity tolerant transgenic plants in finger millet [Eleusine coracana (L.) Gaertn.]. Plant Archives 14:823–829

Jha P, Shashi RA, Agnihotri PK, Kulkarni VM, Bhat V (2011) Efficient Agrobacterium – mediated transformation of Pennisetum glaucum (L.) R. Br. using shoot apices as explant source. Plant Cell Tiss Org Cult 107:501–512

Kikkert JR (1993) The biolistic PDS-1000/he device. Plant Cell Tiss Org Cult 33:221–226

Komari T (1990) Transformation of cultured cells of Chenopodium quinoa by binary vectors that carry a fragment of DNA from the virulence region of pTiBo542. Plant Cell Rep 9:303–306

Komari T, Takakura Y, Ueki J, Kato N, Ishida Y, Hiei Y (2006) Binary vectors and super-binary vectors. Agrobacterium protocols. Methods Mol Biol 343:15–42

Koncz C, Schell J (1986) The promoter of the TL-DNA gene 5 controls the tissue-specific expression of chimaeric genes carried by a novel type of Agrobacterium binary vector. Mol Gen Genet 204:383–396

Kothari SL, Agarwal K, Kumar S (2004) Inorganic nutrient manipulation for highly improved in vitro plant regeneration in finger millet- Eleusine coracana (L.) Gaertn. In Vitro Cell Dev Biol Plant 40:515–519

Kothari SL, Kumar S, Vishnoi RK, Kothari SL, Watanabe KN (2005) Applications of biotechnology for improvement of millet crops: review of progress and future prospects. Plant Biotechnol 22:81–88

Kothari-Chajer A, Sharma M, Kachhwaha S, Kothari SL (2008) Micronutrient optimization results into highly improved in vitro plant regeneration in kodo (*Paspalum scrobiculatum* L.) and finger (*Eleusine coracana* (L.) Gaertn.) millets. Plant Cell Tiss Org Cult 94(2):105–112

Lakkakula S, Stanislaus AC, Jayabalan S, Arockiam SR, Periyasamy R, Manikandan R (2015) Direct plant regeneration from *in vitro* derived shoot apical meristems of finger millet (*Eleusine coracana* (L.) Gaertn.). In Vitro Cell Dev Biol Plant 51:192–200

Lakkakula S, Periyasamy R, Stanislaus AC, Arokiam SR, Subramani P, Ramakrishnan RK, Alagesan S, Manikandan R (2016) Effects of cefotaxime, amino acids and carbon source on somatic embryogenesis and plant regeneration in four Indian genotypes of foxtail millet (*Setaria italica* L.). In Vitro Cell Dev Biol Plant 52:140–153

Lakkakula S, Stanislaus AC, Manikandan R (2017) Improved *Agrobacterium*-mediated transformation and direct plant regeneration in four cultivars of finger millet (*Eleusine coracana* (L.) Gaertn.). Plant Cell Tiss Org Cult 131:547–565

Lambe P, Dinant M, Matagne RF (1995) Differential long-term expression and methylation of the hygromycin phosphotransferase (hph) and β-glucuronidase (GUS) genes in transgenic pearl millet (*Pennisetum americanum*) callus. Plant Sci 108:51–62

Lambe P, Dinant M, Deltour R (2000) Transgenic pearl millet (*Pennisetum glaucum*). In: Bajaj YPS (ed) Biotechnology in agriculture and forestry, transgenic crops I, vol 46. Springer, Berlin, pp 84–108

Last DI, Bretell RIS, Chamberlain DA, Chaundhury AM, Larkin PJ, Marsh EL, Peacock WJ, Dennis ES (1991) p*Emu*: an improved promoter for gene expression in cereal cells. Theor Appl Genet 81:581–588

Latha MA, Rao KV, Reddy VD (2005) Production of transgenic plants resistant to leaf blast disease in finger millet (*Eleusine coracana* (L.) Gaertn.). Plant Sci 169:657–667

Latha MA, Rao KV, Reddy TP, Reddy VD (2006) Development of transgenic pearl millet (*Pennisetum glaucum* (L.) R. Br.) plants resistant to downy mildew. Plant Cell Rep 25:927–935

Lescot M, Déhais P, Thijs G, Marchal K, Moreau Y, Van de Peer Y, Rouzé P, Rombauts S (2002) PlantCARE, a database of plant cis-acting regulatory elements and a portal to tools for in silico analysis of promoter sequences. Nucleic Acids Res 30(1):325–327

Li Z, Upadhyaya NM, Meena S, Gibbs AJ, Waterhouse PM (1997) Comparison of promoters and selectable marker genes for use in Indica rice transformation. Mol Breed 3:1–14

Libiakova G, Jørgensen B, Palmgren G, Ulvskov P, Johansen E (2001) Efficacy of an intron-containing kanamycin resistance gene as a selectable marker in plant transformation. Plant Cell Rep 20(7):610–615

Liu Y, Yu J, Zhao Q, Zhu D, Ao G (2005) Genetic transformation of millet (*Setaria italica*) by *Agrobacterium*. J Agric Biotechnol 13:32–37

Liu Y, Feng X, Xu Y, Yu J, Ao G, Peng Z, Zhao Q (2009) Overexpression of millet ZIP-like gene (*SiPf40*) affects lateral bud outgrowth in tobacco and millet. Plant Physiol Biochem 47:1051–1060

Luciani G, Altpeter F, Yactayo-Chang J, Zhang H, Gallo M, Meagher RL, Wofford D (2007) Expression of in Bahiagrass enhances resistance to fall armyworm. Crop Sci 47(6):2430–2436

Maas C, Simpson CG, Eckes P, Schickler H, Brown JWS, Reiss B, Salchert K, Chet I, Schell J, Reichel C (1997) Expression of intron modified NPT II genes in monocotyledonous and dicotyledonous plant cells. Mol Breed 3:15–28

Mahalakshmi S, Christopher GSB, Reddy TP, Rao KV, Reddy VD (2006) Isolation of a cDNA clone (PcSrp) encoding serine-rich-protein from Porteresia coarctata T. and its expression in yeast and finger millet (Eleusine coracana L.) affording salt tolerance. Planta 224(2):347–359

Maksymiec W (1997) Effect of copper on cellular processes in higher plants. Photosynthetica 34:321–342

Mancini M, Woitovich N, Permingeat HR, Podio M, Siena LA, Ortiz JPA, Pessino SC, Felitti SA (2014) Development of a modified transformation platform for apomixis candidate genes research in *Paspalum notatum* (bahiagrass). In Vitro Cell Dev Biol Plant. https://doi.org/10.1007/s11627-014-9596-2

Martins PK, Nakayama TJ, Ribeiro AP, Cunha BADBD, Nepomuceno AL, Harmon FG et al (2015a) *Setaria viridis* floral-dip: a simple and rapid *Agrobacterium* -mediated transformation method. Biotechnol Rep 6:61–63

Martins PK, Ribeiro AP, Cunha BADB, Kobayashi AK, Molinari HBC (2015b) A simple and highly efficient *Agrobacterium*-mediated transformation protocol for Setaria viridis. Biotechnol Rep 6:41–44

Meyer P, Walgenbach E, Bussmann K, Hombrecher G, Saedler H (1985) Synchronized tobacco protoplasts are efficiently transformed by DNA. Mol Gen Genet MGG 201(3):513–518

Mohanty BD, Gupta SD, Ghosh PD (1985) Callus initiation and plant regeneration in ragi (*Eleusine coracana* Gaertn). Plant Cell Tiss Org Cult 5:147–150

Murashige T, Skoog F (1962) A revised medium for rapid growth and bioassay for tobacco tissue cultures. Physiol Plant 15:473–497

Napoli C, Lemieux C, Jorgensen R (1990) Introduction of a chimeric chalcone synthase gene into petunia results in reversible co-suppression of homologous genes in trans. Plant Cell 2:279–289

Niedz RP, Evens TJ (2007) Regulating plant tissue growth by mineral nutrition. In Vitro Cell Dev Biol Plant 43:370–381

Niedz RP, Sussman MR, Satterlee JS (1995) Green fluorescent protein: an in vivo reporter of plant gene expression. Plant Cell Rep 14(7):403–406

O'Kennedy MM, Burger JT, Botha FC (2004) Pearl millet transformation system using the positive selectable marker gene phosphomannose isomerase. Plant Cell Rep 22:684–690

O'Kennedy MM, Crampton BG, Lorito M, Chakauya E, Breese WA, Burger JT, Botha FC (2011) Expression of a b-1,3-glucanase from a biocontrol fungus in transgenic pearl millet. S Afr J Bot 77:335–345

Ohta S, Mita S, Hattori T, Nakamura K (1990) Construction and expression in tobacco of a β -glucuronidase (GUS) reporter gene containing an intron within the coding region. Plant Cell Physiol 31:805–813

Oldach K, Morgenstern A, Rother S, Girgi M, O'Kennedy M, Lörz H (2001) Efficient in vitro plant regeneration from immature zygotic embryos of pearl millet (*Pennisetum glaucum* (L.) R. Br.) and Sorghum bicolor (L.) Moench. Plant Cell Rep 20(5):416–421

Pan Y, Ma X, Liang H, Zhao Q, Zhu D, Yu J (2015) Spatial and temporal activity of the foxtail millet (Setaria italica) seed-specific promoter pF128. Planta 241:57–67

Prasher DC, Eckenrode VK, Ward WW, Prendergast FG, Cormier MJ (1992) Primary structure of the Aequorea victoria green-fluorescent protein. Gene 111(2):229–233

Plaza-Wuthrich S, Tadele Z (2012) Millet improvement through regeneration and transformation. Biotechnol Mol Biol Rev 7:48–61

Qin FF, Zhao Q, Ao GM, Yu JJ (2008) Co-suppression of Si401, a maize pollen specific *Zm401* homologous gene, results in aberrant anther development in foxtail millet. Euphytica 163(1):103–111

Raineri DM, Bottino P, Gordon MP, Nester EW (1990) *Agrobacterium* – mediated transformation of rice (Oryza sativa L.). Nat Biotechnol 8(1):33–38

Ramadevi R, Rao KV, Reddy VD (2014) *Agrobacteriumtumefaciens*-mediated genetic transformation and production of stable transgenic pearl millet (Pennisetum glaucum [L.] R. Br.). In Vitro Cell Dev Biol Plant 50(4):392–400

Ramegowda Y, Venkategowda R, Jagadish P, Govind G, Hanumanthareddy RR, Makarla U, Guligowda SA (2013) Expression of a rice Zn transporter, OsZIP1, increases Zn concentration in tobacco and finger millet transgenic plants. Plant Biotechnol Rep 7:309–319

Ramineni R, Sadumpati V, Khareedu VR, Vudem DR (2014) Transgenic pearl millet male fertility restorer line (ICMP451) and hybrid (ICMH451) expressing *Brassica juncea* nonexpressor of pathogenesis related genes 1 (*BjNPR1*) exhibit resistance to downy mildew disease. PLoS One 9(3):e90839

Saha P, Blumwald E (2016) Spike-dip transformation of *Setaria viridis*. Plant J 86(1):89–101

Sahrawat AK, Chand S (1999) Stimulatory effect of copper on plant regeneration in indica rice (Oryza sativa L.). J Plant Physiol 154:517–522

Sai NK, Visarada KBRS, Lakshmi YA, Pashupatinath E, Rao SV, Seetharama N (2006) In vitro culture methods in sorghum with shoot tip as the explant material. Plant Cell Rep 25(3):174–182

Sandhu S, Altpeter F (2008) Co-integration, co-expression and inheritance of unlinked minimal transgene expression cassettes in an apomictic turf and forage grass (*Paspalum notatum* Flüggé). Plant Cell Rep 27(11):1755–1765

Sandhu S, Altpeter F, Blount AR (2007) Apomictic bahiagrass expressing the bar gene is highly resistant to glufosinate under field conditions

Sanford JC, Klein TM, Wolf ED, Allen N (1987) Delivery of substances into cells and tissues using a particle bombardment process. J Part Sci Technol 5:27–37

Sen S, Dutta S (2016) A potent bidirectional promoter from the monocot cereal *Eleusine coracana*. Phytochemistry 129:24–35

Sharma KK, Ortiz R (2000) Program for the application of genetic transformation for crop improvement in the semiarid tropics. In Vitro Cell Dev Biol Plant 36:83–92

Sharma M, Kothari-Chajer A, Jagga-Chugh S, Kothari SL (2011) Factors influencing *Agrobacterium* tumefaciens-mediated genetic transformation of *Eleusine coracana* (L.) Gaertn. Plant Cell Tiss Org Cult 105(1):93–104

Shimomura O, Johnson FH, Saiga Y (1962) Extraction, purification and properties of aequorin, a bioluminescent protein from the luminous hydromedusan, Aequorea. J Cell Physiol 59(3):223–239

Smirnova OG, Ibragimova SS, Kochetov AV (2012) Simple database to select promoters for plant transgenesis. Transgenic Res 21(2):429–437

Smith RL, Grando MF, Li YY, Seib JC, Shatters RG (2002) Transformation of bahiagrass (*Paspalum notatum* Flugge). Plant Cell Rep 20:1017–1021

Stachel SE, Messens E, Van Montagu M, Zambryski P (1985) Identification of the signal molecules produced by wounded plant cells that activate T-DNA transfer in *Agrobacterium* tumefaciens. Nature 318(6047):624–629

Stoger E, Vaquero C, Torres E, Sack M, Nicholson L, Drossard J, Williams S, Keen D, Perrin Y, Christou P, Fischer R (2000) Cereal crops as viable production and storage systems for pharmaceutical scFv antibodies. Plant Mol Biol 42(4):583–590

Thakur RP (2008) Pearl millet. In: Satish L, Mawar R, Rathore BS (eds) Disease management in arid land crops. Scientific Publishers, Jodhpur, pp 21–41

Tiecoura K, Kouassi AB, Oulo N, Gonedele Bi S, Dinant M, Ledou L (2015) In vitro transformation of pearl millet (*Pennisetum glaucum* (L). R. BR.): selection of chlorsulfuron resistant plants and long term expression of the gus gene under the control of the emu promoter. Afr J Biotechnol 14:3112–3123

Ueki S, Lacroix B, Krichevsky A, Lazarowitz SG, Citovsky V (2009) Functional transient genetic transformation of Arabidopsis leaves by biolistic bombardment. Nat Protoc 4(1):71

Usami S, Morikawa S, Takebe I, Machida Y (1987) Absence in monocotyledonous plants of the diffusible plant factors inducing T-DNA circularization and *vir* gene expression in *Agrobacterium*. Mol Gen Genet 209:221–226

Vain P, Finer KR, Engler DE, Pratt RC, Finer JJ (1996) Intron-mediated enhancement of gene expression in maize (Zea mays L.) and bluegrass (Poa pratensis L.). Plant Cell Rep 15:489–494

Van Larebeke N, Engler G, Holsters M, Van den Elsacker S, Zaenen I, Schilperoort RA, Schell J (1974) Large plasmid in *Agrobacterium tumefaciens* essential for crown gall-inducing ability. Nature 252(5479):169–170

Vancanneyt G, Schmidt R, O'Connor-Sanchez A, Willmitzer L, Rocha-Sosa M (1990) Construction of an intron-containing marker gene: splicing of the intron in transgenic plants and its use in monitoring early events in *Agrobacterium* tumefaciens plant transformation. Mol Gen Genet 220:245–250

Vasil IK (1982) Plant cell culture and somatic cell genetics of cereals and grasses

Vikrant A, Rashid A (2002) Somatic embryogenesis from immature and mature embryos of a minor millet *Paspalum scrobiculatum* L. Plant Cell Tiss Org Cult 69:71–77

Vikrant A, Rashid A (2003) Somatic embryogenesis from mesocotyl and leaf base segments of *Paspalum scrobiculatum* L., minor millet. In Vitro Cell Dev Biol Plant 39:485–489

Wang MZ, Pan YL, Li C, Liu C, Zhao Q, Ao GM, Yu JJ (2011) Culturing of immature inflorescences and Agrobacterium-mediated transformation of foxtail millet (*Setaria italica*). Afr J
Biotechnol 10:16466–16479

Wang M, Li P, Li C, Pan Y, Jiang X, Zhu D, Zhao Q, Yu J (2014) SiLEA14, a novel atypical LEA
protein, confers abiotic stress resistance in foxtail millet. BMC Plant Biol 14(1):290

Wu LM, Wei YM, Zheng YL (2006) Effects of silver nitrate on the tissue culture of immature
wheat embryos. Russ J Plant Physiol 53(4):530–534

Xiong X, James VA, Zhang H, Altpeter F (2009) Constitutive expression of the barley HvWRKY38
transcription factor enhances drought tolerance in turf and forage grass (*Paspalum notatum*
Flugge). Mol Breeding 25:419–432

Yemets AI, Bayer GY, Blume YB (2013) An effective procedure for in vitro culture of *Eleusine
coracana* (L.) and its application. ISRN Botany. https://doi.org/10.1155/2013/853121

Yilmaz A, Nishiyama MY, Fuentes BG, Souza GM, Janies D, Gray J, Grotewold E (2009)
GRASSIUS: a platform for comparative regulatory genomics across the grasses. Plant Physiol
149(1):171–180

Zhang H, Lomba P, Altpeter F (2007) Improved turf quality of transgenic bahiagrass (*Paspalum
notatum* Flugge) constitutively expressing the ATHB16 gene, a repressor of cell expansion.
Mol Breed 20:415–423

Chapter 12
Transgenic Research on Tomato: Problems, Strategies, and Achievements

Joydeep Banerjee, Saikat Gantait, Sutanu Sarkar,
and Prabir Kumar Bhattacharyya

Abstract Tomato is a climacteric fruit; it is widely consumed as vegetables worldwide either raw or cooked owing to the antioxidative and anticancer properties of lycopene, a dynamic carotenoid pigment of tomato. Nonetheless, since the past few decades, the productivity of tomato is compromised by an array of biotic and abiotic stresses along with deterioration of desirable quality parameters. Consequently, the development of stress-tolerant quality crops is a strategic challenge for agricultural biotechnology. Genetic transformation approach permits to insert defined gene simultaneously avoiding the elimination of any intrinsic genetic attributes unlike the occasion of conventional in situ or true in vitro screening. Till date, a number of attempts have been made to mitigate biotic and abiotic stress on tomato keeping the improvement of quality parameters in mind. Majority of such modifications comprise of the expression of stress-inducible genes, manipulation in the metabolic pathways, or the accumulation of low molecular compounds that function critically in retaining the agility of reactions. In this chapter, we offer an overview of the strategies based on frequently selected target sequences or molecules that are genetically transferred or modified to attain genetically transformed tomatoes tolerant to environmental stresses as well as to improve the quality traits of its fruits.

Keywords Tomato · Abiotic stress · Biotic stress · Genetic transformation · Quality traits

12.1 Introduction

In the recent scenario, horticulture being the fastest-growing sector in agriculture is getting importance worldwide toward food and nutritional security. The global consumption pattern is changing toward non-food grain items in the consumption baskets particularly vegetables and fruits rather than food grains. Consequently, horticulture

J. Banerjee · S. Gantait (✉) · S. Sarkar · P. K. Bhattacharyya
Department of Genetics and Plant Breeding, Faculty of Agriculture, Bidhan Chandra Krishi
Viswavidyalaya, Mohanpur, Nadia, 741252, West Bengal, India

© Springer International Publishing AG, part of Springer Nature 2018
S. S. Gosal, S. H. Wani (eds.), *Biotechnologies of Crop Improvement, Volume 2*,
https://doi.org/10.1007/978-3-319-90650-8_12

is playing a greater role as well as importance among different agricultural sectors and eventually in the global economy. Vegetables are so common in human diet that a meal without a vegetable is supposed to be incomplete in any part of the world. Vegetables having short duration, higher yield, nutritionally rich, and economically viabile, are indispensible in global economy as well as food and nutritional security. Urbanization, increasing per capita income, health awareness, and shifting of farmers toward high-value vegetables for higher income are also important factors for fueling vegetable expansion in the world. Tomato (*Lycopersicon esculentum* Miller. $2n = 2x = 24$), a native to South America more specifically in Andes mountains in Peru-Ecuador-Bolivia region (primary center of origin) and eastern Andes (secondary center of origin) (Swarup 2012), belongs to family Solanaceae and genus *Lycopersicon*. Tomato, a herbaceous sprawling plant, has weak woody stem of 1–3 m in height. The flowers are yellow in color; fruits vary in size and shape from cherry tomatoes, about 1–2 cm in size, to beefsteak tomatoes, about 10 cm or more in diameter. The fruits turn red in color during ripening. Tomato is one of the most important versatile vegetables that contribute to a healthy, well-balanced diet because of its special nutritive value. It has considerable amount of vitamins (B and C), minerals, essential amino acids, sugars, iron, phosphorus, and dietary fibers. Red tomatoes contain lycopene, an antioxidant that may contribute to the protection against carcinogenic substances, but with lower vitamin A content than yellow tomatoes (contains yellow carotenoid pigments). Lycopene plays an important role as quality parameter for the tomato processing industry. The synthesis of lycopene is temperature dependent. Optimum temperature is 16–21 °C and inhibited by >30 °C (Zhang et al. 2008). Besides yellow and red tomato, the tangerine (bright orange beefsteak tomato with prolycopene pigment) is also found. Globally, tomato ranks third vegetable crop after potato and sweet potato but ranks first as canned vegetables. Besides the use of tomato as fresh vegetable, it is used as in a variety of processed products such as soup, ketchup, juice, salad, pickles, sauce, sun-dried tomatoes, salsa, tomato-based powders, and different ready-to-eat products. Huge varietal diversity such as red cherry tomatoes (small, ideal for salad) and black cherry (dark purple varieties) is available in tomato.

During 2014, more than 163 million tons of tomatoes were produced globally which represented the 15% of total global vegetable production. Tomato production had a net value of US$59 billion during 2012, the eighth most important agricultural product worldwide. An amount of US$8.4 billion was involved for fresh tomato exports worldwide in 2015 (FAOSTAT 2016). An increase of 40% in terms of global production of tomato was observed since 2002 (during 2002, production was 116.53 million tons), and as it is seen in the FAO statistics, the increase has been distributed uniformly across the leading ten producing countries. China is in the top position in tomato production with around 50 million metric tons per year (31% of total production), followed by India with around 18 million metric tons (share of 11%), and the USA with nearly 13 million metric tons (share of nearly 8%) (FAOSTAT 2016).

Recently, the growing tomatoes in greenhouses from outdoors are getting attractive by the farmers. Greenhouse tomato production allows farmers to cultivate under optimized conditions. Stability in production; uniformity in shape, size, and quality; higher yields per unit area; and year-round supply to the market attract the farmers for greenhouse tomato cultivation. The Intense is a very unusual tomato that retains its juice even after squeezed which fits the fast-food chains and increase the market value. Marketing includes proper planning, stable production, timely harvesting, attractive packaging, distribution, transportation, storage, and price. To be successful, marketing must be responsive to consumers' demands. Simplistically, it must be customer oriented. Tomato industry is based on two types of marketing which are totally different. To sell in the open market, fresh tomatoes are picked manually, whereas, contractual selling and mechanical harvesting are the common practice in case of processed tomatoes. Dependable regular supply of fresh tomatoes with a planned production schedule is the prerequisite for processing industries. Supply of quality tomato with non-acidic, small locules and thick-walled fresh fruits are the prerequisite of the processing industry. Also, keeping quality is a significant quality factor for processing industries. The network of collection centers, packhouses, and logistic facilities enhances reach and supply chain efficiency and ultimately processing. According to the FAO, an annual increase of more than 25% over last 3 years with an increase of 38% in price during 2014 as compared with the previous year is observed. During 2015, the worldwide trade in fresh tomatoes reached to US$8.4 billion. In 2015, India's global tomato exports were valued at US$67 million, small in comparison with the global trade in fresh tomatoes at US$8.4 billion (or 0.8% of world share) ("World's Top Exports": June 2016). World Processing Tomato Council (WPTC) estimated 41.37 million tons (MT) of tomato were processed globally into value-added products in 2015 which accounts 26% of global production of fresh tomato. Globally, one third (33%) of all tomato processing is undertaken in the USA, mostly in California. The quality of tomato depends upon several factors like cultivar, growing environments and proper harvesting schedule. The characteristics of tomato both physical and chemical also affect the quality of processed product. The availability of the varieties of high yield coupled with good processing qualities to the farmers is the prerequisite to facilitate the farmer involvement in growing tomatoes for processing. The tomato processing deserves the qualities of high total soluble solids (5-6 Brix), pH less than 4.5 (acidity not less than 0.4%), homogeneous red color with a/b color value of at least two along with smooth spotless disease-free surface, fruit weight not less than 50 g, firm flesh, and uniform ripening. (GIAE – India; August, 2016).Varietal development needs attention to combat several biotic and abiotic factors specially heat tolerance without adverse impact on fruit quality characteristics. Breeders have to take the challenge to develop tomato varieties tolerant/resistant to tomato yellow leaf curl diseases caused by whitefly-transmitted begomoviruses, bacterial/fungal wilt, early and late blight, as well as heat-tolerant varieties and other physical quality of the fruit for processing industry. (GIAE – India; August, 2016)

12.2   Transgenic Research on Quality Parameters of Tomato

If you are questioned about the literal meaning of "quality," the answer must be related to appearances, conditions, information, looks, descriptions, illustrations, colors, tastes, flavors, shapes, sizes, textures, and many more attributing parameters, all indicating the degree of excellence in a product. Quality of anything is immensely evaluated by our senses, mostly having the eye-catching potentiality; obviously it is more of an external than internal property and fully dominated by marketability when the things are considered as food. Tomato, i.e., *Solanum lycopersicum* L., represents its fruit berries in the world market as vegetables. Hence it always has to carry qualities of being fruit in the land of the farmers as well as being vegetable in the market for the sellers. The geneticists are doing their best for enhancing the quality parameter of tomato by several genome maneuvering processes, one of which is transgenic approach, and the observable result reflected as the transgenic tomato or we can call genetically engineered or genetically modified (GM) tomato. As the botanical name suggests, ripened tomato berry is rich in antioxidant lycopene, a bright red carotenoid pigment, which is well recognized as a potent anticancerous. When this knowledge was spread in public, the tomato production starts a tremendous hike worldwide. Tomato is one of the members in the Solanaceae family, and it is diploid species with a small genome; approximately 950 megabases genome (0.95 pg/1C) is assembled into 12 tomato chromosomes, and the genome was sequenced by the Tomato Genome Consortium (2012), a multinational team of scientists from 14 countries. Fresh ripened tomato berries are not only adding lots of sweetness, sourness, as well as redness in the food items and salads; its acceptance is noteworthy when used in the varying figures of processed fruit pulp like sauce, soups, or juices. Tomato can be categorized as a model horticultural crop plant for grasping useful genes into its genome. Other important criteria can be mentioned as follows: supplier of highly nutritive berry with many vitamins and fibers but low fats and calories, wider growing adaptability under differing conditions, photoperiod insensitivity, high self-fertility and homozygosity, great reproductive potential, lack of gene duplication, and relatively shorter life-span with large seed production owing to high self-fertility; moreover it is a crossing friendly plant, and via tissue culture techniques, complete plant can be regenerated from any of plant parts.

Before entering into the transgenic research regarding tomato quality, we first have to know what are the quality attributes of tomato berry. In broader sense, shelf life increment of well ripened tomato berry is the unique quality character, and at market level that attribute regulates consumers' visual and sensual demand; we can decompose that umbrella character into many other organoleptic attributes which affect it, viz., good textured with bright red color and well tasted with pleasant aroma and flavor while maintaining all the nutritional values intact when eaten either raw or cooked. Here the transgenic researches on quality parameters tomato berry are discussed.

12.3   Fruit Shelf Life (Ripening, Softening, Texture) and Seedlessness

For seed dispersal plant develops specialized organs that are called fruits, and for rapid seed dispersal fleshy fruits have evolutionary significance, as they are the integral part of diet. Control over fruit ripening and softening gets immense importance, as there is huge reduction of shelf life due to excessive softening of tomato fruits. The shelf life of fruits may be defined as the time when they remain consumable, while stored and softening, shriveling, and rotting of fruits are the key factors that determine the shelf life. Lots of attempts have been done to combat softening of tomato fruit using transgenic approaches by silencing genes that code proteins responsible for cell wall degradation and those influenced the global economy strategically. Fruit texture imposes huge influence on the storage of fruits along with shelf life which ultimately affects consumer preference. To target the small-fruit allele, Liu et al. (2003) made transgenic series of tomato plants having varying number (0 to 4) of *fw2.2* copies; and it was confirmed that a negative fruit-growth regulator is encoded by the gene *fw2.2* by them. Polygalacturonase (PG) enzyme is pectin-degrading enzyme, and it becomes abundant during fruit ripening stage of tomato and has a major role as cell wall hydrolase as well as softening of fruit (Sheehy et al. 1988). Bird et al. (1988) mentioned one *pg* gene per haploid genome of tomato, and the gene has 7 kb length interrupted by eight intervening sequences ranges between 99 and 953 bp. By transferring a transcriptional fusion construct of the putative 1.4 kb promoter and the chloramphenicol acetyltransferase (CAT) gene, they regenerated transgenic tomato plants that expressed CAT in ripe tomato fruit only, not in the other parts like root, leaves, or even in unripe fruits. Giovannoni et al. (1989) proved the key role of PG gene in tomato fruit ripening by transgenic *rin* (ripening inhibitor) fruit; by activating PG gene transcription and other functions, the transcription factor RIN acts as ripening blocker. Smith et al. (1990) transformed tomato plants with a single *pg* antisense gene designed to express anti-PG RNA constitutively. More than 99% reduction in PG enzyme expression was observed in the plants with *anti-pg* gene at homozygous state. In vivo inhibition of pectin degradation activity of PG was proved by their experiment. The antisense gene did not affect combined activity of invertase and pectinesterase, and also ethylene synthesis, accumulation of lycopene, and solubilization of polyuronide were unaffected; these are considered as the parameters of ripening apart from PG. Schuch et al. (1991) transformed tomato cultivar (cv.) Ailsa Craig with the *anti-pg*, and they noticed improvement of fruit firmness, storage life, and transport ability without damage in the PG antisense fruit along with reduced PG activity throughout ripening stage, and they also made tomato juice from those fruits which had significantly higher consistency. Unlike PG, in ripening tomato fruit, expansion is responsible for cell wall disassembly by nonhydrolytic activity. Hightower et al. (1991) expressed antifreeze protein *afa3* in leaves of tomato plants. These proteins are present in the blood of some polar fishes such as *Pseudopleuronectes americanus* (winter flounder) and confer the ability to inhibit recrystallization. Freezing of fruits cause dehydration of cellulose, and thus textural changes occurred that influence qualities of thawed fruits. Hence the work of Hightower et al. (1991) showed a way to improve post-harvest freeze/thaw quality based on the property of antifreeze protein.

To investigate the influence of simultaneous suppression of PG and expansion on fruit texture, Powell et al. (2003) transformed Ailsa Craig tomato plants and hybridized between two transgenic lines that had exhibited either lowered *LePG* or lowered *LeExp1* expression. The fruits from the F_3 plants, having both transgenic constructs at homozygous state, showed significant firmness as well as longer shelf life; the juice prepared from those tomatoes had enhanced viscosity than the non-transformed Ailsa Craig. Smith et al.(2002) reduced fruit softening by silencing β-galactosidase 4 (TBG4) gene in tomato cv. Rutgers. As Bovy et al.(2002) shown the constitutive expression of the gene encoding tomato fruit abscisic stress ripening protein-1 (*TOMASRIP*) during fruit ripening; this gene has been used as an internal reference in many tomato transgenic researches later. Recently Zhu et al. (2014) transformed tomato cv. Ailsa Craig by *SLNAC4* RNAi construct RNA. Three different genes such as *NAM* (no apical meristem), *ATAF* (*Arabidopsis* transcription activation factor), and *CUC* (cup-shaped cotyledon) have the conserved NAC (from first letter of the three genes) domain. RNA interference repressed the NAC domain protein, SlNAC4, and as a result fruit ripening was inhibited along with reduction in ethylene synthesis. This finding suggested the positive regulatory function of the transcription factor SlNAC4 in tomato fruit ripening. Ma et al. (2014) inhibited normal fruit ripening of tomato cv. Zhongshu 6, by overexpressing *SlNAC1*. The overexpression of *SlNAC1* tomato had enhanced abscisic acid (ABA) content, but ethylene production, fruit firmness, and pericarp thickness were reduced in the transformed tomatoes resulting in the early softening of fruits. Earlier Sun et al. (2012b) investigated the role of ABA in tomato fruit (cv. Jia Bao) ripening by suppressing *SlNCED1*; this gene encodes a key enzyme for ABA biosynthesis, 9-cis-epoxycarotenoid dioxygenase (NCED). The transgenic fruit had higher ethylene content and increased accumulation of pectin during ripening; moreover those exhibited considerably fewer seeds, a firmer texture, and an enhanced shelf life, and the juice prepared from those transgenic fruits had enhanced viscosity.

The transgenic tomato story got a hike in 1994 by the company Calgene, USA when they introduced Flavr Savr (also known as CGN-89564) tomato. This transgenic tomato line contains an additional copy of the *pg* gene in antisense orientation, which encodes the PG enzyme. The antisense copy partially inhibits the endogenous *pg* gene transcription, resulting in a reduced concentration of the PG enzyme and in delayed fruit softening; but it was not accepted by the consumers due to its bad taste and its delicate nature hindered the transport. The transgenic tomato lines B, Da, and F were also commercialized during the year in the USA by the company Zeneca and Petoseed, and these lines contain at least one additional copy of the *pg* gene. Da and F lines carry the partial sense *pg* gene, while line B carries a partial antisense *pg* gene; due to the presence of the partial *pg* gene, expression of endogenous PG enzyme is suppressed at the beginning of fruit ripening. The rise in ethylene production level could be used as an earliest indicator of tomato fruit ripening as ethylene synthesis gradually reaches to the peak from early ripening to

Table 12.1 Approved commercialized delayed ripening transgenic tomato

Transgenic tomato line	Gene	Country where commercialized	Company and year of approval
1345-4	Introduction of a truncated *accs*	USA	DNA Plant Technology, 1994
8338	Introduction of *accd*	USA	Monsanto, 1995
351N	Introduction of *sam-k*	USA	Agritope, 1995
Huafan No 1	Introduction of *anti-efe*	China	Huazhong Agricultural University, 1996

Based on Gerszberg et al. (2015)

the middle of the ripening stage (Grierson and Kader 1986). Hence restriction in ethylene synthesis is one of the ideas behind the induction of delayed ripening trait in tomato. A series of delayed ripening transgenic tomato varieties were commercialized in the 1990s mostly in the USA and also in China by different companies (Table 12.1) by the introduction of genes which can reduce ethylene synthesis in tomato fruit, viz., *accs*, *accd*, *sam*, and *anti-efe*. Here you can get a brief account regarding that.

12.3.1 Transgenic Tomato 1345-4

This line contains three copies of the *accs* gene. 1-Aminocyclopropane-1-carboxylate (ACC) is a cyclopropenoid α-amino acid, and it is a precursor of ethylene. We know that ethylene promotes fruit ripening. The *accs* gene encodes the ACC synthase (ACCS) enzyme which catalyzes the synthesis of one ethylene precursor. The truncated version of the enzyme partially suppresses the activity of the natural version of the ACCS, and as a result ethylene accumulation is reduced and fruit ripening is delayed.

12.3.2 Transgenic Tomato 8334

This line has a copy of the *accd* gene that encodes the ACCD (1-aminocyclopropane-1-carboxylic acid deaminase) enzyme which is responsible for catalyzing the degradation process of one ethylene precursor. ACCD has a central role in functional association between plant-bacteria or plant-fungi (Hontzeas et al. 2004; Nascimento et al. 2014); this enzyme converts ACC into α-ketobutyrate and ammonium thus decreasing ACC levels, which ultimately decrease plant ethylene levels. A high concentration of ACC may cause plant growth reduction and finally plant may die (Nascimento et al. 2014). ACCD enzyme was first identified in the bacterium *Pseudomonas* sp. ACP and in the yeast *Hansenula saturnus* (Honma and Shimomura 1978); now the aforesaid yeast is being re-classified as *Cyberlindnera saturnus*.

Klee et al. (1991) cloned ACCD encoding gene and introduced into tomato plants that bore fruits with significant delayed ripening and showed more firmness for at least 6 weeks longer than the non-transgenic fruit. Later Klee (1993) observed significantly slow fruit ripening in the fruits which were detached from the vine in the early stage ripening of transgenic tomato line 5673 which had *accd* from soil bacteria. The fruit that remained in the plants showed rapid ripening. This finding indicates more internal ethylene in attached fruit than the detached one in the transgenic tomato fruit.

12.3.3 Transgenic Tomato 351 N

Its genome carries two or more copies of the *sam-k* gene that encodes an enzyme responsible for hydrolyzing the S-adenosylmethionine (SAM or AdoMet) and ultimately that causes removal of SAM from ethylene biosynthesis. SAM acts as a precursor in the synthesis of ethylene. Good et al. (1994) developed transgenic tomato lines by utilizing a gene from bacteriophage T3 encoding S-adenosylmethionine hydrolase (SAMase), and the resulting transgenic tomato fruit showed a decreased ethylene-synthesizing ability. To regulate SAMase gene expression, they used tomato *E8* gene promoter. E8 is a fruit ripening protein, and it is related to the enzyme catalyzing the last step of ethylene biosynthesis pathway. By using an E8 antisense gene, Peñarrubia et al. (1992a) showed that reduction of E8 protein synthesis during ripening imposes negative effect on ethylene biosynthesis in the transformed tomato plants.

12.3.4 Transgenic Tomato Huafan No 1

Huafan No 1 with long shelf life character was the first GM plant approved for commercialization in China in the year 1996. This line has at least one copy of the *anti-efe* gene conferring delayed ripening character. Yang et al. (2005) used *anti-efe* gene in the construct-specific screening of Huafan No 1. Oxidation of ACC is the last step in ethylene biosynthesis, and this step is catalyzed by the ethylene-forming enzyme (EFE). Antisense RNA of one aminocyclopropane-1-carboxylate oxidase (ACO) gene, i.e, *anti-efe* gene, silences the *aco* gene that encodes an EFE. Previously ethylene production was reduced by using pTOM13 antisense RNA in cv. Ailsa Craig displaying diminished activity of ACO (Hamilton et al. 1990) with longer shelf life but rapid softening (Murray et al. 1993). Total lycopene synthesis, overripening, and fruit spoilage were reduced in EFE-antisense tomato fruit ripened both on and off the plant (Picton et al. 1993). Meli et al. (2010) used RNAi (RNA interference) technology for the α-mannosidase (α-Man) and β-D-N-acetylhexosaminidase (β-Hex), two ripening-specific N-glycoprotein modifying enzymes, and found enhanced firmness as well as 30 days of enhanced shelf life along with reduced softening in the RNAi tomato fruits, and they did not find any other negative effect on phenotype, including yield. They also demonstrated the downregulation of genes responsible for degradation of cell wall and fruit ripening in those transgenic fruits. They also proved excessive fruit softening by overexpressing α-Man or β-Hex. Wang et al. (unpublished) modulated the tomato fruit texture by generating single-gene transgenic lines for cell wall structure-related genes and also by generating multiple gene constructs to understand biological basis of fruit ripening associated with fruit texture; they used CRISPR (Clustered Regularly Interspaced Short Palindromic Repeats) technology and GoldenBraid with RNAi technology to fulfil their purpose. Based on type IIS restriction enzymes and complex multigene construct, GoldenBraid is a standardized assembly system. Their study indicated the efficiency of the CRISPR/Cas9 system to generate DSB (double-stranded break)-induced target gene editing in tomato. Cas9 stands for CRISPR-associated protein-9 nuclease. CRISPR-Cas9 is a molecular biology tool for genome editing where we can remove, add, or alter sections of the DNA sequence. The functions of CRISPR and CRISPR-associated (*Cas*) genes are essential for defense mechanism for bacteria and archaea (Sarrion-Perdigones et al. 2011). Cao et al. (2012) identified a gene (*SlCOBRA-like*) in tomato having homology to *Arabidopsis COBRA*; the *SlCOBRA-like* gene mostly expressed in vegetative parts and premature fruits, but a considerable reduction in its expression occurs during fruit ripening stages. They overexpressed the *SlCOBRA-like* in tomato fruits that showed desirable early developed phenotypes with enhancement in fruit shelf life and firmness. By fruit-specific suppression and overexpression, they confirmed the key function of *SlCOBRA-like* in cell wall architecture and shelf life extension in tomato fruits. Previously Sharkawy et al. (2014) demonstrated the association of PslTIR1 mRNA with higher ethylene and auxin level as well as with the quick loss of fruit firmness in plum. In another experiment, Sharkawy et al. (2016) overexpressed that particular plum auxin receptor gene *PslTIR1* in tomato (cv. Ailsa Craig) and proved that PslTIR1 protein does not stimulate the autocatalytic ethylene synthesis linked to fruit ripening rather it clearly supported the positive regulatory role of auxin in regulating leaf morphology, fruit development, as well as ripening, probably in no connection to the ethylene signaling. They found shelf life reduction of *PslTIR1* tomato fruit measuring some of the associated parameters such as weight loss, penetration strength, and firmness during fruit storage. Though their finding provides a novel and effective way of manipulating fruit quality characters, it also complexes the fruit ripening regulation mechanism.

The cell wall invertase enzyme, *Lycopersicum* Invertase5 (LIN5), is exclusively expressed in flower (ovary, petal, and stamen) and in young fruit; it is a key determinant of the total soluble solid, or Brix value, content in tomato (Zanor et al. 2009). Brix value is an index of the total soluble solids content in the fruit juice. With silencing the expression of LIN5 in tomato cv., Moneymaker Zanor et al. (2009) demonstrated the impact of sugar on fruit hormone and fruit development and the observed reduction in fruit size and seed numbers, accompanied by an increased fruit abortion rate. Another gene silencing approach was carried out by Phan et al. (2007) to establish the role pectin methylesterase in strengthening tomato fruit cell wall; tomato cv. Craigella-Tm-2a was transformed by antisense *Pmeu1*, and the transformed green fruit showed reduced pectin esterase activity along with an enhanced rate of softening during ripening.

12.4 Parthenocarpy

Seedlessness is a desirable trait for processing tomatoes as the tomato processing industry while making tomato paste or puree or ketchup is discarding seeds. Induction of parthenocarpy by ovary-specific expression of the *rolB* gene was accomplished in tomato transgenic; significant increase in jelly fill in the locules and Brix value were observed in the *rolB*-transformed tomato fruits (Carmi et al. 2003). Tomato cv. UC 82 is a popular variety in processing industry, and Rotino et al. (2005) made transgenic UC 82 for the gene *DefH9-RI-iaaM*. This gene produced parthenocarpic fruits, which were with little or no seeds. The presence of the transgene *DefH9-RI-iaaM* did not affect the processing quality of the tomato fruits as revealed by biochemical analysis. Moreover those GM tomatoes possessed higher β-carotene level without altering other technological (color, firmness, dry matter, Brix value, pH) as well as chemical (titratable acidity, organic acids, lycopene, tomatine, total polyphenols, and antioxidant capacity) properties important for processing tomato. Although the fruit number was more in those GM plants but the fruit weight was reduced. Thus the work of Rotino et al. (2005) introduced a strategy to mitigate environmental dispersal of GM seeds by introducing parthenocarpy in *DefH9-Ri-iaaM* GM tomato. Wang et al. (2005) supported the hypothesis of action of IAA9 protein as a repressor in the auxin response pathway by down-regulating *IAA9* in tomato cv. MicroTom and Ailsa Craig. The *IAA9*-inhibited lines bore parthenocarpic fruits. Goetz et al. (2007) overexpressed auxin response factor 8 (ARF8) by transforming tomato cv. Monalbo with *Atarf8-4* from *Arabidopsis thaliana* and found parthenocarpic fruit having increased number and size. On the other hand, silencing of another ARF *Sl-IAA27*, in tomato cv. MicroTom, resulted in higher auxin sensitivity even though *Sl-IAA27* is structural homolog to *Sl-IAA9* (Bassa et al. 2012). RNAi *SlARF7* lines of tomato cv. Moneymaker developed parthenocarpic fruit indicating modifying action of SlARF7 for auxin as well as gibberellin (GA) responses throughout tomato fruit development; the fruits developed by the RNAi *SlARF7* tomato lines displayed similarities with GA-induced fruits as revealed by morphological and molecular analyses (de Jong et al. 2011). Few years ago Schijlen et al. (2007) produced extremely small and parthenocarpic fruits in chalcone synthase (*CHS*) RNAi tomato plants.

12.5 Pigments

12.5.1 Flavonoids

The anthocyanins are the most important pigments in the flavonoids. Anthocyanins are hydrophilic pigments, and its presence is reflected on the red, blue, and purple colors of many fruits and flowers; undoubtedly anthocyanins play an important role for fruit setting as well as seed dispersal by attracting pollinators and seed dispersers. This natural antioxidant synthesis may be induced when plants face stresses. Human gets protection against cardiovascular diseases, cancers, and other chronic disorders by consuming dietary anthocyanin (Tsuda et al. 2003; Wang and Stoner 2008). Tomato peels contain small amounts of other flavonoids, such as flavonols, and risk for cardiovascular cancer and other age-related diseases could be lessened by high intake of flavonoid; flavonoids contain antioxidant as well as anti-inflammatory properties. Few flavonoid classes are found only in few plant species specifically (Schijlen et al. 2006). Butelli et al. (2008) expressed snapdragon (*Antirrhinum majus*), *Delila* (*Del*), and *Rosea1* (*Ros1*) transcription factors into the fruits of a MicroTom, and the transgenic tomato showed enhanced content of the anthocyanin with deep purple color in the pericarp. They increased the life-span of transgenic mice (having natural tendency to develop tumor) by feeding those transgenic tomatoes. Maligeppagol et al. (2013) achieved high (70–100 fold) anthocyanin accumulation by fruit-specific expression of snapdragon *Delila* and *Rosea1* in tomato. Zhang et al. (2013) identified the specific effects of anthocyanins on doubling the shelf life by using virus-induced gene silencing (VIGS) to silence the dihydroflavonol 4-reductase (*SlDFR*), a gene having pivotal function in anthocyanin biosynthesis. Orange sectors of *VIGS-SlDFR*-silenced fruits exhibited similar expression levels like *Del* and *Ros1* to non-silenced, purple sectors. Bassolino et al. (2013) showed accumulation of strong, uniform anthocyanins over tomato fruit peel along with an extended shelf life due to delayed overripening in *Aft/Aft atv/atv* mutant and to induce high anthocyanin production; the mutants were grown with supplemented light as both purple and red regions may be produced on the fruit skin depending on exposure to the light; this anthocyanin accumulation in the skin conferred resistance against the major pathogen *Botrytis cinerea*. The red with low anthocyanin regions showed normal overripening and faster softening than that of the purple regions on the same fruit. The *Aft* (*Anthocyanin fruit*) is a dominant gene, while *atv* (*atroviolacea*) stands for a recessive gene, and *Aft/Aft atv/atv* tomato plants develop extremely purple pigmented fruits (Mes et al. 2008). By expressing *Ros1* together with *Del* in cv. MicroTom, lower level of anthocyanin in the fruit flesh was obtained than that of Butelli et al. (2008).

Earlier Schreiber et al. (2012) constitutively overexpressed *ANT1* gene isolated from wild tomato species *S. chilense*, in cv. Moneymaker background and found the Anthocyanin fruit (AFT) genotypes bearing purple fruits; a MYB transcription factor involved in anthocyanin accumulation is encoded by *ANT1* in *S. chilense*. Transformation with chalcone isomerase encoding gene *chi-a* from *Petunia* showed significantly upregulated flavonol in peel of transformed tomato fruits (Muir et al. 2001; Verhoeyen et al. 2002) and that 65% of flavonols remained in the paste prepared from those tomatoes (Muir et al. 2001). The equal enhancement of flavonol was derived by Colliver et al. (2002) also by transgenic approach, and they mentioned the synergistic action of chalcone synthase and flavonol synthase regarding noticeable upregulation of flavonol biosynthesis in tomato flesh tissues.

Overexpression of the maize regulatory genes *Lc* and *C1* in tomato fruit flesh exhibited synthesis of 20-fold higher flavonol in ripe transgenic tomatoes; this was mainly for the kaempferol glycosides accumulation (Bovy et al. 2002;

Le Gall et al. 2003). Schijlen et al. (2006) isolated five structural flavonoid genes (conferring "novel flavonoid" trait) from different plant sources and transformed the tomato cv. Moneymaker with them. Three of those genes, encoding stilbene synthase (*STS*), chalcone reductase (*CHR*), and flavones synthase (*FNS*), were isolated from grape (*Vitis vinifera* cv. Lavelee), *Medicago sativa*, and *Gerbera hybrida* ,respectively, and the remaining two genes, viz., chalcone synthase (*CHS*) and chalcone isomerase (*CHI A*), were taken from *Petunia hybrida*. Except *STS*, all transgenes inherited with Mendelian pattern and the *STS*-overexpressing tomatoes were seedless with higher stilbene levels. Few *CHS/CHR*- and *CHI/FNS*-overexpressing lines produced pink-red fruits indicating a reduction of yellow-colored flavonoid naringenin chalcone due to increased flux toward deoxychalcones and toward flavones and flavonols, respectively. The transformed tomato fruits exhibited a high level of stilbenes (resveratrol and piceid), flavones (luteolin-7-glucoside and luteolin aglycon), flavonols (quercetin glycosides and kaempferol glycosides), and deoxychalcones (butein and isoliquiritigenin) as revealed by biochemical analysis of peel. In the previous year, Giovinazzo et al. (2005) constitutively expressed the grape *STS* gene in the same tomato cultivar synthesizing fivefold higher stilbenes. The main dietary source of stilbenes is the red wine.

By overexpressing *CHI*, Verhoeyen et al. (2002) developed pink dull tomato fruit. Schijlen et al. (2007) also got similar pink phenotype by RNAi-mediated downregulation of *CHS*. In both the studies, the reduction in enzyme that generates the yellow flavonoid pigment naringenin chalcone (NarCh) was reduced. During fruit development the expression pattern of two genes *SlCHS1* and *SlCHS2* in peel correlates with the expression of other genes associated with flavonoid pathway and NarCh accumulation (Schijlen et al. 2007; Mintz-Oron et al. 2008).

The peel of the *y* mutant (colorless-peel trait) tomato fruit lacks NarCh, and the *y* mutant displays a pink and less glossy fruits during late orange and red stages of fruit development. The flavonoid-related transcription factor SlMYB12 has a key role in tomato peel flavonoid accumulation, and peel-associated expression of *SlMYB12* declines from immature green stage toward the red ripe stage of tomato fruit development. Adato et al. (2009) confirmed the position *SlMYB12* in the *y* mutation locus by downregulating *SlMYB12* in Ailsa Craig *y* mutant along with the recovery of the *y* phenotype overexpression of *SlMYB12*. In another study, accumulation of NarCh was reduced by using VIGS of *SlMYB12* resulting in pink-colored tomato (Ballester et al. 2010).

12.5.2 Carotenoids

The color change in green tomato occurs due to chlorophyll degradation and carotenoid, viz., lycopene (red fruit) and β-carotene (orange fruit) accumulation. Bird et al. (1991) developed yellow ripening tomato fruit by using pTOM5 antisense RNA and proved the role of pTOM5 in the production of phytoene synthase (PSY) which is an important enzyme in carotenoid biosynthesis. No carotenoid accumulation was observed by downregulation of *PSY1* gene (Bartley et al.1992; Fray and Grierson 1993). The yellow peel color of cultivated tomato occurs due to naringenin, the end product of the anthocyanin biosynthesis pathway. Fraser et al. (2002) elevated the lycopene in tomato fruit (cv. Ailsa Craig) through engineering the carotenoid biosynthesis using the fruit-specific expression of a bacterial PSY (*crt*B from *Erwinia uredovora*). The tomato regulatory gene *DET1* was suppressed by RNAi approach resulting in "high pigment" fruit phenotype due to enhancement in carotenoid and flavonoid content (Davuluri et al. 2005). Later Fraser et al. (2007) developed *Psy-1* transgenic tomato plant that bore yellow or pink fruit at the mature green stage of fruit ripening; this finding also suggested pigmentation change associated with *Psy-1* overexpression and the ripening process are not connected. Neily et al. (2011) increased accumulation of polyamine (PA) content by overexpressing the apple spermidine synthase (SPDS) gene (*Md-SPDS1*) in tomato cv. MicroTom; SPDS is a key enzyme in PA biosynthesis. Transcriptional analysis showed the enhancement of carotenoid as well as ethylene production in the transformed tomato fruits as revealed by their study. Their work specifically exhibited higher accumulation of PA in the tomato resulting significant upregulation in PSY in ripe transgenic fruits and finally an enhanced lycopene accumulation in the fruit. Overexpression of *CrtR-b2* (carotene β-hydroxylase 2) in tomato cv. Red Setter developed yellow fruits instead of green due to absence of chlorophyll *a*, and the transformed ripe fruits possessed free violaxanthin and significant quantity of esterified xanthophylls (D'Ambrosio et al. 2011). Earlier lycopene in tomato fruits was fully converted into β-carotene by D'Ambrosio et al. (2004) by transforming tomato plants with *lycopene β-cyclase* (*tlcy-b*). Although Giorio et al. (2008) increased significantly higher xanthophylls content in transformed ripe fruits by overexpressing lycopene β-cyclase (*tlcy-b*) alone rather than co-overexpression of both carotene β-hydroxylase 1 (*CrtR-b1*) and lycopene β-cyclase (*tlcy-b*). Sun et al. (2012a) increased both β-carotene and lycopene in tomato cv. Jia Bao by RNAi-mediated fruit-specific suppression of *SlNCED1*. Using the *SlNAC4*-RNAi in tomato cv. Ailsa Craig, Zhu et al. (2014) reduced the expression of the transcription factor SINAC4 resulting suppression in chlorophyll degradation along with reduction in carotenoid accumulation in the transformed tomato fruit. The tomato cv. Zhongshu 6 was transformed for overexpression of *SINAC1*, and the transformed tomato had the reduced accumulation of total carotenoid and lycopene (Ma et al. 2014).

12.5.3   Folate

Folate deficiency is related to many human birth defects including spina bifida as well as cardiovascular disease; we have to intake folate from dietary supplement due to our inability to synthesize it. Pteridine, p-aminobenzoate (PABA), and glutamate precursors are responsible for folate synthesis (de la Garza et al. 2004, 2007), which produced transgenic tomato fruits with increased folate level due to enhanced production of pteridine and PABA in those transgenic fruits. Later

Waller et al. (2010) increased the folate levels by overexpressing two foreign genes, GTP cyclohydrolase I (GCHI) from mouse and aminodeoxychorismate synthase (ADCS) from *Arabidopsis*; those genes showed association with folate metabolism pathway in tomato.

12.5.4   Flavor and Taste

We judge fruit flavor by combined action of senses of taste and sense of smell, i.e., olfaction, and here orthonasal olfaction (sniffing) and retronasal olfaction (flavor emanating from the oral cavity during eating) play a crucial role by which we can decide whether the food is safe to intake (Small et al. 2004). A complex mixture of volatile and nonvolatile compounds and lipids, amino acids, and carotenoids contributes to the fruit aroma and taste of tomato, and terpenoids are included in volatile precursors. Terpenoid flavor volatiles are derived from carotenoids, and they can strongly influence the food flavor although their availability is generally relatively low (Baldwin et al. 2000; Simkin et al. 2004).

12.5.5   Volatiles

Earlier Speirs et al. (1998) tried to suppress the expression of a tomato alcohol dehydrogenase (ADH) by transforming tomato plant cv. Ailsa Craig with *ADH2* in a sense orientation in constitutive and fruit-specific manner; this ADH 2 has a key role in the interconversion of the aldehyde and alcohol forms of flavor volatiles (Sieso et al. 1976). The transformed ripening fruits showed variable range of ADH activity in case of constitutive expression, but increased level of ADH activity was observed in fruit-specific expression. This indicated a balance between modified level of ADH and some of the flavor generating aldehydes and alcohol volatiles. More intense flavor in the transformed ripe fruit showed enhanced ADH activity and alcohols (Speirs et al. 1998). In ripening tomato fruit, *LeCCD1A and LeCCD1B* encode carotenoid cleavage dioxygenase 1, LeCCD1A, and LeCCD1B, and the latter is highly expressed in ripening fruit; these enzymes contribute to the formation of terpenoid volatiles such as β-ionone and geranylacetone and pseudoionone. The overexpression of antisense RNA for *LeCCD1B* in tomato cv. M82 led to significantly reduced LeCCD1A and LeCCD1B in fruits resulting in huge reduction of aforesaid volatiles (Simkin et al. 2004). MicroTom cultivar develops cherry-type tomatoes, but Moneymaker develops larger globe-type tomatoes. Orzaez et al. (2009) obtained globe-type purple tomatoes in F$_6$ generation from the crossing between *Del/Ros1* MicroTom and non-transformed Moneymaker plants; the transgenic phenotype of *Del/Ros1* fruit is easily scorable. To study the metabolic profiles during tomato fruit ripening, they utilized VIGS as a visual reporter system, and they were able to restore the original red-fruited phenotype by preventing *Del/Ros1* module through VIGS. C6 aldehyde volatile synthesis was reduced by silenced *TomloxC* gene, whereas silencing of gene *phytoene desaturase* reduced two key lycopene-degraded products, viz., geranial and 6-methyl-5-hepten-2-one. Moreover silencing of *SlODO1* gene resulted in increased benzaldehyde accumulation but reduced levels of methyl salicylate and guaiacol in the tomato fruit pericarp; SlODO1 is a tomato homolog of the petunia MYB factor ODORANT1, a positive regulator of benzenoid volatile of flowers. Teiman et al. (2012) used *LoxC*-antisense RNA in tomato cv. M82 to suppress the 13-lipoxygenase which plays an important role in the fatty acid-derived volatile C6 aldehyde and alcohol flavor compounds synthesis pathways; C6 is the most abundant class of volatiles in the fruit. The transformed fruits showed a drastic reduction of C6 volatiles although consumer liked those fruits as the sugar, and acid content was unaffected; this indicated that aroma volatiles and sugar make separate and independent contributions to perceived sweetness. Previously Chen (2004) reduced lipoxygenase TomloxC by using sense and antisense gene construct in tomato cv. Ailsa Craig, and thus they proved the important role of TomloxC in the synthesis of fatty acid-derived flavor compounds. Volatile esters (especially acetate esters) are also important in contributing aroma of many fruits and flowers. Acetate esters are abundant in green tomatos than the red ones, and it has been observed that a negative correlation of acetate esters to human preference of tomato fruits is still persisting; the esterase gene *SlCXE1*, is highly expressed during ripening of the tomato fruit (Goulet et al. 2012). Tomato cv. Flora-Dade was transformed with reduced expression of *SlCXE1*, and the silenced lines possessed high content of acetate esters (Goulet et al. 2012).

12.5.6   Nonvolatiles

The balance between nonvolatile metabolites, i.e., sugars and acidic compounds, has important role for flavor. In tomato fruit citrate and malate are the most abundant acidic metabolites. A wild tomato species *S. pennellii* is deficient in aconitase protein due to the presence of *aco-1* mutant allele, and this species exhibits enhanced citrate and malate content in ripe fruit (Carrari et al. 2003). There lies a robust correlation between the main organic acidic metabolites, viz., carboxylic acids, glutamic acid, and aspartic acid, with tomato fruit flavor (Carli et al. 2009). The expression of two tomato aconitase encoding genes, *SlAco3a* and *SlAco3b*, was silenced by RNAi approach in cv. Moneymaker, and enhanced citrate and malate content was observed in the transgenic ripe fruit, and consequently increase in total carboxylic acid content was observed (Morgan et al. 2013). We intake Vitamin C (L-ascorbic acid or AsA) from fruits and vegetables, but we can't synthesize it due to the presence of a nonfunctional mutation, in one of the key steps of ascorbate biosynthesis, in our genome. Citrus, green chili, aonla, along with tomato, etc. are rich source of AsA as felt from their sour taste. As an antioxidant, ascorbate plays a pivotal role against reactive oxygen species (ROS) produced during physiological processes and biotic and abiotic stresses. The expression of four key ascorbate biosynthesis and recycling enzymes, viz., *SlGME* (encoding GDP-d-mannose-3'5'-epimerase), *SlGalLDH* (encoding L-galactono-1,4-lactone dehydrogenase), *SlAO* (ascorbate oxidase), and *SlMDHAR* (monodehydroascorbate reductase), were reduced in a cherry tomato cv. West Virginia 106 (WVa106) (Garcia et al. 2009); in the *SlGME* RNAi-silenced transgenic lines, ascorbate content was drastically reduced, but in *SlGalLDH* RNAi-silenced line, a little enhancement in total ascorbate content was noticed contrary to the RNAi of *SlGalLDH* by Alhagdow et al. (2007). RNAi-mediated silencing of two tomato genes *SlGME1* and *SlGME2* in cv. WVa106 was conducted by Gilbert et al. (2009). Reduction in ascorbate content leading to ROS accumulation along with huge reduction of fruit firmness was observed in the transformed fruits along with growth abnormality affecting cell division as well as cell expansion. In another study Zhang et al. (2011) overexpressed *SlGME1* and *SlGME2* in cv. Ailsa Craig and observed transgenic lines with significant increase in total AsA in ripe fruits and increased ability of scavenging ROS leading to cold as well as salt stress tolerance. The apoplastic enzyme ascorbate oxidase catalyzes the oxidation of ascorbic acid and controls the cellular ascorbate redox state. Garchery et al. (2013) targeted this enzyme; therefore, they modified tomato cv. WVa106 by using RNAi and decreased ascorbate oxidase activity. The transgenic plants exhibited increased stomatal conductance and enhanced fruit hexose content. Moreover Garchery et al. (2013) established a negative correlation between ascorbate oxidase activity and fruit yield of tomato grown under water deficit condition.

12.5.7 Sweetness

The plant species *Thaumatococcus daniellii* (Benn.) Benth, mainly native to Africa, is renowned for having an intensely sweet protein thaumatin in its fleshy fruit which possesses a specific aftertaste as determined by sensory evaluation, and this thaumatin is also used to develop sweetener. The tomato cv. Beta, carrying the non-ripening (*nor*) mutation, was modified by Bartoszewski et al. (2003) for the trait fruit test by developing transgenic lines that carried thaumatin II cDNA; and those GM tomato produced biologically active thaumatin possessing the same aftertaste as that of African plan species. Monellin is another naturally occurring protein, approximately 10^5 times sweeter than sugar, derived from berries of *Dioscoreophyllum cumminsii*; the plant species is native to West Africa (Peñarrubia et al. 1992b; Reddy et al. 2015). It could be a great alternative to sugar in food for the diabetic people. Peñarrubia et al. (1992b) gave an alternative strategy for enhancing tomato fruit flavor and quality by transferring monellin gene, and transformed tomato plants exhibited significant level of monellin accumulation in fruit. In a study conducted by Reddy et al. (2015), tomato cv. Pusa Ruby was transformed by monellin gene, and enhanced expression of monellin was noticed in the transgenic lines leading to improvement in fruit flavor (sweetness) as well as quality (proteinaceous nature).

Another alternative sweetener miraculin was expressed in tomato cv. Moneymaker (Sun et al. 2007). It is a taste-modifying protein derived from the miracle fruit, red berries of a West African shrub *Richadella dulcifica*. It is not sweet in taste, but it has the ability to modify a sour taste into a sweet one; sour fruits taste sweet if consumed together with this protein. Hence miraculin could be potentially used for diabetic and dietetic people as an alternative low-calorie sweetener and also as a new dietary seasoning. Kim et al. (2010) demonstrated the intense effects, on accumulation of miraculin in transgenic tomato fruits, of gene dosage and genetic background by transforming cv. Moneymaker with a single copy of miraculin gene. The fruit-ripening-specific accumulation of miraculin in tomato fruit cv. MicroTom was carried out by Hiwasa-Tanase et al. (2011) as well as by Hirai et al. (2011).

12.6   Transgenic Research on Abiotic Stress in Tomato

12.6.1   Abiotic Stresses Affecting Tomato Cultivation

12.6.1.1   Drought

Global agricultural production and food security are now under threat owing to the ever-increasing world population, the persistent decline of arable land, and the insufficiency of irrigation water along with the mounting environmental stress. Drought and extreme salinity play the key role in continuous decline of yield potential in major crops. Expression of complex qualitative and quantitative attributes via multiple gene action is observed when the plants are exposed to drought stress. Such changes in morphological traits are associated with an array of alterations at biochemical and physiological level, for instance, the accumulation of compatible solutes or osmolytes (Cortina and Culiáñez-Macià 2005). Deficit of fresh water is a major menace, which has to be faced by the plants when they are exposed to drought or osmotic stress. Such threatening situations can be resulted due to imbalanced transpiration rate in comparison with the rate of water uptake, irregular and low precipitation, and depleted water-holding capability of soil. Additionally, deviations in water potential gradients between soil and plant resulting in loss of cell turgor pressure, alteration of cell volume and protein conformation, and changes in cellular membrane integrity are inferred by drought stress. To combat with such unfavorable circumstances, the plant elicits its cellular and metabolic activities in the form of restriction in photosynthesis, stomata conductance, cell wall and protein synthesis, and eventually cell enlargement (Marco et al. 2015). Prompt biosynthesis of the phytohormone abscisic acid (ABA) and production of compatible solutes by plants along with expression of genes involved in osmotic adjustment and protection/repair of cellular structures are identified as adaptation mechanisms against drought (Bartels and Sunkar 2005; Mahajan and Tuteja 2005; Bhatnagar- Mathur et al. 2008; Janska et al. 2009).

12.6.1.2 Salinity

Virtually, all the traits of the physiological and biochemical of plants are affected by salinity that drastically decreases the yield. To develop salinity-tolerant genotypes by the plant breeders, inordinate efforts have been given to unlock the physiological facets of salinity tolerance of plants because saline soils and saline waters are abundant round the world. Yet, only a fewer number of salinity-tolerant cultivars have been developed despite such prodigious attempts. To utilize the saline soils and saline waters, which are presently unusable, some path-breaking advanced efforts to unravel the salinity-tolerant mechanism are required. While quite a few reasonably salt-tolerant tomato cultivars are there, it has proved challenging to improve elite lines with genes from wild species that confer tolerance owing to the involvement of the multiple number of genes, majority of them have little influence compared to the environment, along with the expense of retrieving the genetic backdrop of the receptor cultivar. Classical breeding approaches to develop exclusive salt-tolerant genotypes exploiting wild species as donors become incompetent during the selection phases. Development of molecular markers strongly associated with the genes prevailing salt tolerance could facilitate the selection of favorable alleles from segregating populations and ultimately integrated into salt-tolerant cultivars: environment can't affect molecular markers (Cuartero et al. 2006).

12.6.1.3 Temperature Stress

Extreme high- or low-temperature stress is usually encountered by plants that grow in tropical or temperate/frigid zones, respectively. Chilling (<20 °C) or freezing (<0 °C) temperatures can tempt formation of ice inside plant cells and thus results in scarcity of water in cells. Additionally, several key enzymatic actions are exceedingly temperature dependent; low temperatures disturb majority of biochemical reactions involving photosynthesis. Even, the fluidity of cell membrane also gets altered. Several responses elicited by chilling stress comprise production of additional energy by induction of primary metabolism and stimulation of molecular chaperones that results alleviation of proteins against freeze-induced denaturation (Chinnusamy et al. 2007; Theocharis et al. 2012).

12.6.2 Transgenic Approach to Mitigate the Abiotic Stress in Tomato

Since tomato is a major crop and shares a large contribution to the global food production, it's an utmost requirement to develop new cultivars having enhanced abiotic stress-tolerant traits. In the present discourse, we have highlighted the key achievements in involvement of transgenic research in tomato with the aim to develop cultivars that can thrive against extreme temperature, salinity, drought, or oxidative stress. One of the classical examples of engineering the tomato plant against a combined stress condition (comprising cold, drought and salinity) is the overexpression of the *mt1D* gene. Transgenic tomato plants were categorized by lessened electrolyte leakage and concurrent rise in lipid peroxidation that was not in the case of wild-type ones. In addition, antioxidant enzymes like superoxide dismutase and catalase activities were significantly increased. "The mannitol-1-phosphate dehydrogenase (mt1D) is in fact responsible for synthesis of Mannitol (a six-carbon, noncyclic sugar-alcohol playing a role in the coenzyme adjustment, free-radical scavenging, storage of energy, and osmoregulation) from fructose in plants" (Gerszberg and HnatuszkoKonka 2017).

12.6.3 Transgenic Approach to Solve Salinity Stress

The demand of transgenic approach to address the crop loss from salinity was triggered when the attempts to develop salt-tolerant cultivars through conventional breeding and in vitro biotechnology (for instance, in vitro selection; Flowers 2004) was found to be inadequate. High-yielding cultivars of elite quality with disease resistance capability was introduced by the plant breeders over the course of time to meet the demands of fruit and vegetable markets as well as food processing industry (Grandillo et al. 1999). It is quite obvious that the newly developed cultivars tolerant to salinity must have some enhance quality attributes and high productivity apart from being salt tolerant. The insertion of genes transporting salt tolerance property to the selected cultivars or parents of existing hybrids, by transformation, is a lucrative solution since, theoretically, susceptible yet high-yielding cultivars must have to be improved to tolerant cultivars, though sustaining each of the valued traits that present cultivars have. Interestingly, the tomato, as a crop, is recommended as an ideal to assay the prospects of marker-assisted selection as well as insertion of alien genes through transformation, as its genetics are well identified in comparison with the other dicotyledons plus multiple transformation approaches can be attempted on this crop (Cuartero et al. 2006). Since 1993 till date, plentiful research articles have been published, and a huge number of scientists have reported the development of salt tolerance either via overexpression of endogenous genes or often, via expression of genes that allegedly express during the course of tolerance (Flowers 2004). As a whole, the acquired outcomes recommend that the expression of distinctive genes in transgenic plants up to a certain extent can intensify salinity tolerance. Conversely, it is unlikely to resolve at the moment that real salt-tolerant cultivars (i.e., with an adequate tolerance based on an agronomic standpoint) have been developed through gene transfer. As Flowers (2004) correctly pointed out, it would be best to escape unwarranted enthusiasm while concluding the decisions on the existing status of this issue. Besides, it would be sensible to consider certain issues, for example, the species involved during transformation, the process for assessing the salinity tolerance, and the intricacies of the trait.

A copious number of reports on transformation pointed at raising the salt tolerance of major grain crops but less for tomato (Arrillaga et al. 1998; Gisbert et al. 2000; Moghaieb et al. 2000; Rus et al. 2001; Zhang and Blumwald 2001; Pineda

2005). Selected scientists inferred that overexpression of two genes, namely, *GlyI* (glyoxalase I) and *GlyII* (glyoxalase II), could outstandingly have enhanced salinity tolerance of tomato, implying the importance of glutathione, indirectly. Mutually decrease in lipid peroxidation and formation of H_2O_2 was detected in transgenic tomato lines exposed to high NaCl level (800 mM). Additionally, non-transformed plants were classified with substantial reduction in chlorophyll a plus b content in contrast to the transgenic lines (Álvarez-Viveros et al. 2013). The glyoxalases are enzymes accountable for detoxification of methylglyoxal and other reactive aldehydes that develop through metabolism. The detoxification activity entails of biphasic feat of glyoxalase I and glyoxalase II exploiting glutathione as a catalytic cofactor (Mustafiz et al. 2010). Remarkable reports from Herbette et al. (2011) revealed that overexpression of glutathione peroxidase (GPx, enzyme that exploits glutathione as a substrate) developed tomato plants extra tolerant to mechanical abiotic stress and low resistant to biotic stress. Wang et al. (2014) offered a sharp substantiation of association within unsaturated fatty acids and salinity tolerance. They developed transgenic tomato by "overexpressing sense and antisense sequences *LeFAD3* –encoding omega-3 fatty acid desaturase that plays an important role in the regulation of the membrane lipid unsaturation." As it transforms 18:2 linoleic acid to 18:3 linolenic acid and presence of it retains the membrane integrated and defends the photosystem, it increases the rate of photosynthesis offering growth vigor and its associated substrates. Consequently, plants carrying the sense sequence and exhibiting superior expression of desaturase flourished and grown more robustly in contrast to the plants carrying the antisense sequence displaying limited expression. In comparison with non-transgenic plants, transgenic tomato lines aggravated the intensities of net photosynthesis, stomatal conductance, and whole plant transpiration under saline conditions following the overexpression of *NtAQP1* gene (originating from *Nicotiana tabacum*) (Sade et al. 2010). Pineda et al. (2012) suggested the modification of plants with *HAL* genes (originating from *Saccharomyces cerevisiae*) to control cation transport procedures (K+ and Na+). Gisbert et al. (2000) presented that tomato plants bearing the *HAL1* gene were categorized by greater salinity tolerance. In view of intracellular cation proportions (K^+ to Na^+), it was observed that transgenic tomato lines displayed the capacity to maintain K^+ in contrast to non-transformed plants exposed to salinity. Hence, overexpression of the yeast gene *HAL5* in tomato increases the salinity tolerance by dropping shoot Na^+ retention for an extended period. The effect of superior expression of Mn superoxide dismutase (Mn-SOD) on salt stress tolerance was examined with the help of transformed tomato plants (Wang et al. 2007). This study implied notably enhanced tolerance to both high salinity and herbicide (methyl viologen) exposure. "The expression of the *FeSOD* gene had a significant influence on changes of cell ultrastructure subcompartments of tomato leaves" (Baranova et al. 2010). Baranova et al. (2014) observed the effect of expression of the *A. thaliana FeSOD1* gene on the dark respiration rate of transgenic tomato regenerants minus salinity in addition to under salinity (chloride and sulfate). The study of Seong et al. (2007) on transgenic tomato revealed that overexpression of the *CaKR1* gene (encoding an ankyrin repeat domain zinc finger) improved resistance to biotic stress and tolerance to oxidative and salt stress.

12.6.4   Transgenic Approach to Solve Chilling Stress

In a significant study, Yu et al. (2009) confirmed that overexpression of tomato omega-3 fatty acid desaturase (*LeFAD3*) gene triggered improved tolerance of tomato plants to cold stress that was accredited to the augmented concentration of the 18:3 fatty acid that alleviated membrane injury. Similarly, overexpression of the *FAD3* gene in genetically modified tomato plants caused in boosted fruit flavor besides increased chilling stress (Dominguez et al. 2010). Further experiments on cold tolerance revealed that the overexpression of *LeFAD7* induced low-temperature tolerance in tomato. Such results could be accredited to deviations in the membrane lipid conformation in tomato (a greater level of trienoic fatty acids in contrast to the dienoic fatty acids content) (Liu et al. 2013). Likewise, it was evidently established that the raised concentration of proline and osmotin in transgenic tomato during chilling stress made the plants further tolerant to cold (Patade et al. 2013). Plants are capable to produce an array of sHSPs (small heat shock proteins) encrypted by multigene families. Kadyrzhanova et al. (1998) exhibited that transcription of the *LeHSP 17.6* gene was heat induced and sustained at an enriched level in ensuing contact to chilling temperature and therefore associated with tolerance to chilling damage. These outcomes are reliable with the findings of Sabehat et al. (1996). They described that safety of tomato following chilling damage provided by pre-storage heat exposure was associated with the stimulation of transcription of HSP17 and HSP 70 mRNAs and with translation of the HSP 17 and HSP 23 proteins that continued during ensuing storage of the fruit at cold temperature. Protection of intracellular redox homeostasis relies on oxidoreductases – glutaredoxins (GRXs). Lately, it has been stated that following the expression of *AtGRX* gene (evolving from *A. thaliana*), the genetically modified tomato plants were adjusted in a superior way to cold stress being at odds with their non-transgenic counterparts. No unfavorable phenotypic variations in growth and development of plants were detected in transgenic lines. (Hu et al. 2015). Earlier, improved tolerance to cold stress in tomatoes were reported via the overexpression of *LetAPX* (Ascorbate peroxidase) (tomato thylakoidal ascorbate peroxidase) gene and this was followed by a substantial decline in GSH and chlorophyll contents as well as APX activities, in contrast with the untransformed plants (Duan et al. 2012). Besides, the transgenic tomato lines were described by reduced MDA, hydrogen peroxide (H_2O_2) level plus ion outflow, greater maximal photochemical efficiency of PSII (Fv/Fm), and superior net photosynthetic rate (Duan et al. 2012). Such outcomes advocate that overexpression of *LetAPX* has a vital function together in lessening photoinhibition and enhancing plant resistance to chilling stress.

12.6.5   Transgenic Approach to Solve Heat Stress

An ample number of strategies were adopted to induce tolerance to heat stress in tomato. An antisense-mediated reduction of *LeFAD7* better the high-temperature tolerance of tomato plants across a high concentration of fatty acids saturation and also alleviated photoinhibition of the photosystem (PS) II (Liu et al. 2010). These outcomes propose that the rise in HT tolerance in tomato plants having antisense expression of *LeFAD7* might be improved by fatty acid flux that results in an array of physiological transformations. Goel et al. (2010) described that transgenic tomato plants having overexpression of the osmotin gene showed greater volume of proline and chlorophyll, relative water content, and leaf expansion in comparison with the control plants in drought stress. Osmotin is a stress-responsive 24-kDa protein that copiously develops in plants throughout abiotic and biotic stresses. Osmotin protein plays an essential part in osmotic control of cells by stimulating synthesis and accumulation of particular solutes into cellular units (Gerszberg and HnatuszkoKonka 2017). Earlier, Cheng et al. (2009) established that tomato plants with overexpression of *SAMDC* gene (SAM decarboxylase catalyzes the synthesis of S-adenosylmethionine, a substrate in polyamine formation) obtained from *Saccharomyces cerevisiae* exhibited superior tolerance to high temperature in contrast to control plants. Genetically transformed lines were depicted by an increased amount of polyamine accumulation (nearly 2.5 times higher than natural conditions). A wide range of plant sHSPs seemingly reveals molecular acclimatization to stress. The regulatory proteins – heat stress transcription factors (HSFs) – control the transcription of HSP encoding genes. These proteins ensue in a dormant form largely in the cytoplasm. Mishra et al. (2002) examined such distinct HSFs. With the altered expression of *HsfA1*, *HsfA2*, or *HsfB1*, their group developed transgenic tomato plants. Studies unveiled that *HsfA1* carried a remarkable part in form of a principal regulator in HSFs A2, B1, and Hsps synthesis. Additionally, in high temperature the posttranscriptional silencing of the *HsfA1* gene too instigates acute deficiencies in thermotolerance and plant growth. Despite such event, Nautiyal et al. (2005) could not detect the *MT-sHSP* in nontransgenic tomatoes at optimal or high temperatures but in genetically transformed tomato plants (carrying *MT-sHSP* gene); in high-temperature stress, anticipated level of thermotolerance was witnessed. Such comparable progressive association was documented by Mahesh et al. (2013). They induce the expression of *MasHSP24.4* gene from wild banana was in a number of tomato explants like root, shoot, and stem at an exposure of 45°C. Consequently, at their regeneration stage, the transgenic tomato lines exhibited superior development and efficiency. Wang et al. (2006) established that transgenic tomato plants undergoing overexpression of the cAPX (ascorbate peroxidase) gene had developed significant tolerance to high-temperature stress along with UV-B exposure stress. Remarkable results in tomato plants were stated by Meng et al. (2015) concerning the overproduction of LeAN2 transcription factor. The mementoes outcomes of their study evidently indicate the enhanced tolerance to high-temperature stress and an augmented concentration of anthocyanins.

12.6.6 Transgenic Approach to Solve Drought Stress

There are interesting reports on genetically transformed tomato plants having overexpression of *PtADC* (a gene from *Poncirus trifoliata*) exhibited increased tolerance for drought and dehydratation stress (Wang et al. 2011). Putrescine is produced by decarboxylation of ornithine; the reaction is induced by ornithine decarboxylase or decarboxylation of arginine in subsidiary pathway; the reaction is induced by arginine decarboxylase. Lately a correlation has been testified between polyamines and the metabolic pathways, and polyamines are found to be engaged in hormonal "cross talks" that are significant to the plant's responses against stress (Alćazar et al. 2010). Likewise, investigations on transgenic overexpression of loss-function mutants offered strong indication of the obstructing character of polyamines in plant reaction to abiotic stress. Trehalose, a disaccharide molecule ubiquitous in diverse groups of organisms (invertebrates, yeast, bacteria) treated with stress situations, is an effectual "osmoprotectant" (Cortina and Culianez-Macia 2005). It is with the virtue of genetic engineering that certain tolerance traits could be found by making alterations to the trehalose metabolism. Few pleiotropic alterations were detected in the tomato plants overexpressing the *ScTPS1* gene (encoding enzyme in trehalose synthesis), where they developed thick shoots, firm dark green leaves, asynchronized rooting system, and sprouted branches as well, besides displaying an exceptional facet regarding their ability to tolerate drought, salinity, and oxidative stresses more than the control plants. In addition, an increased level of starch and chlorophyll was detected in the leaves of genetically modified tomato plants in comparison with that of the untransformed plants (Cortina and Culianez-Macia 2005). Another category of smaller proteins with 25–27 kDa size range, namely, expansins (EXP), also exists in the walls of plant cells that are accountable for optimal expansion of cell wall throughout the plant growth initiating wall stress relaxation and unalterable expansion of cell wall (Xu et al. 2014). Li et al. (2011) reported that an excess of this expansin gene (*TaEXPB23*) stimulated tolerance toward drought in genetically modified tobacco plants. Moreover, research on transgenic overexpression of loss-function mutants provided clear evidence of the preventive role of polyamines in plant response to abiotic stress. Furthermore, few reports indicated that with the presence of *SlAREB* (expression of transcription factor encoding gene) in the leaves of the tomato plant, a combination of stress-responsive genes (*RD29B* gene, *LEA* gene, and trehalose-6-phosphate phosphatase gene) displayed upregulation in terms of their expression. According to proven reports, transgenic tomato plants with an overexpression of *SlAREB1* intended to exhibit an increased tolerance to water and salt stress (Yanez et al. 2009; Hsieh et al. 2010; Orellana et al. 2010).

12.7   Biotic Stresses Affecting Tomato Production

Tomato is grown across the world due to its edible fruits, and it ranks seventh position worldwide in production. As it is taken freshly as well as in processed condition, the production and consumption of it are constantly increasing (Gerszberg et al. 2015). The production of tomato is largely hampered by several biotic stresses including fungi, bacteria, viruses, and nematode infestation. In this section we will discuss about various biotic problems of tomato, their causal organism, and symptoms associated with those diseases. In addition to that, several transgenic researches conducted for controlling different biotic stresses are being described here.

12.7.1   Fungal Diseases of Tomato

12.7.1.1   Late Blight

Late blight is caused by oomycete fungi, *Phytophthora infestans* (Mont.) de Bary, and it is one of the most devastating fungal pathogens of tomato. The disease is facilitated under cool and humid condition, foggy weather, and rainy condition, in the foliar part as well as in the fruits. Although the optimum temperature for sporulation of this pathogen is 18–22°C (64–72°F), it can sporulate in a wide range of temperature from 3 to 26°C (37–79°F). Sporangia can germinate through germ tube at 21–26°C (70–79°F), or it can form zoospores below 18°C (65°F) (Schumann, and D'Arcy. 2000). *P. infestans* has a very short life cycle, and it can be originated from sexual as well as asexual reproduction. Additionally this pathogen is able to produce a number of virulent mutant strains due to its high evolutionary potential, and a number of resistance genes were broken by different new isolates earlier (Panthee and Chen, 2010; Nowicki et al. 2013). Since the last couple of decades, a number of genomics works were started regarding the host as well as the pathogen, and the genome of *P. infestans* was sequenced about a decade ago (Tyler et al. 2006). These genetic and sequence information were found to play significant role in developing modern strategies to combat late blight infestations (Panthee and Chen 2010).

12.7.1.2   Early Blight

Another important damaging disease of tomato is early blight, and the causal organism is *Alternaria solani* Jones and Grout, a necrophytic fungi. It is predominantly occurring in tropical and subtropical areas with frequent rainfall, heavy dew, and high humidity (Agrios 2005). Although this disease causes premature defoliation as well as yield reduction in tomato, but according to some reports, this is the most damaging to tomato causing complete defoliation under severe disease incidence. Several researchers worked on the genetics of the early blight resistance in tomato which is a quantitative trait, but none of the cultivated as well as wild species of tomato were resistant to the tested early blight pathogens (Foolad et al. 2002). To overcome this problem since the last two decades scientists are working on different genomics and transgenic approaches which will be described later in this chapter.

12.7.1.3   Septoria Leaf Spot

Septoria leaf spot (SLS) is another devastating foliar disease of tomato, and it is caused by *Septoria lycopersici* Speg belonging to the phylum Ascomycota. *S. lycopersici* does not directly attack the fruits but initiates infection from the lower leaves and gradually proceeds upward under favorable condition (Gleason and Edmunds 2006). For successful disease development, wet and humid weather conditions are congenial, and according to some reports, due to heavy rainfall, overhead irrigation, and frequent dew as well as moderate temperature, this disease cause severe defoliation of tomato (Gul et al. 2016). Recently due to severity of this disease, several studies were started for knowing the genetics as well as molecular understanding of SLS resistance in tomato through conventional as well as marker assisted selection (Poysa and Tu 1993).

12.7.1.4 Powdery Mildew

Powdery mildew caused by *Oidium lycopersicum*, *Oidium neolycopersici*, and *Leveillula taurica*, belonging to phylum Ascomycota, is another important fungal disease of tomato (Zheng et al. 2013). Although, the common powdery mildew causing fungi possess small host range, but the tomato powdery mildew comprised of a long host range. Disease is facilitated during warm and dry season, and high relative humidity was negatively correlated with the disease severity (Panthee and Chen 2010). This fungus attacks the leaf and causes lesions from green to yellow and finally brown with a light powdery appearance covering of the leaf surface causing death of the leaves (Panthee and Chen 2010). Although the disease not directly attacks the tomato fruit, due to defoliation, it may develop sunscald damage.

12.7.1.5 Fusarium Wilt

Two different diseases of tomato are caused by *Fusarium oxysporum* Schlectend leading to severe crop losses (Agrios 2005), and among them the pathogen F. oxysporum f. sp. lycopersici *F. oxysporum* f. sp. *lycopersici* W. C. Snyder & H. N. Hans is the causal organism of vascular wilt, whereas, the crown and root rot diseases of tomato are caused by *F. oxysporum* f. sp. *radicis-lycopersici* W. R. Jarvis & Shoemaker. Both of them are soil borne, and the optimum growth temperature of soil for vascular wilt infestation is 28°C, whereas the optimum condition for occurrence of crown and root rot of tomato is 18°C and low humidity (Clayton 1923; Kouki et al. 2012). The disease starts at later stages of growth especially during flowering and fruiting stages, whereas wilting starts during hot days causing death of the whole plant (Jones et al. 1991).

12.7.1.6 Verticilium Wilt

Verticillium wilt is another important fungal disease causing severe yield loss in nightshade family members including tomato. The causal organism of Verticilium wilt is a soil-borne fungi of genus *Verticilium* (*V. dahliae* and *V. albo-atrum*) belonging to the phylum Ascomycota (Fradin and Thomma 2006). The disease is progressed in relatively cool temperature, high humidity, as well as high soil moisture condition (Agrios 2005). The symptoms of this disease starts with the formation of yellow blotches on the lower leaves which rapidly turn completely yellow and subsequently wither and drop off. Although the infected plants may survive throughout the growing periods, the disease cause stunted growth and yield reduction (Gleason and Edmunds 2005).

12.7.2 Bacterial Problems of Tomato Production

12.7.2.1 Bacterial Spot

Similar to the fungal infestations, several bacterial diseases also affect tomato plants. One of the economically important bacterial disease is bacterial spot of tomato caused by *Xanthomonas campestris* pv. vesicatoria. The disease causes development of black to brown lesion on leaves; lesions may coalesce to show blighted appearance, subsequent defoliation, as well as lesion formations on fruits which finally cause yield reduction (Louws et al. 2001). It was reported that the lesion on fruit is developed slightly deeper into it, and in addition to that, severe blossom infection may be caused by this organism resulting in flower drop (Gould 2013). The disease causes devastating damage under high humidity and heavy rainfall situation (Tai et al. 1999). Researches regarding the genetics of this disease identified five bacterial races causing disease in tomato (Yang et al. 2005). Recent studies were carried out for incorporating multigenic resistance or quantitative trait resistance in tomato varieties. Transgenic technology has also been used to confer resistance against bacterial spot of tomato, and it will be discussed later in this chapter.

12.7.2.2 Bacterial Speck

Bacterial speck is another important diseases of tomato caused by a bacteria, *Pseudomonas syringae* pv. *tomato*. The disease causes dark brown necrotic specks encircled by yellow halo, and sometimes it causes twisting as well as distortion of leaves (Gould 2013). The disease mostly occurs in the younger and older leaves along with the infection in the stem and fruit. Speck lesions in green fruits are surrounded by dark green halo, and the infections are very superficial. It can cause higher crop loss (around 60%) due to profound abortion of floral bud.

12.7.2.3 Bacterial Canker

The causal organism of bacterial canker is *Clavibacter michiganensis* sub sp. *michiganensis* (CMM), and it is another serious disease of tomato that can spread rapidly creating devastating damage to tomato cultivation worldwide (Li et al. 2013). All aboveground parts of tomato are susceptible to this disease at all growth stages of plant. In tomato seedlings the disease causes small water-soaked spot on leaves resulting stunted growth and wilting and under severe condition causing death of the plant. The infected stems are split and create an open canker, and the vascular system forms reddish-brown discoloration upon this infection (Seebold 2008). *C. michiganensis* subsp. *michiganensis* is a xylem-invading bacterium which can grow at 20–30 °C having optimum growth temperature of 25 °C (Sen et al. 2015). Till now only one successful attempt has been made to combat bacterial canker of tomato by transgenic approach. An endolysin gene (lys) from bacteriophage CMP1 was transferred in tomato genome through Agrobacterium-mediated transformation. No disease incidence was documented in the stably transgene integrated and properly lys-expressed transgenic tomato plants after challenging with *C. michiganensis* subsp. *michiganensis* despite the fact that a small amount of bacteria was available in the xylem sap as well as in leaf extract of those transgenic plants (Wittmann et al. 2016). Another study documented that the inoculation of *C. michiganensis* subsp. *michiganensis* in genetically modified tomato line expressing a bacterial ACC deaminase gene showed significant delay in wilt disease development (Balaji et al. 2008). These transgenic plants were able to synthesize ethylene in reduced amount. Based on the observation on "Never ripe" (Nr) mutant that is nonfunctional in ethylene perception and the ACC deaminase expressing transgenic lines, it was speculated that the ethylene production from the host plant plays a crucial role in disease development by *C. michiganensis* subsp. *michiganensis* (Balaji et al. 2008).

12.7.3 Viral Diseases of Tomato

Viral diseases of tomato are another major concern of tomato growers across the world. Increasing trends of travel, supplying of planting materials from one place to another place as well as large-scale monocropping have made this crop vulnerable to about 136 viral diseases. Major important tomato viral diseases are *Tomato yellow leaf curl virus* (TYLCV), *Tomato spotted wilt virus* (TSWV), *Tomato chlorotic spot virus* (TCSV), *Tobacco mosaic virus (TMV)*, *Tomato mosaic virus* (ToMV), *Tomato ringspot virus* (ToRSV), *Tomato mottle virus* (TMoV), etc. TYLCV is transmitted by whiteflies (*Bemisia tabaci*), and it is one of the major damaging viral diseases of tomato. If the TYLCV virus infects the plant at early growth stages, it causes stunted growth, bushy appearance, and poor yield. Infected leaves show upward curling and crumpling, and yellowing is detected at the leaf edges and in between the leaf veins. TSWV is transmitted by thrips, and stunting growth is also a common problem associated with this viral disease. Additionally, TSWV infection causes numerous dark and small spot formations in leaf, bronze coloration of leaves, deformed and unevenly ripen fruit formations and tips die back as well as wilted appearance. TCSV is transmitted by thrips and cause necrotic or brown spots in upper leaves along with chlorotic spots and mosaic appearance. The disease can progress quickly through bronzing, wilting, as well as leaf deformation and subsequently death of the terminal leaves and stems. TMV infecting tomato does not have a definite mode of transmission, but the virus is transmitted by any means to reach the injured cells of the host plant. The characteristic symptom of TMV infection in tomato is the formation of yellow-green mottling on leaves. The flowers as well as leaves are curled, subnormal in size, and distorted in nature causing a stunted growth of the plant. TEV is primarily found in the Western Hemisphere at all stages of plant growth, and aphid transmits this virus. Infected plants produce shortened internodes and subsequently stunted growth. Leaves of the infected plants show mild mottling, reduced size, and prominent downward curling, whereas the fruit sizes become smaller.

ToMV is mostly transmitted by mechanical injury and improper handling. ToMV-infected younger plants have a characteristic green/yellow mottled appearance along with stunted growth, whereas the infection at later period especially after fruiting stage does not retard the plant growth. The infected fruit may possess necrotic brown patches on them which reduce their market value. ToRSV is caused by a soil-borne virus, and the virus is transmitted by nematode (*Xiphinema americanum*) vector in North America. The infected tomato plants show noticeable curling and necrosis of the terminal part of the actively growing shoots, whereas the basal parts of younger leaves build up brown and clearly distinct necrotic rings. The infected fruits may develop distinct gray to brown corky concentrated rings superficially. TMoV is transmitted by whitefly, and the infected plants have stunted growth and reduced yield. Infected tomato plants show chlorotic mottling on the upper leaves, whereas the middle as well as lower leaves depicts upward curling.

12.7.4   Nematode Infestation in Tomato

Root-knot nematode (*Meloidogyne incognita*) is a potential problem for more than 3000 agricultural crops that initially penetrates the root, subsequently migrates to the vascular tissue, and finally form galls or root knots (Cheng et al. 2015; Dutta et al. 2015). In the infected plants, root-knot or gall formation affects the upward translocation of water as well as nutrients in the root, and which ultimately cause reduced crop yield (Moens et al. 2009).

12.8   Transgenic Research to Mitigate Biotic Stress in Tomato

Through the advent of genomic information in different organisms as well as plants, several reverse genetics approaches have been taken for functional genomics study. Several transgenic researches were conducted by different group of researchers for establishing the function of different pathogenesis-related (PR) or pathogen-derived resistance (PDR) genes either in model organisms or in the targeted species. Different strategies of transgenic research in plant biology field are mentioned below.

12.8.1 Overexpression of Gene Constructs

For boosting the crop production under biotic as well as abiotic stress conditions in a sustainable way, either conventional breeding or transgenic technology could be employed for overexpression of PR or PDR genes. Compared to the conventional breeding, transgenic technology not only broadens the genetic resources by utilizing alien genes from diverse genera, it is less time-consuming, and the deleterious effects of the associated genes can be ignored by this approach. Through this approach either a foreign gene (transgene) or a gene from the related species (cisgene) is expressed in the target organism. The gene is expressed under a constitutive promoter (e.g., CaMV35S and other viral promoters; *nos*, *mas*, and other bacterial promoters; actin, maize ubiquitin, and other plant promoters, etc.) for the expression of the gene of interest (GOI) throughout the transgenic species, whereas the transgene can also be expressed under tissue-specific or stress-inducible promoter for serving some specific need (Benfey and Chua 1990; Banerjee et al. 2015). Since the last two decades, a number of bidirectional promoters have been reported from plant as well as other organisms, which could be efficiently used in the transgenic technology for controlling two different genes at the same time (Yang et al. 2008; Banerjee et al. 2013).

12.8.2 Gene Silencing Approach

The transcript as well as protein expression of a gene can be downregulated by several gene silencing approaches. Through antisense RNA technology, the antisense RNA of a particular RNA sequence is expressed, and it can cause base pairing with the sense RNA strand available in the system and thereby reduces the availability of targeted RNA causing retarded gene expression (Hiatt et al. 1989). Further studies demonstrated RNA interference (RNAi)-mediated gene silencing strategies, and it was a more efficient silencing approach compared to antisense technology (Kadotani et al. 2003). Through RNAi approach, the target gene or gene part is expressed in forward and reverse orientation flanking a linker DNA region to generate a hairpin RNA. With the help of dicer protein, the double-stranded RNA part of the hairpin RNA is chopped into small pieces to produce small interfering RNA (siRNA). Finally using these siRNA, RNA-induced silencing complex (RISC) is formed and that ultimately downregulates the targeted gene expression. After the discovery of another small RNA, namely, microRNA (miRNA), new modified homology-based gene silencing strategy (i.e., artificial microRNA-mediated gene silencing) has been popularly used by many researchers for deregulating the target gene expression (Tiwari et al. 2014; Galvez et al. 2014).

12.8.3 Transgenic Research on Fungal Resistance in Tomato

Since long before transgenic researches started on tomato for generating resistance against *P. infestans*. Stilbenes are belonging to small family of phenylpropanoids and considered as an important components for general defense against plant pathogens. Transgenic tomato plants generated by using two grapevine (*Vitis vinifera*) stilbene synthase genes documented that the generated transgenic lines had significant resistance to *P. infestans* compared to the non-transgenic lines due to expressional upregulation of stilbene synthase and accumulation of its product (phytoalexin trans-resveratrol) in the transgenic lines after pathogen infestation (Thomzik et al. 1997). Further studies identified a late blight resistance gene, *RGA2-blb* (also designated as *Rpi-blb1*), from wild potato species *Solanum bulbocastanum* (Vossen et al. 2003). That resistant gene was used to generate transgenic tomato lines using *Agrobacterium*-mediated gene transfer approach by taking the tomato cultivar Moneymaker. Significant number of primary transformants showed late blight disease resistance as tested for potato as well as tomato *P. infestans* isolates (Vossen et al. 2003). Although in a recent study, RNA interference (RNAi)-mediated gene silencing of *Arabidopsis defense no death1* (*DND1*) gene ortholog in tomato showed improved resistance against late blight and powdery mildew infestation; the *DND1*-silenced tomato plants showed severe dwarfism as well as autonecrosis. Independent transgenic tomato lines were generated by separately overexpressing a maize β-glucanase (M-GLU) gene and an antimicrobial peptide (*Mj*-AMP1) from *Mirabilis jalapa* through *Agrobacterium*-mediated gene transfer approach (Schaefer et al. 2005). Those transgenic lines showed improved resistance against *A. solani* infestation compared to the untransformed control plants. Overexpression of another antimicrobial peptide from *Allium cepa* (Ace-AMP1) showed fungal resistance in transgenic plants (Li et al. 2003; Roy-Barman et al. 2006). Another study documented that the overexpression of complementary DNA (cDNA) of thaumatin-like protein (TLP) from rice in transgenic tomato plants conferred resistance against *A. solani* infestation (Radhajeyalakshmi et al. 2005). *Agrobacterium*-mediated overexpression of a mannitol dehydrogenase (MTD) gene from celery into tomato (NC1 Grape) showed reduced infection to *A. solani* compared to the untransformed tomato plants (Patel et al. 2015). Loss of function of a particular gene, *Mildew resistance Locus O* (*Mlo*), was found to be responsible for resistance to powdery mildew. A couple of research papers documented that the overexpression of *S. lycopersicum* Mlo1 (*SlMlo1*) gene under 35S promoter generated powdery mildew susceptible tomato lines, whereas, the naturally occurring loss-of-function mutant (*ol-2*; 19 bp deletion in the coding DNA sequence) for SlMlo1 gene showed enhanced resistance to *L. taurica* and *O. neolycopersici* causing powdery mildew of tomato (Bai et al. 2008; Zheng et al. 2013). Further studies documented that simultaneous RNAi-mediated gene silencing of SlMlo1, SlMlo5, and SlMlo8 genes had improved resistance to powdery mildew infection compared to the resistance associated with *ol-2* (Zheng et al. 2016). Another study revealed that heterologous expression of yeast Δ-9 desaturase gene in genetically altered transgenic tomato plants enhanced resistance to powdery mildew fungus by inhibiting the fungal spore germination (Wang et al. 2000). A number of fungal species are belonging to the genus *Fusarium* and several transgenic researches were conducted by various researchers to control the infection caused by *Fusarium* sp. in tomato as well as other crops (Jongedijk et al. 1995; Shin et al. 2008; Banerjee et al. 2010). Overexpression of different chitinase genes showed improved resistance in the transgenic tomato plants (Cletus et al. 2013; Girhepuje and Shinde 2011). Jongedijk et al. (1995) documented that overexpression of class I chitinase as well as a class I β-1,3-glucanase gene simultaneously in tomato showed enhanced fungal resistance compared to constitutive expression of either one of them. Later on another group demonstrated that the heterologous expression of class I rice chitinase protein (Chi 11) in tomato showed antifungal activity against *Fusarium sp.* (Abbas et al. 2009). Further studies were conducted to express dual pathogenesis related proteins (tobacco osmotin gene as well as bean chitinase gene) in tomato for governing fusarium wilt resistance (Ouyang et al. 2005). *Agrobacterium*-mediated transformation of tomato was carried out for constitutively expressing radish defensin gene (*Rs-AFP2*) in Bulgarian tomato cultivar Topaz. The transgenic plants showed high level resistance to *F. oxysporum f. sp. lycopersici* as evaluated through detached leaf bioassay experiment (Kostov et al. 2009). Abdallah et al. (2010) raised genetically modified tomato plants by expressing *Medicago sativa* defensin gene (*MsDef1*) under caulimoviral promoter, and the transgenic plants showed fusarium wilt resistance in comparison with the untransformed control plants. Root tissue-specific expression of another defensin (*WD*) gene from wasabi (*Wasabia japonica* Matsum.) into tobacco as well as tomato plants documented significantly improved tolerance to fusarium wilt (Kong et al. 2014). An interesting study demonstrated that the integration of rolA gene from *Agrobacterium rhizogenes*, a soil bacterium into tomato genome, led to the generation of *F. oxysporum f. sp. lycopersici*-resistant/*F. oxysporum f. sp. lyco-persici*-tolerant transgenic tomato plants (Bettini et al. 2016). Overexpression of I-7 gene, a leucine-rich repeat receptor-like protein (LRR-RLP) in tomato background, showed resistance to different races of *F. oxysporum f. sp. lycopersici* (Gonzalez-Cendales et al. 2016).

Transgenic tomato (*L. esculentum*) plants were generated by overexpressing an endochitinase gene (pcht28) from *Lycopersicon chilense* under CaMV 35S promoter (Tabaeizadeh et al. 1999). Those transgenic plants showed significant tolerance to Verticillium wilt caused by different isolates of *V. dahlia*. Later on another group heterologously expressed bacterial ACC deaminase gene in transgenic tomato under different promoters (Robison et al. 2001). Interestingly the transgenic lines overexpressing ACC deaminase under CaMV35S promoter did not show significant reduction of Verticillium wilt infestation, whereas the gene expression under rolD and prb-1b showed strong resistance to *V. dahlia* infestation. An interesting study reported that SERK family members are involved in Verticillium signaling (Fradin et al. 2011). Through virus-induced gene silencing approach in tomato, they documented that SERK1 is responsible for Ve-1-mediated Verticillium resistance. Effector molecules are produced by fungal pathogens for disease establishment in plants and to combat this plant use immune receptors. An immune receptor in tomato (Ve1) is responsible for vascular wilt fungal resistance to race 1 caused by *V. dahliae* and *V. albo-atrum*. de Jonge et al. (2012) established that by transient expression of Ave1 (avirulence on Ve1 of tomato) homologs from *F. oxysporum* and *Cercospora beticola* can govern Ve-1-mediated resistance.

12.8.4 Transgenic Research for Bacterial Resistance in Tomato

Several resistance genes (*Bs1*, *Bs2*, and *Bs3*) were identified from pepper which showed resistance in a "gene for gene" manner against *X. campestris pv. vesicatoria* (*Xcv*). Tai et al. (1999) documented that the transient expression of *Bs2* in tomato showed hypersensitive response, whereas the overexpression of that gene in stable transgenic lines showed strong resistance in the leaf assay against *Xcv* inoculation compared to untransformed tomato leaves (Tai et al. 1999). Further research documented that the overexpression of a nonexpresser of PR genes (*NPR1*) from *Arabidopsis* into transgenic tomato resulted moderate level of enhanced resistance against bacterial spot and improved tolerance to a number of fungal and bacterial pathogens also (Lin et al. 2004). Another group evaluated the bacterial spot disease resistance performance of genetically altered transgenic tomato plants expressing *Bs2* gene from pepper under field condition (Horvath et al. 2012). The transgenic lines demonstrated improved marketable fruit yield compared to the non-transgenic one. Martin et al. (1993) cloned Pto kinase gene from tomato, and transformation of susceptible tomato pants with the cDNA of that gene showed resistance against *Pseudomonas syringae* pv. *tomato*. *Pto* is a disease resistance gene belonging to serine/threonine kinase family which governs race-specific resistance to *Pseudomonas syringae* pv. *tomato* having avrPto gene. Tang et al. (1999) overexpressed *Pto* gene in tomato under Cauliflower mosaic virus 35S (CaMV35S) promoter, and the genetically altered tomato plants showed defense reaction even in absence of Pto-AvrPto. Those transgenic tomato plants conferred disease resistance against *P. s. tomato* without *avrPto* as well as *X. campestris pv vesicatoria* and *Cladosporium fulvum* infestations compared to the non-transgenic one (Tang et al. 1999). Later on another group demonstrated that the overexpression of a transcription factor (*Pti5*), belonging to ethylene response element-binding protein (EREBP) family, in transgenic plants, resulted in improved resistance against *P. syringae pv. tomato* (He et al. 2001). Agrobacterium-mediated transformation of tomato genotype Urfa-2 was carried out for overexpressing the resistance gene against *Pseudomonas syringae* pv. *tomato*. The resultant transgenic plants showed resistance to bacterial speck (race 1 and 0) disease compared to the untransformed plants (Koc et al. 2007). Recently another approach was taken for conferring disease resistance against *P. syringae*. Overexpression of tyramine N-hydroxycinnamoyl transferase (THT), a key enzyme for the production of a secondary metabolite (hydroxycinnamic acid amides; HCAA) in tomato, showed enhanced resistance against *P. syringae* compared to wild-type plants (Campos et al. 2014).

12.8.5   Transgenic Research on Virus Resistance in Tomato

Disease resistance in plants is governed by an incompatible reaction involving the resistant gene (R) from host plant and corresponding avirulence (Avr) gene product from pathogen. Since long time several conventional as well as transgenic researches were conducted for viral disease resistance, and Galvez et al. (2014) nicely presented different approaches for virus resistance in their review. Here we are focusing only on various transgenic researches on viral disease resistance especially on tomato.

12.9   R Gene–Mediated Resistance

Overexpression of some dominant R gene conferred resistance against different viral infection in transgenic tomato plants. Whitham et al. (1996) generated transgenic tomato plants by overexpressing a resistant gene (*N* gene) from tobacco, and the transgenic plants showed resistance to *Tobacco mosaic virus* (TMV) infection. Transformation of susceptible tomato plants was carried out by Lanfermeijer et al. (2003) using 861 amino acid encoding Tm-2^2 gene, and the resultant plants showed resistance against ToMV infection. Another gene from tomato was responsible for resistance against ToMV. In that study it was reported that transgenic expression of Tm-1, a 80 kDa protein coding gene, conferred resistance against ToMV infection, whereas the knockdown of that protein through virus-induced gene silencing approach made the Tm-1-bearing plants sensitive to ToMV infection (Ishibashi et al. 2007). Along with the dominant R gene, some recessive resistance genes from plants demonstrated viral resistance in tomato transgenics. A recessive gene (pvr1) codes for eukaryotic translation initiation factor (eIF4E) was isolated from *Capsicum chinense* and overexpressed in transgenic tomato plants. Generated transgenic plants showed significant resistance against several TEV strains and other potyviruses (Kang et al. 2007).

12.10   PDR-Mediated Resistance

Roger Beachy and his group started working on pathogen-derived resistance (PDR) for combating viral diseases, and later on several research groups worked on PDR for viral disease resistance across plant species (Galvez et al. 2014). Roger Beachy's group expressed the gene responsible for coat protein (CP) formation from *Tobacco mosaic virus* (TMV) into transgenic tobacco, and tomato plants and the transgenic lines showed disease resistance against TMV infection (Abel et al. 1986; Nelson et al. 1988). Later on capsid protein gene (gene VI) from tomato yellow leaf curl virus (TYLCV) was overexpressed in the F1 hybrid (*Lycopersicon esculentum* X *L. pennellii*), and the resultant lines displayed delayed viral symptoms and rapid recovery from TYLCV infection (Kunik et al.

1994). Besides the overexpression of some the full-length genes from virus, Antignus et al. (2004) generated transgenic tomato plants by expressing truncated replication associated with protein gene from the mild strain of TYLCV-Israel. Transformed plants showed resistance against the mild strain of TYLCV-Israel, whereas those plants conferred susceptibility against the severe strain of TYLCV-Israel. Posttranscriptional gene silencing (PTGS)-mediated TYLCV resistance on transgenic tomato was successfully evaluated under field condition by Yang et al. (2004), and the resistant transgenic plants neither showed any viral symptoms nor any viral genomic DNA was detected from them. Very insignificant amount of researches were conducted in transgenic field for conferring resistance against TSWV. Ultzen et al. 1995 demonstrated that the overexpression of the TSWV nucleoprotein (N) gene in transgenic inbred tomato lines showed high-level resistance against TSWV infestation. In addition to that tomato hybrids generated using that N-gene expressing transgenic parental lines also conferred TSWV resistance (Ultzen et al. 1995). An interesting study revealed that certain aptamers could bind effectively to the replication protein (Rep) of different Geminiviridae members. Reyes et al. (2013) identified two peptide aptamers that could bind strongly with Rep protein, and transgenic tomato lines expressing those aptamers separately showed resistance mechanism after inoculation with Geminiviridae members TYLCV and TMoV. Although no TCSV-resistant transgenic tomato plants were reported earlier, broad-spectrum disease resistance to different tospoviruses including TSWV and TCSV were generated in transgenic tobacco plants by overexpressing different linked N genes (Prins et al. 1995). Similarly till now there is no transgenic tomato lines that were generated against ToRSV and TMoV infestation, transgenic tobacco lines were generated by expressing VPg (genome-linked viral protein) and protease (Pro) protein from a raspberry isolates, and the resultant transgenic lines conferred resistance against ToRSV.

Along with the overexpression of pathogen-derived resistance genes, several gene-silencing strategies were employed by different researchers for controlling viral diseases. Hairpin RNA-mediated gene silencing approach was also taken for controlling TYLCV infection in transgenic tomato plants (Fuentes et al. 2006). In that experiment a hairpin construct was generated by putting 726 bp of the TYLCV C1 gene in reverse and forward orientation flanking a linker (intron) DNA, and it was expressed under CaMV35S promoter and NOS terminator. Posttranscriptional gene silencing approach was taken to test the efficacy for resistance against three different tospoviruses including TSWV by using a hairpin RNA construct of L gene from *Watermelon silver mottle virus* (WSMoV). RNAi-mediated gene silenced transgenic tomato plants showed resistance against TSWV (Peng et al. 2014). An interesting study was carried out for generating resistance against tomato chlorotic mottle virus, a DNA virus, through RNAi-mediated gene silencing approach in transgenic *Nicotiana benthamiana* plants. As some of the transgenic lines showed virus susceptibility in spite of producing sufficient siRNA, it was thought that RNAi-mediated viral resistance might not be robust in controlling DNA viruses rather than RNA viruses (Ribeiro et al. 2007). Further research is needed to clarify the efficacy of RNAi-mediated viral disease resistance in DNA and RNA viruses in terms of the production of siRNA, reduction of viral genetic material, and disease resistance.

12.11   Transgenic Research on Nematode Resistance in Tomato

The *Mi* locus of tomato was found to be responsible for the resistance against M. incognita infestation. Further studies identified two genes, namely, Mi-1.1 and Mi-1.2, and a pseudogene having similarity to the disease resistance genes. *Agrobacterium*-mediated transformation documented that the Mi-1.2 gene is responsible for nematode resistance in the susceptible tomato cultivar (Milligan et al. 1998). Later on another group reported that *Bacillus thuringiensis* crystal protein (Cry6) can be ingested by *M. incognita*, and overexpression of Cry6 in tomato roots through A. rhizogenes demonstrated resistance to root-knot nematode (Li et al. 2007). Sooner another group isolated a *Meloidogyne* spp. resistance gene from *Capsicum annuum* (CaMi), and transgenic plants generated by overexpressing CaMi gene documented resistance to root-knot nematode (Chen et al. 2007). In a recent study conducted by Dutta et al. (2015), a RNAi-mediated gene silencing approach was taken for silencing a protease gene from the nematode, namely, cathepsin L cysteine proteinase *(Mi-cpl-1)*. In the transgenic lines about 60–80% reduction in nematode infestation was recorded compared to the untransformed control plants.

12.12   Conclusion

Albeit tomato is a climacteric fruit, it is widely consumed as a vegetable worldwide either raw or cooked. The quality of tomato fruit should be improved considering the flavor, taste, nutrition, color, texture, shelf life, and post-harvest processing. Since the past few decades, the genes implicated to tomato quality parameters have been extensively researched with rapid acceleration in advancement of transgenic technologies. Most of the success stories have been discussed in this text. To combat many human diseases like hepatitis B, HIV, malaria, hemophilia, plague, diarrhea, diphtheria, pulmonary tuberculosis, pertussis (whooping cough), tetanus, inflammatory pain, and even Alzheimer's disease, scientists are expressing many foreign genes in tomato fruits that have the potentiality to be recognized as medicines or edible vaccines. Hence we could say that tomato is a best-studied model fruit crop, with other model plants like *Arabidopsis* or rice, not only for its various useful features but also for its sequenced genome (The Tomato Genome Consortium 2012). Due to the negative opinion of the people toward transgenic plants and foods, all the benefits of the GM tomato fruit are being captivated on the research laboratories and publications instead of reaching into the plate of human being. We the scientific community have a crucial role to educate the people about the benefits of the transgenic including tomato, and then only the world people will be benefitted from the transgenic crops. Genetic transformation approach permits to insert defined gene simultaneously avoiding the elimination of any intrinsic genetic attributes unlike the occasion of conventional in situ or true in vitro screening. Furthermore their efficacy of time and their utility to a broad array of species are the key factors among their recognizable benefits (Gerszberg et al. 2015). As the abiotic stresses are in a manner interconnected in their nature and they generally influence plants concurrently, there are certain issues about the study on genetically stimulated tolerance as well. Therefore, it would possibly be reasonable to acquire an additional multi-target approach for mutable plant responses, in place of aiming on a particular genetic event (influencing one distinct role). Now, the engineering of genes encrypting transcription factors engaged in the control of stress-responsive genes appears to be the ideal policy. Besides the multidirectional action, the overexpression of the transcription factors associated with synchronized reforms of other target sequences. Strategies like this would possibly also tackle the criticism of the inadequate efficiency detected during an individual modification was established (Gerszberg and HnatuszkoKonka 2017). Nonetheless, no matter how effectual the genetic approaches would be, there is one disadvantage quite hard to eliminate, its genetic trait. While the transgenic crops passed through harsher safety assays and policies prior their trade in open market in contrast to non-transgenic crops, they are still under suspicion in several corners of the world causing resilient concerns from public (Eisenstein 2013; Chow et al. 2016; Smart et al. 2016). Although a number of transgenic researches were conducted by various research groups to combat a specific biotic stress or for generating broad-spectrum disease resistance in tomato, till now no successful commercial transgenic variety is available in the market. In consideration of that drawback, further research need to be addressed through on firm trial so the farmers as well as other stakeholders might be interested for commercial acceptance of better transgenic lines in the near future.

References

Abbas DE, Abdallah NA, Madkour MM (2009) Production of transgenic tomato plants with enhanced resistance against the fungal pathogen *Fusarium oxysporum*. Arab J Biotech 12(1):73–84

Abdallah NA, Shah D, Abbas D et al (2010) Stable integration and expression of a plant defensin in tomato confers resistance to fusarium wilt. GM Crops 1(5):344–350

Abel PP, Nelson RS, De B, Hoffmann N, Rogers SG, Fraley RT, Beachy RN (1986) Delay of disease development in transgenic plants that express the tobacco mosaic virus coat protein gene. Science 232:738–743

Adato A, Mandel T, Mintz-Oron S, Venger I, Levy D, Yativ M, Domınguez E, Wang Z, De Vos RC, Jetter R, Schreiber L, Heredia A, Rogachev I, Aharoni A (2009) Fruit-surface flavonoid accumulation in tomato is controlled by a *SlMYB12*-regulated transcriptional network. PLoS Genet 512:e1000777

Agrios GN (2005) Plant pathology, 5th edn. Elsevier, New York

Alcázar R, Altabella T, Marco F, Bortolotti C, Reymond M, Koncz C, Carrasco TAF (2010) Polyamines: molecules with regulatory functions in plant abiotic stress tolerance. Planta 231:1237–1249

Alhagdow M, Mounet F, Gilbert L, Nunes-Nesi A, Garcia V, Just D, Petit J, Beauvoit B, Fernie AR, Rothan C, Baldet P (2007) Silencing of the mitochondrial ascorbate synthesizing enzyme L-galactono-1,4-lactone dehydrogenase (L-GalLDH) affects plant and fruit development in tomato. Plant Physiol 145:1408–1422

Álvarez-Viveros MF, Inostroza-Blancheteau C, Timmermann T, González M, Arce-Johnson P (2013) Overexpression of GlyI and GlyII genes in transgenic tomato (Solanum lycopersicum Mill.) plants confers salt tolerance by decreasing oxidative stress. Mol Biol Rep (4):3281–3290

Antignus Y, Vunsh R, Lachman O, Pearlsman M, Maslenis L, Hananya U, Rosner A (2004) Truncated Rep gene originated from tomato yellow leaf curl virus – Israel [Mild] confers strain-specific resistance in transgenic tomato. Ann Appl Biol 144:39–44

Arrillaga I, Gil Mascarell R, Gisbert C, Sales E, Montesinos C, Serrano R, Moreno V (1998) Expression of the yeast HAL2 gene in tomato increases the in vitro salt tolerance of transgenic progenies. Plant Sci 136:219–226

Bai Y, Pavan S, Zheng Z et al (2008) Naturally occurring broad-spectrum powdery mildew resistance in a central American tomato accession is caused by loss of Mlo function. Mol Plant Microbe Interact 21:30–39

Balaji V, Mayrose M, Sherf O et al (2008) Tomato transcriptional changes in response to *Clavibacter michiganensis* subsp. *michiganensis* reveal a role for ethylene in disease development. Plant Physiol 146(4):1797–1809

Baldwin EA, Scott JW, Shewmaker CK, Schuch W (2000) Flavor trivia and tomato aroma: biochemistry and possible mechanisms for control of important aroma components. Hort Sci 35:1013–1021

Ballester AR, Molthoff J, de Vos R, te Lintel Hekkert B, Orzaez D, Fernandez-Moreno JP, Tripodi P, Grandillo S, Martin C, Heldens J, Ykema M, Granell A, Bovy A (2010) Biochemical and molecular analysis of pink tomatoes: deregulated expression of the gene encoding transcription factor SlMYB12 leads to pink tomato fruit colour. Plant Physiol 1:71–84

Banerjee J, Das N, Dey P, Maiti MK (2010) Transgenically expressed rice germin-like protein1 in tobacco causes hyper-accumulation of H_2O_2 and reinforcement of the cell wall components. Biochem Biophys Res Commun 402 (4):637–643

Banerjee J, Sahoo DK, Dey N, Houtz RL, Maiti IB, Pandey GK (2013) An intergenic region shared by *At4g35985* and *At4g35987* in *Arabidopsis thaliana* is a tissue specific and stress inducible bidirectional promoter analyzed in transgenic Arabidopsis and Tobacco plants. PLoS One 8 (11):e79622

Banerjee J, Sahoo DK, Raha S, Sarkar S, Dey N, Maiti IB (2015) A region containing an *as-1* element of Dahlia Mosaic Virus (DaMV) subgenomic transcript promoter plays a key role in green tissue- and root-specific expression in plants. Plant Mol Biol Rep 33 (3):532–556

Baranova EN, Serenko EK, Balachnina TI, Kosobruhov AA, Kurenina LV, Gulevich AA, Maisuryan AN (2010) Activity of the photosynthetic apparatus and antioxidant enzymes in leaves of transgenic Solanum lycopersicum and Nicotiana tabacum plants, with FeSOD1 gene. Russ Agr Sci 36(4):242–249

Baranova YN, Akanov EN, Gulevich AA, Kurenina LV, Danilova SA, Khaliluev MR (2014) Dark respiration rate of transgenic tomato plants expressing *FeSOD1* gene under chloride and sulfate salinity. Russ Agric Sci 40(1):14–17

Bartels D, Sunkar R (2005) Drought and salt tolerance in plants. Crit Rev Plant Sci 24:23–58

Bartley GE, Viitanen PV, Bacot KO, Scolnik PA (1992) A tomato gene expressed during fruit ripening encodes an enzyme of the carotenoid biosynthesis pathway. J Biol Chem 267:5036–5039

Bartoszewski G, Niedziela A, Szwacka M, Niemirowicz-Szczyt K (2003) Modification of tomato taste in transgenic plants carrying a thaumatin gene from *Thaumatococcus daniellii* benth. Plant Breed (4):347–351

Bassa C, Mila I, Bouzayen M, Audran-Delalande C (2012) Phenotypes associated with down-regulation of *Sl-IAA27* support functional diversity among Aux/IAA family members in tomato. Plant Cell Physiol (9):1583–1595

Bassolino L, Zhang Y, Schoonbeek HJ, Kiferle C, Perata P, Martin C (2013) Accumulation of anthocyanins in tomato skin extends shelf life. New Phytol (3):650–655

Benfey PN, Chua NH (1990) The cauliflower mosaic virus 35S promoter: combinatorial regulation of transcription in plants. Science 250:959–966

Bettini PP, Santangelo E, Baraldi R, Rapparini F, Mosconi P, Crinò P, Mauro ML (2016) *Agrobacterium rhizogenes* rolA gene promotes tolerance to *Fusarium oxysporum* f. sp. *lycopersici* in transgenic tomato plants (*Solanum lycopersicum* L.). J Plant Biochem Biotechnol 25(3):225–233

Bhatnagar-Mathur P, Vadez V, Sharma KK (2008) Transgenic approaches for abiotic stress toler-
ance in plants: retrospect and prospects. Plant Cell Rep 27:411–424
Bird CR, Smith CJS, Ray JA, Moureau P, Bevan MW, Bird AS, Hughes S, Morris PC, Grierson
D, Schuch W (1988) The tomato polygalacturonase gene and ripening-specific expression in
transgenic plants. Plant Mol Biol 11:651–662
Bird CR, Ray JA, Fletcher JD, Boniwell JM, Bird AS, Teulieres C, Blain I, Bramley PM, Schuch
W (1991) Using antisense RNA to study gene function: inhibition of carotenoid biosynthesis
in transgenic tomatoes. Biotechnol 9:635–639
Bovy A, de Vos R, Kemper M, Schijlen E, Almenar Pertejo M, Muir S, Collins G, Robinson S,
Verhoeyen M, Hughes S, Santos-Buelga C, van Tunen A (2002) High-flavonol tomatoes result-
ing from heterologous expression of the maize transcription factor gene *Lc* and *C1*. Plant Cell
14:2509–2526
Butelli E, Titta L, Giorgio M, Mock HP, Matros A, Peterek S, Schijlen EGWM, Hall RD, Bovy
AG, Luo J, Martin C (2008) Enrichment of tomato fruit with health-promoting anthocyanins by
expression of select transcription factors. Nat Biotechnol 26:1301–1308
Campos L, Lisón P, López-Gresa MP et al (2014) Transgenic tomato plants overexpressing tyramine N-hydroxycinnamoyltransferase exhibit elevated hydroxycinnamic acid amide levels and
enhanced resistance to *Pseudomonas syringae*. Mol Plant Microbe Interact 27(10):1159–1169
Cao Y, Tang X, Giovannoni J, Xiao F, Liu Y (2012) Functional characterization of a tomato
COBRA-like gene functioning in fruit development and ripening. BMC Plant Biol 12:211–225
Carli P, Arima S, Fogliano V, Tardella L, Frusciante L, Ercolano MR (2009) Use of network analysis to capture key traits affecting tomato organoleptic quality. J Exp Bot 60:3379–3386
Carmi N, Salts Y, Dedicova B, Shabtai S, Barg R (2003) Induction of parthenocarpy in tomato via
specific expression of the *rolB* gene in the ovary. Planta 217:726–735
Carrari F, Nunes-Nesi A, Gibon Y, Lytovchenko A, Loureiro ME, Fernie AR (2003) Reduced
expression of aconitase results in an enhanced rate of photosynthesis and marked shifts in
carbon partitioning in illuminated leaves of wild species tomato. Plant Physiol 133:1322–1335
Chen G (2004) Identification of a specific isoform of Tomato Lipoxygenase (TomloxC) involved in
the generation of fatty acid-derived flavor compounds. Plant Physiol 136(1):2641–2651
Chen R, Li H, Zhang L et al (2007) CaMi, a root-knot nematode resistance gene from hot pepper
(*Capsium annuum* L.) confers nematode resistance in tomato. Plant Cell Rep 26(7):895–905
Cheng L, Zou Y, Ding S, Zhang J, Yu X, Cao J, Lu G (2009) Polyamine accumulation in transgenic
tomato enhances the tolerance to high temperature stress. J Integr Plant Biol (5):489–499
Cheng X, Liu X, Wang H et al (2015) Effect of emamectin benzoate on root-knot nematodes and
tomato yield. PLoS One 10(10):e0141235
Chinnusamy V, Zhu J, Zhu JK (2007) Cold stress regulation of gene expression in plants. Trends
Plant Sci 12:444–451
Chow S, Norris JF, Bilder B (2016) Insight into the genetically modified foods: from the concerns
of safety to food development (part I). Sci Insigt. https://doi.org/10.15354/si.16.vi010
Clayton EE (1923) The relation of temperature to the Fusarium wilt of the tomato. American J Bot
10(2):71–88
Cletus J, Balasubramanian V, Vashisht D et al (2013) Transgenic expression of plant chitinases to
enhance disease resistance. Biotechnol Lett 35(11):1719–1732
Colliver S, Bovy A, Collins G, Muir S, Robinson S, de Vos CHR, Verhoeyen ME (2002) Improving
the nutritional content of tomatoes through reprogramming their flavonoid biosynthetic pathway. Phytochem Rev 1:113–123
Cortina C, Culianez-Macia FA (2005) Tomato abiotic stress enhanced tolerance by trehalose biosynthesis. Plant Sci 169:75–82
Cuartero J, Boları́n MC, Ası́ns MJ, Moreno V (2006) Increasing salt tolerance in the tomato.
J Exp Bot 57:1045–1058
D'Ambrosio C, Giorio G, Marino I, Merendino A, Petrozza A, Salfi L, Stigliani AL, Cellini F
(2004) Virtually complete conversion of lycopene into b-carotene in fruits of tomato plants
transformed with the *tomato lycopene b-cyclase* (*tlcy-b*) cDNA. Plant Sci 166:207–214

D'Ambrosio C, Stigliani AL, Giorio G (2011) Overexpression of *CrtR-b2* (carotene beta hydroxylase 2) from *S. lycopersicum* L. differentially affects xanthophylls synthesis and accumulation in transgenic tomato plants. Transgenic Res 20:47–60

Davuluri GR, van Tuinen A, Fraser PD, Manfredonia A, Newman R, Burgess D, Brummell DA, King SR, Palys J, Uhlich J, Bramley PM, Pennings HMJ, Bowler C (2005) Fruit-specific RNAi-mediated suppression of DET1 enhances carotenoid and flavonoid content in tomatoes. Nat Biotechnol 23:890–895

de Jong M, Wolters-Arts M, García-Martínez JL, Mariani C, Vriezen WH (2011) The *Solanum lycopersicum* AUXIN RESPONSE FACTOR 7 (SlARF7) mediates cross-talk between auxin and gibberellins signalling during tomato fruit set and development. J Exp Bot (2):617–626

de Jonge R, van Esse HP, Maruthachalam K, Bolton MD, Santhanam P, Saber MK, Zhang Z, Usami T, Lievens B, Subbarao KV, Thomma BPHJ (2012) Tomato immune receptor Ve1 recognizes effector of multiple fungal pathogens uncovered by genome and RNA sequencing. Proc Natl Acad Sci 109(13):5110–5115

de la Garza RID, Quinlivan PE, Klaus SMJ, Basset GJC, Gregory JF, Hanson AD (2004) Folate biofortification in tomatoes by engineering the pteridine branch of folate synthesis. PNAS (38):13720–13725

de la Garza RID, Gregory JF, Hanson AD (2007) Folate biofortification of tomato fruit. PNAS (10):4218–4222

Dominguez T, Hernandez LM, Pennycooke JC, Jimenez P, Martínez-Rivas JM, Sanz C, Stockinger EJ, Sanchez-Serrano JJ, Sanmartin M (2010) Increasing ω-3 desaturase expression in tomato results in altered aroma profile and enhanced resistance to cold stress. Plant Physiol 153:655–665

Duan M, Feng H-L, Wang L-Y, Li D, Meng Q-W (2012) Overexpression of thylakoidal ascorbate peroxidase shows enhanced resistance to chilling stress in tomato. J Plant Physiol 169(9):867–877

Dutta TK, Papolu PK, Banakar P et al (2015) Tomato transgenic plants expressing hairpin construct of a nematode protease gene conferred enhanced resistance to root-knot nematodes. Front Microbiol 6:260

Eisenstein M (2013) Discovery in a dry spell. Nature 501:S7–S9

El-Sharkawy I, Sherif SM, Jones B, Mila I, Kumar PP, Bouzayen M, Jayasankar S (2014) TIR1-like auxin-receptors are involved in the regulation of plum fruit development. J Exp Bot 65:5205–5215

El-Sharkawy I, Sherif S, El-Kayal W, Jones B, Li Z, Sullivan AJ, Jayasankar S (2016) Overexpression of plum auxin receptor *PslTIR1* in tomato alters plant growth, fruit development and fruit shelf-life characteristics. BMC Plant Biol 16:56–67

FAOSTAT Website (http://faostat3.fao.org/home/E)

Flowers TJ (2004) Improving crop salt tolerance. J Exp Bot 55:307–319

Foolad MR, Zhang LP, Khan AA, Niño-Liu D, Lin GY (2002) Identification of QTLs for early blight (*Alternaria solani*) resistance in tomato using backcross populations of a *Lycopersicon esculentum* × *L. hirsutum* cross. Theor Appl Genet 104(6-7):945–958

Fradin EF, Thomma BP (2006) Physiology and molecular aspects of Verticillium wilt diseases caused by V. dahliae and V. albo-atrum. Mol Plant Pathol 7(2):71–86

Fradin EF, Abd-El-Haliem A, Masini L et al (2011) Interfamily transfer of tomato Ve1 mediates Verticillium resistance in Arabidopsis. Plant Physiol 156(4):2255–2265

Fraser PD, Romer S, Shipton CA, Mills PB, Kiano JW, Misawa N, Drake RG, Schuch W, Bramley PM (2002) Evaluation of transgenic tomato plants expressing an additional phytoene synthase in a fruit-specific manner. PNAS (2):1092–1109

Fraser PD, Thomma EMA, Halket JM, Truesdale MR, Yu D, Gerrish C, Bramleya PM (2007) Manipulation of phytoene levels in tomato fruit: effects on isoprenoids, plastids, and intermediary metabolism. Plant Cell 19:3194–3211

Fray RG, Grierson D (1993) Identification and genetic analysis of normal and mutant phytoene synthase genes of tomato by sequencing, complementation and co-suppression. Plant Mol Biol 22:589–602

Fuentes A, Ramos PL, Fiallo E, Callard D, Sa´nchez Y, Peral R, Rodrı´guez R, Pujol M (2006) Intron–hairpin RNA derived from replication associated protein C1 gene confers immunity to Tomato Yellow Leaf Curl Virus infection in transgenic tomato plants. Transgenic Res 15:291–304

Galvez LC, Banerjee J, Pinar H, Mitra A (2014) Engineered plant virus resistance. Plant Sci 228:11–25

Garchery C, Gest N, Do PT, Alhagdow M, Baldet P, Menard G, Rothan C, Massot C, Gautier H, Aarrouf J, Fernie AR, Stevens R (2013) A diminution in ascorbate oxidase activity affects carbon allocation and improves yield in tomato under water deficit. Plant Cell Environ 36:159–175

Garcia V, Stevens R, Gil L, Gilbert L, Gest N, Petit J, Faurobert M, Maucourt M, Deborde C, Moing A, Poessel JL, Jacob D, Bouchet JP, Giraudel JL, Gouble B, Page D, Alhagdow M, Massot C, Gautier H, Lemaire-Chamley M, Rolin D, Usadel B, Lahaye M, Causse M, Baldet P, Rothan C (2009) An integrative genomics approach for deciphering the complex interactions between ascorbate metabolism and fruit growth and composition in tomato. C R Biol (11):1007–1021

Gerszberg A, Hnatuszko-Konka K (2017) Tomato tolerance to abiotic stress: a review of most often engineered target sequences. Plant Growth Regul (online). https://doi.org/10.1007/s10725-017-0251-x

Gerszberg A, Hnatuszko-Konka K, Kowalczyk T, Kononowicz AK (2015) Tomato (*Solanum lycopersicum* L.) in the service of biotechnology. Plant Cell Tiss Org Cult 120:881–902

Gilbert L, Alhagdow M, Nunes-Nesi A, Quemener B, Guillon F, Bouchet B, Faurobert M, Gouble B, Page D, Garcia V, Peti J, Stevens R, Causse M, Fernie AR, Lahaye M, Rothan C, Baldet P (2009) GDP-D-mannose 3,5-epimerase (GME) plays a key role at the intersection of ascorbate and non-cellulosic cell-wall biosynthesis in tomato. Plant J 60:499–508

Giorio G, Stigliani AL, D'Ambrosio C (2008) Over-expression of carotene b-hydroxylase 1 (*CrtR-b1*) and lycopene bcyclase (*Lcy-b*) in transgenic tomato fruits. Acta Hort (789):277–284

Giovannoni JJ, DellaPenna D, Bennett AB, Fischer RL (1989) Expression of a chimeric polygalacturonase gene in transgenic *rin* (ripening inhibitor) tomato fruit results in polyuronide degradation but not fruit softening. Plant Cell 1:53–63

Giovinazzo G, D'Amico L, Paradiso A, Bollino R, Sparvoli F, DeGara L (2005) Antioxidant metabolite profiles in tomato fruit constitutively expressing the grapevine stilbene synthase gene. Plant Biotechnol J 3:57–69

Girhepuje PV, Shinde GB (2011) Transgenic tomato plants expressing a wheat endochitinase gene demonstrate enhanced resistance to *Fusarium oxysporum* f. sp. *lycopersici*. Plant Cell Tiss Org Cult 105(2):243–251

Gisbert C, Rus AM, Boları´n MC, Coronado JM, Arrillaga I, Montesinos C, Caro M, Serrano R, Moreno V (2000) The yeast HAL1 gene improves salt tolerance of transgenic tomato. Plant Physiol 123:393–402

Gleason ML, Edmunds BA (2005) Tomato diseases and disorders. vinesgardens.org

Gleason ML, Edmunds BA (2006) Tomato diseases and disorders. University Extension PM 1266, Iowa State University, Ames

Goel D, Singh AK, Yadav V, Babbar SB, Bansal KC (2010) Overexpression of osmotin gene confers tolerance to salt and drought stresses in transgenic tomato (Solanum lycopersicum L.). Protoplasma 245:133–141

Goetz M, Hooper LC, Johnson SD, Rodrigues JC, Vivian-Smith A, Koltunov AM (2007) Expression of aberrant forms of auxin response factor 8 stimulates parthenocarpy in Arabidopsis and tomato. Plant Physiol 2:336–351

Gonzalez-cendales Y, Catanzariti AM, Baker B, Mcgrath DJ, Jones DA (2016) Identification of I-7 expands the repertoire of genes for resistance to Fusarium wilt in tomato to three resistance gene classes. Mol Plant Pathol 17(3):448–463

Good X, Kellogg JA, Wagoner W, Langhoff D, Matsumura W, Bestwick RK (1994) Reduced ethylene synthesis by transgenic tomatoes expressing S-adenosylmethionine hydrolase. Plant Mol Biol 26:781–790

Gould WA (2013) Tomato production, processing and technology, 3rd edn. Woodhead Publishing, Sawston, Cambridge, UK

Goulet C, Mageroy MH, Lam NB, Floystad A, Tieman DM, Klee HJ (2012) Role of an esterase in flavor volatile variation within the tomato clade. PNAS 109:19009–19014

Grandillo S, Zamir D, Tanksley SD (1999) Genetic improvement of processing tomatoes: a 20 years perspective. Euphytica 110:85–97

Grierson D, Kader AA (1986) Fruit ripening and quality. In: Atherton JG, Rudich J (eds) The tomato crop: a scientific basis for improvement. Chapman and Hall, London, pp 241–280

Gul Z, Ahmed M, Khan ZU, Khan B, Iqbal M (2016) Evaluation of tomato lines against Septoria leaf spot under field conditions and its effect on fruit yield. Agric Sci 7(04):181–186

Hamilton AJ, Lycett GW, Grierson D (1990) Antisense gene that inhibits synthesis of the hormone ethylene in transgenic plants. Nature 346:284–287

He P, Warren RF, Zhao T et al (2001) Overexpression of Pti5 in tomato potentiates pathogen-induced defense gene expression and enhances disease resistance to *Pseudomonas syringae* pv. *tomato*. Mol Plant Microbe Interact 14(12):1453–1457

Herbette S, Tourvieille de Labrouhe D, Drevet JR, Roeckel-Drevet P (2011) Transgenic tomatoes showing higher glutathione peroxidase antioxidant activity are more resistant to an abiotic stress but more susceptible to biotic stresses. Plant Sci 180:548–553

Hiatt WR, Kramer M, Sheehy RE (1989) The application of antisense RNA technology to plants. Genetic engineering: principles and methods. J. K. Setlow. Boston, MA, Springer US, pp 49–63

Hightower R, Baden C, Penzes E, Lund P, Dunsmuir P (1991) Expression of antifreeze proteins in transgenic plants. Plant Mol Biol 17:1013–1021

Hirai T, Kim YW, Kato K, Hiwasa-Tanase K, Ezura H (2011) Uniform accumulation of recombinant miraculin protein in transgenic tomato fruit using a fruit-ripening-specific E8 promoter. Transgenic Res 20:1285–1292

Hiwasa-Tanase K, Nyarubona M, Hirai T, Kato K, Ichikawa T, Ezura H (2011) High-level accumulation of recombinant miraculin protein in transgenic tomatoes expressing a synthetic *miraculin* gene with optimized codon usage terminated by the native miraculin terminator. Plant Cell Rep 30:113–124

Honma M, Shimomura T (1978) Metabolism of 1-aminocyclopropane-1- carboxylic acid. Agric Biol Chem 42:1825–1831

Hontzeas N, Zoidakis J, Glick BR, Abu-Omar MM (2004) Expression and characterization of 1-aminocyclopropane-1-carboxylate deaminase from the rhizobacterium *Pseudomonas putida* UW4: a key enzyme in bacterial plant growth promotion. Biochim Biophys Acta 1703:11–19

Horvath DM, Stall RE, Jones JB et al (2012) Transgenic resistance confers effective field level control of bacterial spot disease in tomato. PLoS One 7(8):e42036. https://doi.org/10.1371/journal.pone.0042036

Hsieh TH, Li CW, Su RC, Cheng CP, Sanjaya TYC, Chan MT (2010) A tomato bZIP transcription factor, SlAREB, is involved in water deficit and salt stress response. Planta 231:1459–1473

Hu Y, Wu Q, Sprague SA, Park J, Oh M, Rajashekar CB, Koiwa H, Nakata PA, Cheng N, Hirschi KD, Frank F White FF, Park S (2015) Tomato expressing Arabidopsis glutaredoxin gene AtGRXS17 confers tolerance to chilling stress via modulating cold responsive components. Hort Res 2:15051

Ishibashi K, Masuda K, Naito S et al (2007) An inhibitor of viral RNA replication is encoded by a plant resistance gene. Proc Natl Acad Sci U S A 104(34):13833–13838

Janska A, Marsik P, Zelenkova S, Ovesna J (2009) Cold stress and acclimation – what is important for metabolic adjustment? Plant Biol 12:395–405

Jones JB, Jones JP, Stall RE, Zitter TA (eds) (1991) Compendium of tomato diseases. The American Phytopathological Society, St. Paul, pp 31–42

Jongedijk E. et al. (1995) Synergistic activity of chitinases and β-1,3-glucanases enhances fungal resistance in transgenic tomato plants. The Methodology of Plant Genetic Manipulation: Criteria for Decision Making: Proceedings of the Eucarpia Plant Genetic Manipulation Section Meeting held at Cork, Ireland from September 11 to September 14, 1994. AC Cassells and PW Jones. Dordrecht, Springer Netherlands, pp. 173–180

Kadotani N, Nakayashiki H, Tosa Y, Mayama S (2003) RNA Silencing in the Phytopathogenic fungus Magnaporthe oryzae. Mol Plant Microbe Interact 16(9):769–776

Kadyrzhanova DK, Vlachonasios KE, Ververidis P, Dilley DR (1998) Molecular cloning of a novel heat induced/chilling tolerance related cDNA1 in tomato fruit by use of mRNA differential display. Plant Mol Biol 36:885–895

Kang BC, Yeam I, Li H et al (2007) Ectopic expression of a recessive resistance gene generates dominant potyvirus resistance in plants. Plant Biotechnol J 5(4):526–536

Kim YW, Hirai T, Kato K, Hiwasa-Tanase K, Ezura H (2010) Gene dosage and genetic background affect miraculin accumulation in transgenic tomato fruits. Plant Biotechnol 27:333–338

Klee HJ (1993) Ripening physiology of fruit from transgenic tomato (*Lycopersicon esculentum*) plants with reduced ethylene synthesis. Plant Physiol 102:911–916

Klee HJ, Hayford MB, Kretzmer KA, Barry GF, Kishore GM (1991) Control of ethylene synthesis by expression of a bacterial enzyme in transgenic tomato plants. Plant Cell 3:1187–1193

Koc NK, Kayim M, Yetisir H et al (2007) The improvement of resistance to bacterial speck in transgenic tomato plants by Agrobacterium tumefaciens mediated transformation. Russian J Plant Physiol 54(1):89–96

Kong K, Ntui VO, Makabe S et al (2014) Transgenic tobacco and tomato plants expressing Wasabi defensin genes driven by root-specific LjNRT2 and AtNRT2.1 promoters confer resistance against *Fusarium oxysporum*. Plant Biotechnol 31:89–96

Kostov K, Christova P, Slavov S et al (2009) Constitutive expression of a radish defensin gene Rs-AFP2 in tomato increases the resistance to fungal pathogens. Biotechnol Biotechnol Equip 23(1):1121–1125

Kouki S, Saidi N, Rajeb AB et al (2012) Control of Fusarium wilt of tomato caused by Fusarium oxysporum f. sp. radicis-lycopersici using mixture of vegetable and Posidonia oceanica compost. Appl Environ Soil Sci 239639:1–11. https://doi.org/10.1155/2012/239639

Kunik T, Salomon R, Zamir D, Navot N, Zeidan M, Michelson I, Gafni Y, Czosnek H (1994) Transgenic tomato plants expressing the tomato yellow leaf curl virus capsid protein are resistant to the virus. Nat Biotechnol 12:500–504

Lanfermeijer FC, Dijkhuis J, Sturre MJG et al (2003) Cloning and characterization of the durable tomato mosaic virus resistance gene Tm-2² from *Lycopersicon esculentum*. Plant Mol Biol 52(5):1039–1051

Le Gall G, DuPont MS, Mellon FA, Davis AL, Collins GJ, Verhoeyen ME, Colquhoun IJ (2003) Characterization and content of flavonoid glycosides in genetically modified tomato (*Lycopersicon esculentum*) fruits. J Agric Food Chem 51:2438–2446

Li X, Gasic K, Cammue B et al (2003) Transgenic rose lines harboring an antimicrobial protein gene, Ace-AMP1, demonstrate enhanced resistance to powdery mildew (Sphaerotheca pannosa). Planta 218(2):226–232

Li XQ, Wei JZ, Tan A et al (2007) Resistance to root-knot nematode in tomato roots expressing a nematicidal *Bacillus thuringiensis* crystal protein. Plant Biotechnol J 5(4):455–464

Li F, Xinga S, Guoa Q, Zhaoa M, Zhanga J, Gaoa Q, Wangb G, Wanga W (2011) Drought tolerance through over-expression of the expansin gene TaEXPB23 intransgenic tobacco. J Plant Physiol 168:960–966

Li S, Jin X, Chen J et al (2013) Inhibitory activities of venom alkaloids of Red Imported Fire Ant against *Clavibacter michiganensis* subsp. *michiganensis in vitro* and the application of piperidine alkaloids to manage symptom development of bacterial canker on tomato in the greenhouse. Int J Pest Manage 59(2):150–156

Lin WC, Lu CF, Wu JW et al (2004) Transgenic tomato plants expressing the Arabidopsis NPR1 gene display enhanced resistance to a spectrum of fungal and bacterial diseases. Transgenic Res 13(6):567–581

Liu J, Cong B, Tanksley SD (2003) Generation and analysis of an artificial gene dosage series in tomato to study the mechanism by which the cloned quantitative trait locus *fw2.2* controls fruit size. Plant Physiol (1):292–299

Liu X, Yang JH, Li B, Yang XM, Meng QW (2010) Antisense expression of tomato chloroplast omega-3 fatty acid desaturase gene (LeFAD7) enhances the tomato high-temperature tolerance through reductions of trienoic fatty acids and alterations of physiological parameters. Photosyntheyhica (1):59–66

Liu XY, Teng YB, Meng QW (2013) Enhancement of low-temperature tolerance in transgenic tomato plants overexpressing Lefad7 through regulation of trienoic fatty acids. Photosyntheyhica (2):238–244

Louws FJ, Wilson M, Cuppels DA et al (2001) Field control of bacterial spot of tomato and pepper and bacterial speck of tomato using a plant activator. Plant Dis 85:481–488

Ma N, Feng H, Meng X, Li D, Yang D, Wu C, Meng Q (2014) Overexpression of tomato *SlNAC1* transcription factor alters fruit pigmentation and softening. BMC Plant Biol 14:351–364

Mahajan S, Tuteja N (2005) Cold, salinity and drought stresses: an overview. Arch Biochem Biophys 444:139–158

Mahesh U, Mamidala P, Rapolu S, Aragao FJL, Souza MT, Rao PJM, Kirti PB, Nanna RS (2013) Constitutive overexpression of small HSP24.4 gene in transgenic tomato conferring tolerance to high-temperature stress. Mol Breed 32:687–697

Maligeppagol M, Chandra GS, Prakash M, Navale PM, Deepa H, Rajeev PR, Asokan R, Babu KP, Babu CCB, Rao VK, Kumar KNK (2013) Anthocyanin enrichment of tomato (*Solanum lycopersicum* L.) fruit by metabolic engineering. Curr Sci 1:72–80

Marco F, Bitrián M, Carrasco P, Rajam MV, Alcázar R, Tiburcio AF (2015) In: Bahadur B et al. (eds) Plant biology and biotechnology: Volume II: Plant genomics and biotechnology, Springer India, New Delhi, pp 579–609

Martin G, Brommonschenkel S, Chunwongse J, Frary A, Ganal M, Spivey R, Wu T, Earle E, Tanksley S (1993) Map-based cloning of a protein kinase gene conferring disease resistance in tomato. Science 262 (5138):1432–1436

Meli VS, Ghosh S, Prabha TN, Chakraborty N, Chakraborty S, Datta A (2010) Enhancement of fruit shelf life by suppressing *N*-glycan processing enzymes. PNAS 107:2413–2418

Meng X, Wang JR, Wang GD, Liang XQ, Li XD, Meng QW (2015) An R2R3-MYB gene, LeAN2, positively regulated the thermotolerance in transgenic tomato. J Plant Physiol 175:1–8

Mes PJ, Boches P, Myers JR, Durst R (2008) Characterization of tomatoes expressing anthocyanin in the fruit. J Amer Soc Hort Sci 133:262–269

Milligan SB, Bodeau J, Yaghoobi J et al (1998) The root knot nematode resistance gene *Mi* from tomato is a member of the leucine zipper, nucleotide binding, leucine-rich repeat family of plant genes. Plant Cell 10(8):1307–1319

Mintz-Oron S, Mandel T, Rogachev I, Feldberg L, Lotan O, Yativ M, Wang Z, Jetter R, Venger I, Adato A, Aharoni A (2008) Gene expression and metabolism in tomato fruit surface tissues. Plant Physiol 147:823–851

Mishra SK, Tripp J, Winkelhaus S, Tschiersch B, Theres K, Nover L, Scharf KD (2002) In the complex family of heat stress transcription factors, HsfA1 has a unique role as master regulator of thermotolerance in tomato. Genes Dev 16:1555–1567

Moens M, Perry RN, Starr JL (2009) *Meloidogyne* spp.-a diverse group of novel and important plant parasite. In: Perry RN, Moens M, Starr JL (eds) Root-knot nematodes. CAB International, Wallingford, Oxfordshire, pp 1–17

Moghaieb REA, Tanaka N, Saneoka H, Hussein HA, Yousef SS, Ewada MAF, Aly MAM, Fujita K (2000) Expression of betaine aldehyde dehydrogenase gene in transgenic tomato hairy roots leads to the accumulation of glycine betaine and contributes to the maintenance of the osmotic potential under salt stress. Soil Sci Plant Nutr 46:873–883

Morgan MJ, Osorio S, Gehl B, Baxter CJ, Kruger NJ, Ratcliffe RG, Fernie AR, Sweetlove LJ (2013) Metabolic engineering of tomato fruit organic acid content guided by biochemical analysis of an introgression line. Plant Physiol (1):397–407

Muir SR, Collins GJ, Robinson S, Hughes SG, Bovy AG, de Vos CH, van Tunen AJ, Verhoyen ME (2001) Overexpression of petunia chalcone isomerase in tomato results in fruit containing increased levels of flavonols. Nat Biotechnol (5):470–474

Murray AJ, Hobson GE, Schuch W, Bird CR (1993) Reduced ethylene synthesis in EFE anti-sense tomatoes has differential effects on fruit ripening processes. Post harvest Biol Technol 2:301–313

Mustafiz A, Sahoo KK, Singla-Pareek SL, Sopory SK (2010) Metabolic engineering of glyoxalase pathway for enhancing stress tolerance in plants. In: Ramanjulu S (ed) Plant stress tolerance, volume 639 of the series methods mol biol, pp 95–118

Nascimento FX, Rossi MJ, Soares CRFS, McConkey BJ, Glick BR (2014) New Insights into 1-aminocyclopropane-1-carboxylate (ACC) deaminase phylogeny, evolution and ecological significance. PLoS One 9:e99168

Nautiyal PC, Shono M, Egawa Y (2005) Enhanced thermotolerance of the vegetative part of MT-sHSP transgenic tomato line. Sci Hortic 105:393–409

Neily MH, Matsukura C, Maucourt M, Bernillon S, Deborde C, Moing A, Yin YG, Saito T, Mori K, Asamizua E, Rolin D, Moriguchi T, Ezura H (2011) Enhanced polyamine accumulation alters carotenoid metabolism at the transcriptional level in tomato fruit over-expressing spermidine synthase. J Plant Physiol 168:242–252

Nelson RS, McCormick SM, Delannay X, Dubé P, Layton J, Anderson EJ, Kaniewska M, Proksch RK, Horsch RB, Rogers SG, Fraley RT, Beachy RN (1988) Virus tolerance, plant performance of transgenic tomato plants expressing coat protein from tobacco mosaic virus. Nat Biotechnol 6:403–409

Nowicki M, Kozik EU, Foolad MR (2013) Late blight of tomato. In: Varshney RK, Tuberosa R (eds) Translational genomics for crop breeding. John Wiley & Sons Ltd., USA, pp 241–265

Orellana S, Yanez M, Espinoza A, Verdugo I, Gonzales E, Ruiz-Lara S, Cacaretto JA (2010) The transcription factor SlAREB1 confers drought, salt stress tolerance and regulates biotic and abiotic stress-related genes in tomato. Plant Cell Environ 33:2191–2208

Orzaez D, Medina A, Torre S, Fernandez-Moreno JP, Rambla JL, Fernandez-del-Carmen A, Butelli E, Martin C, Granell A (2009) A visual reporter system for virus-induced gene silencing in tomato fruit based on anthocyanin accumulation. Plant Physiol (3):1122–1134

Ouyang B, Chen YH, Li HX et al (2005) Transformation of tomatoes with osmotin and chitinase genes and their resistance to Fusarium wilt. J Hortic Sci Biotech 80(5):517–522

Panthee DR, Chen F (2010) Genomics of fungal disease resistance in tomato. Curr Genomics 11(1):30–39

Patade VY, Khatri D, Kumari M, Grover A, Gupta SM, Ahmed Z (2013) Cold tolerance in osmotin transgenic tomato (Solanum lycopersicum L.) is associated with modulation in transcript abundance of stress responsive genes. SpringerPlus 2:117

Patel TK, Krasnyanski SF, Allen GC et al (2015) Progeny of selfed plants from tomato breeding line 'NC1 grape' overexpressing mannitol dehydrogenase (MTD) have increased resistance to the early blight fungus, Alternaria solani. Plant Health Progress. https://doi.org/10.1094/PHP-RS-15-0022

Peñarrubia L, Aguilar M, Margossian L, Fischer RL (1992a) An antisense gene stimulates ethylene hormone production during tomato fruit ripening. Plant Cell 4:681–687

Peñarrubia L, Kim R, Giovannoni J, Kim SH Fischer RL (1992b) Production of the sweet protein monellin in transgenic plants. Nat Biotechnol 10:561–564

Peng J-C, Chen T-C, Raja JAJ et al. (2014) Broad-spectrum transgenic resistance against distinct tospovirus species at the genus level. PLoS One 9(5):e96073

Phan TD, Bo W, West G, Lycett GW, Tucker GA (2007) Silencing of the major salt-dependent isoform of pectinesterase in tomato alters fruit softening. Plant Physiol 144:1960–1967

Picton S, Barton SL, Bouzayen M, Hamilton AJ, Grierson D (1993) Altered fruit ripening and leaf senescence in tomatoes expressing an antisense ethylene-forming enzyme transgene. Plant J 3:469–481

Pineda B (2005) Ana´lisis functional de diversos genes relacionados con la tolerancia a la salinidad y el estre´s hı´drico en plantas transge´nicas de tomate (Lycopersicon esculentum Mill). PhD thesis, Universidad Polite´cnica de Valencia

Pineda B, Garcia-Abellan JO, Perez F, Moyano E, Garcia Sogo B, Campos JF, Angosto T, Morales B, Capel J, Moreno V, Lozano R, Bolarin MC, Atares A (2012) Tomato: genomic approaches for salt and drought stress tolerance. In: Tuteja N, Gill SS, Tiburcio A, Tuteja R (eds) Improving crop resistance to abiotic stress, Vol 1 & Vol 2. Wiley-VCH Verlag GmbH & Co. KGaA, Weinheim, Germany

Powell ALT, Kalamaki MS, Kurien PA, Gurrieri S, Bennett AB (2003) Simultaneous transgenic suppression of LePG and LeExp1influences fruit texture and juice viscosity in a fresh market tomato variety. J Agric Food Chem 51:7450–7455

Poysa V, Tu JC (1993) Response of cultivars and breeding lines of *Lycopersicon* spp. to Septoria lycopersici. Canadian Plant Dis Survey 73:9–13

Prins M, de Haan P, Luyten R et al (1995) Broad resistance to tospoviruses in transgenic tobacco plants expressing three tospoviral nucleoprotein gene sequences. Mol Plant Microbe Interact 8:85–91

Radhajeyalakshmi R, Velazhahan R, Balasubramanian P et al. (2005) Overexpression of thaumatin-like protein in transgenic tomato plants confers enhanced resistance to *Alternaria solani*. Arch Phytopathol Plant Protect 38(4):257–265

Reddy CS, Vjayalakshmi M, Kaul T, Islam T, Reddy MK (2015) Improving flavour and quality of tomatoes by expression of synthetic gene encoding sweet protein monellin. Mol Biotechnol 57:448–453

Reyes MI, Nash TE, Dallas MM et al (2013) Peptide aptamers that bind to geminivirus replication proteins confer a resistance phenotype to tomato yellow leaf curl virus and tomato mottle virus infection in tomato. J Virol 87(17):9691–9706

Ribeiro SG, Lohuis H, Goldbach R et al (2007) Tomato chlorotic mottle virus is a target of RNA silencing but the presence of specific short interfering RNAs does not guarantee resistance in transgenic plants. J Virol 81(4):1563–1573

Robison MM, Shah S, Tamot B et al (2001) Reduced symptoms of Verticillium wilt in transgenic tomato expressing a bacterial ACC deaminase. Mol Plant Pathol 2(3):135–145

Rotino G, Acciarri N, Sabatini E, Mennella G, Scalzo RL, Maestrelli A, Molesini B, Pandolfini T, Scalzo J, Mezzetti B, Spena A (2005) Open field trial of genetically modified parthenocarpic tomato: seedlessness and fruit quality. BMC Biotechnol 5(1):32

Roy-Barman S, Sautter C, Chattoo BB (2006) Expression of the lipid transfer protein Ace-AMP1 in transgenic wheat enhances antifungal activity and defense responses. Transgenic Res 15(4):435–446

Rus AM, Estan˜ MT, Gisbert C, Garcia-Sogo B, Serrano R, Caro M, Moreno V, MC B´n (2001) Expressing the yeast HAL1 gene in tomato increases fruit yield and enhances K+/Na+ selectivity under salt stress. Plant Cell Environ 24:875–880

Sabehat A, Weiss D, Lurie S (1996) The correlation between heat shock protein accumulation and persistence and chilling in tomato fruit. Plant Physiol 110:531–537

Sade N, Gebretsadik M, Seligmann R, Schwartz A, Wallach R, Moshelion M (2010) The role of tobacco aquaporin1 in improving water use efficiency, hydraulic conductivity, and yield production under salt stress. Plant Physiol 152:245–254

Sarrion-Perdigones A, Falconi EE, Zandalinas SI, Juárez P, Fernández-del-Carmen A, Granell A, Orzaez D (2011) GoldenBraid: an iterative cloning system for standardized assembly of reusable genetic modules. PLoS One 6:e21622

Schaefer SC, Gasic K, Cammue B (2005) Enhanced resistance to early blight in transgenic tomato lines expressing heterologous plant defense genes. Planta 222:858–866

Schijlen E, de Vos CHR, Jonker H, van den Broeck H, Molthoff J, van Tunen A, Martens S, Bovy A (2006) Pathway engineering for healthy phytochemicals leading to the production of novel flavonoids in tomato fruit. Plant Biotechnol J 4:433–444

Schijlen EGWM, de Vos CHR, Martens S, Jonker HH, Rosin FM, Molthoff JF, Tikunov YM, Angenent GC, van Tunen AJ, Bovy AG (2007) RNA interference silencing of chalcone synthase, the first step in the flavonoid biosynthesis pathway, leads to parthenocarpic tomato fruits. Plant Physiol 144:1520–1530

Schreiber G, Reuveni M, Evenor D, Oren-Shamir M, Ovadia R, Sapir-Mir M, Bootbool-Man A, Nahon S, Shlomo H, Chen L, Levin I (2012) *ANTHOCYANIN1* from Solanum chilense is more efficient in accumulating anthocyanin metabolites than its *Solanum lycopersicum* counterpart in association with the ANTHOCYANIN FRUIT phenotype of tomato. Theor Appl Genet 124:295–307

Schuch W, Kanczler J, Robertson D, Hobson G, Tucker G, Grierson D, Bright S, Bird C (1991) Fruit quality characteristics of transgenic tomato fruit with altered polygalacturonase activity. Hort Sci 26:1517–1520

Schumann GL, D'Arcy CJ (2000) Late blight of potato and tomato. The Plant Health Instructor. https://doi.org/10.1094/PHI-I-2000-0724-01

Seebold K (2008) Bacterial canker of tomato. Plant pathology fact sheet. Cooperative Extension Service. University of Kentucky, Agricultural and Natural Resources

Sen Y, van der Wolf J, Visser RG et al (2015) Bacterial canker of tomato: current knowledge of detection, management, resistance, and interactions. Plant Dis 99:4–13

Seong ES, Cho HS, Choi D, Joung YH, Lim CK, Hur JH, Wang MH (2007) Tomato plants over-expressing CaKR1 enhanced tolerance to salt and oxidative stress. Biochem Biophys Res Commun 363:983–988

Sheehy RE, Kramer M, Hiatt WR (1988) Reduction of polygalacturonase activity in tomato fruit by antisense RNA. PNAS 85:8805–8809

Shin S, Mackintosh CA, Lewis J et al (2008) Transgenic wheat expressing a barley class II chitinase gene has enhanced resistance against *Fusarium graminearum*. J Exp Bot 59(9):2371–2378

Sieso V, Nicolas M, Seck S, Crouzet J (1976) Constituants volatils de la tomate: mise en evidence et formation par voie enzymatique du *trans*-hexene-2-ol. Agric Biol Chem 40:2349–2353

Simkin AJ, Schwartz SH, Auldridge M, Taylor MG, Klee HJ (2004) The tomato carotenoid cleavage dioxygenase 1 genes contribute to the formation of the flavor volatiles b-ionone, pseudoionone, and geranylacetone. Plant J 40:882–892

Small DM, Voss J, Mak YE, Simmons KB, Parrish T, Gitelman D (2004) Experience-dependent neural integration of taste and smell in the human brain. J Neurophysiol 92:1892–1903

Smart R, Blum M, Wesseler J (2016) Trends in approval times for genetically engineered crops in the United States and the European Union. J Agric Econ. https://doi.org/10.1111/1477-9552.12171

Smith CJ, Watson CF, Morris PC, Bird CR, Seymour GB, Gray JE, Arnold C, Tucker GA, Schuch W, Harding S, Griersonr D (1990) Inheritance and effect on ripening of antisense polygalacturonase genes in transgenic tomatoes. Plant Mol Biol 14:369–379

Smith DL, Abbott JA, Gross KC (2002) Down-regulation of tomato β-galactosidase 4 results in decreased fruit softening. Plant Physiol 129:1755–1762

Speirs J, Lee E, Holt K, Yong-Duk K, Scott NS, Loveys B, Schuch W (1998) Genetic manipulation of alcohol dehydrogenase levels in ripening tomato fruit affects the balance of some flavor aldehydes and alcohols. Plant Physiol 117:1047–1058

Sun HJ, Kataoka H, Yano M, Ezura H (2007) Genetically stable expression of functional miraculin, a new type of alternative sweetener, in transgenic tomato plants. Plant Biotechnol J 5:768–777

Sun L, Yuan B, Zhang M, Wang L, Cui M, Wang Q, Leng P (2012a) Fruit-specific RNAi-mediated suppression of *SlNCED1* increases both lycopene and β-carotene contents in tomato fruit. J Exp Bot 63:3097–3108

Sun L, Sun Y, Zhang M, Wang L, Ren J, Cui M, Wang Y, Ji K, Li P, Li Q, Chen P, Dai S, Duan C, Wu Y, Ping Leng P (2012b) Suppression of 9-cis-epoxycarotenoid dioxygenase, which encodes a key enzyme in abscisic acid biosynthesis, alters fruit texture in transgenic tomato. Plant Physiol 158:283–298

Sun K, Wolters AM, Loonen AE et al (2016) Down-regulation of Arabidopsis DND1 orthologs in potato and tomato leads to broad-spectrum resistance to late blight and powdery mildew. Transgenic Res 25:123–138

Swarup V (2012) Vegetable Science and Technology. Kalyani Publishers, B/I/1292, Rajendra Nagar, Ludhiana-141008, India

Tabaeizadeh Z, Agharbaoui Z, Harrak H et al (1999) Transgenic tomato plants expressing a *Lycopersicon chilense* chitinase gene demonstrate improved resistance to *Verticillium dahliae* race 2. Plant Cell Rep 19(2):197–202

Tai TH, Dahlbeck D, Clark ET et al (1999) Expression of the Bs2 pepper gene confers resistance to bacterial spot disease in tomato. Proc Natl Acad Sci U S A 96(24):14153–14158

Tang X, Xie M, Kim YJ et al (1999) Overexpression of Pto activates defense responses and confers broad resistance. Plant Cell 11(1):15–29

The Tomato Genome Consortium (2012) The tomato genome sequence provides insights into fleshy fruit evolution. Nature 485:635–641

Theocharis A, Clément C, Barka E (2012) Physiological and molecular changes in plants grown at low temperatures. Planta 235:1091–1105

Thomzik JE, Stenzel K, Stöcker R, Schreier PH, Hain R, Stahl DJ (1997) Synthesis of a grapevine phytoalexin in transgenic tomatoes (*Lycopersicon esculentum* Mill.) conditions resistance against *Phytophthora infestans*. Physiol Mol Plant Pathol 51(4):265–278

Tieman D, Bliss P, McIntyre LM, Blandon-Ubeda A, Bies D, Odabasi AZ, GR R´g, van der Knaap E, Taylor MG, Goulet C, Mageroy MH, Snyder DJ, Colquhoun T, Moskowitz H, Clark DG, Sims C, Bartoshuk L, Klee HJ (2012) The chemical interactions underlying tomato flavor preferences. Curr Biol 22:1035–1039

Tiwari M, Sharma D, Trivedi PK (2014) Artificial microRNA mediated gene silencing in plants: progress and perspectives. Plant Mol Biol 86(1-2):1–18

Tsuda T, Horio F, Uchida K, Aoki H, Osawa T (2003) Dietary cyanidin 3-O-β-D-glucoside-rich purple corn color prevents obesity and ameliorates hyperglycemia in mice. J Nutr 133:2125–2130

Tyler BM et al (2006) *Phytophthora* genome sequences uncover evolutionary origins and mechanisms of pathogenesis. Science 313:1261–1266

Ultzen T, Gielen J, Venema F, Westerbroek A, de Haan P, Tan M-L, Schram A, van Grinsven M, Goldbach R (1995) Resistance to tomato spotted wilt virus in transgenic tomato hybrids. Euphytica 85 (1-3):159–168

van der Vossen E, Sikkema A, Bt H et al (2003) An ancient R gene from the wild potato species Solanum bulbocastanum confers broad-spectrum resistance to Phytophthora infestans in cultivated potato and tomato. Plant J 36:867–882

Verhoeyen ME, Bovy A, Collins G, Muir S, Robinson S, de Vos CHR, Colliver S (2002) Increasing antioxidant levels in tomatoes through modification of the flavonoid biosynthetic pathway. J Exp Bot 53:2099–2106

Waller JC, Akhtar TA, Lara-Nunez A, Gregory JF, McQuinn RP, Giovannoni JJ, Hanson AD (2010) Developmental and feedforward control of the expression of folate biosynthesis genes in tomato fruit. Mol Plant (1):66–77

Wang LS, Stoner GD (2008) Anthocyanins and their role in cancer prevention. Cancer Lett 269:281–290

Wang C, Chin CK, Gianfagna T (2000) Relationship between cutin monomers and tomato resistance to powdery mildew infection. Physiol Mol Plant Pathol 57(2):55–61

Wang H, Jones B, Li Z, Frasse P, Delalande C, Regad F, Chaabouni S, Latche A, Pech JC, Bouzayen M (2005) The tomato Aux/IAA transcription factor IAA9 is involved in fruit development and leaf morphogenesis. Plant Cell (10):2676–2692

Wang Y, Wisniewski M, Meilan R, Cui M, Fuchigami L (2006) Transgenic tomato (Lycopersicon esculentum) overexpressing cAPX exhibits enhanced tolerance to UV-B and heat stress. J App Hort 8:87–90

Wang Y, Wisniewski M, Meilan R, Uratsu SL, Cui M, Dandekar A, Fuchigami L (2007) Ectopic expression of Mn-SOD in Lycopersicon esculentum leads to enhanced tolerance to salt and oxidative tress. J App Hort 9:3–8

Wang BQ, Zhang QF, Liu JH, Li GH (2011) Overexpression of PtADC confers enhanced dehydration and drought tolerance in transgenic tobacco and tomato: effect on ROS elimination. Biochem Biophys Res Commn 413:10–16

Wang HS, Yu C, Tang XF, Zhu ZJ, Ma NN, Meng QW (2014) A tomato endoplasmic reticulum (ER)-type omega-3 fatty acid desaturase (LeFAD3) functions in early seedling tolerance to salinity stress. Plant Cell Rep 33:131–142

Wang D, Seymour GB, Fray R, Foster T (Unpublished) Modulation of Tomato Fruit Texture by Silencing Cell Wall Structure-Related Genes Divisions of Plant and Crop and Food Science, University of Nottingham, Loughborough LE12 5RD

Whitham S, Mccormick S, Baker B (1996) The N gene of tobacco confers resistance to tobacco mosaic virus in transgenic tomato. Proc Natl Acad Sci U S A 93:8776–8781

Wittmann J, Brancato C, Berendzen KW et al (2016) Development of a tomato plant resistant to *Clavibacter michiganensis* using the endolysin gene of bacteriophage CMP1 as a transgene. Plant Pathol 65:496–502

Xu Q, Xu X, Shi Y, Xu J, Huang B (2014) Transgenic tobacco plants overexpressing a grass PpEXP1 gene exhibit enhanced tolerance to heat stress. PLoS One. https://doi.org/10.1371/journal.pone.0100792

Yanez M, Caseres S, Orellana S, Bastias A, Verdugo I, Luiz-Lara S, Casaretto JA (2009) An abiotic stress-responsive bZIP transcription factor from wild and cultivated tomatoes regulates stress-related genes. Plant Cell Rep (10):1497–1507

Yang Y, Sherwood TA, Patte CP, Hiebert E, Polston JE (2004) Use of tomato yellow leaf curl virus Rep gene sequences to engineer TYLCV resistance in tomato. Phytopathol 94:490–496

Yang L, Shen H, Pan A, Chen J, Huang C, Zhang D (2005) Screening and construct-specific detection methods of transgenic Huafan No 1 tomato by conventional and real-time PCR. J Sci Food Agric 85:2159–2166

Yang MQ, Taylor J, Elnitski L (2008) Comparative analyses of bidirectional promoters in vertebrates. BMC Bioinformatics 28:1471–2105

Yu C, Wang H-S, Yang S, Tang X-F, Duan M, Meng Q-W (2009) Overexpression of endoplasmic reticulum omega-3 fatty acid desaturase gene improves chilling tolerance in tomato. Plant Physiol Biochem 47(11-12):1102–1112

Zanor MI, Osorio S, Nunes-Nesi A, Carrari F, Lohse M, Usadel B, Kuhn C, Bleiss W, Giavalisco P, Willmitzer L, Sulpice R, Zhou YH, Fernie AR (2009) RNA interference of LIN5 in tomato confirms its role in controlling Brix content, uncovers the influence of sugars on the levels of fruit hormones, and demonstrates the importance of sucrose cleavage for normal fruit development and fertility. Plant Physiol (3):1204–1218

Zhang HX, Blumwald E (2001) Transgenic salt-tolerant tomato plants accumulate salt in foliage but not in fruit. Nat Biotechnol 19:765–768

Zhang T, Shi J, Wang Y, Xue SJ (2008) Cultivar and agricultural management on lycopene and vitamin C contents in tomato fruits. In: Preedy VR, Watson RR (eds) Tomatoes and tomato products. Science Publishers, Enfield, pp 27–45

Zhang C, Liu J, Zhang Y, Cai X, Gong P, Zhang J, Wang T, Li H, Ye Z (2011) Overexpression of *SlGMEs* leads to ascorbate accumulation with enhanced oxidative stress, cold, and salt tolerance in tomato. Plant Cell Rep 30:389–398

Zhang Y, Butelli E, Stefano RD, Schoonbeek H, Magusin A, Pagliarani C, Wellner N, Hill L, Orzaez D, Granell A, Jones JDG, Martin C (2013) Anthocyanins double the shelf life of tomatoes by delaying overripening and reducing susceptibility to gray mold. Curr Biol 23:1094–1100

Zheng Z, Nonomura T, Appiano M et al. (2013) Loss of function in Mlo orthologs reduces susceptibility of pepper and tomato to powdery mildew disease caused by *Leveillula taurica*. PLoS One 8(7):e70723

Zheng Z, Appiano M, Pavan S et al. (2016) Genome-wide study of the tomato SlMLO gene family and its functional characterization in response to the powdery mildew fungus *Oidium neolycopersici*. Front Plant Sci 7:380

Zhu M, Chen G, Zhou S, Tu Y, Wang Y, Dong T, Hu Z (2014) A new tomato NAC (NAM/ATAF1/2/CUC2) transcription factor, SlNAC4, functions as a positive regulator of fruit ripening and carotenoid accumulation. Plant Cell Physiol 55:119–135

Chapter 13
Genetic Transformation in Eucalyptus

Shuchishweta Vinay Kendurkar and Mamatha Rangaswamy

Abstract *Eucalyptus*, commonly known as eucalypts, has over 700 species and is native to Australia and the neighboring islands of Timor and Indonesia. Due to their superior growth, adaptability to specific environments, and desirable wood properties, *Eucalyptus* species have become the most valuable and widely planted hardwoods in the world. The main theme to attempt genetic transformation in trees is the improvement of productivity and quality. The potential of production of trees with novel traits is one of the most distinct benefits of genetic transformation. There are three prerequisites for successful genetic transformation of a cell or tissue: introduction of the DNA into the cell, its integration into the host genome, and the controlled expression of the introduced DNA. Common methods for genetic transformation are usually divided into indirect or direct transformation. Biological methods using bacteria are referred to as indirect, while direct methods are physical which are based on the penetration of the cellular wall. Indirect transformation methods introduce plasmids/independent circular molecules of DNA that are found in bacteria, separate from the bacterial chromosome into the target cell by means of bacteria capable of transferring genes to higher plant species. The most popular used microorganisms are *Agrobacterium tumefaciens* and *Agrobacterium rhizogenes*. Direct transfer includes electroporation and microprojectile/biolistics/particle bombardment.

Keywords *Eucalyptus* · Genetic transformation · Direct transfer · Indirect transfer · *Agrobacterium* · Electroporation · Biolistic

S. V. Kendurkar (✉) · M. Rangaswamy
Plant Tissue Culture Division, National Chemical Laboratory, Pune, India

© Springer International Publishing AG, part of Springer Nature 2018
S. S. Gosal, S. H. Wani (eds.), *Biotechnologies of Crop Improvement, Volume 2*,
https://doi.org/10.1007/978-3-319-90650-8_13

13.1   Introduction

Forestry has been enormously benefited from the development and implementation of improved silvicultural forest management practices and breeding techniques. Establishing forest plantations to meet the ever-increasing demand for tree products has been a long-standing tradition in the tropics (Evans 1999). These have contributed significantly to the improvement of forest tree species in the past and will continue to have a substantial impact on the genetic gain and productivity of economically important tree species by providing better germplasm and improved management practices for plantation forests (Turnbull 1999). Conventional breeding methods of most of the tree species are often constrained by the long reproductive cycles and difficulty in achieving significant improvements toward the complex traits such as wood properties, disease/pest control, and tolerance to abiotic stresses. The state of food and agriculture reported that biotechnology is considered to be much more than genetic engineering (FAO 2004). Biotechnology is an adjunct to the long-established traditional tree improvement practices and utilizes fundamental discoveries in the field of plant tissue culture for clonal forestry, gene transfer techniques, molecular biology, and genomics. These new discoveries provide an extended platform for the improvement of traits that have previously been considered impractical via conventional breeding methods. Biotechnology also provides exciting opportunities to further expand our understanding of genome organization and functioning of genes associated with complex value-added traits and to transfer such genes into economically important tree species. This will lead to the development and deployment of trees ready to meet the future demand of the world's ever-increasing population for timber and other forest products while preserving natural forests for future generations (Vikas et al. 2015).

Genetic engineering has the potential to boost global wood production in many ways (Sutton 1999; Sedjo 2001). Applications for plantation forests currently include resistance to biodegradable herbicides, altered lignin properties for reduced downstream processing costs or improved burning, resistance to selected pests, altered reproductive mechanisms for faster breeding or genetic containment, phytoremediation of polluted sites, and the production of novel chemicals or pharmaceuticals (Strauss et al. 1999, 2001; Yanchuck 2001). It might also be possible to manipulate wood quality traits, photosynthetic efficiency, and tolerance to abiotic stresses such as drought (Fenning and Gershanzon 2002), economic benefits resulting from the introduction of forest biotechnology (Sedjo 2001), and ecological issues associated with the deployment of genetically modified forest tree species (Van Frankenhuyzen and Beardmore 2004).

Conventional tree improvement programs involve selection, genetic crosses, and recurrent testing, have been used around the world to improve plantation forestry yield, and have certainly proven useful. Unfortunately, the recognition of the potential of biotechnology in the forest sector is much more limited. However, it is becoming an increasingly important component of the processing sector, such as pulp and paper production, and it also plays an important role in various stages of the production chain, from planting to harvesting. One of the first applications of biotechnology in forestry was the inoculation of seedlings with symbiotic organisms (specifically mycorrhizae) with the objective of increasing seedling growth. Since then, tremendous progress has been made in the field of forest biotechnology, which currently focuses on main areas such as propagation, genetic transformation, transgenic approaches, abiotic stress resistance, biotic stress resistance, modification of lignin, RNAi interference, marker-assisted selection and QTL mapping, and future directions in forest tree genomics research. The directed desirable gene transfer from one organism to other and subsequent stable integration and expression of foreign gene into the genome are referred as genetic transformation. The transferred gene is known as transgene, and the organisms that develop after a successful gene transfer are known as transgenic. The most widely used term is GMO, i.e., genetically modified organisms. Transgenic plants are generated by introducing foreign DNA into a plant, and regenerated plants contain the foreign DNA. Transgenic plants are the plants that carry the stably integrated foreign genes (Chawla 2000).

13.1.1   Origin of Eucalyptus Species and Their Hybrids

Over 700 *Eucalyptus* species, commonly known as eucalypts, are native to Australia and the neighboring islands of Timor and Indonesia (Groves 1994; Ladiges 1997; Myburg et al. 2007). Eucalypts grow across a wide range of soil types and climatic environments, ranging from lowland tropical forests to temperate high elevations that regularly experience freezing temperatures. Natural *Eucalyptus* forests cover over 40 million hectares (Eldridge et al. 1994).

Eucalypts are among the fastest growing woody plants in the world with mean annual increments up to 100 m^3/ha. Due to their superior growth, adaptability to specific environments, and desirable wood properties, *Eucalyptus* species have become the most valuable and widely planted hardwoods in the world with ~11.8 million hectares planted in 90 countries (FAO 2007). They are widely grown as exotic plantation species in tropical and subtropical regions of Africa, South America, and Asia. They are also planted in some temperate regions of Europe, South America, North America, and Australia. Four eucalypt species, *Eucalyptus grandis*, *E. urophylla*, *E. camaldulensis*, and *E. globulus* together with various hybrids with these species, account for about 80% of the eucalypt plantations worldwide (Eldridge et al. 1993; Grattapaglia and Kirst 2008), selected mainly based on their good growth and form and adaptability in different regions. *E. globulus* is the premier species for temperate zone plantations in Portugal, Spain, Chile, and Australia. *E. grandis* is the most widely used species in plantation forestry worldwide in tropical and subtropical areas. It is planted as a pure species but also utilized as a parental species in hybrid breeding (Myburg et al. 2007).

Eucalypts have been historically classified into two genera (*Angophora* Cav and *Eucalyptus* L'Her) that belong to the Myrtaceae family of angiosperms (Briggs and Johnson 1979). Over the years, several classifications of the genus *Eucalyptus* have been proposed. Among these, a comprehensive and informal classification proposed by Pryor and Johnson (1971) has been widely used by taxonomists and ecologists. This classification recognizes seven subgenera within *Eucalyptus* (*Corymbia*, *Blakella*, *Eudesmia*, *Gaubaea*, *Idiogenes*, *Monocalyptus*, and *Symphyomyrtus*). An eighth subgenus, *Telocalyptus*, was subsequently added to this list by Johnson (1976).

13.1.2   Cytogenetics of Eucalyptus

All examined species of the genus *Eucalyptus* are diploid (2n = 22) with haploid chromosome number of $n = 11$ (Myburg et al. 2007). There are no confirmed reports of natural polyploidization in *Eucalyptus* (Eldridge et al. 1994; Grattapaglia and Bradshaw 1994; Potts and Wiltshire 1997). The chromosomes of *Eucalyptus* species are extremely small in size (2–6 μm) with diploid nuclear (2C) DNA content ranging from 0.77 to 1.47 pg (Grattapaglia and Bradshaw 1994).

Natural hybridization among different subgenera and sections within the genus *Eucalyptus* is rare, and hybrid viability decreases with increasing taxonomic distance between parents (Griffin et al. 1988a; Potts and Dungey 2004). Even among closely related species, hybridization rates are generally very low (Volker 1995).

13.1.3   Need for Transformation

The main theme to attempt genetic transformation in trees is the improvement of productivity and quality. The potential of production of trees with novel traits is one of the most distinct benefits of genetic transformation. The first successful transformation in trees was achieved by Fillatti et al. (1987) in *Populus*. Since then steady progress has been witnessed in many new inventions and techniques over the past two decades, which have been reviewed extensively (Table 13.1). The idea of using several species, most of which belong to the genera *Eucalyptus*, *Pinus*, *Picea*, and *Populus*, and rubber for molecular farming of desired products is also gaining momentum (Merkle and Dean 2000; Pena and Seguin 2001; Herschbach and Kopriva 2002; Diouf 2003; Gallardo et al. 2003; Gartland et al. 2003).

Transgenic technology is undoubtedly a powerful complementary tool available to the molecular breeder. Considering that industrial *Eucalyptus* forests are almost exclusively clonal, transgenics will most likely have an increasing role not only in wood quality improvement but in resolving problems related to pest and pathogen susceptibility and/or abiotic stress tolerance (e.g., frost, drought) owing to monoculture that might limit the expansion or survival of existing plantations. The introduction of genes that confer traits that do not display variation within the *Eucalyptus* gene pool or impossible to be attained by the natural recombination processes might radically modify the ways that forests are planted or that forest products are derived.

Table 13.1 Summary of the *Eucalyptus* transformation studies

Sl. no	Plant species	Tissue/organ used for transformation	Genes transferred	Marker gene	Promotors	Mode of transformation	Reference
1	*E. gunnii*	Protoplasts	NA	NA	NA	Electroporation using PEG	Teulieres et al. (1991)
2	*E. citriodora*	Protoplasts	NA	NA	NA	Electroporation using PEG	Manders et al. (1992)
3	*E. globulus*	Six-day-old embryos	uid II			Biolistic	Rochange et al. (1995)
4	*E. globulus*	Zygotic embryos	uid II	nptII		Particle gun bombardment	Serrano et al. (1996)
5	*E. camaldulensis*	Leaf explants from *in vitro*-grown plants	*uid II*	*nptII*	CaMV355	*A. tumefaciens*-mediated gene transfer	Mullins et al. (1997)
6	*E. camaldulensis*	Hypocotyl segments from *in vitro* seedlings	*uid II*	*nptII*	CaMV355	*A. tumefaciens*-mediated gene transfer	Ho et al. (1998)
7	*E. camaldulensis*	Cotyledon and hypocotyl	*uid II*	*nptII*	CaMV 35S	*A. tumefaciens*-mediated gene transfer	Moralejo et al. (1998)
8	*E. grandis*	NA	*Glyphosate resistance gene*	*NA*	NA	NA	Llewellyn (1999)
9	*E. camaldulensis*	Cotyledon and hypocotyl	*cry3A*	*har*	CaMV 35S	*A. tumefaciens*-mediated gene transfer	Harcourt et al. (2000)
10	*E. camaldulensis*	Leaf explants from *in vitro* shoot cultures	*Populus tremuloides* C4H+ gus	*nptII*	CaMV 35S	*A. tumefaciens*-mediated gene transfer	Chen et al. (2001)
11	*E. grandis x E. urophylla*	Seedling	*uid II* *Lhcb1*2*	*nptII* *nptII*	CaMV 35S	*A. tumefaciens*-mediated gene transfer	Gonzalez et al. (2002)
12	*E. camaldulensis*	NA	*Antisense Ntlim1*	*NA*	NA	NA	Kawaoka et al. (2003)

(continued)

Table 13.1 (continued)

Sl. no	Plant species	Tissue/organ used for transformation	Genes transferred	Marker gene	Promotors	Mode of transformation	Reference
13	*E. grandis* x E. *urophylla*	Leaf explants from seedling-derived shoot cultures	*CAD antisense cDNA*	*nptII*	CaMV 35S	*A. tumefaciens*-mediated gene transfer	Tournier et al. (2003)
14	*E. camaldulensis*	Leaf explants from *in vitro*-grown seedlings	CAD gene	*nptII*	CaMV 35S	*A. tumefaciens*-mediated gene transfer	Valerio et al. (2003)
15	*E. globulus*	Apical stem segments	*uid II*			*A. tumefaciens*-mediated gene transfer	Antanas et al. (2005)
16	*E. camaldulensis*	Mature leaf, cotyledon, and hypocotyl	*uid II*	*nptII*	CaMV 35S	*A. tumefaciens*-mediated gene transfer	Chen et al. (2007)
17	*E. occidentalis*	Cotyledon and hypocotyl	*uid II*	*nptII*	CaMV 35S	*A. tumefaciens*-mediated gene transfer	Southerton (2007)
18	*E. tereticornis*	Cotyledon and hypocotyl explants	*uid II*	*nptII*	CaMV 35S	*A. tumefaciens*-mediated gene transfer	Prakash and Gurumurthi (2009)
19	*E. saligna*	Leaf explants from *in vitro*-grown seedlings	*P5CSF129A uidA*	*nptII*	CaMV 35S	*A. tumefaciens*-mediated gene transfer	Dibax et al. (2010)
20	*E. tereticornis*	Healthy actively growing callus	â-Glucuronidase and hygromycin phosphotransferase			Particle bombardment	Nair and Vijayalakshmi (2010)
21	*E. tereticornis*	Fully expanded leaves from elongated microshoots	*uid II*	*nptII*	CaMV 35S	*A. tumefaciens*-mediated gene transfer	Aggarwal et al. (2011)
	E. grandis, E. urophylla, E. globulus		EgCesA3	*nptII*	CaMV 35S	*A. tumefaciens*-mediated gene transfer	Marques et al. (2011)
22	*E. grandis* x E. *urophylla*	Leaf from micro-cutting	*EguCBF1a/b*	*nptII* and *gfp*	CaMV 35S	*A. tumefaciens*-mediated gene transfer	Navarro et al. (2011)

No.	Species	Explant	Gene/construct		Method	Reference
23	E. grandis, E. grandis x E. nitens, E. grandis x E. camaldulensis, E. grandis x E. urophylla	Leaf explants from in vitro-grown plants	uidA gene (β-glucuronidase; GUS) and hpt (hygromycin phosphotransferase)		A. tumefaciens-mediated gene transfer	Raj et al. (2011)
24	E. globulus	Hypocotyl sections with or without shoot apex from in vitro-germinated seedlings	(codA) with a GUS reporter gene	nptII	A. tumefaciens-mediated gene transfer	Matsunaga et al. (2012)
25	E. camaldulensis	Leaf explants from in vitro-grown seedlings	pCAMBIA3301 (35S, GUS, NOS)		Agrobiolistics transformation A. tumefaciens-mediated gene transfer	Mendonca et al. (2013)
26	E. camaldulensis	Cotyledonary leaves, plantlet leaves, and hypocotyls of young in vitro plants and calli	pGA482 with uidA (GUS) gene and nptII		A. tumefaciens-mediated gene transfer	Ahad et al. (2014)
27	E. globulus	Microshoots	uidA (β-glucuronidase) the gfp (green fluorescent protein) gene and the nptII (neomycin phosphotransferase)		A. tumefaciens-mediated gene transfer	de la Torre et al. (2014)
28	E. urophylla x E. grandis	Axillary shoots with short internodes			Sonication-assisted agrobacterium-mediated transformation method	Makouanzi et al. (2014)
29	E. grandis	Hypocotyl of 14-day-old seedling	Eucalyptus cinnamoyl-CoA reductase1 (EgCCR1) gene		Hairy root induction	Plasencia et al. (2016)
30	E. urophylla		14GLU4 and CAD gene			Chen et al. (2016a, b)
31	E. urophylla x E. grandis	Hypocotyl segments from aseptically grown seedlings	aiiA	PPP3	A. tumefaciens-mediated gene transfer	Ouyang and Li (2016)

In spite of the recognized economic importance of Eucalyptus in world forestry, very little has been published on transgenic experiments in species of the genus. Several reports have documented the production of transformed callus, tissue, and root organs; however, reports on transformed plants are scarce (MacRae and van Staden 1990; Machado et al. 1997).

A marked genotype effect has been observed on the efficiency of regeneration and consequently, stable transformation. This fact has prompted several groups to first identify Eucalyptus "lab rats," i.e., easily regenerable genotypes, and only after that develop improved protocols to generate large numbers of independent transformation events. This research is carried out primarily by private companies. The most representative and complete work published on Eucalyptus transformation was carried out by a French-Brazilian group where an easily regenerable plant was selected after screening around 300 plantlets of an E. grandis x E. urophylla hybrid. This plant exhibited the best compromise between short- and long-term GUS expression levels, regeneration, and micropropagation efficiency under selection after transformation. Following this selection step, stably transformed plants were obtained for some reporter genes and a cinnamyl alcohol dehydrogenase (CAD) antisense cDNA from E. gunnii at an efficiency of 10% (120 transformed and confirmed plants from 1200 explants), thus demonstrating the efficiency of the protocol (Tournier et al. 2003).

Recently, Chen et al. (2016a) also described a basic Agrobacterium-mediated genetic transformation protocol through organogenesis for the production of transgenic plants using E. camaldulensis. More importantly, modifications of the protocol for mature tissues derived from elite trees and other Eucalyptus species were also described. Efficient transformation protocols have also been developed in Japan, where an E. camaldulensis has been used (Kawazu et al. 1996) as well as different labs in the USA with E. grandis and E. urophylla (Hinchee and Chiang, "personal communication"). However, these have not yet been published although they constitute important components of patent applications (Grattapaglia and Kirst 2008).

The current information on Eucalyptus transgenesis points to a very promising future as far as the technical possibility of generating stably transformed Eucalyptus plants is concerned. However, according to Grattapaglia and Kirst (2008), some strategic issues in the adoption of transgenic technology have been a matter of recent debate, including:

(a) What is the magnitude of the attainable gain and cost/benefit relationship by manipulating lignification or cellulose genes, when compared to the exploitation of the genetic variation in Eucalyptus by hybridization and intensive selection?

(b) What are the specific biosafety and intellectual property issues relevant to transgenic eucalypts and the time and investment necessary to solve them to actually be able to plant transgenic trees on a large scale?

(c) What is the speed by which breeding programs generate new and better clones for several adaptability traits (growth, pest resistance, clonability, etc.) compared to the time needed for regulatory approval of every new transgenic clone?

(d) What is the life-span of a patent in the local regulation as compared to the time needed to effectively make returns on the patent from the planted forest before the patent goes into public domain?

(e) What are the market issues that the company has to consider in adopting transgenics both in relation to public perception and forest certification processes?

All these and other issues will have to be carefully considered without overlooking that, just as occurred in annual crops such as soybean, maize, and cotton, the use of transgenics could become a major technology divide and represent the necessary condition for a forest-based industry to continue competitive in the world scenario (Grattapaglia and Kirst 2008).

13.2   Modes of Transformation

Common methods for genetic transformation are usually divided into indirect or direct transformation (Qayyum et al. 2009). Biological methods using bacteria are referred to as indirect, while direct methods are physical which are based on the penetration of the cellular wall. Indirect transformation methods introduce plasmids/independent circular molecules of DNA that are found in bacteria, separate from the bacterial chromosome into the target cell by means of bacteria capable of transferring genes to higher plant species (Broothaerts et al. 2005). The most popular used microorganisms are *Agrobacterium tumefaciens* and *Agrobacterium rhizogenes*, two soil native bacteria (Bevan et al. 1983; Zupan and Zambryski 1997; Patnaik and Khurana 2001; Smith 2003; Dhar et al. 2011).

There are three prerequisites for successful genetic transformation of a cell or tissue: introduction of the DNA into the cell, its integration into the host genome, and the controlled expression of the introduced DNA (Lindsey and Jones 1990).

The application of genetic transformation technology is dependent on the availability of efficient systems for the transfer of foreign genetic material into cells capable of giving rise to fertile plants. For this reason a wide array of techniques has been developed for DNA transfer into plant cells. Various methods for delivering foreign DNA into eucalyptus have been studied. They include *Agrobacterium*-mediated transfer, electroporation, and microprojectile/biolistics/particle bombardment.

13.2.1   Indirect Transformation

13.2.1.1   *Agrobacterium*-Mediated Gene Transfer

Introduction of DNA into a cell or tissue can be achieved in many ways, the most commonly used method being based on the natural gene transfer system of the soil bacterium *A. tumefaciens*. So far, *Agrobacterium*-mediated transformation is the most followed and reported method in eucalyptus.

The host range of *A. tumefaciens* was initially thought to include only dicotyledonous species; however, T-DNA insertion has subsequently been demonstrated in some monocotyledonous plant species (Chan et al. 1992; Hooykaas and Schilperoort 1992; Delbreil and Jullien 1993; Ritchie 1991). *Agrobacterium* also has the ability to insert a particular segment of foreign DNA into the plant genome, under the control of genetic elements within the bacterium, such as the promoter. The promoter is a regulatory element in the immediate vicinity of the transcription start site (Beilmann et al. 1992) and is the DNA region which binds RNA polymerase and directs the enzyme to the correct transcriptional site so that RNA synthesis can begin. Unless placed under the control of suitable promoter elements, bacterial genes are not transcribed after integration into the plant genome. *Agrobacterium* T-DNA is the exception, since it contains its own promoter elements (Fraley et al. 1983). Genetic transformation methods rely on random insertion of DNA into the plant genome (Potrykus et al. 1985). Successful integration is therefore influenced by factors such as DNA conformation, concentration, and the type of vector used. Different promoters have different efficiencies and transcriptional levels.

First report of *Agrobacterium*-mediated transformation was on *E. globulus* and *E. gunnii* by Chirqui et al. (1992). According to Mullins et al. (1997), indirect gene transfer approach is preferred for long-living tree species as it is known to reduce the insertion of multiple copies of the transgene, which can lead to gene silencing. Reliable regeneration protocols for *E. camaldulensis* using leaf explants from well-grown plants have been developed by Mullins et al. (1997) using 24 clones. Identical protocols were also successful in the regeneration of some clones of *E. microtheca*, *E. ochrophloia*, *E. grandis*, and *E. marginata* but at lower frequencies. Co-cultivation of *E. camaldulensis* leaf explants with *A. tumefaciens* strains carrying a kanamycin resistance gene and the reporter gene glucuronidase (*gus*), followed by selection on kanamycin, allowed the selection of transformed shoots that could be rooted on selective media. Transformation of the plants was verified by staining for the *gus* enzyme in various plant tissues and *npt*II assays and by Southern blotting on isolated DNA using specific probes for both the *gus* and selectable marker genes. Transformed tissue was obtained with five clones of *E. camaldulensis* tested and a number of *A. tumefaciens* strains. However, only one clone regenerated transformed whole plants reliably.

Machado et al. (1997) evaluated the susceptibility of *E. grandis* × *E. urophylla* hybrids to 12 *A. tumefaciens* wild strains. Different degrees of virulence are tested using these different stains, indicating the possibility of transforming eucalyptus hybrids with *Agrobacterium*-derived vectors. The ability of *A. tumefaciens* to infect eucalyptus varied across species and genotypes.

An efficient system for transformation of *E. camaldulensis* and production of transgenic plants was developed by Ho et al. (1998) by co-cultivation of hypocotyl segments with *A. tumefaciens* containing a binary Ti plasmid vector harboring *npt*II and *gus* genes. Histochemical assay revealed the expression of the *gus* gene in leaf, stem, and root tissues of transgenic plants. Insertion of the *gus* gene in the nuclear genome of transgenic plants was verified by Southern hybridization analysis, further confirming the integration and expression of T-DNA in these plants.

Krimi et al. (2006) reported that *E. occidentalis* was more susceptible to this bacterium than *E. camaldulensis* and *E. cladocalyx*. Sonication-assisted *Agrobacterium* transformation (SAAT) system was also employed in the production of transgenic eucalyptus from *E. grandis* × *E. urophylla* hybrid (Gonzalez et al. 2002). The report indicated that germinated seeds and seedlings showed high percentage of transient *gus* expression when sonicated for 30 s and presonication greatly enhanced the efficiency of transformation. The efficiency of the method was also assessed using a chimeric construct containing the Lhchl2 gene of the 28 kDa chlorophyll a/b binding pea protein from the LHC11 antenna. Using this construct, four stable transformants were generated and confirmed with genomic blotting.

The latest advancement in eucalyptus transformation has been reported by Gred Bossinger group from University of Melbourne. They reported *Agrobacterium*-mediated *in vivo* transformation of wood-producing stem segments in eucalyptus. Unlike the earlier three methods which led to the generation of stable transgenic plants, this procedure is more involved in generating transgenic sectors in growing eucalyptus plants.

Studies were carried out with a view to develop *A. tumefaciens*-mediated genetic transformation protocol in eucalyptus using leaf and apical meristem explants derived from *in vitro*-grown microshoot cultures (Figs. 13.1 and 13.2). The explants were subjected to kanamycin sensitivity assay using a range of 0–50 mg/l concentration of kanamycin by preculturing for 0–3 days on regeneration medium. Different concentrations of *A. tumefaciens* strain LBA 4404 harboring the binary

Fig. 13.1 Regeneration in *Eucalyptus tereticornis* Sm. Leaf explants from mature tree-derived microshoots

Fig. 13.2 *Agrobacterium tumefaciens*-mediated transformation of *Eucalyptus tereticornis* Sm

vector pCAMBIA2301 containing the *uidA* and kanamycin resistance were used for infection (O.D 0.3–0.6. at 600 nm) of the explants by co-culturing for 2, 3, and 4 days. Optimum concentration of cefotaxime in selective regeneration medium was determined by well diffusion method to control the growth of *Agrobacterium*. Co-cultivated explants were transferred to selective regeneration medium. Histochemical assay was performed for leaves dissected from apical meristem and regenerating shoot from leaf explants. Leaves harvested from these cultures were tested positive for GUS assay. Beyond 10 mg/l concentration of kanamycin, explants turned brown and in few cases necrosis occurred. Explants precultured for 3 days turned brown. *A. tumefaciens* at OD 0.6 and co-cultivated for 3 days showed hyperinfection and led to the death of explants. In selective regeneration medium concentration below 300 mg/l did not control growth of *Agrobacterium* (Kendurkar, Khan, and coworkers, "personal communication", 2012).

13.2.2 Direct Transformation

The cellular wall is the natural barrier that all methods of genetic transformation have to overcome to achieve DNA penetration into the cell. Direct methods originated in the 1980s due to the big interest in modifying crops, almost impossible to be manipulated by *Agrobacterium* (Neumann et al. 1982; Paszkowski et al. 1984, Vasil 2005). They offer an alternative for integrating multiple copies of a desired gene with minimal cellular toxicity at random sites into the genome (Qayyum et al. 2009). Their disadvantages involve problems with plant regeneration and a low transient expression of transgenes. Direct methods are electroporation (Zimmermann and Vienken 1982; Luciano et al. 1987; Chand et al. 1988; Toriyama et al. 1988; Tada et al. 1990; Spörlein and Koop 1991; Cheah 2001; Danilova 2007; Hjouj and Rubinsky 2010) and biolistics (Han and Yang 2000; Kothari et al. 2005; Castellanos-Hernández et al. 2009).

A wide diversity of sources of transgenes and regulatory elements, and intended traits, have been tested, including expression of reporter genes; insect, disease, and herbicide resistance; modified wood properties; modified flowering and fertility; and modified growth rate and stature. Differentiation of transformed cells is a prerequisite to obtaining transgenic plants, and two systems are being used in forest trees: organogenesis and embryogenesis. Such transformation procedures, including the use of selectable markers and screening methods, are well established. It is possible to introduce one or more perfectly characterized new characters without, in theory, adversely affecting the overall genetic makeup of the plant. This approach also offers the possibility of overcoming the genetic barrier between species, in a relatively shorter time frame than through conventional tree breeding. Harfouche et al. (2011) suggest that major obstacles to efficient production of transgenic trees are:

(i) Difficulties in plant regeneration from *Agrobacterium*-infected or particle-bombarded explants
(ii) Incomplete development beyond the *in vitro* stage of rooted plants for establishing field trials
(iii) Transgene instability during the long life-span of forest trees, including transgene silencing and somaclonal variation

Once transgenesis is performed at the cell level, *in vitro* culture techniques can be used to regenerate the entire tree.

13.2.2.1   Electroporation

Electroporation is a popular technique of genetic transformation because it is simple, quick, and highly efficient for a wide variety of tissues (Zimmermann and Vienken 1982; Luciano et al. 1987; Chand et al. 1988; Toriyama et al. 1988; Tada et al. 1990; Spörlein and Koop 1991; Cheah 2001; Danilova 2007). It is commonly used to transport biochemical substances like lipids, proteins, ribonucleic acid (RNA), and DNA to the cell interior. The method enhances the formation of pores on the cell surface due to a polarity alteration on the membrane, caused by an electrical field. This phenomenon can be observed with a microscope. It has been mainly applied to transform plant protoplasts, i.e., cells without a wall, of various cellular types like corn with efficiency of 90 transgenic plants recovered from 1440 maize embryos (6.2%) and wheat with an efficiency of 3 plants from 1080 embryos (0.3%) (Tada et al. 1990; Spörlein and Koop 1991; Cheah 2001; Danilova 2007).

An electrical field (alternate or pulsed) applied to a cellular suspension induces a dipolar moment inside the cells and a potential difference through the plasmatic membrane (Zimmermann and Vienken 1982; Toriyama et al. 1988; Zhang and Wu 1988; Kubiniec et al. 1990; Weaver and Chizmadzhev 1996; Escoffre et al. 2009). The induced voltage can lead to cell permeabilization due to an electrical imbalance in the plasma membrane when the potential difference is bigger than 0.5 V at normal conditions of pressure and temperature (there is a membrane voltage threshold from 0.5 to 1 V). It has been shown that the pulse length, type, and duration have a strong effect on the transformation efficiency (Abdul-Baki et al. 1990; Joersbo and Brunstedt 1990; Kubiniec et al. 1990; Weaver and Chizmadzhev 1996). The effects of this electrical imbalance are reversible only when the electrical pulse lasts less than 100 µs (Turgut-Kara and Ari 2010). Under these circumstances, DNA can be introduced into the cells without changing the cellular functions or the membrane integrity (Djuzenova et al. 1996). It has been proposed (Sukhorukov et al.1998; Saulis et al. 1991) that the membrane permeabilization is due to the transitory force of electro-deformation produced by the electrostatic interaction of the dipoles generated on the cells due to the applied electrical field. Several physical factors such as transmembrane potential created by the imposing pulsed electric field, extent of membrane permeation, duration of the permeated state, mode and duration of molecular flow, global and local (surface) concentrations of DNA, form of DNA, tolerance of cells to membrane permeation, and the heterogeneity of the cell population may affect the electro-transfection efficiency (Gallie et al. 1989; Dekeyser et al. 1990; Izawa et al. 1991; Hui 1995; Weaver 1995; Escoffre et al. 2009).

Unfortunately, this technique has a low efficiency and can only be applied to protoplasts, using laborious protocols for the regeneration after genetic transformation. In this technique mixture containing cells and DNA is exposed to very high-voltage electrical pulses (4000–8000 V/cm) for very brief time periods (few milliseconds). It results in formation of transient pores in the plasma membrane, through which DNA seems to enter inside the cell and then nucleus. Initially, transformation of *E. saligna* using electroporation has been reported by Kawasu et al. (1990). Further, *cat* and *gus* gene transient expression has been studied in the protoplasts of *E. gunnii* and *E. citriodora* obtained via electroporation and polyethylene glycol (PEG) treatment (Teulieres et al. 1991; Manders et al. 1992).

13.2.2.2 Biolistic Method

Biolistics, also known as "particle bombardment" or "gene gun technique," consists on the acceleration of high-density carrier particles, approximately two microns in diameter (which is smaller than a plant cell), covered with genes that pass through the cells, leaving the DNA inside (Hartman et al. 1994; Huang and Zhang 1999; Han and Yang 2000; Huang et al. 2002; Kothari et al. 2005; Antony Ceasar and Ignacimuthu 2009). It was designed at Cornell University in 1987 to handle the genetic transformation of cereals (Sanford et al. 1987); however, it can be used on many species. The technique can be employed for nuclear and chloroplast transformation (Boynton et al. 1988). Cells, protoplasts, organized tissues like meristems (a group of non-differentiated cells with active mitosis), embryos, or callus can be used as a target (Sanford et al. 1993). Originally the biolistic method was developed with the aim to transform monocotyledons, which are recalcitrant to transformation with *Agrobacterium*. Comparison of *Agrobacterium* and biolistics in terms of transformation efficiency, transgene copy number, expression, inheritance, and physical structure of the transgenic loci using fluorescence in situ hybridization shows that, in general, *Agrobacterium* offers significant advantages over biolistics (Southgate et al. 1995; Snyder et al. 1999; Taylor and Fauquet 2002; Travella et al. 2005). Nevertheless, biolistics is the most accepted direct technique for genetic transformation of plants because it can be used for many species, subcellular organelles, bacteria, fungi, and even animal cells, because it has a short processing time and low costs involved in the production of transgenic plants and due to its simplicity for introduction of multiple genes or chimeric DNA (DNA from two different species). Furthermore, it does not need a vector of a specific sequence and does not depend on the electrophysiological properties of the cell, like the electrical potential and the structural components of the cellular membrane. However, the transformation parameters must be optimized to each biological target employed (Sanford et al. 1993).

In plant research, the major applications of biolistics include transient gene expression studies, production of transgenic plants, and inoculation of plants with viral pathogens (Taylor and Fauquet 2002). Only 50% of the tissue under bombardment survives to obtain a transformed plant. The method has a transformation efficiency of 0.002 with a genetic translational degree from 17 to 36% of the relative activity during events in one bombardment, while there is up to 70% of activity in the genetic expression during events in multiple bombardments (Oard et al. 1990). Particle bombardment has been used to genetically transform several plants. Microprojectile bombardment or biolistic method employs high velocity metal particles to deliver biologically active DNA into plant cells. The concept has been described by Sanford (1990). Following the original observation by Klein et al. (1987), tungsten particles could be used to introduce macromolecules such as RNA and DNA into epidermal cells of onion with subsequent transient expression of enzymes encoded by these compounds. Christou et al. (1991) demonstrated that the process could be used to deliver biologically active DNA into living cells and result in the recovery of stable transformants.

The first attempt on optimization of biological and physical parameters for particle bombardment in *E. globules* was carried by Rochange et al. (1995). Based on transient *gus* expression studies, they observed that both gun powder and compressed helium gas device exhibited similar transformation efficiency and reported that 6-day-old cultured embryos were best suited for genetic transformation of eucalyptus. Later, Serrano et al. (1996) for the first time reported successful regeneration of single *E. globulus* plant. Stable transformation after biolistic DNA delivery was investigated using 6-day-old cultured zygotic embryos as target material, and whole plants were recovered through organogenesis after particle gun bombardment of a linear T-DNA fragment harboring *gus* and *nptII* genes. After 2 months, neoformed *gus*-positive calli were obtained from the "T-DNA"-bombarded embryos. *Gus*-expressing calli were also recovered after selection with kanamycin following bombardment, and integration of both genes into the *Eucalyptus* genome was confirmed by Southern blot analysis.

Biolistic transformation was also carried on hypocotyls and cotyledons of *E. grandis* × *E. urophylla* hybrids by Sartoretto et al. (2002). They reported that *gus*-expressing calli couldn't regenerate into transgenic shoots and that the tissue culture conditions favorable for regeneration hinder the regeneration of transgenic tissues and vice versa. Genetic transformation of ITC 3, a superior clone of *E. tereticornis*, was attempted by Nair and Vijayalakshmi (2010) with transformation vectors pAHC 25 and pHX4 carrying â-glucuronidase and hygromycin phosphotransferase, respectively. However, *gus* expression could not be achieved. However, there are no reports of successful transformation work in these lines where eucalyptus-specific transformation parameters have been optimized.

At CSIR-NCL, genetic transformation studies were carried out using apical meristems and leaves dissected from *in vitro*-grown microshoots of elite clones of *Eucalyptus tereticornis* Sm. Biolistic DNA delivery system and *Agrobacterium*-mediated transformation systems were studied in depth. Critical parameters for plant regeneration and transformation were standardized. Efficient regeneration was observed in 1–2-week-old precultured meristems and leaves. Transformation efficiency was enhanced using particle bombardment followed by co-cultivation with *Agrobacterium tumefaciens*. The *Agrobacterium* strain GV 2260 harboring the binary vector pCAMBIA1301 coated on the gold microprojectiles containing *uid A* and hptII genes produced transformants which were screened on selection media containing hygromycin (20 mg/l). Confirmation of the putative hygromycin-resistant transformants was done by the GUS assay. The transformation frequency observed was 43.11%. Further confirmation of GUS-positive, hygromycin-resistant transformants was carried out by polymerase chain reaction. This regeneration system allows effective transformation and direct regeneration of *E. tereticornis* Sm. from apical meristems and leaf explants. These results showed that gene transfer by high-velocity microprojectiles is a rapid and direct means for transforming intact plant cells and tissues (Kendurkar, Khan, and coworkers, "personal communication", 2012).

13.2.3   Factors Affecting the Transformation

The nopaline synthase (nos) and the 35S transcript promoters from the cauliflower mosaic virus (CaMV) are commonly used promoters (Hensgens et al. 1992). Chimeric genes that function in plants may thus be produced, and these are often marked with bacterial antibiotic resistance genes. An important limitation in the use of gene transfer technology is the fact that there is considerable inter-transformant variability in expression levels of introduced genes (Dunsmuir et al. 1988; Hooykaas and Schilperoort 1992). The factors governing the expression of foreign genes and causing this variability in plants remain an enigma, although factors such as DNA methylation (Hooykaas and Schilperoort 1992; Ottaviani et al. 1993), DNA copy number, and position in the hI lst genome have been suggested to affect expression of the introduced gene (Gao et al. 1991), and these factors may also function over long distances, such as chromatin folding (Dunsmuir et al. 1988).The correlation between copy number and gene expression in transformants has been reported to be positive (Gendloff et al. 1990), indeterminate (Shirsat et al. 1989), or negative (Hobbs et al. 1990). This is further complicated by the fact that the introduction of additional copies of naturally occurring genes may have a repressive effect on gene expression (Hobbs et al. 1993).

The finding that transgenes can influence expression of each other as well as the expression of resident genes in transgenic plants (referred to as co-suppression) indicates that there is still much to be learnt about the nuclear processes involved in gene regulation and genome maintenance (Kooter and Mol 1993). The advances that have been made in the chemical regulation of transgene expression explants (Ward et al. 1993), allows the manipulation of levels or gene expression in order to gain an understanding of the functions of individual genes. Although the constitutive expression of inserted genes is adequate at present, other inserted genes may be useful only if placed under exogenous control or regulation (Ward et al. 1993).

13.2.3.1   *In Vitro* Culture

In vitro culture technique involves propagating plant tissues (units as small as a cell) in a controlled environment free of microorganisms. Approximately 34% of all biotechnology activities reported in forestry over the past 10 years was related to propagation (Chaix and Monteuuis 2004; Wheeler et al. 2003). An entire tree can be regenerated from a single cell. *In vitro* culture can be used to reproduce seedlings and to cryopreserve cell lines from which it will be possible to regenerate other copies of the same seedlings in the future. During *in vitro* plant culture, regeneration occurs via two main pathways: organogenesis and somatic embryogenesis. Organogenesis is the regeneration of plants through organ formation on an explant or from cell masses and for somatic embryogenesis; it is done through the formation of embryo-like structures. Organogenesis has been the method of choice for species such as poplar and eucalyptus; embryogenesis has been used very successfully with conifers. Both processes provide the means to clonally propagate large numbers of elite trees for research and reforestation. One drawback of somatic embryogenesis is that it is fully applicable only using juvenile material as initial explants (embryos but difficult to carry out with needles). To capture maximum gains, a two-step procedure must be established. Firstly, while testing new lines produced with replicated clonal trees, tissue lines must be cryopreserved. Secondly, once the best clone has been identified after a few years of testing, cryopreserved tissues of the best lines are put back into *in vitro* culture for tree multiplication and propagation. *In vitro* culture is also essential to genetic engineering or transgenesis work because it provides the material on which the technology can be carried out (Park et al. 1998).

Selection of the type of the explants, namely, mature tree-derived shoots or seeds, plays a very important aspect for the success of the transformation. Mature tree-derived shoot shows high degree of genetic integrity and is most responsive to various processes in comparison with juvenile explants.

13.2.3.2 Selection of Vectors for Gene Cloning in Plants

An important part of recombinant DNA technology is the selection of a suitable vector (carrier) into which DNA sequences can be inserted (Sambrook et al. 1989). In the context of plant genetic engineering, a vector may be defined as an agent that will facilitate one or more steps in the overall process of placing foreign genetic material into plants or their constituent parts (Mantell et al. 1985). The term "plant gene vector" applies to potential carriers for the transfer of genetic information both between plants and from other organisms, such as bacteria, to plants (Mantell et al. 1985). Likely candidates for vectors are those biological systems where entry of the nucleic acid usually occurs pathogenically (Grierson and Covcy 1984), such as the T-DNA (transferred DNA) of the *Agrobacterium* Ti plasmid.

13.2.4   Genes/Traits of Interest

13.2.4.1   Cellulose/Lignin Content

Pulp and paper production in particular requires treatment to separate lignins from cellulose, a procedure that is costly, energy-consuming, and polluting. The isolation of genes encoding enzymes involved in lignification (Boerjan et al. 2003) has made it possible to envisage controlled modulation of lignins through the downregulation of gene expression (e.g., by the antisense RNA strategy). Significant results in changing lignin quantity and/or quality, which both affect wood processing, have already been obtained in different species (Boudet et al. 2003). Improvement by conventional breeding is not as rapid as envisaged to meet the increasing demand for good-quality planting materials. Genetic engineering potentially offers to the eucalyptus breeder opportunities to add new genes, including those that do not occur naturally in the *Eucalyptus* genome, without the involvement of sexual reproduction. Genetic modification of lignin content having economic importance in pulp and paper production has been reported using genetic engineering (Li et al. 1997; Hu and Wang 1999).

During chemical pulping of wood, one of the most expensive and environmentally hazardous processes is to separate lignin from cellulose and hemicellulose (Pilate et al. 2002). The production of plant material with lower contents of lignin would mean a significant reduction of cost and pollution to the paper industry. It is now possible to develop transgenic trees that have lower lignin content but do not have unfavorable physiological characteristics. Biochemical pathways in lignin synthesis have been the subject of numerous investigations, and several genes responsible for the enzymes involved have been characterized (Tzfira et al. 1998; Merkle and Dean 2000). One of the approaches to obtain reduced lignin forest trees has been the downregulation of lignin biosynthesis pathways (Hu and Wang 1999). The main genes involved with genetic transformation targeting lignin reduction are 4-coumarate-coenzyme A ligase (4CL) (Hu and Wang 1999), cinnamyl alcohol dehydrogenase (CAD – the final enzyme in the biosynthesis of lignin monomers) (Baucher et al. 1996), and caffeate/5-hydroxyferulate O-methyltransferase (COMT enzyme involved in syringyl lignin synthesis) (Lapierre et al. 1999). After cellulose, lignin is the most abundant organic compound in the biosphere and makes up 15–35% of the dry weight of trees. By manipulating the expression of these genes, it has been possible to modify the lignin content or structure.

The huge economic importance of *Eucalyptus* wood has been a driving force to delineate the lignin pathway in this genus. More than 20 years ago, the gene encoding of the cinnamyl alcohol dehydrogenase (CAD), which catalyzes the last step in monolignol biosynthesis, was cloned in eucalyptus after tobacco among all plant species (Grima-Pettenati et al. 1993). The gene encoding cinnamoyl-CoA reductase (CCR), which catalyzes the penultimate step in monolignol biosynthesis, was first cloned in eucalyptus (Lacombe et al. 1997) and subsequently used to clone its orthologs in other plant species such as poplar (Leple et al. 1998), maize (Pichon et al. 1998), and *Arabidopsis* (Lauvergeat et al. 2001). Other SCW-related genes, including transcription factors, have been cloned in eucalyptus but, due to the lack of an efficient stable transformation system, their functional characterization had been achieved mainly using heterologous systems such as poplar (Feuillet et al. 1995; Hawkins et al. 1997; Samaj et al. 1998; Lauvergeat et al. 2002; Legay et al. 2010), tobacco (Goicoechea et al. 2005; Lacombe et al. 2000), and *Arabidopsis* (Baghdady et al. 2006; Creux et al. 2008; Foucart et al. 2009; Hussey et al. 2011; Legay et al. 2010).

The recent availability of the *E. grandis* genome (Myburg et al. 2014) has allowed genome-wide characterization of many gene families, notably those involved in the lignin biosynthetic pathway (Carocha et al. 2015) as well as transcription factor families containing members known to regulate SCW formation such as the R2R3-MYB (Soler et al. 2015), NAC (Hussey et al. 2015), ARF (Yu et al. 2014), and Aux/IAA (Yu et al. 2015) among others. These studies have underscored many new candidates potentially regulating wood formation that need to be functionally characterized. Although stable transformation protocols have been established for several *Eucalyptus* species (Tournier et al. 2003; Girijashankar 2011; de la Torre et al. 2014), they are not suitable for medium-/high-throughput functional characterization of genes because they are tedious and time-consuming and present low efficiencies. For these reasons, only very few functional studies have been performed in transgenic *Eucalyptus* (Girijashankar 2011).

A procedure for *A. tumefaciens*-mediated genetic transformation of a juvenile *E. camaldulensis* clone was reported by Valério et al. (2003). CAD antisense full-length cDNAs from *E. gunnii* or *Nicotiana tabacum* were introduced under the control of the CaMV 35S DE promoter. From 44 individual transgenic shoots selected by PCR analysis, 32% exhibited a significant reduction of CAD activity, up to 83%. The use of the heterologous tobacco CAD cDNA construct was less efficient (up to 65% reduction). Transcript levels in three lines obtained using the homologous eucalyptus cDNA confirmed the under-expression of the CAD gene, and Southern blot data indicated a low transgene copy number ranging between 1 and 3. The most downregulated plant contained a single transgene copy. Therefore, for the first time in eucalyptus, genetically modified plantlets exhibiting a strong inhibition of CAD activity associated with decreased transcription were recovered. Five transgenic lines went through a wood chemical analysis which showed no differences in lignin quantity (through Fourier-transform infrared spectroscopy), composition (through analytical pyrolysis), or pulp yield (through kraft pulping) compared to control trees. Despite the downregulation of the CAD gene in this *Eucalyptus* species of economic interest, the lack of significant changes in lignin profiles indicates that probably the trees were not sufficiently suppressed in CAD throughout development to exhibit obvious modifications in lignin and pulping. This raises the problem of the requirements for an efficient modulation of lignification in trees such as eucalyptus.

An efficient procedure to stably introduce genes into an economically important pulp tree (*E. grandis* × *E. urophylla*) was carried out by Tournier et al. (2003). Seedlings were selected according to their regeneration (adventitious organogenesis) and transformation capacity. After cloning, the best genotype out of 250 tested was transformed via *A. tumefaciens*. A cinnamyl alcohol dehydrogenase (CAD) antisense cDNA from *E. gunnii* was transferred, under the control of the 35S CaMV promoter with a double enhancer sequence, into a selected genotype. According to kanamycin resistance and PCR verification, out of 120 transformants that were generated, 58% significantly inhibited for CAD activity and 9 exhibited the highest downregulation, ranging from 69% to 78% (22% residual activity). Southern blot hybridization showed a low transgene copy number, ranging from 1 to 4, depending on the transgenic line. Northern analyses on the 5–16 and 3–23 lines (respectively, one and two insertion sites) demonstrated the antisense origin of CAD gene inhibition. With, respectively, 26% and 22% of residual CAD activity, these two lines were considered as the most interesting and transferred to the greenhouse for further analyses.

Transformation protocol for *E. tereticornis* Sm. using cotyledon and hypocotyl explants was developed by Prakash and Gurumurthi (2009). Precultured cotyledon and hypocotyls explants were co-cultured with *A. tumefaciens* strain LBA 4404 harboring the binary vector pBI121 containing the *uidA* and neomycin phosphotransferase II genes for 2 days. They were transferred to selective regeneration medium containing 0.5 mg/l 6-benzylaminopurine (BAP), 0.1 mg/l naphthalene acetic acid, 40 mg/l kanamycin, and 300 mg/l cefotaxime. After two passages, the putatively transformed regenerants were transferred to MS liquid medium containing 0.5 mg/l BAP and 40 mg/l kanamycin on paper bridges for further development and elongation. The elongated kanamycin-resistant shoots were subsequently rooted on the MS medium supplemented with 1.0 mg/l indole-3-butyric acid and 40 mg/l kanamycin. A strong β-glucuronidase activity was detected in the transformed plants by histochemical assay. Integration of T-DNA into the nuclear genome of transgenic plants was confirmed by polymerase chain reaction and Southern hybridization.

A fast, reliable, and efficient protocol to obtain and easily detect co-transformed *E. grandis* hairy roots using fluorescent marker eucalyptus hairy roots was carried out by Plasencia et al. (2016) with an average efficiency of 62%. It was further demonstrated that co-transformed hairy roots are suitable for protein subcellular localization, gene expression patterns through RT-qPCR and promoter expression, as well as the modulation of endogenous gene expression. Downregulation of the cinnamoyl-CoA reductase1 (EgCCR1) gene, encoding a key enzyme in lignin biosynthesis, led to transgenic roots with reduced lignin levels and thinner cell walls. This gene was used as a proof of concept to demonstrate the function of genes involved in secondary cell wall biosynthesis and wood formation.

13.2.4.2   Proline Biosynthesis

Organogenesis and *A. tumefaciens*-mediated transformation of *E. saligna* with *P5CS* gene which encodes pyrroline-5-carboxylate synthetase (*P5CS*), the key enzyme in proline biosynthesis, was carried out by Dibax et al. (2010). After selection of the most responsive genotype, shoot organogenesis was induced on leaf explants cultured on a callus induction medium followed by subculture on a shoot induction medium. Shoots were elongated and transferred to a rooting medium and transplanted to pots and acclimatized in a greenhouse. For genetic transformation, a binary vector carrying *P5CSF129A* and *uidA* genes, both under control of the 35SCaMV promoter, was used. Leaves were co-cultured with *A. tumefaciens* in the dark for 5 d. The explants were transferred to the selective callogenesis inducing medium containing kanamycin and cefotaxime. Calli developed shoots that were cultured on an elongation medium for 14 d and finally multiplied. The presence of the transgene in the plant genome was demonstrated by PCR and confirmed by Southern blot analysis. Proline content in the leaves was four times higher in transformed than in untransformed plants while the proline content in the roots was similar in both types of plants.

Procedure for the *A. tumefaciens*-mediated T-DNA delivery into the elite clone(s) of *E. tereticornis* using leaf explants from microshoots has been developed by Agarwal et al. (2011). Among two strains of *A. tumefaciens,* namely, EHA105 and LBA4404 (harboring pBI121 plasmid), strain EHA105 was found to be more efficient. Method of injury to tissue, presence of acetosyringone in co-cultivation medium, and photoperiod during co-cultivation also influenced the expression of transient *gus* activity. Stable transformation was confirmed on the basis of GUS activity and PCR amplification of DNA fragments specific to *uidA* and *nptII* genes.

13.2.4.3   Biotic Resistance/Resistance to Insects and Pathogens

The use of genetic engineering to improve tree resistance to insects and microbial pests has been the subject of investigation in several laboratories. In several parts of the world, fungal and bacterial infestations cause substantial forest losses. These losses are very often underestimated, as compared to the damage caused by insects, because the damage is less visible. However, it is possible to induce resistance by introducing genes associated with the production of antifungal or antibacterial proteins like endochitinase and PPO (Seguin 1999). Genetically engineered insect resistance can be environmentally beneficial because of the reduced need for synthetic insecticides. Forest trees play host to a wide range of fungal, bacterial, and viral pathogens. Trees engineered for disease resistance can provide both environmental and commercial benefits. Enhanced disease resistance has been achieved using a variety of genes derived from plants and microorganisms, with varying degrees of success (Jia et al. 2010). Testing for disease resistance in a natural setting is imperative, and multi-year field trials will be needed to verify the durability of resistance against ever-evolving pathogen populations.

13.2.4.4 Abiotic Stress Resistance

Drought, which is often associated with osmotic or salinity stress, is a major factor involved in the decrease in forest productivity. Enhancing drought and salinity tolerance is of particular importance when reforesting marginal arid and semiarid areas, which are prone to radiation. Molecular control of plant response to abiotic stress is complex, usually involving coordinated expression of several genes. The use of known abiotic stress-associated genes from other species to enhance tolerance in forest trees has been limited. However, recent studies in genomics, transcriptomics, and proteomics in several forest tree species, as well as release of the draft *E. grandis* genomic sequence (www.eucagen.org), have provided new tools for improving abiotic stress tolerance in trees (Harfouche et al. 2011). Overexpression of a pepper ERF/AP2 transcription factor, CaPF1, in eastern white pine resulted in a significant increase in tolerance to drought, freezing, and salt stress. The increased tolerance was associated with polyamine biosynthesis. Moreover, overexpression of the choline oxidase (codA) gene from *Arthrobacter globiformis* resulted in increased tolerance to NaCl in several lines of *E. globules* (Yu et al. 2009).

13.3   Future Prospects

Biotechnology application in forestry has made tremendous strides in the past decades. Many tree species engineered for expression of a variety of traits are already under extensive cultivation in many parts of the word. The status of biotechnology in India is very encouraging with many opportunities. All modern biotechnologies require large research and development investments. The allocation of funds, through either private or public agencies, needs to achieve a balance between building scientific capabilities and knowledge and supporting more applied, well-proven forestry technologies. From a genetic perspective, concerns that biotechnology is "unnatural" ignore the dynamic changes in the genetic code that occur within and across species genomes through modification of transposable genes or elements by virus vectors and through mutation. Tissue culture or genetically modified trees will be substantial managers with time to evaluate many of the issues being faced in agriculture; the economic realities of relatively long generations will continue to be a major challenge for investors in biotechnology in forest trees. It appears that genetic modification will therefore become a reality only for particularly novel and valuable traits in short-rotation species in intensively managed plantations. Other forest management decisions with potentially more serious ecological consequences, including large-scale species introductions or inappropriate use of provenances or improved trees from even conventional breeding, need to be evaluated by foresters, managers, and regulatory agencies in the same way as the products of modern biotechnology. However, while those in the forefront of any technology will promote its potential benefits, in the end it will be the economic and regulatory systems of governing bodies at the national and global levels that must evaluate the technology's relevance and appropriateness (Kumar et al. 2015).

13.4   Conclusions

For the world to be supplied with wood, it needs long-term sustainable, high-yielding, and short-rotation plantation forests. Biotechnology is essential to achieve this goal. The logic of plantation forests will undoubtedly play a major role in achieving global sustainability. The only real problem is the damage done to the Earth's natural forests before the contribution of plantation forests is recognized. Genetic engineering is becoming a routine method in forestry. The possibilities in agriculture are more because of the broader knowledge available, shorter rotation period, and background breeding information; forest trees are also clearly in the focus of research. The methods to transfer genes into the genome of trees offer ample opportunities in the field of breeding research. Based on the Indian government's aim to enlarge the total area covered by forests to 23% by 2010 and to 33% by 2020, biotechnology is going to play a central role to tackle specific challenges (Kumar et al. 2015). Research is still going on regarding the problems related to transgenic trees with respect to concerning all aspects of biosafety including efforts to prevent the escape of transgenes into natural populations. Approval of the commercial use of transgenic trees and their easy vegetative propagation by cuttings may result in the rapid distribution of transgenic plant material in the near future.

References

Abdul-Baki AA, Saunders JA, Matthews BF, Pittarelli GW (1990) DNA uptake during electroporation of germinating pollen grains. Plant Sci 70(2):181–190

Aggarwal D, Kumar A, Reddy MS (2011) *Agrobacterium tumefaciens* mediated genetic transformation of selected elite clone(s) of *Eucalyptus tereticornis*. Acta Physiol Plant 33(5):1603–1611

Ahad A, Maqbool A, Malik KA (2014) Optimization of *Agrobacterium tumefaciens* mediated transformation in *Eucalyptus camaldulensis*. Pak J Bot 76(2):735–774

Antanas VS, Van Beveren K, Leitch MA, Bossinger G (2005) *Agrobacterium*-mediated *in vitro* transformation of wood-producing stem segments in Eucalypts. Plant Cell Rep 23:617–624

Antony Ceasar S, Ignacimuthu S (2009) Genetic engineering of millets: current status and future prospects. Biotechnol Lett 31:779–788

Baghdady A, Blervacq AS, Jouanin L, Grima-Pettenati J, Sivadon P, Hawkins S (2006) *Eucalyptus gunnii* CCR and CAD2 promoters are active in lignifying cells during primary and secondary xylem formation in *Arabidopsis thaliana*. Plant Physiol Biochem 44(11–12):674–683

Baucher M, Chabbert B, Pilate G, Doorsselaere JV, Tollier MT, Petit-Conil M, Cornu D, Monties B, Van Montagu M, Inze D, Jouanin L, Boerjan W (1996) Red xylem and higher lignin extractability by down-regulating a cinnamyl alcohol dehydrogenase in poplar. Plant Physiol 112:1479–1490

BeiLmann A, Albrecht K, Schultze S, Wanner G, Pfitzner UM (1992) Activation of a truncated PR-1 promotor by endogenous enhancers in transgenic plants. Plant Mol Biol 18:65–78

Bevan MW, Flavell RB, Chilton MD (1983) A chimaeric antibiotic resistance gene as a selectable marker for plant cell transformation. Nature 304(5922):184–187

Boerjan W, Ralph J, Baucher M (2003) Lignin biosynthesis. Annu Rev Plant Biol 54:519–546

Boudet AM, Kajita S, Grima-Pettenati J, Goffner D (2003) Lignins and lignocellulosics: a better control of synthesis for new and improved uses. Trends Plant Sci 8(12):576–581

Boynton JE, Gillham NW, Harris EH, Hosler JP, Johnson AM, Jones AR et al (1988) Chloroplast transformation in Chlamydomonas with high velocity microprojectiles. Science 240(4858):1534–1538

Briggs BG, Johnson LAS (1979) Evolution in the Myrtaceae – evidence from inflorescence structure. Proc Linnean Soc NSW 102:157–256

Broothaerts W, Mitchell HJ, Weir B, Kaines S, Smith LMA, Yang W et al (2005) Gene transfer to plants by diverse species of bacteria. Nature 433(7026):629–633

Carocha V, Soler M, Hefe C, Cassan-Wang H, Fevereiro P, Alexander A (2015) Genome-wide analysis of the lignin toolbox of *Eucalyptus grandis*. New Phytol 206:1–17

Castellanos-Hernández OA, Rodríguez-Sahagun A, Acevedo-Hernández GJ, Rodríguez-Garay B, Cabrera-Ponce JL, Herrera-Estrella LR (2009) Transgenic Paulownia elongata plants using biolistic-mediated transformation. Plant Cell Tiss Org Cult 9(2):175–181

Chaix G, Monteuuis O (2004). Biotechnology in the forestry sector. Food and Agriculture Organization. In: Preliminary review of biotechnology in forestry, including genetic modification. FAO-Division des ressources forestières. Rome: FAO, pp 19–56

Chan MT, Lee TM, Chang HH (1992) Transformation of Indica Rice (*Oryza saliva* L.) mediated by *Agrobacterium tumefaciens*. Plant Cell Physiol 33:577–583

Chand PK, Ochatt SJ, Rech EL, Power JB, Davey MR (1988) Electroporation stimulates plant regeneration from protoplasts of the woody medicinal species *Solarium dulcamara* L. J Exp Bot 39(9):1267–1274

Chawla HS (2000) Introduction to plant biotechnology. Enfield, NH, USA

Cheah KT (2001) Methods for producing genetically modified plants, genetically modified plants, plant materials and plant products produced thereby. United States Patent No 6255559 [Issued July 2001]

Chen ZZ, Chang SH, Ho CK, Chen YC, Tsai JB, Chiang VL (2001) Plant production of transgenic *Eucalyptus camaldulensis* carrying the *Populus tremuloides* cinnamate 4-hydroxylase gene. Taiwan J For Sci 16:249–258

Chen ZZ, Ho CK, Ahn IS, Chiang VL (2007) Eucalyptus. In: Wang K (ed) Methods in molecular biology. Vol. 344: Agrobacterium protocols Vol. II. Humana Press, Inc., Totowa, pp 125–134

Chen BW, Xiao YF, Li JJ, Liu HL, Qin ZH, Gai Y, Jiang XN (2016a) Identification of the CAD gene from *Eucalyptus urophylla* GLU4 and its functional analysis in transgenic tobacco. Genet Mol Res 15(4):1–21

Chen W, Li S, Yu H, Liu X, Huang L, Wang Q, Liu H, Cui Y, Tang Y, Zhang P, Wang C (2016b) ER adaptor SCAP translocates and recruits IRF3 to perinuclear microsome induced by cytosolic microbial DNAs. PLoS Pathog 12(2):1–22

Chriqui D, Adam S, Caissard JC, Noin M, Azim A (1992) Shoot regeneration and *Agrobacterium*-mediated transformation of *E. globulus* and *E. gunnii*. In: Schonau APG (ed) IUFRO symposium on intensive forestry: the role of Eucalyptus. Proceedings. Pretoria, South Africa: South African Institute of Forestry, pp 70–80

Christou P, Ford TL, Kofron M (1991) Production of transgenic rice (*Oryza Sativa* L.) plants from agronomically important indica and japonica varieties via electric discharge particle acceleration of exogenous DNA into immature zygotic embryos. Nat Biotechnol 9(10):957–962

Creux NM, Ranik M, Berger DK, Myburg AA (2008) Comparative analysis of orthologous cellulose synthase promoters from *Arabidopsis*, *Populus* and *Eucalyptus*: evidence of conserved regulatory elements in angiosperms. New Phytol 179:722–737

Danilova SA (2007) The technologies for genetic transformation of cereals. Russ J Plant Physiol 54:569–581

de la Torre F, Rodrıguez R, Jorge G, Villar B, Alvarez-Otero R, Grima- Pettenati J, Gallego PP (2014) Genetic transformation of *Eucalyptus globulus* using the vascular-specific EgCCR as an alternative to the constitutive CaMV35S promoter. Plant Cell Tiss Org Cult 117:77–84

Dekeyser RA, Claes B, De Rycke RMU, Habets ME, Van Montagu MC, Caplan AB (1990) Transient gene expression in intact and organized rice tissues. Plant Cell 2(7):591–602

Delbreil B, Jullien M (1993) Agrobacterium mediated transformation of *Asparagus officinalis* L. Long term embryogenic callus and regeneration of transgenic plants. Plant Cell Rep 13:372–376

Dhar MK, Kaul S, Kour J (2011) Towards the development of better crops by genetic transformation using engineered plant chromosomes. Plant Cell Rep 30(5):799–806

Dibax R, Quisen RC, Bona C, Quoirin M (2010) Plant regeneration from cotyledonary explants of *Eucalyptus camaldulensis* Dehn and histological study of organogenesis in vitro. Braz Arch Biol Technol 53(2):311–331. Mar/Apro

Diouf D (2003) Genetic transformation of forest trees. African J Biotechnol 2:328–333

Djuzenova CS, Zimmermann U, Frank H, Sukhorukov VL, Richter E, Fuhr G (1996) Effect of medium conductivity and composition on the uptake of propidium iodide into electropermeabilized myeloma cells. Biochim Biophys Acta-Biomembr 1284:143–152

Dunsmuir P, Bond D, Lee K, Gidoni D, Townsend J (1988) Stability of introduced genes and stability in expression. In: Gelvin SB, Schilperoort RA, Verma DPS (eds) Plant molecular biology manual. Kluwer Academic Publishers, Belgium, pp 1–7

Eldridge K, Davidson J, Harwood C, Van WG (1993) Eucalypt domestication and breeding. Oxford University Press, New York electroporation in high electric fields. EMBO J 1(7):841–845

Eldridge K, Davidson C, Harwood C, Van Wyk G (1994) Eucalyptus domestication and breeding. Oxford University Press, New York

Escoffre JM, Portet T, Wasungu L, Teissié J, Dean D, Rols MP (2009) What is (still not) known of the mechanism by which electroporation mediates gene transfer and expression in cells and tissues? Mol Biotechnol 41(3):286–295

Evans J (1999) Planted forests of the wet and dry tropics: their variety, nature, and significance. In: Planted forests: contributions to the quest for sustainable societies. Springer, Netherlands, pp 25–36

FAO (2004) The state of food and agriculture 2003–2004. Agricultural biotechnology: meeting the needs of the poor? FAO agriculture series, no. 35 Rome

FAO (2007) The state of food and agriculture. Paying farmers for environmental services. Food and agriculture organization of the united nations, Rome

Fenning TM, Gershenzon J (2002) Where will the wood come from? Plantation forests and the role of biotechnology. Trends Biotechnol 20:291–296

Feuillet C, Lauvergeat V, Deswarte C, Pilate G, Boudet A, Grima-Pettenati J (1995) Tissue-specific and cell-specific expression of a cinnamyl alcohol dehydrogenase promoter in transgenic poplar plants. Plant Mol Biol 27:651–667

Fillatti JJ, Selmer J, McCown B, Hassig B, Comai L (1987) *Agrobacterium* mediated transformation and regeneration of *Populus*. Mol Gen Genet 206:192–199

Foucart C, Jauneau A, Gion JM, Amelot N, Martinez Y, Panegos P, Grima-Pettenati J (2009) Over expression of EgROP1, a *Eucalyptus* vascular-expressed Rac-like small GTPase, affects secondary xylem formation in *Arabidopsis thaliana*. New Phytol 183:1014–1029

Fraley RT, Rogers SG, Horsch RD, Sanders PR, Flick S (1983) Expression of bacterial genes in plant cells. Proc Natl Acad Sci USA 80:4803–4807

Gallardo F, Fu J, Jing ZP, Kirby EG, Caovas FM (2003) Genetic modification of amino acid metabolism in woody plants. Plant Physiol Biochem 41:587–594

Gallie DR, Lucas WJ, Walbot V (1989) Visualizing mRNA expression in plant protoplasts: factors influencing efficient mRNA uptake and translation. Plant Cell 1(3):301–311

Gao J, Lee M, An G (1991) The stability of foreign protein production in genetically modified plant cells. Plant Cell Rep 10:533–536

Gartland KMA, Crow RM, Fenning TM, Gartland JS (2003) Genetically modified trees: production, properties and potential. J Arboric 29(5):259–266

Gendloff EH, Bowen B, Buchholz WG (1990) Quantification of chloramphenicol acetyl transferase in transgenic tobacco plants by ELISA and correlation with gene copy number. Plant Mol Biol 14:575–583

Girijashankar V (2011) Genetic transformation of *Eucalyptus*. Physiol Mol Biol Plants 17:9–23

Goicoechea M, Lacombe E, Legay S, Mihaljevic S, Rech P, Jauneau A, Lapierre C (2005) EgMYB2, a new transcriptional activator from *Eucalyptus* xylem, regulates secondary cell wall formation and lignin biosynthesis. Plant J 43:553–567

Gonza'lez ER, de Andrade A, Bertolo AL, Coelho G, Tozelli R, Prado VA, Veneziano MT, Labate CA (2002) Production of transgenic *Eucalyptus grandis*, *E. urophylla* using the sonication-assisted *Agrobacterium transformation* (SAAT) system. Funct Plant Biol 29:97–102

Grattapaglia D, Bradshaw HD (1994) Nuclear-DNA content of commercially important *Eucalyptus* species and hybrids. Can J For Res 24(5):1074–1078

Grattapaglia D, Kirst M (2008) *Eucalyptus* applied genomics: from gene sequences to breeding tools. New Phytol 179(4):911–929

Grierson D, Covey SN (1984) Prospects for the genetic engineering of plants. In: Plant molecular biology, 1st edn. Blackie & Son Limited, Glasgow/London, pp 147–159

Griffin AR, Burgess IP, Wolf L (1988a) Patterns of natural and manipulated hybridisation in the genus *Eucalyptus* L'Hérit. A review. Aust J Bot 36:41–66

Grima-Pettenati J, Feuillet C, Goffner D, Borderies G, Boudet AM (1993) Molecular cloning and expression of a *Eucalyptus gunnii* cDNA clone encoding cinnamyl alcohol dehydrogenase. Plant Mol Biol 21:1085–1095

Groves RH (1994) Australian vegetation, 2nd edn. Cambridge University Press, Cambridge

Han KY, Yang J (2000) Genetic transformation of orchids. United States Patent No 6020538 [Issued February 2000]

Harcourt RL, Kyozuka J, Floyd RB, Bateman KS, Tanaka H, Decroocq V, Llewellyn DJ, Zhu X, Peacock WJ, Dennis ES (2000) Insect- and herbicide-resistant transgenic *Eucalyptus*. Mol Breed 6:307–315

Harfouche A, Meilan R, Altman A (2011) Tree genetic engineering and applications to sustainable forestry and biomass production. Trends Biotechnol 29:9–17

Hartman CL, Lee L, Day PR, Tumer NE (1994) Herbicide resistant turfgrass (*Agrostis palustris huds*) by biolistic transformation. Bio/Technol 12(9):919–923

Hawkins S, Samaj J, Lauvergeat V, Boudet A, Grima-Pettenati J (1997) Cinnamyl alcohol dehydrogenase: identification of new sites of promoter activity in transgenic poplar. Plant Physiol 113:321–325

Hensgens LAM, Fornerod MWJ, Rueb S, Winkler AA, van der Veen S, Schilperoort RA (1992) Translation controls the expression level of a chimaeric reporter gene. Plant Mol Biol 20:921–938

Herschbach C, Kopriva S (2002) Transgenic trees as tools in tree and plant physiology. Trees 16:250–261

Hjouj M, Rubinsky B (2010) Magnetic resonance imaging characteristics of nonthermal irreversible electroporation in vegetable tissue. J Membr Biol 236(1):137–146

Ho CK, Chang SH, Tsay JY, Tsai CJ, Chiang VL, Chen ZZ (1998) *Agrobacterium tumefaciens* mediated transformation of *Eucalyptus camaldulensis* and production of transgenic plants. Plant Cell Rep 17:675–680

Hobbs SLA, Kpoda P, DeLong CMO (1990) The effect of T-DNA copy number, position and methylation on reporter gene expression in tobacco transformants. Plant Mol Biol 15:851–864

Hobbs SLA, Warkentin TD, DeLong CMO (1993) Transgene copy number can be positively or negatively associated with transgene expression. Plant Mol Biol 21:17–26

Hooykaas PJJ, Schilperoort RA (1992) *Agrobacterium* and plant genetic engineering. Plant Mol Biol 19:15–38

Hu CY, Wang L (1999) In-planta soybean transformation technologies developed in China: procedure, confirmation and field performance. In Vitro Cell Dev Biol Plant 35(5):417–420

Huang M, Zhang L (1999) Association of the movement protein of alfalfa mosaic virus with the endoplasmic reticulum and its trafficking in epidermal cells of onion bulb scales. Mol Plant Microbe Interact 12(8):680–690

Huang J, Wu L, Yalda D, Adkins Y, Kelleher SL, Crane M et al (2002) Expression of functional recombinant human lysozyme in transgenic rice cell culture. Transgenic Res 11(3):229–239

Hui SW (1995) Effects of pulse length and strength on electroporation efficiency. In: Nickoloff JA (ed) Methods in molecular biology. Plant cell electroporation and electrofusion protocols. Humana Press Inc., Totowa, pp 29–40

Hussey SG, Mizrachi E, Spokevicius AV, Bossinger G, Berger DK, Myburg AA (2011) SND2, a NAC transcription factor gene, regulates genes involved in secondary cell wall development in *Arabidopsis* fibres and increases fibre cell area in *Eucalyptus*. BMC Plant Biol 11:173

Hussey SG, Saedi MN, Hefer CA, Myburg AA, Grima-Pettenati J (2015) Structural, evolutionary and functional analysis of the NAC domain protein family in *Eucalyptus*. New Phytol 206:1337–1350

Izawa T, Miyazaki C, Yamamoto M, Terada R, Iida S, Shimamoto K (1991) Introduction and transposition of the maize transposable element Ac in rice (*Oryza sativa* L.). Mol Gen Genet 227(3):391–396

Jia Z, Sun Y, Yuan L, Tian Q, Luo K (2010) The chitinase gene (Bbchit1) from *Beauveria bassiana* enhances resistance to *Cytospora chrysosperma* in *Populus tomentosa* Carr. Biotechnol Lett 32:1325–1332

Joersbo M, Brunstedt J (1990) Direct gene transfer to plant protoplasts by electroporation by alternating, rectangular and exponentially decaying pulses. Plant Cell Rep 8(12):701–705

Johnson LAS (1976) Problems of species and genera in Eucalyptus (Myrtaceae). Plant Syst Evol 125:155–167

Kawaoka A, Nanto K, Sugita K, Endo S, Watanabe KY, Matsunaga EH (2003) Production and analysis of lignin modified transgenic *eucalyptus*. Tree Biotechnology Symposium, Sweden

Kawasu T, Doi K, Ohta T, Shinohara Y, Ito K, Shibata M (1990) Transformation of *Eucalyptus saligna* using electroporation. Abstract In: 7th international congress on plant tissue and cell culture, Amsterdam. IAPTC 24–29 June, pp 64

Kawazu T, Dol K, Tatemichi y, Ito K, Shibata M (1996) Regeneration of transgenic plants by nodule culture systems in *Eucalyptus camaldulensis*. In: Proc IUFRO Conf-"Ttee improvement for sustainable tropical forest", pp 492–497

Klein TM, Wolf ED, Wu R, Sanford JC (1987) High-velocity microprojectiles for delivering nucleic acids into living cells. Nature 327(6117):70–73

Kooter JM, Mol JNM (1993) Trans-inactivation of gene expression in plants. Curr Sci 4:166–171

Kothari SL, Kumar S, Vishnoi RK, Kothari SL, Watanabe KN (2005) Applications of biotechnology for improvement of millet crops: review of progress and future prospects. Plant Biotechnol 22:81–88

Krimi Z, Raio A, Petit A, Nesme X, Dessaux Y (2006) *Eucalyptus occidentalis* plantlets are naturally infected by pathogenic *Agrobacterium tumefaciens*. Eur J Plant Pathol 116:237–246

Kubiniec RT, Liang H, Hui SW (1990) Effects of pulse length and pulse strength on transfection by electroporation. Biotechniques 8(1):16–20

Kumar V, Rout S, Tak MK, Deepak KR (2015) Application of biotechnology in forestry: current status and future perspective. Nat Environ Pollut Technol 14(3):1–9

Lacombe E, Hawkins S, Van Doorsselaere J, Piquemal J, Goffner D, Poeydomenge O, Boudeta M (1997) Cinnamoyl CoA reductase, the first committed enzyme of the lignin branch biosynthetic pathway: cloning, expression and phylogenetic relationships. Plant J 11:429–441

Lacombe E, Van Doorsselaere J, Boerjan W, Boudet A, Grima-Pettenati J (2000) Characterization of cis-elements required for vascular expression of the cinnamoyl CoA reductase gene and for protein-DNA complex formation. Plant J 23:663–676

Ladiges PY (1997) Phylogenetic history and classification of Eucalypt. In: Willaim J, Woinarski J (eds) Eucalypt ecology: individual to ecosystems. Cambridge University Press, Cambridge, pp 16–29

Lapierre C, Pollet B, Petit-Conil M, Toval G, Romero J, Pilate G, Leple JC, Boerjan W, Ferret V, De Nadai V, Jouanin L (1999) Structural alterations of lignins in transgenic poplars with depressed cinnamyl alcohol dehydrogenase or caffeic acid O-methyltransferase activity have an opposite impact on the efficiency of industrial kraft pulping. Plant Physiol 119:153–163

Lauvergeat V, Lacomme C, Lacombe E, Lasserre E, Roby D, Grima-Pettenati J (2001) Two cinnamoyl-CoA reductase (CCR) genes from *Arabidopsis thaliana* are differentially expressed during development and in response to infection with pathogenic bacteria. Phytochemistry 57:1187–1195

Lauvergeat V, Rech P, Jauneau A, Guez C, Coutos-Thevenot P, Grima-Pettenati J (2002) The vascular expression pattern directed by the *Eucalyptus gunnii* cinnamyl alcohol dehydrogenase EgCAD2 promoter is conserved among woody and herbaceous plant species. Plant Mol Biol 50:497–509

Legay S, Sivadon P, Blervacq AS, Pavy N, Baghdady A, Tremblay L, Levasseur C (2010) EgMYB1, an R2R3 MYB transcription factor from *Eucalyptus* negatively regulates secondary cell wall formation in Arabidopsis and poplar. New Phytol 188:774–786

Leple JC, Grima-Pettenati J, Van Montagu M, Boerjan W (1998) A cDNA encoding cinnamoyl-CoA reductase from *Populus trichocarpa* (PGR98–121). Plant Physiol 117:1126–1126

Li SL, You ZG, Liang DX, Li SR (1997) Extraction and separation of allelochemicals in wheat and its herbicidal efficacy on Imperata cylindrical. Acta phytophyl Sun 24:81–84

Lindsey K, Jones MGK (1990) Selection of transformed cells. In: Dix PJ (ed) Plant cell line selection procedures and applications. YCH, Weinheim/New York/Basel/Cambridge, pp 317–339

Llewellyn DJ (1999) Herbicide tolerant forest trees. In: Jain SM, Minocha SC (eds) Molecular biology of woody plants, vol 2. Kluwer Academic Publ. Dordrecht, The Netherlands, pp 439–466

Luciano CS, Rhoads RE, Shaw JG (1987) Synthesis of potyviral RNA and proteins in tobacco mesophyll protoplasts inoculated by electroporation. Plant Sci 51(2–3):295–303

Machado LO, de Andrade GM, Cid LPB, Penchel RM, Brasileiro ACM (1997) *Agrobacterium* strain specificity and shooty tumour formation in eucalypt *Eucalyptus grandis × E. urophylla*. Plant Cell Rep 16:299–303

MacRae S, von Staden J (1990) In vitro culture of Eucalyptus grandis. Effect of glling agents on propagation. J Plant Physiology 137:249–251

Makouanzi G, Bouvet JM, Denis M, Saya A, Mankessi F, Vigneron P (2014) Assessing the additive and dominance genetic effects of vegetative propagation ability in *Eucalyptus*—influence of modeling on genetic gain. Tree Genet Genomes 10(5):1243–1256

Manders G, dos Santos AVP, d'Utra Vaz FB, Davey MR, Power JB (1992) Transient gene expression in electroporated protoplasts of *Eucalyptus citriodora* hook. Plant Cell Tiss Org Cult 30(1):69–75

Mantell SH, Matthews A, McKee RA (1985) Vectors for gene cloning in plants. In: Principles of plant biotechnology: an introduction to genetic engineering in plants. Blackwell Scientific Publications, Oxford/London/Edinburgh/Boston/Palo Alto/Melbourne, p 6288

Marques S, Garcia-Gonzalo J, Borges JG, Botequim B, Oliveira MM, Tomé J, Tomé M (2011) Developing post –fire Eucalyptus globules stand damage and tree mortality models for enhanced forest planning in Portugal. Silva Fennica 45(1):69–83

Matsunaga E, Nanto K, Oishi M, Ebinuma H, Morishita Y, Sakurai N, Suzuki H, Shibata D, Shimada T (2012) Agrobacterium-mediated transformation of Eucalyptus globulus using explants with shoot apex with introduction of bacterial choline oxidase gene to enhance salt tolerance. Plant Cell Rep 31:225–235

Mendonça EG, Stein VS, Balieiro FP, Lima CDF, Santos BR, Paiva LV (2013) Genetic trans-
    formation of *Eucalyptus camaldulensis* by agrobalistic method. Revista Árvore, Viçosa-MG
    37(3):419–429
Merkle SA, Dean JF (2000) Forest tree biotechnology. Curr Opin Biotechnol 11:298–300
Moralejo M, Rochange F, Boudet AM, Teulieres C (1998) Generation of transgenic *Eucalyptus
    globulus* plantlets through *Agrobacterium tumefaciens* mediated transformation. Aust J Plant
    Physiol 25:207–212
Mullins KV, Llewellyn DJ, Hartney VJ, Strauss S, Dennis ES (1997) Regeneration and transforma-
    tion of *Eucalyptus camaldulensis*. Plant Cell Rep 16(11):787–791
Myburg AA, Potts BM, Marques C, Kirst M, Gion JM, Grattapaglia D, Grima-Pettenat J (2007)
    Eucalypts. In: Kole C (ed) Genome mapping and molecular breeding in plants. Springer Berlin
    Heidelberg, Berlin, Heidelberg, pp 115–160
Myburg AA, Grattapaglia D, Tuskan GA, Hellsten U, Hayes RD, Grimwood J, Jl J (2014) The
    genome of *Eucalyptus grandis*. Nature 509:356–362
Nair SG, Vijayalakshmi C (2010) Genetic transformation of ITC-3, a superior clone of *Eucalyptus
    tereticornis*. Indian J Agric Res 44(3):229–232
Navarro M, Ayax C, Martinez Y, Laue J, Koyal WEI, Marque C, Tuelieres C (2011) Two EguCBF1
    genes over expressed in *Eucalyptus* display a different impact on stress tolerance and plant
    development. Plant Biotechnol J 9:50–63
Neumann E, Schaefer-Ridder M, Wang Y (1982) Gene transfer into mouse lyoma cells by electro-
    poration in high electric fields. EMBO J 1(7):841–845
Oard JH, Paige DF, Simmonds JA, Gradziel TM (1990) Transient gene expression in maize, rice
    and wheat cells using an air gun apparatus. Plant Physiol 92(2):334–339
Ottaviani MP, Smits T, Hansisch ten Cate CH (1993) Differntial methylation and expression of
    the beta- glucuronidase and neomycin phosphotranferase genes in transgenic potato cv Bintje.
    Plant Sci 88:73–81
Ouyang LJ, Li LM (2016) Effects of an inducible aiiA gene on disease resistance in *Eucalyptus
    urophylla* x *Eucalyptus grandis*. Transgenic Res 25:441–452. https://doi.org/10.1007/
    s11248-016-9940-x
Park YS, Barrett JD, Bonga JM (1998) Application of somatic embryogenesis in high-value clonal
    forestry: deployment, genetic control, and stability of cryopreserved clones. In Vitro Cell Dev
    Biol Plant 34(3):231–293
Paszkowski J, Shillito RD, Saul M, Mandák V, Hohn T, Hohn B et al (1984) Direct gene transfer
    to plants. EMBO J 3(12):2717–2722
Patnaik D, Khurana P (2001) Wheat biotechnology: a mini-review. Electron J Biotechnol
    4(2):38–66
Pena L, Seguin A (2001) Recent advances in the genetic transformation of trees. Trends Biotechnol
    19:5000–5506
Pichon M, Courbou I, Beckert M, Boudet AM, Grima-Pettenati J (1998) Cloning and characteriza-
    tion of two maize cDNAs encoding cinnamoyl-CoA reductase (CCR) and differential expres-
    sion of the corresponding genes. Plant Mol Biol 38:671–676
Pilate G, Guiney E, Holt K, Petit-Conil M, Lapierre C, Leple JC, Pollet B, Mila I, Webster EA,
    Marstorp HG, Hopkins DW, Jouanin L, BoerjanW SW, Cornu D, Halpin C (2002) Field and
    pulping performances of transgenic trees with altered lignification. Nat Biotechnol 20:607–612
Plasencia A, Soler M, Dupas A, Ladouce N, Martins G, Martinez Y, Lapierre C, Franche C, Truchet
    I, Pettenati JG (2016) *Eucalyptus* hairy roots, a fast, efficient and versatile tool to explore func-
    tion and expression of genes involved in wood formation. Plant Biotechnol J 14:1381–1393
Potrykus I, Saul MW, Petruska J, Paszkowski J, Shillito RD (1985) Direct gene transfer to cells of
    a graminaceous monocot. Mol Gen Genet 199(2):183–188
Potts BM, Dungey HS (2004) Interspecific hybridization of *Eucalyptus*: key issues for breeders
    and geneticists. New For 27(2):115–138
Potts BM, Wiltshire RJE (1997) Eucalypt genetics and genecology. In: Williams J, Woinarski J (eds)
    Eucalypt ecology: individuals to ecosystems. Cambridge University Press, Cambridge, pp 56–91

Prakash MG, Gurumurthi K (2009) Genetic transformation and regeneration of transgenic plants from precultured cotyledon and hypocotyl explants of *Eucalyptus tereticornis* Sm. Using *Agrobacterium tumefaciens*. In Vitro Cell Dev Biol Plant 45(4):429–434

Pryor LD, Johnson LAS (1971) A classification of the eucalypts. Australian National University Press, Canberra

Qayyum A, Bakhsh A, Kiani S, Shahzad K, Ali Shahid A, Husnain T et al (2009) The myth of plant transformation. Biotechnol Adv 27(6):753–763

Raj D, Veale A, Ma C, Strauss SH, Myburg AA (2011) Optimization of a plant regeneration and genetic transformation protocol for *Eucalyptus* clonal genotypes. BMC Proceedings 5(7):132

Ritchie GA (1991) The commercial use of conifer rooted cuttings in forestry: a world overview. New Forests 5:247–275

Rochange F, Serrano L, Marque C, Teulieres C, Boudet AM (1995) DNA delivery into *Eucalyptus globulus* zygotic embryos through biolistics: optimization of the biological and physical parameters of bombardment for two different particle guns. Plant Cell Rep 14:674–678

Samaj J, Hawkins S, Lauvergeat V, Grima-Pettenati J, Boudet A (1998) Immunolocalization of cinnamyl alcohol dehydrogenase 2 (CAD 2) indicates a good correlation with cell-specific activity of CAD 2 promoter in transgenic poplar shoots. Planta 204:437–443

Sambrook J, Fritsch EF, Maniatis T (1989) Molecular cloning: a laboratory manual, 2nd edn. Cold Spring Harbour Laboratory Press, U.S.A.

Sanford JC (1990) Biolistic plant transformation. Physiol Plant 79(1):206–209

Sanford JC, Klein TM, Wolf ED, Allen N (1987) Delivery of substances into cells and tissues using a particle bombardment process. J Part Sci Technol 5:27–37

Sanford JC, Smith FD, Russel JA (1993) Optimizing the biolistic process for different biological applications. Methods Enzymol 217:483–509

Sartoretto LM, Barrueto CLP, Brasileiro ACM (2002) Biolistic transformation of *Eucalyptus grandis* × *E. urophylla* callus. Funct Plant Biol 29:917–924

Saulis G, Venslauskas MS, Naktinis J (1991) Kinetics of pore resealing in cell membranes after electroporation. Bioelectrochem Bioenerg 26(1):1–13

Sedjo RA (2001) From foraging to cropping: the transition to plantation forestry, and implications for wood supply and demand. Unasylva 52:24–27

Séguin A (1999) Transgenic trees resistant to microbial pests. For Chron 75:303–304

Serrano L, Rochange F, Sembalt JP, Marque C, Teulieres C, Boudet AM (1996) Genetic transformation of *Eucalyptus globulus* through biolistics: complementary development procedures for organogenesis from zygotic embryos and stable transformation of corresponding proliferating tissue. J Exp Bot 47:285–290

Shirsat AH, Wilford N, Croy RRD (1989) Gene copy number and levels of expression intransgenic plants of a seed specific gene. Plant Sci 61:75–80

Smith KR (2003) Gene therapy: theoretical and bioethical concepts. Arch Med Res 34:247–268

Snyder GW, Ingersoll JC, Smigocki AC, Owens LD (1999) Introduction of pathogen defense genes and a cytokinin biosynthesis gene into sugar beet (*Beta vulgaris* L.) by *Agrobacterium* or particle bombardment. Plant Cell Rep 18(10):829–834

Soler M, Camargo ELO, Carocha V, Cassan-Wang H, San Clemente H, Savelli B, Hefer C (2015) The *Eucalyptus grandis* R2R3-MYB transcription factor family: evidence for woody growth-related evolution and function. New Phytol 206:1364–1377

Southerton SG (2007) Early flowering induction and *Agrobacterium transformation* of hardwood tree species *Eucalyptus occidentalis*. Funct Plant Biol 34:707–713

Southgate EM, Davey MR, Power JB, Marchant R (1995) Factors affecting the genetic engineering of plants by microprojectile bombardment. Biotechnol Adv 13(4):631–651

Spörlein B, Koop H-U (1991) Lipofectin: direct gene transfer to higher plants using cationic liposomes. Theor Appl Genet 83(1):1–5

Strauss SH, Boerjan W, Cairney J, Campbell M, Dean J, Ellis D, Jouanin L, Sundberg B (1999) Forest biotechnology makes its position known. Nat Biotechnol 17:1145

Strauss SH, DiFazio SP, Meilan R (2001) Genetically modified poplars in context. Forest Chron 77:271–279

Sukhorukov VL, Mussauer H, Zimmermann U (1998) The effect of electrical deformation forces on the electropermeabilization of erythrocyte membranes in low- and high-conductivity media. J Membr Biol 163:235–245

Sutton WRJ (1999) The need for planted forests and the example of radiate pine. New For 17:95–109

Tada Y, Sakamoto M, Fujiyama T (1990) Efficient gene introduction into rice by electroporation and analysis of transgenic plants: use of electroporation buffer lacking chloride ions. Theor Appl Genet 80(4):475–480

Taylor NJ, Fauquet CM (2002) Microparticle bombardment as a tool in plant science and agricultural biotechnology. DNA Cell Biol 21(12):963–977

Teulieres C, Grima Pettenati J, Curie C, Teissie J, Boudet AM (1991) Transient foreign gene expression in polyethylene/glycol treated or electropulsated *Eucalyptus gunnii* protoplasts. Plant Cell Tiss Org Cult 25:125–132

Toriyama K, Arimoto Y, Uchimiya H, Hinata K (1988) Transgenic rice plants after direct gene transfer into protoplasts. Nat Biotechnol 6(9):1072–1074

Tournier V, Grat S, Marque C, Kayal WE, Penchel R, de Andrade G, Boudet AM, Teulières C (2003) An efficient procedure to stably introduce genes into an economically important pulp tree (*Eucalyptus grandis x E. urophylla*). Transgenic Res 12:403–411

Travella S, Ross SM, Harden J, Everett C, Snape JW, Harwood WA (2005) A comparison of transgenic barley lines produced by particle bombardment and *Agrobacterium*-mediated techniques. Plant Cell Rep 23(12):780–789

Turgut-Kara N, Ari S (2010) The optimization of voltage parameter for tissue electroporation in somatic embryos of *Astragalus chrysochlorus* (Leguminosae). Afr J Biotechnol 9(29):4584–4588

Turnbull JW (1999) Eucalypt plantations. New For 17:37–52

Tzfira T, Zuker A, Altman A (1998) Forest-tree biotechnology: genetic transformation and its application to future forests. Trends Biotechnol 16:439–446

Valério L, Carter D, Rodrigues CJ, Tournier V, Gominho J, Marque C, Boudet AM, Maunders M, Pereira H, Teulières C (2003) Down regulation of cinnamyl alcohol dehydrogenase, a lignification enzyme, in *Eucalyptus camaldulensis*. Mol Breeding 12(2):157–167

Van Frankenhuyzen K, Beardmore T (2004) Current status and environmental impact of transgenic forest trees. Can J For Res 34:1163–1180

Vasil IK (2005) The story of transgenic cereals: the challenge, the debate, and the solution: a historical perspective. In Vitro Cell Dev Biol Plant 41(5):577–583

Vikas K, Rout S, Tak MK, Deepak KR (2015) Application of biotechnology in forestry: current status and future perspective. Nat Environ Pollut Technol 14:645–653

Volker PW (1995) Evaluation of *E nitens* x *globules* for commercial forestry. IUFRO:1–4

Ward ER, Ryals JA, Miflin BJ (1993) Chemical regulation of transgene expression in plants. Plant Mol Biol 22:361–366

Weaver JC (1995) Electroporation theory. In: Nickoloff JA (ed) Methods in molecular biology. Plant cell electroporation and electrofusion protocols. Humana Press Inc., Totowa, pp 3–28

Weaver JC, Chizmadzhev YA (1996) Theory of electroporation: a review. Biochem Bioenergetics 41:135–160

Wheeler MA, Byrne M, McComb JA (2003) Little genetic differentiation within the dominant forest tree, *Eucalyptus marginata* (Myrtaceae) of South- Western Australia. Silvae Genetica 52(5–6):254–259

Yanchuck AD (2001) The role and implications of biotechnological tools in forestry. Unasylva 52:53–61

Yu X, Kikuchi A, Matsunaga E, Morishita Y, Nanto K, Sakurai N, Suzuki H, Shibata D, Shimada T, Watanabe NK (2009) Establishment of the evaluation system of salt tolerance on transgenic woody plants in the special netted-house. Plant Biotechnol 26:135–141

Yu H, Soler M, Mila I, San Clemente H, Savelli B, Dunand C, Paiva JAP (2014) Genome-wide characterization and expression profiling of the auxin response factor (ARF) gene family in *Eucalyptus grandis* PLoS One 9(9): e0108906. https://doi:10.1371/journal.pone.0108906

Yu H, Soler M, San Clemente H, Mila I, Paiva JAP, Myburg AA, Bouzayen M (2015) Comprehensive genome-wide analysis of the Aux/ IAA gene family in *Eucalyptus*: evidence for the role of EgrIAA4 in wood formation. Plant Cell Physiol 56:700–714

Zhang W, Wu R (1988) Efficient regeneration of transgenic plants from rice protoplasts and correctly regulated expression of the foreign gene in the plants. Theor Appl Genet 76(6):835–840

Zimmermann U, Vienken J (1982) Stable transformation of maize after gene transfer by electroporation. J Membr Biol 67:165–182

Zupan J, Zambryski P (1997) The *Agrobacterium* DNA transfer complex. Crit Rev Plant Sci 16:279–295

Chapter 14
Transgenic Manipulation of Glutamine Synthetase: A Target with Untapped Potential in Various Aspects of Crop Improvement

Donald James, Bhabesh Borphukan, Dhirendra Fartyal, V. M. M. Achary, and M. K. Reddy

Abstract Glutamine synthetase (GS) plays a key role in the nitrogen (N) metabolism in higher plants. N is a major limiting nutrient in crop production, and most of it is lost due to volatilization or leaching which has deleterious effects on the environment. Hence, GS is considered a prime target for transgenic approaches to increase nitrogen use efficiency (NUE) which is paramount for sustainable agriculture. The current status of such attempts at increasing NUE utilizing GS, their outcomes, constraints, and future prospects have been discussed in detail. GS is also modulated by various abiotic stresses including salt and drought which have adverse effects on crop production. Modulation of GS by various abiotic stresses and transgenic approaches utilizing GS for tolerance, their results, limitations, and possibilities of further advancement have been reviewed. GS is also the target of the commonly used herbicide glufosinate (Basta). Herbicide-tolerant transgenic crops have become a necessity for modern agriculture, given the labor and expenditure involved in traditional weed control practices. In the light of public resentment and biosafety concerns of utilizing bacterial genes for herbicide tolerance in food crops, the overexpression of mutant GS enzymes as an alternative strategy for developing glufosinate-resistant crops has been discussed. This chapter also examines the inconsistent results of overexpression of GS genes for various applications in view of intricate regulation of GS due to its critical role in metabolism.

Keywords Glutamine synthetase · Nitrogen use efficiency · Abiotic stress · Herbicide tolerance · Glufosinate · Transgenic overexpression · Regulation

D. James · B. Borphukan · D. Fartyal · V. M. M. Achary · M. K. Reddy (✉)
International Centre for Genetic Engineering and Biotechnology, New Delhi, India
e-mail: reddy@icgeb.res.in

© Springer International Publishing AG, part of Springer Nature 2018
S. S. Gosal, S. H. Wani (eds.), *Biotechnologies of Crop Improvement, Volume 2*,
https://doi.org/10.1007/978-3-319-90650-8_14

14.1 Glutamine Synthetase: A Key Enzyme Involved in Nitrogen Metabolism of Higher Plants

Glutamine synthetase (GS, EC: 6.3.1.2), also referred to as glutamate-ammonia ligase, is the key enzyme responsible for primary nitrogen (N) assimilation in higher plants (Miflin and Habash 2002; Bernard and Habash 2009). Along with glutamate synthase, also known as glutamine oxoglutarate aminotransferase (GOGAT, EC 1.4.1.14 and EC 1.4.7.1), GS takes part in the GS-GOGAT cycle (Fig. 14.1) which serves as the cornerstone of N metabolism. GS is primarily responsible for assimilating ammonia[1] (NH_4^+), a cytotoxic and reactive metabolite, produced from the fixation of atmospheric N, and direct nitrate or ammonia uptake from soil (Hirel and Lea 2001). GS is also responsible for the reassimilation of NH_4^+ produced during various cellular metabolic processes including photorespiration and protein degradation. GS has a high affinity for NH_4^+ and catalyzes the ATP- and Mg^{2+}-dependent condensation of NH_4^+ with glutamate to form glutamine. The enzyme GOGAT then converts glutamine and 2-oxoglutarate into two glutamate molecules, thus recycling glutamate for further NH_4^+ assimilation (Fig. 14.1). The glutamate produced by the GS/GOGAT cycle, through the action of various aminotransferases, serves as the N donor for biosynthesis of amino acids, which then act as precursors for all nitrogenous biomolecules including proteins, enzymes, secondary metabolites, cytochrome/phytochrome, chlorophyll, and nucleic acids (Forde and Lea 2007; Bernard

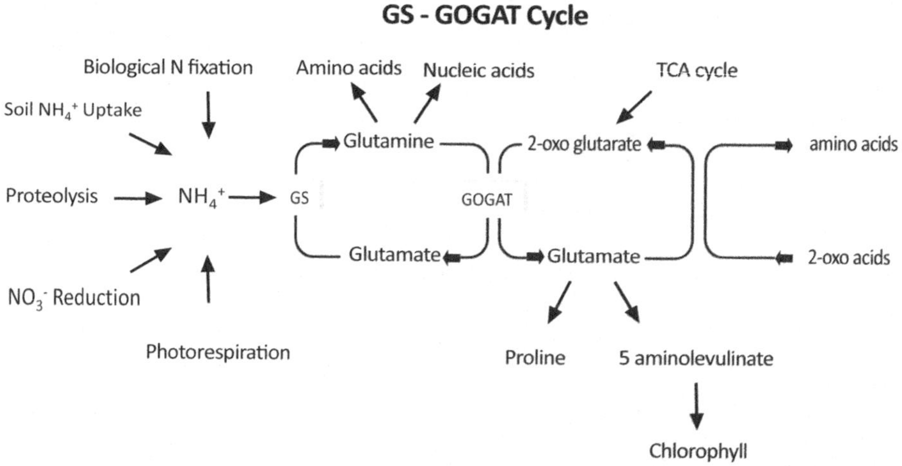

GS - GOGAT Cycle

Fig. 14.1 The glutamine synthetase-glutamine oxoglutarate aminotransferase (GS-GOGAT) cycle serves as the cornerstone of N metabolism in higher plants. (Modified from Donn and Kocher, 2002)

---

[1] In aqueous solution, ammonia (NH_3) is in equilibrium with its cationic form ammonium (NH_4^+), and at typical cytosolic pH ~99% is present as NH_4^+. Both the charged and uncharged species have been abbreviated as NH_4^+ in this chapter.

and Habash 2009). A single N atom can move through this GS/GOGAT cycle multiple times after soil uptake, assimilation, and remobilization till its eventual filling as storage proteins in seeds (Hirel et al. 2007). This organic N is transported between compounds mainly through the function of glutamine-amide transferases and transaminases, but a considerable portion is also released as NH_4^+ and reassimilated by GS activity. For example, in legume plants, ureide compounds like allantoin play a key role in moving nitrogen, and their N is released as NH_4^+ through the action of the enzyme urease. Similarly, in cereals, asparagine, a highly efficient N transport compound in lieu of its high carbon-to-nitrogen (C/N) ratio, is metabolized into aspartate and NH_4^+ by asparaginase (Miflin and Habash 2002). In other instances, glutamine, the end product of GS, can itself be the main form of organic N transport, as is the case in rice phloem and in the xylem of poplar (Bernard and Habash 2009). Therefore, during the growth cycle of a plant, N is repeatedly released as NH_4^+ and recycled into biomolecules via GS. Moreover, NH_4^+, which is released during photorespiration in C3 plants and can exceed that produced by primary N assimilation by as much as ten times, is also reassimilated through the action of GS (Keys et al. 1978; Leegood et al. 1995). Also, in the root nodules of leguminous plants, GS has a key function of assimilating the ammonia being released by N-fixing microbes (Atkins 1987). Glutamine synthetase thus occupies a central role in the plant N metabolic pathway and controls crop growth and productivity (Miflin and Habash 2002; Tabuchi et al. 2005; Kichey et al. 2006; Habash et al. 2007; Bernard and Habash 2009; Lothier et al. 2011; Brestic et al. 2014; Simons et al. 2014; Thomsen et al. 2014; Wang et al. 2015).

14.2   Glutamine Synthetase: Evolution, Structure, and Roles of Multiple Isoforms

14.2.1   Glutamine Synthetase and Its Evolution

GS is considered to be one of the oldest existing and functioning genes and thereby acts as a good molecular clock for phylogenetic analyses (Pesole et al. 1991; Kumada et al. 1993). The GS gene superfamily in living organisms can be classified into three distinct classes, GSI, GSII, and GSIII, based on their gene sequence, molecular weight, and quaternary structure (Woods and Reid 1993). They are primarily distinguished by the length of their proteins: GSI with 360 amino acids, GSII with 450, and GSIII with 730 on average (van Rooyen et al. 2011). GSI is typically considered to be limited to prokaryotes and GSII to eukaryotes; however, the variable distribution of these classes within the domains of life is not uncommon (Ghoshroy et al. 2010). For example, GSII genes have been found in certain prokaryotic organisms, such as *Rhizobium*, *Streptomyces*, and *Frankia* (Biesiadka and Legocki 1997). This was initially considered to be a horizontal gene transfer event from plant to symbiotic bacteria, but Kumada et al. (1993) reasoned that ancient gene duplication provided a much better explanation. Conversely, studies have predicted the existence of

GSI-like genes in eukaryotes like the plants *Arabidopsis thaliana*, *Saccharum* sp., and *Medicago truncatula* (Mathis et al. 2000; Nogueira et al. 2005; Seabra et al. 2010). However, the exact function and roles of these GSI-like enzymes in plants are yet to be understood although recent studies have proposed a possible role in nitrogen and biotic stress signaling (Silva et al. 2015; Doskočilová et al. 2011). Thus, it was surmised that the GS gene family evolved anciently and prior to the divergence of eukaryotes and prokaryotes and that the GSI and GSII classes underwent paralogous evolution (Ghoshroy et al. 2010; Mathis et al. 2000).

The GSII class is by far the most abundant type of GS in higher plants. Ghoshroy et al. (2010) provided phylogenetic evidence that GSII genes were transferred by a non-endosymbiotic horizontal gene transfer event from Eubacteria to the Chloroplastida early in plant evolution. The third class, GSIII, was identified more recently and is found in many diverse organisms like cyanobacteria (Reyes and Florencio 1994) and anaerobic bacteria (Goodman and Woods 1993; Amaya et al. 2005) but is predicted to be much more widespread.

14.2.2 Structure of GS Enzymes

The GS enzymes are usually homo-oligomers composed of a closed two ring structure (with active sites formed between the monomers) and arranged in a dihedral symmetry. GSI enzymes were determined to be dodecameric oligomers with a hexagonal 622 symmetry (van Rooyen et al. 2011). A high-resolution atomic crystal structure of the type II GS from maize (*Zea mays*) (PDB ID: 2d3c) presented a decameric structure (Unno et al. 2006), unlike the dodecameric structure formed by bacterial GSI (Almassy et al. 1986; Krajewski et al. 2005). Crystal structures from *Medicago truncatula* (Seabra et al. 2009; Torreira et al. 2014), yeast (He et al. 2009), and mammals (Krajewski et al. 2008) have dispelled the earlier notion of an octameric form and confirmed the decameric structure of GSII enzymes with a 522 symmetry (van Rooyen et al. 2011). The decameric GSII enzyme in plants is made up of two pentameric rings of subunits facing each other and containing ten active sites, each of which is formed in between two adjacent subunits in the pentamer (Unno et al. 2006) (Fig. 14.2). It is worth mentioning that the first crystal structure of a GSIII enzyme revealed a surprising inversion in the inter-ring interface which explains the strikingly low (around 10%) global sequence identity of GSIII class with other GS enzymes (van Rooyen et al. 2011). It also suggests that the double-ringed architecture, seen conserved throughout the GS gene family, does not have a role in its regulation. It was observed that, although the quaternary structure differed considerably, the GS active site structure and catalytic fold are conserved in all types of GS enzymes, and slight differences in ligand binding can be exploited for designing type-specific inhibitors. The overall "funnel-shaped pocket" geometry of the active site is also a structural feature seen conserved among GS enzymes (Eisenberg et al. 2000; Unno et al. 2006; van Rooyen et al. 2011). Although the three classes have structural similarities, they differ considerably at the amino acid sequence level, as well as in their regulation and response to feedback inhibitors

Fig. 14.2 (a) Top view of ribbon representation of the decameric GS enzyme from *Zea mays* (PDB ID: 2d3c) showing PPT (red spheres) bound to the active sites. (b) Side view of the GS showing the two pentameric rings facing each other, with active sites (denoted by red arrows) occurring between two adjacent monomeric subunits (each color represents one monomer)

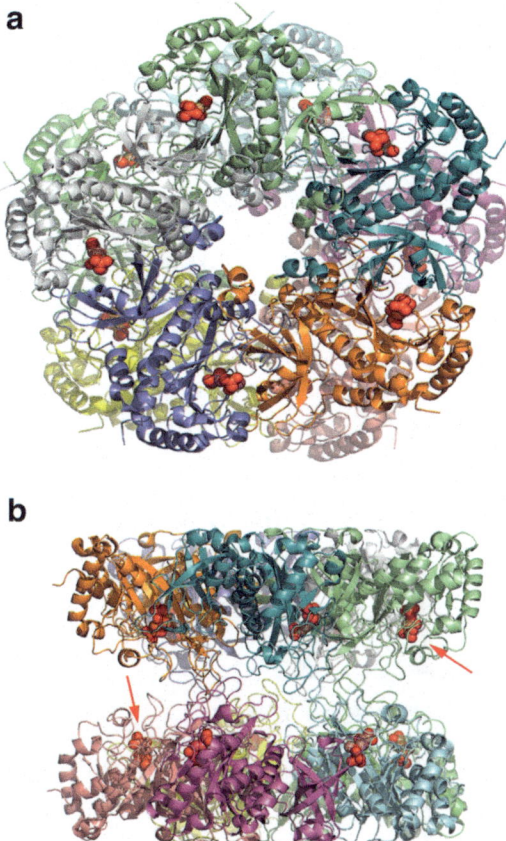

(Eisenberg et al. 2000). The *Zea mays* GS was also crystallized in complex with phosphinothricin (PPT) (PDB id: 2d3c), a structural analogue of the GS substrate glutamate and also a potent herbicide due to its inhibition of GS activity (Unno et al. 2006) (Fig. 14.2). This structure has helped in the understanding of the inhibitory mechanism of PPT and further serves as a guide for the characterization of new potential mutations conferring tolerance to PPT as is discussed in Sect. 1.5.3. The characterization of atomic structures of GS in other crop plants is of special significance to further advance its roles in plant productivity and agronomic utility.

14.2.3   GS Isoforms in Plants

In vascular plants, two isoforms of the GSII class of GS, which is the predominant class in plants, were originally differentiated by chromatography (O'Neal and Joy 1973; McParland et al. 1976; Mann et al. 1979; McNally et al. 1983). Based on their size differences and distinct subcellular localization, they were classified as the cytosolic GS1 and the chloroplastic GS2 isoforms (McNally et al. 1983).

The cytosolic GS1 isoform (polypeptide molecular weight (MW) ~ 38–40 kDa) is responsible for primary assimilation of inorganic N availed from soil in the form of nitrate or ammonia and the reassimilation of NH_4^+ released by protein degradation in senescing leaves, whereas the larger chloroplast localized GS2 isoform (polypeptide MW ~ 42–45 kDa) is responsible for reassimilation of NH_4^+ released during photorespiration and nitrate (NO_3^-) reduction in plastids (Wallsgrove et al. 1987; Leegood et al. 1995). In *Arabidopsis*, the GS2 gene was shown to be dual targeted to both the chloroplast and mitochondria, although this has not been reported from other species (Taira et al. 2004). In most plants, there exists usually a single gene for the chloroplastic GS2 and as many as three to five genes of the cytosolic GS1 isoform. However, multiple GS2 isoforms have been identified in alfalfa (*Medicago sativa*) and wheat (*Triticum aestivum*) (Zozaya- Hinchliffe et al. 2005; Bernard et al. 2008). The multigenic status of GS suggests a complex role for GS isoforms in various aspects of plant N metabolism (Fig. 14.3). For example, in rice (*Oryza sativa*), one gene encodes the plastidic GS2 (*OsGS2*), and three genes encode cytosolic GS1 (*OsGS1;1*, *OsGS1;2* and *OsGS1;3*). *OsGS1;1* is ubiquitous but expressed more in the shoots, *OsGS1;2* is expressed mostly in the root, *OsGS1;3* is limited to the spikelets, and *OsGS2* is abundant in the leaf (Tabuchi et al. 2005). In wheat, there are seven genes coding for three different forms of GS1. *TaGS1a*, *TaGS1b*, and *TaGS1c* encode *TaGS1; 1*, *TaGSr1*, and *TaGSr2* code for the *TaGS1;2* (also known

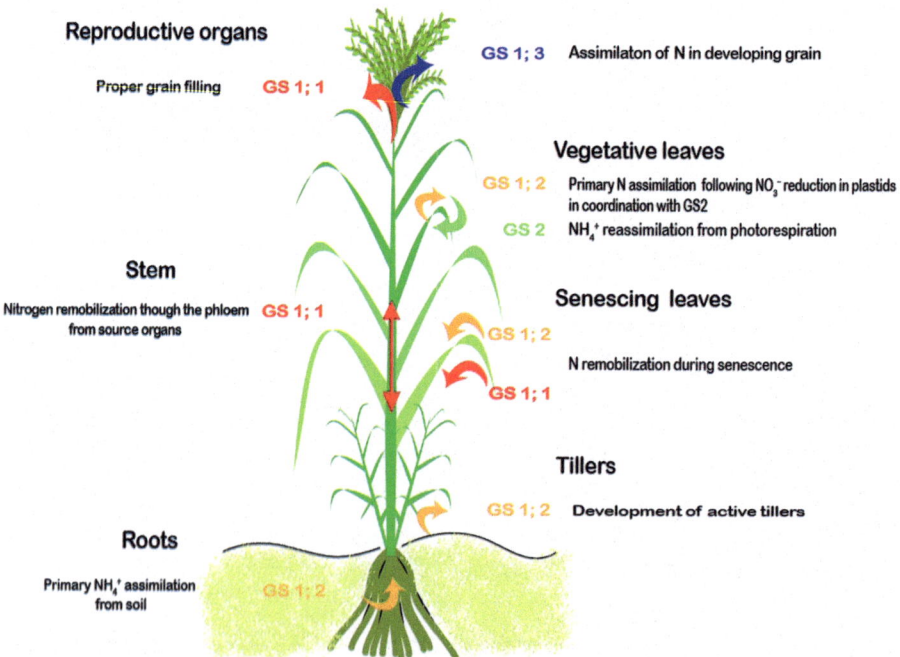

Fig. 14.3 Tissue localizations of various GS isoforms and associated roles in plant N metabolism. (Modified from Thomsen et al. 2014)

as GSr) and *TaGSe1* and *TaGSe2* code for *TaGS1;3* (also referred to as GSe). Furthermore, three genes coding for the chloroplastic isoform GS2 (*TaGS2a*, *TaGS2b*, and *TaGS2c*) are known (Bernard et al. 2008; Wang et al. 2015).

In maize, there are five isoenzymes of GS1 (*ZmGS1;1*,[2] *ZmGS1;2*, *ZmGS1;3*, *ZmGS1;4*, and *ZmGS1;5*) and one chloroplastic GS2 (*ZmGS2*) (Martin et al. 2006). Similarly, in barley, five isozymes of cytosolic GS1 have been identified (*HvGS1;1*, *HvGS1;2*, *HvGS1;3*, *HvGS1;4*, *HvGS1;5*) along with one chloroplastic GS2 (*HvGS2*) (Baima et al. 1989; Goodall et al. 2013; Avila-Ospina et al. 2015). *Arabidopsis* (*Arabidopsis thaliana*) has five putative GS1 genes, namely, *AtGS1;1*, *AtGS1;2*, *AtGS1;3*, *AtGS1;4*, and *AtGS1;5*, and one GS2 (*AtGS2*) (Ishiyama et al. 2004a). Phylogenetic analysis of GSII genes in plants has shown that the cytosolic and chloroplastic GS isoforms form two sister groups (Doyle 1991; Pesole et al. 1991) and that gene duplication followed by independent evolution, through the alteration of promoter and regulatory regions of existing genes, led to the emergence of such varied isoforms in plant species (Biesiadka and Legocki 1997).

14.3 Glutamine Synthetase: Roles in Nitrogen Use Efficiency and Crop Yield Improvement

14.3.1 The Need for Nitrogen Use Efficient Crops

Crop growth and development requires N abundantly, and hence N is generally the most limiting nutrient for crop productivity and yield. Since the production of N fertilizers is energy-intensive and its cost is dependent on the price of energy, it is an expensive nutrient to supply, and commercial N fertilizers contribute to a majority share in the cost of crop production. The "Green Revolution" primarily relied on the use of high fertilizer-responsive dwarf varieties which did not lodge on being supplied high amounts of N. In the last half a century, the amount of synthetic N fertilizers applied to crops has increased significantly, consequently leading to a dramatic increase in crop yields which has sustained our agriculture (Mulvaney et al. 2009). There is a growing consensus that such unwarranted use of nitrogen fertilizers is not a sustainable practice, because most of the nitrogenous fertilizer supplied, more than 50% according to some estimates (Freney 2013), is lost due to volatilization or leached into water bodies, thereby causing eutrophication and severe damage to the environment (Socolow 1999). Poor interconversion of nitrogen fertilizers and their partial capture also lead to nitrous oxide emissions which add to global warming (Bouwman et al. 2002). Therefore, developing crop plants with better nitrogen use efficiency (NUE), which is defined as the biomass/grain yield per unit N accessible for uptake (Brauer and Shelp 2010), is a major challenge we currently face for

---

[2] Some authors use Gln as an alternate abbreviation for the GS gene. Hence, *ZmGS1;1* is *Gln1–1* of maize, *AtGS1;1* is *Gln1;1* of Arabidopsis, etc. The format used throughout this chapter is consistent but may vary in literature.

sustainable agriculture. Given its critical role in N assimilation and recycling, GS has been a prime target in consideration for improvement of NUE through transgenic routes (Brauer and Shelp 2010; Swarbreck et al. 2011).

14.3.2 Significance of GS in NUE and Yield

Expression and Mutational Analysis Several gene expression and mutational (knockout/knockdown) analyses of the various GS isogenes have shown that they play specific and critical roles in the development and yield of crops. In rice, knockout of cytosolic *OsGS1;1*, which is localized predominantly in the vasculature of mature leaves, resulted in a drastic reduction in shoot growth and grain filling along with severe metabolic imbalances. It is noteworthy that in these knockout mutants, the isogenes *OsGS1;2* and *OsGS1;3* could not compensate for the loss of *OsGS1;1*, which signify a non-overlapping and non-redundant role for each of them in rice (Tabuchi et al. 2005; Kusano et al. 2011). Also, knockout mutants of *OsGS1;2* isoform, which is expressed mainly in surface cells of roots, showed a drastic reduction in active tiller number and therefore panicle number at harvest, signifying their role in the primary assimilation of ammonium ions taken up by rice roots from the soil. The *OsGS1;1* isoform, also present to some extent in the roots, was unable to compensate for this function (Funayama et al. 2013; Ohashi et al. 2015).

Similarly, the GS1 isogenes *ZmGln1;3* and *ZmGln1;4*, which are predominantly localized in leaves, play a role in proper development of cobs in maize, in relation to kernel number and kernel size, respectively (Martin et al. 2006; Canas et al. 2010). The *ZmGln1;3* isoform is expressed constitutively in the mesophyll cells, signifying a function in the synthesis of glutamine from NO_3^- reduction, whereas expression of *ZmGln1;4* is limited to the bundle sheath cells and is upregulated during senescence, suggesting a role in reassimilation of NH_4^+ released during degradation of Rubisco (Martin et al. 2006). The cytosolic *HvGS1;1* isoform in barley is localized abundantly in vascular tissues of the stem, and its response to changes in N supply shows that it is important in transport and remobilization of N, similar to the role of *OsGS1;1* in rice, whereas *HvGS1;2* was the major GS isoform expressed in leaf mesophyll cells and the cortex and pericycle of roots, and the increase in its expression in leaves under increasing N levels suggests its function in the primary assimilation of N. The *HvGS1;3* isoform was predominantly and specifically expressed during grain development. Interestingly, *HvGS1;3* expression was also seen to be enhanced in the roots of plants grown under high NH_4^+, which implied a possible role in the protection against NH_4^+ toxicity in the roots (Goodall et al. 2013). Furthermore, the GS2 genes of durum wheat are also known to be involved in determining grain protein content (GPC) (Gadaleta et al. 2011). In wheat, a recent study examined the GS expression and enzymatic activity in different tissues and developmental stages of ten durum wheat genotypes in relation to its GPC. It was observed that the genotypes which had the highest GS expression and activity had a high GPC and vice versa and the genotypes with lower expression and activity had low GPC (Nigro et al. 2016).

In addition, *Arabidopsis* cytosolic GS genes, *AtGS-1;1* and *AtGS-1;2*, are known to have specific roles to play during germination and seed development. In particular, during seed germination, *AtGS-1;1* influenced primary root development in response to N provided exogenously, whereas *AtGS-1;2* was found to be necessary for remobilization of N in germinating seeds and for seed yield (Guan et al. 2015). It was observed that shoot growth, total GS1 activity, and amino acid contents were markedly impaired in *AtGS1;2* knockout mutants which demonstrated that *AtGS1;2* is the main isozyme contributing to GS1 activity in *Arabidopsis* (Guan et al. 2016). Also, *AtGS1;2* was found to be essential for nitrogen assimilation under ample N supply and could be upregulated in roots to relieve ammonium toxicity (Lothier et al. 2011; Guan et al. 2016; Konishi et al. 2016).

QTL Studies Several large population-based quantitative trait loci (QTL) studies have also implicated the crucial role of GS in yield and NUE components. Studies in maize (Hirel et al. 2001; Gallais and Hirel 2004; Hirel et al. 2007) and rice (Yamaya et al. 2002; Obara et al. 2004) have shown QTLs relating to grain parameters co-localizing at previously mapped GS loci. In maize, QTLs for the leaf GS activity have been shown to coincide with QTLs for yield and thus shows the putative role of GS in maize kernel yield (Hirel et al. 2001). In maize, the *ZmGS1;4* locus was coincident with a QTL for the thousand kernel weight trait, and the *ZmGS1;3* locus was coincident with two QTLs for thousand kernel weight and yield. In rice, Obara et al. (2001) reported QTLs associated with putative cytosolic GS activity to be colocated in the vicinity of QTLs for physiological and agronomical traits affected by nitrogen recycling such as spikelet number and panicle weight. In addition, a structural gene for GS1 was mapped in the QTL region for one-spikelet weight signifying its role in grain development during senescence, most likely through its nitrogen export capacity (Obara et al. 2001). Subsequently, Obara et al. (2004) also reported that a putative QTL for GS1 protein content colocated at the QTL regions for spikelet number, soluble protein content, and panicle weight. In wheat, large-effect QTLs for grain N % and total grain weight were found to be associated with the GS1 locus (Habash et al. 2007). On the same lines, Fontaine et al. (2009) reported the correlation of GS with QTLs for physiological and agronomic traits linked to better NUE in wheat. Furthermore, QTLs for total GS activity in flag leaf positively correlated to QTLs for stem and grain N in wheat, but lower correlations were associated with loci for grain yield (Habash et al. 2007; Bernard et al. 2008). Also, the cytosolic *HvGS1;1* of barley was found to be located near to a genetic marker HVM074 which is associated with a major QTL for grain protein content (See et al. 2002). In *Arabidopsis*, QTLs for NUE identified under contrasting nitrate availability co-localized with the cytosolic *AtGS1;2* gene (Loudet et al. 2003). Moreover, a meta-QTL analysis of three robust independent QTL studies associated with NUE in wheat found four meta-QTLs to be associated with different paralogues of the glutamine synthetase (GS) gene highlighting the critical role of GS in NUE (Quraishi et al. 2011).

14.3.3 Transgenic Overexpression of GS: Effects on NUE and Yield

Physiological, molecular, and genetic analyses from diverse plants evidently show the key role of GS in determining many aspects of nitrogen use efficiency trait and yield in crops and its potential utilization for sustainable agriculture. Therefore, several genetic engineering studies have attempted modulating the levels of the GS enzyme in different plants with the aim of understanding its roles in plant growth and/or NUE (Eckes et al. 1989; Hemon et al. 1990; Hirel et al. 1992; Temple et al. 1993, 1994, 1998; Su et al. 1995; Vincent et al. 1997; Brugiere et al. 1999; Gallardo et al. 1999; Limami et al. 1999; Ortega et al. 2001; Fuentes et al. 2001; Oliveira et al. 2002; Carvalho et al. 2003; Jing et al. 2004; Fei et al. 2003, 2006; Fu et al. 2003; Harrison et al. 2003; Sun et al. 2005a, 2005b; Huang et al. 2005; Martin et al. 2006; Seger et al. 2009; Cai et al. 2009; Brauer et al. 2011; Wang et al. 2013; Zhu et al. 2014; He et al. 2014; Bao et al. 2014; Zhu et al. 2015) (Table 14.1). The theory behind many of these attempts was that an enhanced GS activity should increase growth as a result of increased N uptake, while under low N conditions or during senescence, increased GS activity will expedite the remobilization of N from vegetative parts to the seeds (Habash et al. 2001, Oliveira et al. 2002, Fuentes et al. 2001, Thomsen et al. 2014).

An *Arabidopsis* transgenic overexpressing the GS1 gene from algae (*Dunaliella viridis*) showed significantly increased leaf size (29%), stem length (26%), root length (26%), silique number (30%), and fresh weight (22–46% at different NO_3^- concentrations) compared to the wild type. Also, these lines had higher total N, total amino acid, and soluble protein content than the wild-type controls (Zhu et al. 2014). Also, transgenic *Arabidopsis* lines overexpressing bacterial GSI genes from *Klebsiella* sp. D1–5 and *Lactococcus* sp. showed enhanced fresh weight (12%) and dry weight (13%) compared to the wild-type plants at different concentrations of NO_3^- supplied. Further characterization showed that the transgenic lines had higher soluble protein concentration (7–11%), higher total nitrogen content (5–8%), and total amino acid content (4–8%), thereby showing an improved NUE compared to wild-type controls (Zhu et al. 2015). Similarly, overexpression of *Arabidopsis* Dof1 (a transcription factor regulating GS gene expression), *AtGS1;4*, and *AtGS2* in separate transgenic tobacco (*Nicotiana tabacum*) lines, all under a light-responsive Rubisco small subunit promoter (*rbcS*), enhanced nitrogen assimilation under low-nitrogen conditions. In all these transgenic tobacco lines, leaf surface area, plant length, total protein content, total amino acid content, chlorophyll content, and also glucose and sucrose contents in leaves were significantly higher than in wild-type controls. In addition, they had higher activities of many C-N metabolic enzymes like pyruvate kinase (PK), phosphoenolpyruvate carboxylase (PEPC), and nitrate reductase (NR) (Wang et al. 2013).

Transgenic tobacco (*Nicotiana tabacum*) plants of many GS1 genes driven by the *CaMV* 35S promoter have displayed an increase in GS activity and GS1 protein along with reduced free ammonia content and/or increased fresh weight and growth

Table 14.1 Transgenic modulation of GS: outcomes with regard to various aspects of crop improvement

GS isoform; source	Transgenic plant (control used[a])	Construct features	Alteration in GS activity[b]	Alteration in biomass/yield	Alteration in NUE[c]-associated phenotype	Associated phenotypes	Special comments	References
GS1 alfalfa (*M. sativa*)	Tobacco (UT)	*CaMV 35S* Leaf- and stem-specific *ST-LS1* promoters	↑ Fivefold (SA)	N/A	N/A	Normal and fertile plants; PPT resistance ↑ 20-fold	First reported transgenic OX of GS in plants	Eckes et al. (1989)
γ-GS1 French bean (*P. vulgaris*)	Tobacco (UT)	*CaMV 35S*;	24–27% (of TGA)	N/A	N/A	N/A	GS targeted to mitochondria using β-F 1 ATPase mitochondrial targeting sequence	Hemon et al. (1990)
GS1(root) soybean (*G. max*)	Tobacco; *Lotus* (UT)	*CaMV 35S*	25–30% (of TGA)	N/A	N/A	N/A	Induced native cytosolic GS expression	Hirel et al. (1992)
GS1 alfalfa (*M. sativa*)	Tobacco (UT)	*CaMV 35S* Sense and antisense constructs	↑ 25% (TGA in leaves) (sense) ↓ 40% (TGA in leaves) (antisense)	N/A	N/A	↑ 45% TSP (Sense) ↓ 40% TSP (antisense)	Silencing due to heterologous antisense GS transcript occurs at the level of translation	Temple et al. (1993)

(continued)

Table 14.1 (continued)

GS isoform; source	Transgenic plant (control used[a])	Construct features	Alteration in GS activity[b]	Alteration in biomass/yield	Alteration in NUE[c]-associated phenotype	Associated phenotypes	Special comments	References
GS2 tobacco (*N. tabacum*)	Tobacco	Rubisco small subunit promoter (*rbcS*)	↑ 2–2.3-fold (SA of GS2 in leaves)	↑ 20–30% TFW; ↑ 7–14% TDW	N/A	↓ NH_3; ↑ FAA (Gln and Glu)	----	Migge et al. (2000)
glnA *Azospirillum brasilense*	Rice (Wt)	*CaMV 35S* Rice actin1 (*Act1*)	N/A	N/A	↑ Growth (under LN)	N/A	----	Su et al. (1995)
GS1 soybean (*G. max*)	*Lotus* (UT)	*CaMV 35S*	↑ 80% (TGA in shoots)	No change	↑ Growth (under HN) ↓ NH_4^+ uptake	↑ FAA ↑ NH_4^+ (shoots: under HN) ↑ NH_4^+; ↓ Carbohydrate content (roots: under HN)	Premature flowering and senescence	Vincent et al. (1997)
GS1 soybean (*G. max*)	*Lotus* (UT/VC)	*A. rhizogenes RolD* root-specific promoter	↑ 25–30% (TGA in roots)	RFW and SFW ↓ 30–50%	↓ NO_3^- uptake	↑ NO_3^- (shoot) ↓ NO_3^- (roots)	-ve correlation b/w root GS expression and plant biomass	Limami et al. (1999)

Gene	Host	Promoter/construct	Expression	Nodule FW		NH4+	Comment	Reference
GS1 soybean (*G. max*)	*Lotus* (UT)	Nodule-specific *LBC3* promoter of soybean; antisense construct	↓ 50% (TGA in nodules)	↑ Twofold RFW and SFW ↑ 20%	N/A	↑ NH$_4^+$ in nodules	↓ GS activity in root nodules of *Lotus* correlated with ↑ biomass accumulation	Harrison et al. (2003)
GS1 (gln-α) French bean (*P. vulgaris*)	Wheat (A)	Rubisco small subunit promoter (*rbcS*)	↑ Threefold (TGA in flag leaves at late dough stage)	↑ RDW and TDW ↑ Grain N content	N/A	N/A	OX reduced native GS2 activity by 50%	Habash et al. (2001)
GS1(glnβ1) soybean (*G. max*)	Alfalfa (UT/Wt)	*CaMV* 35S	No change	N/A	N/A	N/A	↑ GS transcript level but mRNA instability or protein turnover led to ↓ GS protein and activity	Ortega et al. (2001)
GS1 soybean (*G. max*)	Tobacco; (UT)	*CaMV* 35S	↑ 4–13-fold (TGA)	↑ 70% SDW, ↑ 100% RDW under LN No change under HN	↑ Growth (LN) No change under HN	50% ↑ leaf area under LN	Maintenance of photosynthetic rate under LN improved growth	Fuentes et al. (2001)

(continued)

Table 14.1 (continued)

GS isoform; source	Transgenic plant (control used)[a]	Construct features	Alteration in GS activity[b]	Alteration in biomass/yield	Alteration in NUE[c]-associated phenotype	Associated phenotypes	Special comments	References
GS1; GS2 Pea (*P. sativum*)	Tobacco; (UT)	*CaMV 35S*	↑ 26–50% (TGA) in GS1 lines; GS activity ↓ 93% reduced in GS2 OX lines	↑ TFW, TDW, TSP under both HN and LN for GS1 OX lines; ↓ 55% biomass for GS2 OX lines	↑ Growth (LN and HN)	Sixfold ↓ NH_4^+ (leaves)	Improved growth phenotype was light dependent in GS1 OX lines; GS2 OX lines were co-suppressed	Oliveira et al. (2002)
GS1 *Medicago truncatula*	*Medicago truncatula* (UT)	Nodule-specific *LgHB* promoter *Mtbl*; sense and antisense constructs	↑ Twofold (TGA; sense) ↓ Fourfold (TGA; antisense) (in root nodules)	N/A	N/A	N/A	GS negatively regulated AS and GDH expression	Carvalho et al. (2003)
GS1 conifer (*Pinus* sp.)	Poplar (UT)	*CaMV 35S*	↑ 42% (SA)	↑ 76% Ht (2 months); ↑ 21% Ht (6 months)	N/A	↑ TSP; ↑ Chlorophyll (leaves) ↑ Leaf area ↑ Numbers of nodes and leaves	---	Gallardo et al. (1999), Fu et al. (2003)

GS1 conifer (*Pinus* sp.)	Poplar (UT)	*CaMV* 35S	N/A	↑ 21% Ht (first yr. - FT); ↑ 36% Ht (second yr. -FT); ↑ 41% Ht (third yr - FT)	N/A	↑ 21% protein content in bark in transgenic trees after first yr. of growth	Field trials of poplar transgenics developed by Gallardo et al. 1999	Jing et al. (2004)
GS1 conifer (*Pinus* sp.)	Poplar (UT)	*CaMV* 35S	↑ 62% (TGA under LN); ↑ 81% (TGA under HN)	↑112 LDW, ↑80% Ht (under LN); ↑ 26 LDW, ↑ 35% Ht (under HN)	↑ NUE	↓ 85% NH_4^+ (leaves) ↑ 230% Gln ↑ 20% FAA (under HN); ↑ 31% Gln ↑ 15% FAA (under LN)	Characterization of polar transgenics developed by Gallardo et al. 1999, under varying nitrogen regimens	Man et al. (2005)
GS1 conifer (*Pinus* sp.)	Poplar (Wt)	*CaMV* 35S	N/A	↑ Fivefold growth rate over controls after spraying with 275 g ha^{-1} PPT equivalent herbicide	N/A	↑ PPT resistance; 20-45% of transformed plantlets viable after 14 days in100 μM PPT in vitro	Herbicide tolerance study of polar transgenics developed by Gallardo et al. 1999	Pascual et al. (2008)

(continued)

Table 14.1 (continued)

GS isoform; source	Transgenic plant (control used)[a]	Construct features	Alteration in GS activity[b]	Alteration in biomass/yield	Alteration in NUE[c]-associated phenotype	Associated phenotypes	Special comments	References
GS1 conifer (*Pinus* sp.)	Poplar (Wt)	*CaMV* 35S	N/A	N/A	N/A	↑Tolerance to drought stress; ↑chlorophyll; ↑photorespiratory capacity	Drought tolerance studies of polar transgenics developed by Gallardo et al. 1999. Copper and miRNA-mediated differential regulation of genes involved in ROS protection confer drought tolerance in *GS1* OX Tg poplar	El Khatib et al. (2004), Molina-Rueda et al. (2013, 2015)
GS1;3 maize (*Z. mays*)	Maize (A)	Cassava vein mosaic virus (*CsVMV*) promoter	↑Two- to threefold (TGA in leaves)	↑42% grain yield (under LN); No change in SDW	↑42% UtE	Grain yield ↑ due to ↑30% in kernel number	Strong correlation determined b/w leaf GS activity and grain yield; knockout mutants of *ZmGS1;3* and *ZmGS1;4* reported to have reduced kernel number and size, respectively	Martin et al. (2006)

GS1 soybean (*G. max*)	Pea (Wt)	*CaMV 35S* Nodule-specific promoter (*LBC*) Root-specific promoter (*rolD*)	↑ 0–12-fold (TGA) (various tissues)	↓ 23% to ↑ 33% TDW (under LN and HN (inconsistent results)	↓ 29% to ↑ 30% UI (under LN and HN (Brauer and Shelp 2010)	N/A	Inconsistent GS activity and phenotypes	Fei et al. (2003, 2006)
GS1 + GS2 (concurrently) pea (*P. sativum*)	Rice (Wt)	Rice actin1 (*Act1*) Maize ubiquitin (*Ubi*) promoters	N/A	↑ FW (LN)	N/A	↑ Growth (LN); ↑ PPT resistance (up to 0.3% v/v leaf painting)	---	Sun et al. (2005a, b)
GS1 + GS2 (concurrently) pea (*P. sativum*)	Wheat (Wt)	Rice actin1 (*Act1*) Maize ubiquitin (*Ubi*) promoters	N/A	N/A	N/A	↑ PPT resistance (up to 0.3% v/v leaf painting)	---	Huang et al. (2005)

(continued)

Table 14.1 (continued)

GS isoform; source	Transgenic plant (control used)[a]	Construct features	Alteration in GS activity[b]	Alteration in biomass/yield	Alteration in NUE[c]-associated phenotype	Associated phenotypes	Special comments	References
OsGS1;1 OsGS1;2 (Rice) (O. sativa) glnA (E.coli)	Rice (Wt) Separate transgenics of each gene	CaMV 35S	↑18–25% (TGA under ON) ↑36–46% (TGA under LN)	↓7–39% grain yield No change in SDW	↓NUE$_{grain}$	OsGS1;2 OX Tg plants resistant to PPT (0.5% v/v spraying) but sensitive to cold, drought, salt OsGS1;1 OX and glnA OX Tg rice plants showed no significant tolerance to abiotic stresses	All three types of transgenic OX lines led to decrease in yield C-N metabolic imbalance and poor N transport from stem to leaf were identified as the reason for poor growth phenotype and reduced yield	Cai et al. (2009), Bao et al. (2014)
OsGS1;2 rice (O. sativa)	Rice (A)	Maize ubiquitin (Ubi)	↑ 1.7–2.8-fold (TGA in leaves)	No change in vegetative yield; ↑ percent spikelets filled (under both HN and LN)	↑HI; ↑NHI; ↑UtE compared to azygous controls	No change in UpE hence not N use efficient under LN ↑UtE under growth chamber conditions; not under greenhouse conditions	Reappraisal study; transformation effects significantly masked OX phenotype Highlights the need for using azygous controls under field conditions in assessing true potential of GS OX Tg plants	Brauer et al. (2011)

OsGS2 rice (O. sativa)	Rice (Wt)	CaMV 35S Catalase first intron GS2 with transit peptide	N/A	N/A	N/A	OsGS2 OX Tg rice was significantly tolerant to salt stress (150 mM NaCl)	Enhanced photorespiratory capacity in OsGS2 OX Tg lines led to salt tolerance · Hoshida et al. (2000)
OsGS1;1 Rice (O. sativa)	Rice (Wt)	CaMV 35S	N/A	N/A	N/A	OsGS1;1 OX Tg rice was tolerant to cadmium toxicity (100 µM CdCl$_2$)	Modulation of enzymes involved in protection from membrane lipid peroxidation led to tolerance · Lee et al. (2013)
SbGS1;2 Sorghum (S. bicolor)	Sorghum (A)	CaMV 35S Maize ubiquitin (Ubi)	↑ 1.3–1.6-fold (TGA in leaves); ↑ 1.3–2.2-fold (TGA in roots)	↑ Twofold SDW (under ON) No change in SDW, RDW, grain yield (under LN)	No change	N/A	GS OX maybe affected by environmental and transformational influences · Urriola and Rathore (2015)

(continued)

Table 14.1 (continued)

GS isoform; source	Transgenic plant (control used)[a]	Construct features	Alteration in GS activity[b]	Alteration in biomass/yield	Alteration in NUE[c]-associated phenotype	Associated phenotypes	Special comments	References
AtGS1;4 AtGS2 AtDof1.7 (A. thaliana)	Tobacco (Wt) Separate transgenics of each gene as well as double transgenics (Dof + GS1;4 and Dof + GS2)	Rubisco small subunit promoter (rbcS)	↑ 140–170% (TGA in leaves)	↑ FW, DW, plant Ht (under LN)	↑ NUE; ↑ growth under LN More so in double transgenics	↑ Chlorophyll content; ↑ Shoot N content; ↑ TSP ↑ PEPC and PK activity (Dof OX Tg)	NUE highest in double transgenics	Wang et al. (2013)
DvGS1 green algae (Dunaliella viridis)	Arabidopsis (Wt)	CaMV 35S	↑ 29% (TGA in leaves)	↑ 22–46% TFW; ↑ 12–2% TDW; ↑ 26–40% TSP	↑ NUE	↑ Leaf area; ↑ silique weight; ↑root length	Results are promising but have to be independently verified	Zhu et al. (2014)
GSI type genes from Klebsiella Lactococcus	Arabidopsis (Wt)	CaMV 35S	↑ 8–14% (TGA in leaves)	↑ 12–13% FW and DW (under LN and HN)	↑ NUE	↑ Total N content; ↑ TSP; ↑ FAA	Results are promising but have to be independently verified	Zhu et al. (2015)

| Mutant *OsGS1;1* Rice (*O. sativa*) | *Arabidopsis* (Wt) | Double *CaMV* 35S (D35S) promoter | N/A | N/A | N/A | ↑ Resistance to PPT (150 mg/L spray PPT); (upto 40 mg/L in vitro) | DNA shuffling identified a R295K mutant rice *GS1;1* which conferred PPT tolerance to *Arabidopsis*. Result is promising and has to be further verified in rice under field conditions | Tian et al. (2015) |

Modified from Brauer and Shelp (2010)

A azygous (null segregants of individual transgenics lines), VC vector control (transformed with plant transformation vectors without gene), GS (glutamine synthetase), OX (overexpression), ↑ increase, ↓ decrease, PPT (phosphinothricin), TSP (total soluble proteins), FAA (free amino acids), LN (low nitrogen), N (nitrogen), HN (high nitrogen), ON (optimum nitrogen), TFW (total fresh weight), TDW (total dry weight), SFW (shoot fresh weight), RFW (root fresh weight), SDW (shoot dry weight), RDW (shoot dry weight), PEPC (phosphoenolpyruvate carboxykinase), PK (pyruvate kinase)

[a]Controls used for comparison against transgenic plants; UT: untransformed control (untransformed plants obtained after tissue culture)

[b]Activity of GS measured as total GS activity (TGA) or specific activity (SA)

[c]NUE (nitrogen use efficiency: yield/biomass per unit N supplied); NUE_{grain} (grain nitrogen use efficiency: spikelet yield per unit N supplied); UtE (Utilization N use efficiency: biomass or yield/shoot N content); HI (harvest index: grain weight/biomass weight); NHI (nitrogen harvest index: spikelet N content/shoot N content); UpE (uptake efficiency shoot N per unit of N supplied); UI (usage index: square of SDW per unit of shoot N)

(Eckes et al. 1989; Hemon et al. 1990; Hirel et al. 1992, 1997; Temple et al. 1993; Lam et al. 1995; Migge et al. 2000; Fuentes et al. 2001; Oliveira et al. 2002). Tobacco transgenics overexpressing the alfalfa GS1 under a *CaMV* 35S constitutive promoter showed no change in photosynthesis or growth under optimum nitrogen fertilization conditions. However, under N starvation these transgenics had 50% more leaf area, ~70% more shoot, and ~100% larger root dry weight than their respective low N controls. N deprivation in these GS-overexpressing lines had no significant effect on photosynthetic ability and levels of chlorophyll and amino acids, while controls exhibited a 30–50% decrease in photosynthesis under limiting N conditions (Fuentes et al. 2001). These results suggest that the transgenic lines exhibited relatively better N assimilation and supply more N to the vital photosynthetic machinery under N stress. In contrast, Oliveira et al. (2002) reported greater dry matter accumulation under *abundant N conditions* in tobacco-overexpressing GS1 lines than in controls. Further experiments showed significant differences in shoot fresh mass under both abundant and limiting N levels in these transgenic plants. Also, these GS overexpression lines exhibited higher photorespiratory rates in leaves as measured by increased CO_2 evolution, as well as lower free ammonia concentrations (an 88% decrease) compared with non-transgenic controls. The growth increases in these GS-overexpressing transgenics are hypothesized to be due to increased photosynthesis and N assimilation (Oliveira et al. 2002). Whereas the study by Oliveira et al. (2002) reported considerable biomass increase in the tobacco GS overexpression lines at both luxury and limiting N, the study by Fuentes et al. (2001) reports differences only under limiting N levels. The inconsistency in these results may be due to the use of different GS isogenes and promoters and/or differences in growing methods. Overexpression of the tobacco GS2 gene in tobacco under an *rbcS* promoter exhibited a significant decrease in the leaf NH_4^+ content (~3.7 times) and increase in free amino acids such as glutamine (~2.3 times) and glutamate (~2.5 times). Leaf soluble protein content per unit fresh weight remained unchanged, and a mechanism downstream of the synthesis of the primary organic metabolites of N assimilation was attributed to limiting the leaf protein accumulation. However, overexpression of GS2 increased the rate of growth in the transgenic tobacco which had higher fresh weight than the wild-type control plants grown under similar conditions (Migge et al. 2000).

Poplar trees (*Populus tremula*) overexpressing a GS1 gene from pine (*Pinus sylvestris*) exhibited an increase in the levels of total soluble protein and chlorophyll content in leaves of transformed trees. Moreover, the mean net growth in height of GS-overexpressing trees in comparison to untransformed controls showed a 76% increase in height at 2 months and a 21% increase at 6 months (Gallardo et al. 1999; Fu et al. 2003). A study of the same transgenics, under field trials, showed that the trees reached heights that were on average 21, 36, and 41% taller than control trees after the first, second, and third year of growth, respectively. Analyses of stem diameter and protein contents of the bark showed higher levels of nitrogen accumulation in the stem of transgenics (Jing et al. 2004). Further characterization of these poplar transgenic lines also demonstrated higher leaf dry mass under limiting N conditions than under non-limiting N conditions. These plants also showed 85% less free NH_4^+

than leaves of non-transgenic control plants. Also, under high N conditions, leaves of transgenics plants showed enhanced levels of free glutamine and total free amino acids. Moreover, ^{15}N-enrichment experiments in these transgenics showed that 27% more labeled N was taken up into biomolecules in transgenic lines than in non-transgenic controls, thereby demonstrating increased N assimilation efficiency in GS transgenic lines (Man et al. 2005).

In wheat (*Triticum aestivum*) plants, overexpressing the GS1 gene (*gln-α*) from common bean (*Phaseolus vulgaris*) under the rice Rubisco promoter led to an increase in root dry mass, grain yield, nitrogen uptake efficiency (UpE), and earlier flower and seed development but does not affect shoot dry mass (Habash et al. 2001). The yield differences are due to increases in grain mass and grain N content, but not grain number. However, these results were not reproducible using the same wheat transgenics in an independent study (Schjoerring 2005). Overexpression of *ZmGln1;3* in maize (*Zea mays*) using a constitutive promoter resulted in an increase of 30% in kernel number, ultimately leading to a higher yield in transgenic plants compared with controls, but no significant changes were observed in shoot dry matter production or NUE (Martin et al. 2006). Similarly, overexpression of two maize GS1 genes (*ZmGln 1;3* and *ZmGln 1;4*) in separate maize transgenics showed improved nitrogen assimilation, NUE, and yield (up to 20%) as inferred from the enhanced yield-related traits such as ear weight, ear diameter, ear length, hundred-kernel weight, and grain weight per ear (He et al. 2014). In sorghum (*Sorghum bicolor*), transgenic plants overexpressing a sorghum cytosolic GS gene (*SbGln1;2*) under the control of the maize ubiquitin (*Ubq*) promoter showed enhanced tillering, whereas vegetative shoot biomass increased twofold when grown under optimal nitrogen conditions (Urriola and Rathore 2015).

In rice, ectopic overexpression of the *OsGS1;2* isoform using a maize *Ubq* promoter resulted in increased spikelet yield, harvest index, N harvest index (spikelet N content/shoot N content), and UtE relative to their azygous controls in plants grown under high N conditions (Brauer et al. 2011). Intriguingly, the overexpression lines did not differ from azygous controls in vegetative yield or shoot N content. Thus, overexpression of *OsGS1;2* improved N partitioning in rice during grain filling leading to an improved UtE. However, it did not show a better NUE phenotype under limiting N and therefore is improbable to result in the use of less N under field conditions (Brauer et al. 2011). Individual transgenic rice plants with the *CaMV* 35S promoter driving a rice GS1;1 (*OsGS1;1*), a rice GS1;2 (*OsGS1;2*), and an *E.coli* glnA gene exhibited an increased metabolic level as seen by increases in leaf total GS activities and soluble protein concentrations and higher total amino acids and total nitrogen contents in whole plants grown hydroponically under both limiting and non-limiting N conditions. However, in plants grown to maturity under limiting N conditions in the field, both total amino acids in grains and grain yield production were decreased in all these GS-overexpressing transgenics compared with wild-type plants (Cai et al. 2009). Further systematic study of the *OsGS1;1-* and *OsGS1;2*-overexpressing lines was done by analyzing the carbon-nitrogen metabolic status, growth phenotype, and gene expression profiles of these transgenic rice plants (Bao et al. 2014). This study revealed that these transgenics exhibited a poor plant growth phenotype, lower yield, and decreased carbon/nitrogen (C/N) ratio in the stem due to the accumulation of nitrogen. A detailed metabolite profiling and gene expression analysis under different N environments demonstrated significant changes in free amino acids, individual sugars, organic acids, and gene expression patterns of C and N metabolic pathway-related genes between transgenic lines and wild-type controls. Also, the chlorophyll content, photosynthetic rate, soluble proteins, and carbohydrates varied inconsistently in both the overexpressing transgenic plants (Bao et al. 2014). A transgenic rice plant simultaneously overexpressing both the plastidic and cytosolic GS isoforms under the control of rice actin 1 (*Act1*) and maize *Ubq* promoters displayed increased fresh weight and was tolerant to nitrogen deficiency (Sun et al. 2005a).

14.3.4 Possible Reasons for Inconsistent Results of GS Overexpression

GS1 overexpression has in some cases led to higher yields and shoot biomass and/ or better NUE (Habash et al. 2001; Oliveira et al. 2002; Fuentes et al. 2001; Martin et al. 2006), however, various studies have reported nil or negative effects on yield and/or NUE (Temple et al. 1993; Vincent et al. 1997; Limami et al. 1999; Ortega et al. 2001; Fei et al. 2003, 2006; Cai et al. 2009; Brauer et al. 2011; Bao et al. 2014). One possible reason for this inconsistency is that GS activity is increased only slightly. For example, 18–23% increase in GS activity in tobacco does not manifest in an altered phenotype, whereas tobacco plants with more than a 26% increase in GS activity display significant biomass increases (Hirel et al. 1992; Temple et al. 1993).

Also, the lack of biomass improvement from increased GS activity may be due to species-specific regulation, as is the case in overexpression of GS in *Lotus* which has consistently resulted in negative phenotypes such as decreased or no alteration in biomass, accelerated senescence, and pollen sterility in comparison to controls (Vincent et al. 1997; Limami et al. 1999; Suarez et al. 2003). Intriguingly, it was observed that limiting GS activity using an antisense construct of GS improved fresh weight in *Lotus* (Harrison et al. 2003). A similar scenario manifested in GS1 overexpression lines of pea (*Pisum sativum*) which showed inconsistent biomass variation (Fei et al. 2003, 2006).

Another reason for these inconsistent results might be due to the fact that most of these studies had a relatively unsophisticated transcriptional control of the transgene expression, because of the use of the constitutive *CaMV* 35S promoter. It is well understood that different GS isogenes have varying cellular-, tissue-, and developmental stage-specific, non-redundant roles to play in N sensing, assimilation, transport, and remobilization, and also much variation occurs in GS activity among different species (Forde et al. 1989; Kamachi et al. 1991, 1992; Li et al. 1993; Finnemann and Schjoerring 2000; Habash et al. 2001; Tabuchi et al. 2007; Bernard et al. 2008; Orsel et al. 2014; Ohashi et al. 2015). Thus, the use of more sophisticated and refined strategies which would combine gene stacking with tissue and/or developmental stage-specific expression is likely to provide better outcomes (Thomsen et al. 2014).

Moreover, most of these experiments have been done only on individual transgenics in and under greenhouse conditions. It has been observed that the metabolic roles of GS depend on environmental conditions, metabolic status of various tissues in the plant, and the plant in general and can even vary during different times of the day (Miflin and Habash 2002). Since environmental conditions and soil N regimens play an integral role in response of GS, studies under field conditions, utilizing larger transgenic populations in replications, and analyses at appropriate times of the day over multiple developmental stages will be a more robust approach to further our understanding.

Using wild-type controls offer a good comparison in NUE and yield experiments but are unlikely to take into account transformation effects such as somaclonal variation, positional effect, and the high energy expenses of overexpression (Miki et al. 2009). It was observed that transformation effects had a significant effect on the productivity of the GS-overexpressing transgenic rice (Brauer et al. 2011). In comparison, using azygous controls (progeny of transgenic parent lines that have lost the transgene through segregation) do take in into account transformation effects but do not allow comparison to benchmark cultivars in production. Hence, the use of both wild-type and azygous controls would take into consideration the impact of transformation effects as well assess the possibility of use of these novel transgenics in improving crop production under field conditions (Brauer and Shelp 2010; Brauer et al. 2011).

14.3.5 Intricate Regulation of GS Might Limit Simple Overexpression Strategies

It is noteworthy to mention that the GS-GOGAT cycle is estimated to use around 15% of the ATP pool of a cell (Harper et al. 2010). Given the high energy expenditure involved, the presence of a complex regulatory system may also account for the inconsistent results of overexpression studies. Regulation of plant GSs studied till date occurs at several levels: tissue-, organ-, and phenological stage-specific transcription of different isoforms in response to various environmental cues; stability and posttranscriptional processing; translational regulation; subcellular targeting, modification, and processing; assembly of the holoenzyme; posttranslational modification; and also enzyme degradation via oxidative turnover (Kamachi et al. 1991; Li et al. 1993; Temple et al. 1993; Cren and Hirel 1999; Ortega et al. 1999; Hirel et al. 2001; Riedel et al. 2001; Tobin and Yamaya 2001; Miflin and Habash 2002; Ishiyama et al. 2004b; Lima et al. 2006a; Tabuchi et al. 2007; Bernard and Habash 2009; Orsel et al. 2014; Thomsen et al. 2014; Seabra and Carvalho 2015). Transcriptional regulation of GS has been found to be influenced by light intensity, flux of various N and C metabolites, abundance of C skeletons available for amino acid synthesis, as well as the external availability and form of N (NH_4^+ or NO_3^-) (Thomsen et al. 2014). Glutamate, the substrate for GS and more importantly glutamate-to-glutamine ratio, has been shown to control GS expression levels (Watanabe et al. 1997; Masclaux-Daubresse et al. 2005). Also, the promoters of different GS isogenes contain many regulatory elements such as light regulatory elements (LREs), elements conferring tissue-specific expression, as well as many known transcription factor binding sites (Castro-Rodriguez et al. 2011). Various transcription factors such as MYB transcription factor CCA1 (circadian clock-associated1), OsMYB55, Dof (DNA binding with one finger), and NLP7 (NIN-like protein 7) have been implicated in the control of GS expression and/or regulating cellular C-N balance and NUE (Gutiérrez et al. 2008; Gutiérrez 2012; Castro-Rodriguez et al. 2011; Yanagisawa et al. 2004; Castaings et al. 2009; Marchive et al. 2013; Konishi and Yanagisawa, 2013; Thomsen et al. 2014; Yu et al. 2016). These transcription factors might play a role in the alteration of GS expression as well as coordinating C and N metabolism. The exact mechanisms involved in sensing of C and N metabolite flux levels for transcriptional regulation of GS are yet to be unraveled (Thomsen et al. 2014). Also, many QTL studies have indicated regions in the genome that are associated with GS activity, but do not co-localize with known GS genes. The genes in these regions may contain novel regulatory factors of GS, and their functional analysis may further enhance our understanding of the complex regulation of GS in plants (Miflin and Habash 2002).

It was observed that the soybean GS1 gene, *GmglnβI*, when introduced as a transgene is subject to posttranscriptional regulation via its 3′ UTR which is involved in both transcript turnover and translational repression. This regulation was found to be influenced by N metabolites such as glutamine and cellular C/N ratios (Ortega et al. 2006; Simon and Sengupta-Gopalan 2010). This regulation was proposed to be likely universal in nature though further confirmation in other species is needed. Furthermore, evidence suggesting a posttranscriptional control of the transgene was seen in rice GS-overexpressing lines (Brauer et al. 2011). The rice transgenic lines from different transformation events had a 5–15-fold increased level of GS expression than in the wild type and 2–3 times the GS activity. These results are also consistent with other studies of GS overexpression in pea (Fei et al. 2003), *Arabidopsis* (Migge et al. 2000), alfalfa (Ortega et al. 2001), and tobacco (Oliveira et al. 2002). It was also reported that the 5′UTR of GS1 may function as a translational enhancer although the exact mechanism is still unknown (Ortega et al. 2012). Thus including the 5′ UTR region and discarding the 3′UTR region in future overexpression studies might enhance efficiency of the GS transgene expression.

Posttranslational regulation of GS1 activity is known to be controlled by phosphorylation-mediated protein turnover, nitration by nitric oxide (NO), and oxidative turnover (Thomsen et al. 2014). GS1 phosphorylation is influenced by N fixation and light suggesting that this form of regulation is vital for coordination of N assimilation upon sensing external cues (Lima et al. 2006a). A Ca^{2+}-dependent protein kinase-related kinase (CRK), AtCRK3, was found to specifically interact with and phosphorylate the cytosolic glutamine synthetase, *AtGS1;1*, of *Arabidopsis* and may regulate nitrogen remobilization during senescence (Li et al. 2006). Also, 14–3-3 proteins, which bind to GS upon phosphorylation, have been found to regulate the activity of GS2 through posttranslational mechanism and may have a significant role in N sensing and signal transduction, but further characterization is required to understand the exact mechanisms (Moorhead et al. 1999; Finnemann and Schjoerring 2000; Riedel et al. 2001; Lima et al. 2006a, b). Very recently, ACR11, a uridylyltransferase-like protein, was reported to be an activator of GS2 in *Arabidopsis*, although its specific function remains unknown (Osanai et al. 2017). Interestingly, although homologs of the signal transduction protein PII which has been shown to link C and N signaling in prokaryotes have been identified in rice and *Arabidopsis*, its direct regulation of GS in plants has not been observed (Uhrig et al. 2009). The oxidative turnover-mediated inactivation of GS was observed in soybean in response to exogenous application of ammonia, whereas the NO-based posttranslational inactivation of GS was inferred to be in order to channel metabolites to boost antioxidant defenses. Furthermore, various small RNA molecules have been implicated in the control of C-N metabolism and may help in improving NUE, but till date the miRNAs that directly regulate specific GS genes are yet to be identified (Fischer et al. 2013; Thomsen et al. 2014).

14.3.6 Future Strategies for Utilizing GS in NUE and Yield Enhancement

Future overexpression strategies using GS for improving NUE will have to take into account the intricate regulation of GS and its intimate interaction with the C-N metabolic pathway to overcome potential metabolic bottlenecks. Concomitant overexpression of other genes of the C-N metabolic pathway will play an integral part of such attempts in order to prevent imbalances in metabolism caused by limitations in substrate and/or inhibition by end products, thereby downregulating GS1 transgene expression or activity. These genes can include enzymes such as GOGAT, which supplies the substrate for GS; asparagine synthetase (AS); PEPc (phosphoenolpyruvate carboxylase) and ICDH (isocitrate dehydrogenase) which provides C skeletons like 2-OG (2 oxoglutatrate); GDH (glutamine dehydrogenase) which is also responsible for NH_4^+ assimilation to a lesser extent; ammonia and NO_3 transporters (NRTs and AMTs); amino acid transporters such as alanine aminotransferase (AlaAT) or aspartate aminotransferase (AspAT) which are involved in the transport of N; transcription factors and regulatory proteins like *Dof*, CCA, MYB, PII protein, NLP7, and ENOD93–1; and heteromeric G proteins associated with DEP1 (dense and erect panicle 1) which have been reported to regulate the C-N metabolic pathway (Lancien et al. 2000; Good et al. 2004; Yanagisawa et al. 2004; Gutiérrez et al. 2008, Gutiérrez 2012; Rueda- Lopez et al. 2008; Castaings et al. 2009; Masumoto et al. 2010; Castro-Rodriguez et al. 2011; Konishi and Yanagisawa 2013; Marchive et al. 2013; Sun et al. 2014; Thomsen et al. 2014; Yu et al. 2016).

In fact, such strategies are already being utilized and show much promise. For example, concomitant overexpression of *Arabidopsis Dof1.7* and *AtGS1;3* in tobacco (*Nicotiana tabacum*) utilizing an in vitro gene pyramiding approach based on Gateway™ cloning showed a considerably higher NUE as compared to only *AtGln1;4*-overexpressing and wild-type tobacco controls (Wang et al. 2013). In contrast, concurrent overexpression of a cytosolic GS1 from soybean and sucrose phosphate synthase (SPS) from maize in transgenic tobacco by crossbreeding of single transformants showed that the GS1 activity was lower in the co-transformants compared to only GS1 transformants, specifically under low-nitrogen conditions (Seger et al. 2014).

Thomsen et al. (2014) have detailed the specific areas where the overexpression of GS in a concerted manner might help in improving NUE and yield. Targeted GS overexpression can be optimized for synchronization of N fluxes with the early spike development stages to sustain a high number of fertile florets that may lead to increased number of grains per spike and thereby yield. For example, in wheat and barley, the flux of N to the early spike development is important to reduce floret abortion and increase the number of fertile florets per spike leading to higher yields (Ferrante et al. 2010). Also, overexpression of specific GS isogenes to improve N filling in grains during generative growth stages may also lead to better yields. Furthermore, tissue- and organ-specific overexpression of amino acid transporters such as AlaAT or AspAT and in tandem with regulated GS overexpression might help in diversion of glutamine and increased phloem export of amino acids which can overcome end product inhibition and the associated metabolic imbalances. Also, this strategy may lead to improved xylem N loading in the stem and ensure enhanced N transport in roots, all having beneficial effects on the NUE trait. It is worthy to mention that tissue-specific overexpression of a barley *AlaAT* gene under the control of a rice root epidermis-specific promoter *OsAnt1* showed a significantly improved NUE (Shrawat et al. 2008). In fact, Arcadia Biosciences of California, USA, has developed this NUE trait in rice and canola, and extensive field trials have reported that the NUE rice lines, when grown under half the recommended dose of nitrogen fertilizer, outyielded the control varieties by 22% in the first year and by 30% in the second year of trials. Similarly a team of researchers at the Nanjing Agricultural University, China, working in collaboration with scientists at the John Innes Centre, UK, found that constitutive overexpression of the *OsNRT2.3b* gene in rice enhanced its pH-buffering capability, thereby increasing N, Fe, and P uptake (Fan et al. 2016). In extensive field trials, the overexpression of *OsNRT2.3b* enhanced nitrogen use efficiency (NUE) and grain yield by 40%, which is an extraordinary advance of using a single gene transgenic approach to increase NUE and yield in rice. Hence synchronized overexpression of these genes with GS is a worthwhile attempt to explore if it can possibly outperform single transformants and improve NUE and yield even further.

So far, it has been difficult to determine whether the varied regulation of the transgene GS activity and the inconsistent results of yield and NUE improvement are due to transgenic positional effects, transformational effects due to tissue culture, environmental influence, or species-specific internal regulations of the C-N metabolic pathway within the transgenic plants. Given the high energy expenditure involved and the intricately regulated interplay between carbon and nitrogen metabolic pathways, it might be a combination of such factors that lead to this variation in phenotype. Nonetheless, preliminary results are sufficiently encouraging to suggest that further attempts at transgenic manipulation of GS, for crop yield improvement and improved NUE, are warranted.

14.4   Glutamine Synthetase: Transgenesis for Tolerance to Various Abiotic Stresses

Abiotic stresses including salinity, drought, cold, high temperatures, flooding, and metal toxicity in soils have adverse impact on plant growth and yield under field conditions and have become an impediment in realizing global food security for the burgeoning population. Therefore developing climate-resilient crops which can face such adverse environmental conditions is a need to be addressed with utmost urgency (Gill et al. 2014). These crops will not only have to grow under stressed conditions but also yield as much as they do under normal field environments, if not more. The critical role of genes involved in N uptake, assimilation, remobilization, and partitioning under abiotic stress will also have to be considered if we are to improve crop yield especially under field conditions where overlapping episodes of multiple stresses are usually common (Goel and Singh 2015; Suzuki et al. 2014).

14.4.1   GS Is Differentially Modulated in Various Abiotic Stresses

Genes involved in N metabolism, including GS, are known to be significantly modulated during various stress responses in different plants (Goel and Singh 2015; Wang et al. 2012). This is possibly because nitrogen assimilation is more responsive to moderate stress than photosynthetic assimilation, and abiotic stresses result in the increase of cellular processes like photorespiration (high light, salt, and oxidative stress) and proteolysis (generally all stresses) which can evolve NH_4^+ (Teixeira and Fidalgo 2009; Wingler et al. 1999). The capability of crop plants to grow under low N supply may also be linked to their tolerance to other stresses (Bernard and Habash 2009). For example, a relative of *A. thaliana*, namely, *Thellungiella halophila*, was found to be tolerant to nitrogen limitation and salt stress, as a result of its improved N uptake and assimilation (Kant et al. 2008).

Different studies have given varying results in the modulation of GS with respect to various stresses and in different tissues. Although generally, in leaves during drought or salt stress, the GS2 activity and protein levels decline, while the cytosolic GS shows an increase or maintains the same level (Bauer et al. 1997; Lutts et al. 1999; Santos et al. 2004). For example, in drought-stressed tomato, it was reported that cytosolic GS gene expression increased in leaves, whereas the plastidic isoform showed no change (Bauer et al. 1997). Similarly, in sunflower plants GS1 mRNA and protein contents increased under salt and drought stress, while plastidic GS2 activity decreased in leaves under salt stress (Santos et al. 2004). Salinity also led to increased GS activity in mulberry (Ramanjulu et al. 1994). Similarly in foxtail millet, Veeranagamallaiah et al. (2007) reported increased GS protein and proline accumulation in salt-tolerant cultivar than in salt-susceptible cultivar under moderate salt stress. Similarly in salt-stressed cowpea and cashew, the total leaf GS activity was increased by salt stress, whereas in the roots it was slightly lowered (Silveira et al. 2001; Silveira et al. 2003).

In roots, the response is vague, with results from potatoes and rice seedlings showing a decrease in total activity of GS under salt stress (although the total GS activity was higher in the salt-resistant cultivar in rice) (Lutts et al. 1999; Teixeira and Pereira 2007) and reports in sunflower and rice observing an increased GS1 protein and/or activity (Santos et al. 2004; Yan et al. 2005). A suppression subtractive hybridization (SSH) study to identify early salt stress-responsive genes in tomato roots identified both GS isoforms to be differentially regulated. The downregulation of GS expression under severe salt stress was confirmed by microarray analysis (Ouyang et al. 2007). In potato, Teixeira and Pereira (2007) noted that the total GS activity in leaves decreased under salt and drought stresses but were enhanced in growing tubers under drought stress conditions. But interestingly, the decrease observed in GS activity in leaves was not due to a decrease in GS protein content, thereby signifying a posttranslational mechanism of GS regulation at work. In contrast, in tomato seedlings under salinity stress, GS activity was inhibited in the leaves but was enhanced in the roots (Debouba et al. 2006), whereas it was reported to be enhanced in leaves by Hossain et al. (2012). The discrepancies in results could be due to the fact that there GS exists as several isoforms in plants which have nonredundant functions and measurement of total GS activity may mask isoform-specific differences. Another reason could be the phenological stage-specific regulation of the various GS isoforms in different species. The variation in stress severity may also affect the results as it has been shown that chloroplastic GS2 undergoes oxidative degradation by hydroxyl ROS (reactive oxygen species) under higher levels of stress (Ishida et al. 2002, Palatnik et al. 1999). Thus under severe stress conditions, the total GS activity in leaves would likely reduce due to lower GS2 activity as in most plants GS2 forms the major isoform in leaves (Lancien et al. 2000). Indeed, in wheat, it was observed that at a lower salinity stress (150 mM NaCl), the total GS activity increased, while at higher salinity (300 mM NaCl), GS activity decreased drastically (Wang et al. 2007). In another study, GS activity in roots of salt-stressed wheat plants increased slightly, whereas in shoots it decreased by approx. 30% as compared to control plants. While the activity of GS1 isoform grew slightly under saline stress, the drop in the total GS activity in shoots of salt-stressed plants was found to be due to the drop in activity of GS2 isoform (Kwinta and Cal 2005).

Nagy et al. (2013) observed that in well-watered wheat plants, the total GS activity in the older leaves was lower than in younger flag leaf in both drought-sensitive and drought-tolerant cultivars. But under drought, the flag leaf of the sensitive cultivars had lower GS activity than the older leaves. However, in drought-tolerant genotypes, it maintained high GS activity in the flag leaf even under drought stress. They proposed that the untimely senescence of the flag leaf and increasing ratio of cytoplasmic to chloroplastic GS due to accelerated senescence induced by water deficit are indicators of drought stress sensitivity and can be used as markers of drought stress tolerance (Nagy et al. 2013). Recently, a comparative proteomic study of wheat cultivars under drought stress found that the chloroplastic GS2 was significantly upregulated in a drought-tolerant cultivar as compared to the sensitive cultivar (Cheng et al. 2016). Also, a recent gene expression analysis under salinity and drought stress between contrasting durum wheat genotypes showed that the most tolerant genotype exhibited the highest GS activity and was the only one showing enhanced expression of both GS1 and GS2 under stress conditions compared with control (Yousfi et al. 2015).

In *Arabidopsis*, studies revealed *AtGS1;1*, *AtGS1;3*, and *AtGS1; 4* to be upregulated by abiotic stresses, and it was further observed that the *AtGS1;1* double mutant of *Arabidopsis* was susceptible to salt, cold, and oxidative stresses (Ji 2011). Moreover, a proteomic study in *Arabidopsis* based on 2D gel analysis of proteins showed differential GS1 accumulation in response to cold treatment (Kwon et al. 2007). In tea (*Camellia sinensis*), GS expression and activity were increased by cadmium and salt stress but decreased by copper, aluminum, drought, cold, and heat stress (Rana et al. 2008). Cadmium-treated tomato plants displayed a decrease in chloroplastic GS protein and mRNA and an increase in cytosolic GS transcripts and proteins (Chaffei et al. 2004).

Sahu et al. (2001) reported based on their studies on GS activities of salt-tolerant and susceptible rice cultivars that higher activity of glutamine synthetase and consequently higher metabolic activity in the tolerant cultivar might be a biochemical adaptation for salt tolerance in rice. They suggested the use of the wild rice relative *Porteresia coarctata*, which has a high level of salt tolerance, as a potential source for GS in overexpression studies.

Wang et al. (2012) reported that the expression levels of *OsGS1;1*, *OsGS1;2*, and *OsGS1;3* in young leaves of rice under salt stress remained the same, while their expression in old leaves was significantly upregulated. *OsGS2* expression levels were downregulated both in young and old leaves more so in the latter. They concluded that in old leaves, downregulation of *OsGS2* might be due to reduced photorespiration following chloroplastic injury under severe salt stress. In a recent comparative study on the expression and activity of various GS isoforms in rice under drought stress, it was inferred that a relatively maintained *OsGS2* level and the overexpression of *OsGS1;1* may contribute to the enhanced drought tolerance characteristics of the drought-tolerant cultivar Khitish (Singh and Ghosh 2013). Moreover, Lu et al. (2005) have reported increased GS activity in rice roots subjected to low-temperature stress. Furthermore, a recent QTL analysis in potato revealed that the cytosolic GS is essential for improving photosynthetic efficiency and water use efficiency (WUE). It was observed that GS activity was more enhanced in the high WUE bulk population than in the low bulk population (Kaminski et al. 2015).

14.4.2 Transgenic Overexpression of GS for Abiotic Stress Tolerance: Outcomes and Future Outlook

The results discussed in Sect. 1.4.1 have given further credence that manipulating GS through transgenic approaches may help in achieving better tolerance and yield of crop plants under various adverse abiotic stresses. Indeed, transgenics overexpressing GS have shown improved tolerance against various abiotic stresses (Table 14.1). Kozaki and Takeba (1996) showed that transgenic tobacco plants overexpressing GS2 had substantially enhanced photorespiration rates and displayed tolerance to high light intensity. Their results further strengthened the role of photorespiration as a protective mechanism against photooxidation of the photosynthetic apparatus under abiotic stresses. Although the role of photorespiration as a photoprotective mechanism is controversial due to the significant loss of CO_2 associated with it (Ogren 1984), photorespiration might function as a potential route for the dissipation of excess light energy or reducing power (Osmond and Grace 1995; Willekens et al. 1997; Wingler et al. 2000). It was observed that barley mutants lacking GS2 showed no photorespiratory capacity which along with further lines of evidence showed that the rate-limiting step in photorespiration could be the reassimilation of NH_4^+ catalyzed by chloroplastic GS2 (Wallsgrove et al. 1987; Hausler et al. 1994; Kozaki and Takeba 1996).

Furthermore, the role of GS2 in photorespiration was studied by overexpression of the rice chloroplast *GS2* gene in transgenic rice, which showed an increased photorespiration capacity and salt tolerance of rice (Hoshida et al. 2000). After a salt stress treatment of 150 mM NaCl for 2 weeks, the control plants completely lost photosystem II (PSII) activity, but transgenic plants with higher GS2 activity retained more than 90% of PSII activity. Additionally, in the presence of isonicotinic acid hydrazide, a potent inhibitor of photorespiration, transgenic plants became salt sensitive like the control plants. This was the first direct evidence indicating a protective role of photorespiration against salt stress, and it was noted that increased photorespiration conferred tolerance to salt in rice plants. Preliminary results also suggested cold tolerance in the transformants (Hoshida et al. 2000). Moreover, ectopic overexpression of the cytosolic pea GS1 in transgenic tobacco plants showed the exact opposite phenotypes as compared to the barley GS2 mutants reported by Wallsgrove et al. (1987). These transgenic tobacco plants showed increased photorespiration, increased Ser/Gly ratio, and severe reduction in the levels of free ammonium (Oliveira et al. 2002). Also, ectopic overexpression of a pine cytoplasmic GS1 gene in transgenic poplar increased photorespiratory activity and conferred enhanced tolerance to drought stress (el Khatib et al. 2004). Further characterization of these transgenics revealed that changes in the N metabolism in these poplar transgenics led to the differential regulation of genes involved in conferring protection against ROS hence improving drought tolerance (Molina-Rueda et al. 2013; Molina-Rueda and Kirby 2015). Lee et al. (2013) reported that the overexpression of the cytosolic *OsGS1;1* gene in rice under the control of the *CaMV* 35S promoter improved tolerance to cadmium stress by modulating the enzymes responsible for protection against oxidative damage. In contrast, Cai et al. (2009) reported that overexpression of *OsGS1;2* gene in rice resulted in higher sensitivity to salt, drought, and cold stresses, while the phenotype of the *OsGS1;1* overexpression line did not differ significantly from the wild type under these stresses.

The accumulation of proline, glycine betaine, γ-amino butyric acid (GABA), and polyamines is a common response to various abiotic stresses (Vinocur and Altman 2005). They act as osmolytes and/or by protecting cellular membranes and proteins. Transgenic tobacco plants expressing an antisense form of the phloem-specific cytosolic tobacco GS1 gene showed more sensitivity to salt stress and had significantly less proline accumulation than wild-type plants. Based on a labeled $^{15}NH_4^+$ pulse-chase kinetic study, it was observed that the decrease in proline production was directly related to glutamine availability in the leaves of these tobacco transgenic plants. Therefore, it was inferred that GS plays a key role in regulating proline production in the phloem and that higher GS activity is essential to synthesizing proline under water stress (Brugière et al. 1999). These results were in line with the observations of Larher et al. (1998), who reported that the application of a GS inhibitor prevented the conversion of amino acids into proline in rapeseed leaf discs. Moreover, similar results were observed in GS2 mutants of *Lotus* where proline content during drought was significantly lower than in WT plants. These mutants also had multiple changes in transcriptomic and physiological levels which led to a compromised recovery following re-watering after severe drought stress (Diaz et al. 2010).

Interestingly, in a study conducted on the highly desiccation-tolerant "resurrection plant" *Sporobolus stapfianus*, it was noted that the desiccation tolerance was correlated to higher total GS enzyme activity and maintenance of an elevated chloroplastic GS protein content during stress. Surprisingly, desiccation tolerance was not seen associated with accumulation of proline and GABA, but rather in the preferential accumulation of asparagine and arginine, which would also need a higher GS activity. It was argued that the large accumulation of asparagine and arginine during desiccation could serve as vital C and N reservoirs necessary during rehydration (Martinelli et al. 2007).

Furthermore, it can be reasoned that stress-induced increase in ammonia production from proteolysis, phenylalanine ammonia lyase activity, and photorespiration (Lutts et al. 1999) could induce higher GS-GOGAT activity and thereby supply precursors for proline and other stress specific N compounds, though further research is essential to validate this hypothesis. It is probable that, since chloroplastic GS2 activity is downregulated by oxidation under severe stresses, cultivars maintaining an optimum GS2 level even under such conditions may have improved tolerance. Based on the current understanding, the prevention of oxidative turnover of GS under severe stresses may play a role in alleviating them and thereby lead to better yields. The overexpression of a mutant GS resistant to oxidative modulation is a probable approach worth implementing. Thus, a more comprehensive understanding of the regulation of GS and the functions of various isoforms in stress would be needed before consistent results can be obtained. But the prospect of manipulating GS to enhance yield and NUE under abiotic stress conditions is lucrative enough to continue such directed efforts.

14.5 Overexpressing Mutant GS: An Alternative Approach for Developing Herbicide-Resistant Crops in Modern Agriculture.

14.5.1 GS Is the Target of the Common Herbicide Glufosinate

Over the past 40 years or so, the crucial role of GS in the recycling and detoxification of NH_4^+ released by metabolic processes such as photorespiration, catabolism of amino acids and nucleic acids, and nitrate reduction in plants has been established. Due to the toxic and volatile nature of ammonia, and limited availability of N, the role of GS in higher plants is vital. With the discovery of potential inhibitors of GS, it was proposed that GS was a suitable target for developing novel herbicides.

A team at the University of Tübingen in the 1960s discovered a tripeptide from *Streptomyces viridochromogenes* which had inhibitory effects in bacteria. It consisted of two alanine residues linked to a unique amino acid which was named L-phosphinothricin (PPT; 2-amino-4-(hydroxymethylposphinyl) butanoic acid). Due to its structural analogy to glutamate, it was tested and demonstrated to inhibit GS (Bayer et al. 1972). Independently, a Japanese team at Meiji Seika Kaisha Company discovered an antibiotic from *Streptomyces hygroscopicus* which showed biological activity comparable to PPT and named it bialaphos (Kondo et al. 1973). PPT was introduced as a nonselective herbicide for broad-spectrum post-emergent weed control in 1984 under the common name glufosinate ammonium, whereas the tripeptide developed by at Meiji Seika Kaisha was introduced as a foliar herbicide under the trade name Herbiace (Donn and Kocher 2002). Today, glufosinate ammonium is a widely used post-emergent broad-spectrum herbicide, and the synthetic racemic mixture of active L-PPT and inactive D-PPT forms is marketed under various trade names (Basta®, Liberty®, Ignite®, Challenge, Finale).

Glufosinate (PPT), as was discussed earlier, is a structural analogue of glutamate which occupies the substrate pocket and blocks glutamate binding to GS (Gill and Eisenberg 2001) (Fig. 14.4). Inhibition of GS by glufosinate can lead to the buildup of large amounts of ammonia in plant cells (Wild and Manderscheid 1984; Tachibana et al. 1986). Inhibition of the GS by glufosinate in plants is manifested by ammonia accumulation, inhibition of photosynthesis, inhibition of amino acid synthesis, and severe damage to plant tissues, which eventually result in death of the plants (Tachibana et al. 1986; Donn and Kocher 2002).

14.5.2 Glufosinate Herbicide-Resistant Crops: Current Status

Due to its broad-spectrum nature, minimum residual activity in soil, and comparatively low toxicity for nontarget organisms, the generation of glufosinate herbicide-tolerant crops was found to be conducive for post-emergent weed control in annual

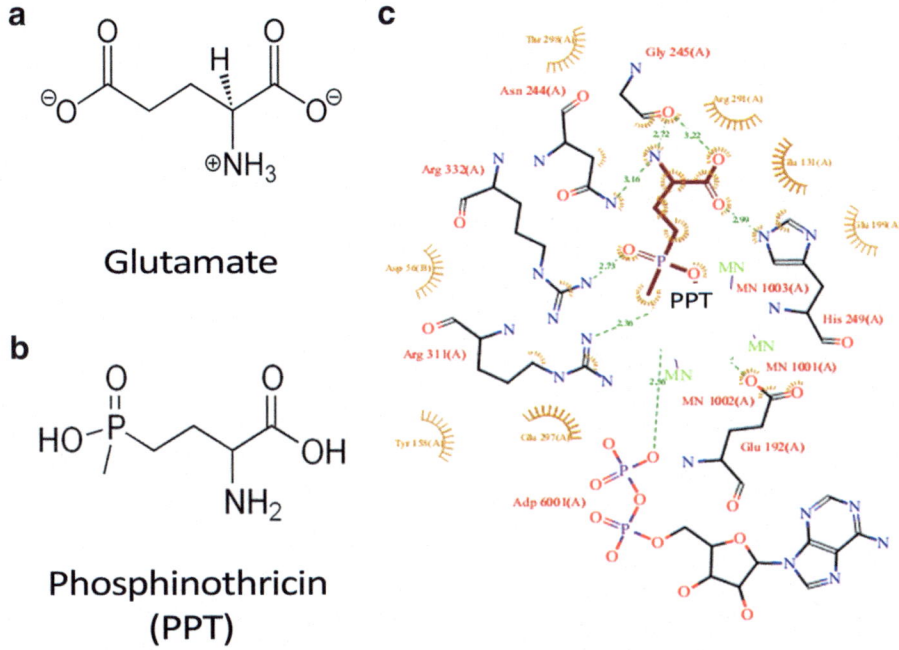

Fig. 14.4 (**a**) Phosphinothricin (PPT) is a structural analogue of (**b**) glutamate, the substrate of GS, and occupies the substrate pocket thereby blocking binding of glutamate. (**c**) The amino acid residues of GS involved in binding with PPT

field crops (Donn and Kocher 2002). The development of herbicide-resistant/ herbicide-tolerant crops by genetic engineering offers the farming community a novel option for weed control. These crops allow nonselective, broad-range herbicides to be used as selective herbicides, effectively controlling a wider range of weed species than a selective herbicide would, without injuring the crop.

Commercial herbicide-tolerant crops are usually developed by three common routes:

1. Expressing a gene for detoxification of the herbicide
2. Overexpressing the herbicide's target protein or modifying it for resistance
3. Making physiological or physical barriers that limit the entry of the herbicide into the plant

As an example of the first approach, a gene conferring resistance against glufosinate by detoxifying it via acetylation was discovered from *S. hygroscopicus* and named *bar* (bialaphos resistance) gene (Thompson et al. 1987), whereas the respective gene from *S. viridochromogenes* was named as *pat* (phosphinothricin acetyltransferase) (Strauch et al. 1988). The synthetic plant codon-optimized variants of these genes have been vastly utilized in the generation of glufosinate-tolerant plants for commercial purpose, as well as in plant transformation studies as selectable markers (Eckes et al. 1989). Since 1997, several glufosinate-tolerant genetically modified (GM) crops utilizing these genes, such as Liberty Link ™ maize, canola, cotton, soybean, and sugarbeet from Bayer AG Ltd., are commercially available for cultivation.

14.5.3   Overexpression of Mutant GS as an Alternate Strategy for Developing Glufosinate-Resistant Crops

Given the better performance and instant commercial success of the first route based on detoxification using the *pat* and *bar* enzymes, attempts using the second alternative route of overexpressing GS or studying possible mutations for resistance have been largely neglected. However, many studies have shown the potential of this approach. A cell culture of tobacco grown in PPT showed increased PPT resistance due to a 20-fold overexpression of the GS enzyme (Ishida et al. 1989). Also, Donn et al. (1984) reported that alfalfa (*Medicago sativa*) suspension cultures selected on increasing PPT concentrations resulted in the amplification of GS gene that led to enhanced GS expression and resistance to PPT. Transgenic tobacco plants overexpressing this alfalfa GS gene driven by a *CaMV* 35S promoter showed a 5-fold increase in specific GS activity and a corresponding 20-fold increase in resistance to PPT (Eckes et al. 1989). Also, the alfalfa GS gene could complement a mutant GS gene (*glnA*) in *E. coli* (DasSarma et al. 1986). Many such auxotrophic *E. coli* strains are now available which can be utilized to select for PPT-resistant plant GS (JW3841−1 strain available at the Coli Genetic Stock Centre, Yale University, USA, or the ET8894 strain).

Moreover, several overexpression studies of GS in crops have been reported to show tolerance to PPT. In rice, transgenic plants overexpressing the *OsGS1;2* gene under a *CaMV* 35S promoter showed resistance to 10 mg/l of Basta in vitro and 0.5% (v/v) solution of Basta applied as a foliar spray. However, it was seen that *OsGS1;1* overexpression did not result in Basta tolerance (Cai et al. 2009). Sun et al. (2005b) reported that simultaneous overexpression of *PsGS1* and *PsGS2* in rice plants conferred resistance to 0.3% Basta solution painted on leaves. Similar results were seen in wheat plants simultaneously overexpressing both cytosolic and chloroplastic GS isoforms which tolerated up to 0.3% Basta when painted on leaves (Huang et al. 2005). Also, transgenic poplar overexpressing a cytosolic pine GS showed considerable tolerance to a foliar application of PPT with enhanced growth in transgenic over controls (Pascual et al. 2008).

Although, when compared to resistance levels in plants obtained using the PPT detoxifying *bar* or *pat* genes, the overexpression of GS as a strategy for developing PPT-resistant crops was found to show unsuitably low resistance for commercial viability, therefore, it is reasonable that overexpression of mutant GS resistant to PPT might be a better approach. Several initial attempts to mutagenize GS for PPT tolerance failed, although potential mutations which could result in PPT resistance were identified (Table 14.2) (Donn and Kocher 2002). Recent studies in cell lines of

Table 14.2 Mutations of GS identified which may confer tolerance to glufosinate herbicide

Mutation (residue positions)	Organism	Resistance to[a]	Reference
207 is Gly (X207G) 245 is anything other than Gly (X245 ≠ G) Gly 245 can preferentially be Ser/Cys/ Arg (G245S; G245C; G245R) Arg 332 to Lys (R332K)	*M. sativa*	PPT	Goodman et al. (1990) (US patent US4975374 A)
Glu 297 to Ala (G297A) (known as the glutamate loop)	*P. vulgaris*	MSX	Clemente and Márquez (1999)
Glu 304 to Ala/asp (Glu loop) (E304A; E304D) Ala 305 Glu (also a part of Glu loop) (A305E)	*B. subtilis*	MSX	Wray and Fisher (2010)
His 249 Tyr (H249Y)	*G. max*	PPT	Pornprom et al. (2009)
Arg 295 to Lys (R295K)	*O. sativa*	PPT	Tian et al. (2015)

[a]PPT (phosphinothricin); MSX (methionine sulfoximine)

soybean and mung bean which were found to be resistant to PPT reported multiple mutations that might confer tolerance to PPT (Pornprom et al. 2008, 2009). A histidine at 249 to tyrosine (H249Y) mutation was proposed to be the major determinant of PPT tolerance (Pornprom et al. 2009). Maize cell suspension cultures selected for resistance to PPT identified several mutations incorporated in GS of resistant cell lines which highlighted the role of mutational changes in GS leading to resistance (Chompoo and Pornprom, 2008). Moreover, site-directed mutagenesis of the glutamate 297 residue (part of the "glutamate loop" seen conserved in GS) to alanine (G297A) of *Phaseolus vulgaris* GS was found to confer resistance to L-methionine sulfoximine (MSX), which is also an inhibitor of GS similar to PPT (Clemente and Márquez 1999). Given the close structural relation of MSX to PPT, and the fact that their binding patterns are similar (Ronzio and Meister 1968; Manderscheid and Wild 1986), the G297A mutation should also show potential resistance to PPT.

Due to the overlap of binding sites for PPT and glutamate, it is likely that point mutations leading to a mutant enzyme with lowered binding affinity for PPT will also reduce binding affinity for glutamate and hence reduce GS enzymatic activity (Donn and Kocher 2002). But, the utilization of enzyme engineering strategies and allele mining for natural variants might help overcome this limitation. Moreover, the presence of multiple GS isoforms in plants means that mutations would have to be introduced in multiple, if not all GS isoforms for improved resistance. Also, GS being a multimeric enzyme, there is a chance that the mutant and native subunits may form mixed enzyme complexes which are likely to be inhibited by PPT (De Block et al. 1987). But, very recently, a DNA shuffling of the *OsGS1;1* gene of rice under selective pressure of high concentrations of PPT identified an arginine at 295 position to lysine (R295K) mutation as responsible for conferring PPT resistance.

Heterologous complementation studies in a GS mutant *Saccharomyces cerevisiae* and transgenic overexpression of the *OsGS1;1* R295K mutant gene in wild-type *Arabidopsis* confirmed its ability to confer high levels of tolerance to PPT (Tian et al. 2015). More importantly, it was observed that the substrate binding affinity decreased only slightly for the *OsGS1;1* R295K mutant in comparison to native *OsGS1;1* enzyme under in vitro conditions, providing encouragement for further improvement using such techniques.

14.5.4 Future Prospects in Developing Glufosinate-Resistant Crops

Keeping in mind biosafety concerns and general public resentment of the transformation of heterologous bacterial genes such as *pat* and *bar* into food crops, overexpression of mutant GS genes may be a promising alternative strategy for developing future crops resistant to glufosinate. Techniques like DNA shuffling and directed evolution mutagenesis can be further exploited in developing PPT-resistant GS in crop plants but would require structural characterization of other crop GSs, as currently only the structure of maize GS has been elucidated (Unno et al. 2006). Moreover, such structures if crystallized bound with PPT in complex can be utilized in molecular dynamic simulation studies to characterize possible GS mutants and their significance in conferring tolerance to PPT. With the availability of recent technologies like in vitro gene pyramiding using Gateway™ vectors, pyramiding multiple PPT-resistant GS isoforms for better tolerance also looks to be a worthwhile approach, although one should keep in mind the regulation of GS as discussed in Sect. 1.3.5. Furthermore, mining of various allelic forms of GS from different species and biochemical and structural analysis of their tolerance to PPT along with the application of mutational approaches to engineer PPT-resistant GS mutants can help in developing the commercial viability of this approach.

14.6 Conclusion

Given the rise of herbicide resistance in weeds, including to glufosinate (Jalaludin et al. 2010), stacked trait transgenic crops with resistance to multiple herbicides have gained importance over crops with single herbicide resistance trait. But development of multiple herbicide-tolerant crops has been challenging due to the inherent limitations in transforming large constructs or crossbreeding of traits developed in different cultivars (Que et al. 2010). With the advent of genome editing, the use of technologies like CRISPR in mutational editing and/or overexpression of mutated herbicide target genes for developing cisgenic herbicide-resistant crops looks promising. In fact, CRISPR-based genome editing technology has already been used to make chlorsulfuron herbicide-resistant rice by introducing several discrete point mutations in the acetolactate synthase (ALS) gene (Sun et al. 2016). The use of genome editing technologies has special relevance in the case of GS as it can encompass the concurrent improvement of NUE, abiotic stress, and herbicide resistance. Thus, further study of the regulation, allelic variation, and structure of plant glutamine synthetase will pay rich dividends and lead to realizing the full potential of this vital and ancient gene in various aspects of crop improvement.

References

Almassy RJ, Janson CA, Hamlin R et al (1986) Novel subunit–subunit interactions in the structure of glutamine synthetase. Nature 323:304–309

Amaya KR, Kocherginskaya SA, Mackie RI et al (2005) Biochemical and mutational analysis of glutamine synthetase type III from the rumen anaerobe *Ruminococcus albus*. J Bacteriol 187:7481–7491. https://doi.org/10.1128/JB.187.21.7481-7491

Atkins CA (1987) Metabolism and translocation of fixed nitrogen in the nodulated legume. In: van Diest A (ed) Plant and soil interfaces and interactions. Springer Netherlands, Dordrecht, pp 157–169

Avila-Ospina L, Marmagne A, Talbotec J et al (2015) The identification of new cytosolic glutamine synthetase and asparagine synthetase genes in barley (*Hordeum vulgare* L.), and their expression during leaf senescence. J Exp Bot 66:2013–2026. https://doi.org/10.1093/jxb/erv003

Baima S, Haegi A, Stroman P et al (1989) Characterization of a cDNA clone for barley leaf glutamine synthetase. Carlsb Res Commun 54:1–9

Bao A, Zhao Z, Ding G et al (2014) Accumulated expression level of cytosolic glutamine synthetase gene (*OsGS1;1 or OsGS1;2*) alter plant development and the carbon-nitrogen metabolic status in rice. PLoS One 9:e95581. https://doi.org/10.1371/journal.pone.0095581

Bauer D, Biehler K, Fock H et al (1997) A role for cytosolic glutamine synthetase in the remobilization of leaf nitrogen during water stress in tomato. Physiol Plant 99:241–248. https://doi.org/10.1111/j.1399-3054.1997.tb05408.x

Bayer E, Gugel KH, Hagele K et al (1972) Phosphinothricin und Phosphinothricylalanyl-alanin. Helv Chim Acta 55:224–239. https://doi.org/10.1002/hlca.19720550126. (In German)

Bernard SM, Habash DZ (2009) The importance of cytosolic glutamine synthetase in nitrogen assimilation and recycling. New Phytol 182:608–620

Bernard SM, Møller ALB, Dionisio G et al (2008) Gene expression, cellular localisation and function of glutamine synthetase isozymes in wheat (*Triticum aestivum L.*). Plant Mol Biol 67:89–105. https://doi.org/10.1007/s11103-008-9303-y

Biesiadka J, Legocki AB (1997) Evolution of the glutamine synthetase gene in plants. Plant Sci 128:51–58. https://doi.org/10.1016/S0168-9452(97)00137-4

Bouwman AF, Boumans LJM, Batjes NH (2002) Emissions of N_2O and NO from fertilised fields: summary of available measurement data. Global Biogeochem Cycles 16:1058

Brauer EK, Shelp BJ (2010) Nitrogen use efficiency: re-consideration of the bioengineering approach. Botany 88:103–109

Brauer EK, Rochon A, Bi YM et al (2011) Reappraisal of nitrogen use efficiency in rice overexpressing glutamine synthetase. Physiol Plant 141:361–372. https://doi.org/10.1111/j.1399-3054.2011.01443.x

Brestic M, Zivcak M, Olsovska K et al (2014) Reduced glutamine synthetase activity plays a role in control of photosynthetic responses to high light in barley leaves. Plant Physiol Biochem 81:74–83

Brugière N, Dubois F, Limami A et al (1999) Glutamine synthetase in the phloem plays a major role in controlling proline production. Plant Cell 11:1995–2012. https://doi.org/10.1105/tpc.11.10.1995

Cai H, Zhou Y, Xiao J et al (2009) Overexpressed glutamine synthetase gene modifies nitrogen metabolism and abiotic stress responses in rice. Plant Cell Rep 28:527–537. https://doi.org/10.1007/s00299-008-0665-z

Canas RA, Quilleré I, Lea PJ et al (2010) Analysis of amino acid metabolism in the ear of maize mutants deficient in two cytosolic glutamine synthetase isoenzymes highlights the importance of asparagine for nitrogen translocation within sink organs. Plant Biotechnol J 8:966–978

Carvalho HG, Lopes-Cardoso IA, Lima LM et al (2003) Nodule-specific modulation of glutamine synthetase in transgenic *Medicago truncatula* leads to inverse alterations in asparagine synthetase expression. Plant Physiol 133:243–252. https://doi.org/10.1104/PP.102.017830

Castaings L, Camargo A, Pocholle D et al (2009) The nodule inception-like protein 7 modulates nitrate sensing and metabolism in Arabidopsis. Plant J 57:426–435. https://doi.org/10.1111/j.1365-313X.2008.03695.x

Castro-Rodríguez V, García-Gutiérrez A, Canales J et al (2011) The glutamine synthetase gene family in *Populus*. BMC Plant Biol 11:119. https://doi.org/10.1186/1471-2229-11-119

Chaffei C, Pageau K, Suzuki A et al (2004) Cadmium toxicity induced changes in nitrogen Management in *Lycopersicon esculentum* leading to a metabolic safeguard through an amino acid storage strategy. Plant Cell Physiol 45:1681–1693. https://doi.org/10.1093/pcp/pch192

Cheng L, Wang Y, He Q et al (2016) Comparative proteomics illustrates the complexity of drought resistance mechanisms in two wheat (*Triticum aestivum* L.) cultivars under dehydration and rehydration. BMC Plant Biol 16:188. https://doi.org/10.1186/s12870-016-0871-8

Chompoo J, Pornprom T (2008) RT-PCR based detection of resistance conferred by an insensitive GS in glufosinate-resistant maize cell lines. Pestic Biochem Physiol 90:189–195. https://doi.org/10.1016/j.pestbp.2007.11.007

Clemente MT, Márquez AJ (1999) Site-directed mutagenesis of Glu-297 from the α-polypeptide of *Phaseolus vulgaris* glutamine synthetase alters kinetic and structural properties and confers resistance to L-methionine sulfoximine. Plant Mol Biol 40:835–845. https://doi.org/10.1023/A:1006257323624

Cren M, Hirel B (1999) Glutamine synthetase in higher plants: regulation of gene and protein expression from the organ to the cell. Plant Cell Physiol 40:1187–1193. https://doi.org/10.1093/oxfordjournals.pcp.a029506

DasSarma S, Tisher E, Goodman HM (1986) Plant glutamine synthetase complements Glu a mutation in *Escherichia coli*. Science 232:1242–1244

De Block M, Botterman J, Vandewiele M et al (1987) Engineering herbicide resistance in plants by expression of a detoxifying enzyme. EMBO J 6:2513–2518. https://doi.org/10.1016/j.molcatb.2014.03.001

Debouba M, Gouia H, Suzuki A et al (2006) NaCl stress effects on enzymes involved in nitrogen assimilation pathway in tomato "*Lycopersicon esculentum*" seedlings. J Plant Physiol 163:1247–1258. https://doi.org/10.1016/j.jplph.2005.09.012

Díaz P, Betti M, Sánchez DH et al (2010) Deficiency in plastidic glutamine synthetase alters proline metabolism and transcriptomic response in *Lotus japonicus* under drought stress. New Phytol 188:1001–1013. https://doi.org/10.1111/j.1469-8137.2010.03440.x

Donn G, Köcher H (2002) Inhibitors of glutamine Synthetase. In: Böger P et al (eds) Herbicide Classes in Development. Springer, Berlin/Heidelberg, pp 87–101

Donn G, Tischer E, Smith JA et al (1984) Herbicide-resistant alfalfa cells: an example of gene amplification in plants. J Mol Appl Genet 2:621–635

Doskočilová A, Plíhal O, Volc J et al (2011) A nodulin/glutamine synthetase-like fusion protein is implicated in the regulation of root morphogenesis and in signalling triggered by flagellin. Planta 234:459–476. https://doi.org/10.1007/s00425-011-1419-7

Doyle JJ (1991) Evolution of higher plant glutamine synthetase genes tissue specificity as a criterion for predicting orthology. Mol Biol Evol 8:366–377

Eckes P, Schmitt P, Daub W et al (1989) Overproduction of alfalfa glutamine synthetase in transgenic tobacco plants. Mol Gen Genet 217:263–268

Eisenberg D, Gill HS, Pfluegl GMU et al (2000) Structure–function relationships of glutamine synthetases. Biochim Biophys Acta Protein Struct Mol Enzymol 1477:122–145. https://doi.org/10.1016/S0167-4838(99)00270-8

El-Khatib RT, Hamerlynck EP, Gallardo F et al (2004) Transgenic poplar characterized by ectopic expression of a pine cytosolic glutamine synthetase gene exhibits enhanced tolerance to water stress. Tree Physiol 24:729–736. https://doi.org/10.1093/treephys/24.7.729

Fan X, Tang Z, Tan Y et al (2016) Overexpression of a pH-sensitive nitrate transporter in rice increases crop yields. Proc Natl Acad Sci U S A 113:7118–7123. https://doi.org/10.1073/pnas.1525184113

Fei H, Chaillou S, Hirel B et al (2003) Overexpression of a soybean cytosolic glutamine synthetase gene linked to organ-specific promoters in pea plants grown in different concentrations of nitrate. Planta 216:467–474. https://doi.org/10.1007/s00425-002-0873-7

Fei H, Chaillou S, Hirel B et al (2006) Effects of the overexpression of a soybean cytosolic glutamine synthetase gene (GS15) linked to organ-specific promoters on growth and nitrogen accumulation of pea plants supplied with ammonium. Plant Physiol Biochem 44:543–550. https://doi.org/10.1016/j.plaphy.2006.09.007

Ferrante A, Savin R, Slafer GA (2010) Floret development of durum wheat in response to nitrogen availability. J Exp Bot 61:4351–4359. https://doi.org/10.1093/jxb/erq236

Finnemann J, Schjoerring JK (2000) Post-translational regulation of cytosolic glutamine synthetase by reversible phosphorylation and 14-3-3 protein interaction. Plant J 24:171–181. https://doi.org/10.1046/j.1365-313X.2000.00863.x

Fischer JJ, Beatty PH, Good AG et al (2013) Manipulation of MicroRNA expression to improve nitrogen use efficiency. Plant Sci 210:70–81

Fontaine JX, Ravel C, Pageau K et al (2009) A quantitative genetic study for elucidating the contribution of glutamine synthetase, glutamate dehydrogenase and other nitrogen-related physiological traits to the agronomic performance of common wheat. Theor Appl Genet 119:645–662. https://doi.org/10.1007/s00122-009-1076-4

Forde BG, Lea PJ (2007) Glutamate in plants: metabolism, regulation, and signalling. J Exp Bot 58:2339–2358

Forde BG, Day HM, Turton JF et al (1989) Two glutamine Synthetase genes from Phaseolus vulgaris L. display contrasting developmental and spatial patterns of expression in transgenic *Lotus corniculatus* plants. Plant Cell 1:391–401

Freney JR (2013) Management practices to increase efficiency of fertilizer and animal nitrogen and minimize nitrogen loss to the atmosphere and groundwater, CSIRO Plant Industry, Australia, http://www.fftc.agnet.org/files/lib_articles/20130422100045/tb186.pdf. Accessed 19 Mar 2017

Fu J, Sampalo R, Gallardo F et al (2003) Assembly of a cytosolic pine glutamine synthetase holoenzyme in leaves of transgenic poplar leads to enhanced vegetative growth in young plants. Plant Cell Environ 26:411–418. https://doi.org/10.1046/j.1365-3040.2003.00972.x

Fuentes SI, Allen DJ, Ortiz-Lopez A et al (2001) Over-expression of cytosolic glutamine synthetase increases photosynthesis and growth at low nitrogen concentrations. J Exp Bot 52:1071–1081. https://doi.org/10.1093/jexbot/52.358.1071

Funayama K, Kojima S, Tabuchi M et al (2013) Cytosolic glutamine Synthetase1 ; 2 is responsible for the primary assimilation of ammonium in Rice roots. 54:934–943. https://doi.org/10.1093/pcp/pct046

Gadaleta A, Nigro D, Giancaspro A et al (2011) The glutamine synthetase (GS2) genes in relation to grain protein content of durum wheat. Funct Integr Genomics 11:665–670. https://doi.org/10.1007/s10142-011-0235-2

Gallais A, Hirel B (2004) An approach to the genetics of nitrogen use efficiency in maize. J Exp Bot 55:295–306

Gallardo F, Fu J, Cantón FR et al (1999) Expression of a conifer glutamine synthetase gene in transgenic poplar. Planta 210:19–26

Ghoshroy S, Binder M, Tartar A et al (2010) Molecular evolution of glutamine synthetase II: phylogenetic evidence of a non-endosymbiotic gene transfer event early in plant evolution. BMC Evol Biol 10:198. https://doi.org/10.1186/1471-2148-10-198

Gill HS, Eisenberg D (2001) The crystal structure of phosphinothricin in the active site of glutamine synthetase illuminates the mechanism of enzymatic inhibition. Biochemistry 40:1903–1912. https://doi.org/10.1021/bi002438h

Gill SS, Gill R, Tuteja R et al (2014) Genetic engineering of crops: a ray of hope for enhanced food security. Plant Signal Behav 9:e28545. https://doi.org/10.4161/psb.28545

Goel P, Singh AK (2015) Abiotic stresses downregulate key genes involved in nitrogen uptake and assimilation in *Brassica juncea* l. PLoS One 10:e0143645. https://doi.org/10.1371/journal.pone.0143645

Good AG, Shrawat AK, Muench DG (2004) Can less yield more? Is reducing nutrient input into the environment compatible with maintaining crop production? Trends Plant Sci 9:597–605

Goodall AJ, Kumar P, Tobin AK (2013) Identification and expression analyses of cytosolic glutamine synthetase genes in barley (*Hordeum vulgare* L.). Plant Cell Physiol 54:492–505. https://doi.org/10.1093/pcp/pct006

Goodman HJK, Woods DR (1993) Cloning and nucleotide sequence of the *Butyrivibrio fibrisolvens* gene encoding a type III glutamine synthetase. J Gen Microbiol 139:1487–1493. https://doi.org/10.1099/00221287-139-7-1487

Goodman H, DasSarma S, Tischer E et al (1990) Expression of wild type and mutant glutamine synthetase in foreign hosts. US Patent US4975374 A, 4 Dec 1990

Guan M, Moller IS, Schjoerring JK (2015) Two cytosolic glutamine synthetase isoforms play specific roles for seed germination and seed yield structure in Arabidopsis. J Exp Bot 66:203–212. https://doi.org/10.1093/jxb/eru411

Guan M, de Bang TC, Pedersen C et al (2016) Cytosolic glutamine Synthetase Gln1;2 is the main Isozyme contributing to GS1 activity and can be up-regulated to relieve ammonium toxicity. Plant Physiol 171:1921–1933. https://doi.org/10.1104/pp.16.01195

Gutiérrez RA (2012) Systems biology for enhanced plant nitrogen nutrition. Science (80-) 336:1673–1675. https://doi.org/10.1126/science.1217620

Gutiérrez RA, Stokes TL, Thum K et al (2008) Systems approach identifies an organic nitrogen-responsive gene network that is regulated by the master clock control gene CCA1. Proc Natl Acad Sci U S A 105:4939–4944. https://doi.org/10.1073/pnas.0800211105

Habash DZ, Massiah AJ, Rong HL et al (2001) The role of cytosolic glutamine synthetase in wheat. Ann Appl Biol 138:83–89. https://doi.org/10.1111/j.1744-7348.2001.tb00087.x

Habash DZ, Bernard S, Schondelmaier J et al (2007) The genetics of nitrogen use in hexaploid wheat: N utilisation, development and yield. Theor Appl Genet 114:403–419. https://doi.org/10.1007/s00122-006-0429-5

Harper CJ, Hayward D, Kidd M, Wiid I, van Helden P (2010) Glutamate dehydrogenase and glutamine synthetase are regulated in response to nitrogen availability in Myocbacterium smegmatis. BMC Microbiol 10:138

Harrison J, Pou de Crescenzo MA, Sene O et al (2003) Does lowering glutamine synthetase activity in nodules modify nitrogen metabolism and growth of *Lotus japonicus*? Plant Physiol 133:253–262. https://doi.org/10.1104/pp.102.016766

Häusler RE, Lea PJ, Leegood RC (1994) Control of photosynthesis in barley leaves with reduced activities of glutamine synthetase or glutamate synthase - II. Control of electron transport and CO2 assimilation. Planta 194:418–435

He YX, Gui L, Liu YZ et al (2009) Crystal structure of Saccharomyces cerevisiae glutamine synthetase Gln1 suggests a nanotube-like supramolecular assembly. Proteins Struct Funct Bioinf 76:249–254. https://doi.org/10.1002/prot.22403

He C, Liu C, Liu Q et al (2014) Over-expression of glutamine synthetase genes Gln1-3/Gln1-4 improved nitrogen assimilation and maize yields. Maydica 59:50–256. http://www.maydica.org/articles/59_250.pdf

Hemon P, Robbins MP, Cullimore JV (1990) Targeting of glutamine synthetase to the mitochondria of transgenic tobacco. Plant Mol Biol 15:895–904. https://doi.org/10.1007/BF00039428

Hirel B, Lea PJ (2001) Ammonia Assimilation. In: Lea PJ et al (eds) Plant Nitrogen. Springer, Berlin/Heidelberg, pp 79–99. (ISBN 978-3-662-04064-5)

Hirel B, Marsolier MC, Hoarau J et al (1992) Forcing expression of a soybean root glutamine synthetase gene in tobacco leaves induces a native gene encoding cytosolic enzyme. Plant Mol Biol 20:207–218

Hirel B, Phillipson B, Murchie E et al (1997) Manipulating the pathway of ammonia assimilation in transgenic non-legumes and legumes. Z Pflanzenernähr Bodenk 160:283–290

Hirel B, Bertin P, Quilleré I et al (2001) Towards a better understanding of the genetic and physiological basis for nitrogen use efficiency in maize. Plant Physiol 125:1258–1270. https://doi.org/10.1104/PP.125.3.1258

Hirel B, Le Gouis J, Ney B et al (2007) The challenge of improving nitrogen use efficiency in crop plants: towards a more central role for genetic variability and quantitative genetics within integrated approaches. J Exp Bot 58:2369–2387

Hoshida H, Tanaka Y, Hibino T et al (2000) Enhanced tolerance to salt stress in transgenic rice that overexpresses chloroplast glutamine synthetase. Plant Mol Biol 43:103–111

Hossain MA, Uddin MK, Ismail MR et al (2012) Responses of glutamine Synthetase-glutamate synthase cycle enzymes in tomato leaves under salinity stress. Int J Agric Biol 14:509–515

Huang QM, Liu WH, Sun H et al (2005) Agrobacterium tumefaciens mediated transgenic wheat plants with glutamine synthetases confer tolerance to herbicide. J Plant Ecol 29:338–344. (in Chinese)

Ishida Y, Hiyoshi T, Sano M et al (1989) Selection and characterization of a herbicide-tolerant cell line of tobacco (Nicotiana tabacum L.). Plant Sci 63:227–235. https://doi.org/10.1016/0168-9452(89)90248-3

Ishida H, Anzawa D, Kokubun N et al (2002) Direct evidence for non-enzymatic fragmentation of chloroplastic glutamine synthetase by a reactive oxygen species. Plant Cell Environ 25:625–631. https://doi.org/10.1046/j.1365-3040.2002.00851.x

Ishiyama K, Inoue E, Watanabe-Takahashi A et al (2004a) Kinetic properties and ammonium-dependent regulation of cytosolic isoenzymes of glutamine synthetase in Arabidopsis. J Biol Chem 279:16598–16605

Ishiyama K, Inoue E, Tabuchi M et al (2004b) Biochemical background and compartmentalized functions of cytosolic glutamine synthetase for active ammonium assimilation in rice roots. Plant Cell Physiol 45:1640–1647

Jalaludin A, Ngim J, Bakar BHJ et al (2010) Preliminary findings of potentially resistant goosegrass (Eleusine indica) to glufosinate-ammonium in Malaysia. Weed Biol Manag 10:256–260. https://doi.org/10.1111/j.1445-6664.2010.00392.x

Ji Y (2011) The role of cytosolic glutamine synthetases in abiotic stress and development in Arabidopsis thaliana. Dissertation, University of Saskatchewan

Jing ZP, Gallardo F, Pascual MB et al (2004) Improved growth in a field trial of transgenic hybrid poplar overexpressing glutamine synthetase. New Phytol 164:137–145

Kamachi K, Yamaya T, Mae T et al (1991) A role for glutamine Synthetase in the remobilization of leaf nitrogen during natural senescence in Rice leaves. Plant Physiol 96:411–417. https://doi.org/10.1104/pp.96.2.411

Kamachi K, Yamaya T, Hayakawa T et al (1992) Changes in cytosolic glutamine Synthetase polypeptide and its mRNA in a leaf blade of Rice plants during natural senescence. Plant Physiol 98:1323–1329. https://doi.org/10.1104/pp.98.4.1323

Kaminski KP, Kørup K, Andersen MN et al (2015) Cytosolic glutamine synthetase is important for photosynthetic efficiency and water use efficiency in potato as revealed by high-throughput sequencing QTL analysis. Theor Appl Genet 128:2143–2153. https://doi.org/10.1007/s00122-015-2573-2

Kant S, Bi YM, Weretilnyk E et al (2008) The Arabidopsis halophytic relative Thellungiella halophila tolerates nitrogen-limiting conditions by maintaining growth, nitrogen uptake, and assimilation. Plant Physiol 147:1168–1180

Keys AJ, Bird IF, Cornelius MJ et al (1978) Photorespiratory nitrogen cycle. Nature 275:741–743

Kichey T, Heumez E, Pocholle D et al (2006) Combined agronomic and physiological aspects of nitrogen management in wheat highlight a central role for glutamine synthetase. New Phytol 169:265–278. https://doi.org/10.1111/j.1469-8137.2005.01606.x

Kondo Y, Shomura T, Ogawa Y et al (1973) Isolation and physicochemical and biological characterization of SF-1293 substances. Sci Rep Meiji Seika Kaisha 13:34–43

Konishi M, Yanagisawa S (2013) Arabidopsis NIN-like transcription factors have a central role in nitrate signalling. Nat Commun 4:1617. https://doi.org/10.1038/ncomms2621

Konishi N, Ishiyama K, Beier MP et al (2016) Contributions of two cytosolic glutamine synthetase isozymes to ammonium assimilation in Arabidopsis roots. J Exp Bot 19:erw454. https://doi.org/10.1093/jxb/erw454

Kozaki A, Takeba G (1996) Photorespiration protects C 3 plants from photooxidation. Nature 384:557–560. https://doi.org/10.1038/384557a0

Krajewski WW, Jones TA, Mowbray SL (2005) Structure of mycobacterium tuberculosis glutamine synthetase in complex with a transition-state mimic provides functional insights. Proc Natl Acad Sci U S A 102:10499–10504. https://doi.org/10.1073/pnas.0502248102

Krajewski WW, Collins R, Holmberg-Schiavone L et al (2008) Crystal structures of mammalian glutamine Synthetases illustrate substrate-induced conformational changes and provide opportunities for drug and herbicide design. J Mol Biol 375:217–228. https://doi.org/10.1016/j.jmb.2007.10.029

Kumada Y, Benson DR, Hillemann D et al (1993) Evolution of the glutamine synthetase gene, one of the oldest existing and functioning genes. Proc Natl Acad Sci U S A 90:3009–3013

Kusano M, Tabuchi M, Fukushima A, Funayama K, Diaz C, Kobayashi M, Hayashi N, Tsuchiya YN, Takahashi H, Kamata A, et al (2011) Metabolomics data reveal a crucial role of cytosolic glutamine synthetase 1;1 in coordinating metabolic balance in rice. Plant J 66: 456–466

Kwinta J, Cal K (2005) Effects of salinity stress on the activity of glutamine synthetase and glutamate dehydrogenase in triticale seedlings. Pol J Environ Stud 14:125–130

Kwon SJ, Kwon SI, Bae MS et al (2007) Role of the methionine sulfoxide reductase MsrB3 in cold acclimation in Arabidopsis. Plant Cell Physiol 48:1713–1723. https://doi.org/10.1093/pcp/pcm143

Lam HM, Coschigano K, Schultz C et al (1995) Use of Arabidopsis mutants and genes to study amide amino acid biosynthesis. Plant Cell 7:887. https://doi.org/10.1105/tpc.7.7.887

Lancien M, Gadal P, Hodges M (2000) Update on biochemistry enzyme redundancy and the importance of 2-Oxoglutarate in higher plant ammonium assimilation. Plant Physiol 123:817–824. https://doi.org/10.1104/pp.123.3.817

Larher F, Aziz A, Deleu C et al (1998) Suppression of the osmoinduced proline response of rapeseed leaf discs by polyamines. Physiol Plant 102:139–147. https://doi.org/10.1034/j.1399-3054.1998.1020118.x

Lee HJ, Abdula SE, Jang DW et al (2013) Overexpression of the glutamine synthetase gene modulates oxidative stress response in rice after exposure to cadmium stress. Plant Cell Rep 32:1521–1529. https://doi.org/10.1007/s00299-013-1464-8

Leegood RC, Lea PJ, Adcock MD et al (1995) The regulation and control of photorespiration. J Exp Bot 46:1397–1414

Li M, Villemur R, Hussey PJ et al (1993) Differential expression of six glutamine synthetase genes in Zea mays. Plant Mol Biol 23:401–407. https://doi.org/10.1007/BF00029015

Li RJ, Hua W, Lu YT (2006) Arabidopsis cytosolic glutamine synthetase AtGLN1;1 is a potential substrate of AtCRK3 involved in leaf senescence. Biochem Biophys Res Commun 342:119–126. https://doi.org/10.1016/j.bbrc.2006.01.100

Lima L, Seabra A, Melo P et al (2006a) Phosphorylation and subsequent interaction with 14-3-3 proteins regulate plastid glutamine synthetase in *Medicago truncatula*. Planta 223:558–567. https://doi.org/10.1007/s00425-005-0097-8

Lima L, Seabra A, Melo P et al (2006b) Post-translational regulation of cytosolic glutamine synthetase of Medicago truncatula. J Exp Bot 57:2751–2761. https://doi.org/10.1093/jxb/erl036

Limami A, Phillipson B, Ameziane R et al (1999) Does root glutamine synthetase control plant biomass production in *Lotus japonicus* L.? Planta 209:495–502. https://doi.org/10.1007/s004250050753

Lothier J, Gaufichon L, Sormani R et al (2011) The cytosolic glutamine synthetase GLN1;2 plays a role in the control of plant growth and ammonium homeostasis in Arabidopsis rosettes when nitrate supply is not limiting. J Exp Bot 62:1375–1390. https://doi.org/10.1093/jxb/erq299

Loudet O, Chaillou S, Merigout P et al (2003) Quantitative trait loci analysis of nitrogen use efficiency in Arabidopsis. Plant Physiol 131:345–358. https://doi.org/10.1104/pp.102.010785

Lu B, Yuan Y, Zhang C et al (2005) Modulation of key enzymes involved in ammonium assimilation and carbon metabolism by low temperature in rice (*Oryza sativa* L.) roots. Plant Sci 169:295–302. https://doi.org/10.1016/j.plantsci.2004.09.031

Lutts S, Majerus V, Kinet J-M (1999) NaCl effects on proline metabolism in rice (*Oryza sativa*) seedlings. Physiol Plant 105:450–458. https://doi.org/10.1034/j.1399-3054.1999.105309.x

Man HM, Boriel R, El-Khatib R et al (2005) Characterization of transgenic poplar with ectopic expression of pine cytosolic glutamine synthetase under conditions of varying nitrogen availability. New Phytol 167:31–39. https://doi.org/10.1111/j.1469-8137.2005.01461.x

Manderscheid R, Wild A (1986) Studies on the mechanism of inhibition of phosphinothricin of glutamine synthetase isolated from *Triticum aestivum* L. J Plant Physiol 123:135–142. https://doi.org/10.1016/S0176-1617(86)80134-1

Mann AF, Femten PA, Stewart GR (1979) Identification of two forms of glutamine synthetase in barley (Hordeum vulgare L.). Biochem Biophys Res Comm 88:515–521

Marchive C, Roudier F, Castaings L et al (2013) Nuclear retention of the transcription factor NLP7 orchestrates the early response to nitrate in plants. Nat Commun 4:1713. https://doi.org/10.1038/ncomms2650

Martin A, Lee J, Kichey T et al (2006) Two cytosolic glutamine Synthetase isoforms of maize are specifically involved in the control of grain production. Plant Cell 18:3252–3274. https://doi.org/10.1105/tpc.106.042689

Martinelli T, Whittaker A, Bochicchio A et al (2007) Amino acid pattern and glutamate metabolism during dehydration stress in the "resurrection" plant *Sporobolus stapfianus*: a comparison between desiccation-sensitive and desiccation-tolerant leaves. J Exp Bot 58:3037–3046. https://doi.org/10.1093/jxb/erm161

Masclaux-Daubresse C, Carrayol E, Valadier MH (2005) The two nitrogen mobilisation- and senescence-associated GS1 and GDH genes are controlled by C and N metabolites. Planta 221:580–588. https://doi.org/10.1007/s00425-004-1468-2

Masumoto C, Miyazawa S-I, Ohkawa H et al (2010) Phosphoenolpyruvate carboxylase intrinsically located in the chloroplast of rice plays a crucial role in ammonium assimilation. Proc Natl Acad Sci U S A 107:5226–5231. https://doi.org/10.1073/pnas.0913127107

Mathis R, Gamas P, Meyer Y, Cullimore JV (2000) The presence of GSI-like genes in higher plants: support for the paralogous evolution of GSI and GSII genes. J Mol Evol 50:116–122

McNally SF, Hirel B, Gadal P et al (1983) Evidence for a specific isoform content related to their possible physiological role and their compartmentation within the leaf. Plant Physiol 72:22–25

McParland RH, Guevara JG, Becker RR et al (1976) The purification and properties of the glutamine synthetase from the cytosol of Soyabean root nodules. Biochem J 153:597–606

Miflin BJ, Habash DZ (2002) The role of glutamine synthetase and glutamate dehydrogenase in nitrogen assimilation and possibilities for improvement in the nitrogen utilization of crops. J Exp Bot 53:979–987

Migge A, Carrayol E, Hirel B, Becker TW (2000) Leaf-specific overexpression of plastidic glutamine synthetase stimulates the growth of transgenic tobacco seedlings. Planta 210:252–260. https://doi.org/10.1007/PL00008132

Miki B, Abdeen A, Manabe Y, MacDonald P (2009) Selectable marker genes and unintended changes to the plant transcriptome. Plant Biotechnol J 7:211–218. https://doi.org/10.1111/j.1467-7652.2009.00400.x

Molina-Rueda JJ, Kirby EG (2015) Transgenic poplar expressing the pine GS1a show alterations in nitrogen homeostasis during drought. Plant Physiol Biochem 94:181–190

Molina-Rueda JJ, Tsai CJ, Kirby EG (2013) The Populus superoxide dismutase gene family and its responses to drought stress in transgenic poplar overexpressing a pine cytosolic glutamine Synthetase (GS1a). PLoS One 8:e56421

Moorhead G, Douglas P, Cotelle V et al (1999) Phosphorylation-dependent interactions between enzymes of plant metabolism and 14-3-3 proteins. Plant J 18:1–12. https://doi.org/10.1046/j.1365-313X.1999.00417.x

Mulvaney RL, Khan SA, Ellsworth TR (2009) Synthetic nitrogen fertilizers deplete soil nitrogen: a global dilemma for sustainable cereal production. J Environ Qual 38:2295–2314

Nagy Z, Nemeth E, Guoth A et al (2013) Metabolic indicators of drought stress tolerance in wheat: glutamine synthetase isoenzymes and Rubisco. Plant Physiol Biochem 67:48–54. https://doi.org/10.1016/j.plaphy.2013.03.001

Nigro D, Fortunato S, Giove SL, Paradiso A, Gu YQ, Blanco A, de Pinto MC, Gadaleta A (2016) Glutamine synthetase in Durum Wheat: Genotypic Variation and Relationship with Grain Protein Content. Front Plant Sci 7: 971

Nogueira EM, Olivares FL, Japiassu JC et al (2005) Characterization of glutamine synthetase genes in sugarcane genotypes with different rates of biological nitrogen fixation. Plant Sci 169:14

O'Neal TD, Joy KW (1973) Glutamine synthetase of pea leaves: purification, stabilization and pH optima. Arch Biochem Biophys 159:113–122

Obara M, Kajiura M, Fukuta Y et al (2001) Mapping of QTLs associated with cytosolic glutamine synthetase and NADH-glutamate synthase in rice (Oryza sativa L.). J Exp Bot 52:1209–1217

Obara M, Sato T, Sasaki S et al (2004) Identification and characterization of a QTL on chromosome2 for cytosolic glutamine synthetase content and panicle number in rice. Theor Appl Genet 110:1–11. https://doi.org/10.1007/s00122-004-1828-0

Ogren WL (1984) Photorespiration: pathways, regulation, and modification. Annu Rev Plant Physiol 35:415–442. https://doi.org/10.1146/annurev.pp.35.060184.002215

Ohashi M, Ishiyama K, Kusano M et al (2015) Lack of cytosolic glutamine synthetase1;2 in vascular tissues of axillary buds causes severe reduction in their outgrowth and disorder of metabolic balance in rice seedlings. Plant J 81:347–356. https://doi.org/10.1111/tpj.12731

Oliveira IC, Brears T, Knight TJ et al (2002) Overexpression of cytosolic glutamine synthetase: relation to nitrogen, light, and photorespiration. Plant Physiol 129:1170–1180. https://doi.org/10.1104/pp.020013

Orsel M, Moison M, Clouet V et al (2014) Sixteen cytosolic glutamine synthetase genes identified in the Brassica napus L. genome are differentially regulated depending on nitrogen regimes and leaf senescence. J Exp Bot 65:3927–3947. https://doi.org/10.1093/jxb/eru041

Ortega JL, Roche D, Sengupta-Gopalan C (1999) Oxidative turnover of soybean root glutamine synthetase. In vitro and in vivo studies. Plant Physiol 119:1483–1496. https://doi.org/10.1104/PP.119.4.1483

Ortega JL, Temple SJ, Sengupta-Gopalan C (2001) Constitutive overexpression of cytosolic glutamine synthetase (GS1) gene in transgenic alfalfa demonstrates that GS1 may be regulated at the level of RNA stability and protein turnover. Plant Physiol 126:109–121. https://doi.org/10.1104/PP.126.1.109

Ortega JL, Moguel-Esponda S, Potenza C et al (2006) The 3′ untranslated region of a soybean cytosolic glutamine synthetase (GS1) affects transcript stability and protein accumulation in transgenic alfalfa. Plant J 45:832–846. https://doi.org/10.1111/j.1365-313X.2005.02644.x

Ortega JL, Wilson OL, Sengupta-Gopalan C (2012) The 5′ untranslated region of the soybean cytosolic glutamine synthetase β1 gene contains prokaryotic translation initiation signals and acts as a translational enhancer in plants. Mol Gen Genomics 287:881–893. https://doi.org/10.1007/s00438-012-0724-6

Osanai T, Kuwahara A, Otsuki H et al (2017) ACR11 is an activator of plastid-type glutamine Synthetase GS2 in Arabidopsis thaliana. Plant Cell Physiol 6:29668. https://doi.org/10.1093/pcp/pcx033x

Osmond CB, Grace SC (1995) Perspectives on photoinhibition and photorespiration in the field: quintessential inefficiencies of the light and dark reactions of photosynthesis? J Exp Bot 46:1351–1362. https://doi.org/10.1093/jxb/46.special_issue.1351

Ouyang B, Yang T, Li H et al (2007) Identification of early salt stress response genes in tomato root by suppression subtractive hybridization and microarray analysis. J Exp Bot 58:507–520. https://doi.org/10.1093/jxb/erl258

Palatnik JF, Carrillo N, Valle EM (1999) The role of photosynthetic electron transport in the oxidative degradation of Chloroplastic glutamine Synthetase. Plant Physiol 121:471–478

Pascual MB, Jing ZP, Kirby EG et al (2008) Response of transgenic poplar overexpressing cytosolic glutamine synthetase to phosphinothricin. Phytochemistry 69:382–389. https://doi.org/10.1016/j.phytochem.2007.07.031

Pesole G, Bozzettit MP, Lanave C et al (1991) Glutamine synthetase gene evolution: a good molecular clock. Proc Natl Acad Sci U S A 88:522–526

Pornprom T, Pengnual A, Udomprasert N et al (2008) The role of altered glutamine Synthetase in conferring resistance to Glufosinate in Mungbean cell selections. Thai J Agric Sci 41:3–4

Pornprom T, Prodmatee N, Chatchawankanphanich O (2009) Glutamine synthetase mutation conferring target-site-based resistance to glufosinate in soybean cell selections. Pest Manag Sci 65:216–222. https://doi.org/10.1002/ps.1671

Que Q, Chilton M-DM, de Fontes CM et al (2010) Trait stacking in transgenic crops: challenges and opportunities. GM Crops 1:220–229. https://doi.org/10.4161/gmcr.1.4.13439

Quraishi UM, Abrouk M, Murat F et al (2011) Cross-genome map based dissection of a nitrogen use efficiency ortho-metaQTL in bread wheat unravels concerted cereal genome evolution. Plant J 65:745–756. https://doi.org/10.1111/j.1365-313X.2010.04461.x

Ramanjulu S, Veeranjaneyulu K, Sudhakar C (1994) Short-term shifts in nitrogen metabolism in mulberry *Morus alba* under salt shock. Phytochemistry 37:991–995. https://doi.org/10.1016/S0031-9422(00)89515-1

Rana NK, Mohanpuria P, Yadav SK (2008) Cloning and characterization of a cytosolic glutamine synthetase from *Camellia sinensis* (L.) O. Kuntze that is upregulated by ABA, SA, and H2O2. Mol Biotechnol 39:49–56. https://doi.org/10.1007/s12033-007-9027-2

Reyes JC, Florencio FJ (1994) A new type of glutamine synthetase in cyanobacteria: the protein encoded by the glnN gene supports nitrogen assimilation in *Synechocystis* sp. strain PCC 6803. J Bacteriol 176:1260–1267. https://doi.org/10.1128/JB.176.5.1260-1267.1994

Riedel J, Tischner R, Mäck G (2001) The chloroplastic glutamine synthetase (GS-2) of tobacco is phosphorylated and associated with 14-3-3 proteins inside the chloroplast. Planta 213:396–401. https://doi.org/10.1007/s004250000509

Ronzio RA, Meister A (1968) Phosphorylation of methionine sulfoximine by glutamine synthetase. Proc Natl Acad Sci USA 59:164–170

Rueda-López M, Crespillo R, Cánovas FM et al (2008) Differential regulation of two glutamine synthetase genes by a single Dof transcription factor. Plant J 56:73–85. https://doi.org/10.1111/j.1365-313X.2008.03573.x

Sahu AC, Sahoo SK, Sahoo N (2001) NaCl-stress induced alteration in glutamine synthetase activity in excised senescing leaves of a salt-sensitive and a salt-tolerant rice cultivar in light and darkness. Plant Growth Regul 34:287–292. https://doi.org/10.1023/A:1013395701308

Santos C, Pereira A, Pereira S et al (2004) Regulation of glutamine synthetase expression in sunflower cells exposed to salt and osmotic stress. Sci Hortic (Amsterdam) 103:101–111. https://doi.org/10.1016/j.scienta.2004.04.010

Schjoerring JK (2005) EU Research Project SUSTAIN: Developing Wheat with Enhanced Nitrogen Use Efficiency Towards a Sustainable System of Production, QLK5-CT-2001-01461, CORDIS http://cordis.europa.eu/result/report/rcn/44834_en.html

Seabra AR, Carvalho HG (2015) Glutamine synthetase in Medicago truncatula, unveiling new secrets of a very old enzyme. Front Plant Sci 6:578. https://doi.org/10.3389/fpls.2015.00578

Seabra AR, Carvalho H, Pereira PJB (2009) Crystallization and preliminary crystallographic characterization of glutamine synthetase from *Medicago truncatula*. Acta Crystallogr Sect F Struct Biol Cryst Commun 65:1309–1312. https://doi.org/10.1107/S1744309109047381

Seabra AR, Vieira CP, Cullimore JV, Carvalho HG (2010) Medicago truncatula contains a second gene encoding a plastid located glutamine synthetase exclusively expressed in developing seeds. BMC Plant Biol 10:183. https://doi.org/10.1186/1471-2229-10-183

See D, Kanazin V, Kephart K, Blake T (2002) Mapping genes controlling variation in barley grain protein concentration. Crop Sci 42:680–685. https://doi.org/10.2135/cropsci2002.6800

Seger M, Ortega JL, Bagga S et al (2009) Repercussion of mesophyll-specific overexpression of a soybean cytosolic glutamine synthetase gene in alfalfa (Medicago sativa L.) and tobacco (*Nicotiana tabacum* L.). Plant Sci 176:119–129. https://doi.org/10.1016/j.plantsci.2008.10.006

Seger M, Gebril S, Tabilona J et al (2014) Impact of concurrent overexpression of cytosolic gluta-mine synthetase (GS1) and sucrose phosphate synthase (SPS) on growth and development in transgenic tobacco. Planta 241:69–81. https://doi.org/10.1007/s00425-014-2165-4

Shrawat AK, Carroll RT, DePauw M et al (2008) Genetic engineering of improved nitrogen use efficiency in rice by the tissue-specific expression of alanine aminotransferase. Plant Biotechnol J 6:722–732. https://doi.org/10.1111/j.1467-7652.2008.00351.x

Silva LS, Seabra AR, Leitão JN, Carvalho HG (2015) Possible role of glutamine synthetase of the prokaryotic type (GSI-like) in nitrogen signaling in Medicago truncatula. Plant Sci 240:98–108. https://doi.org/10.1016/j.plantsci.2015.09.001

Silveira JAG, Melo AR, Viégas R, Oliveira JT (2001) Salinity-induced effects on nitrogen assimila-tion related to growth in cowpea plants. Environ Exp Bot 46:171–179. https://doi.org/10.1016/S0098-8472(01)00095-8

Silveira JAG, Viégas R, da Rocha IMA et al (2003) Proline accumulation and glutamine synthetase activity are increased by salt-induced proteolysis in cashew leaves. J Plant Physiol 160:115–123. https://doi.org/10.1078/0176-1617-00890

Simon B, Sengupta-Gopalan C (2010) The 3' untranslated region of the two cytosolic glutamine synthetase (GS1) genes in alfalfa (Medicago sativa) regulates transcript stability in response to glutamine. Planta 232:1151–1162. https://doi.org/10.1007/s00425-010-1247-1

Simons M, Saha R, Amiour N et al (2014) Assessing the metabolic impact of nitrogen availabil-ity using a compartmentalized maize leaf genome-scale model. Plant Physiol 166:1659–1674. https://doi.org/10.1104/pp.114.245787

Singh KK, Ghosh S (2013) Regulation of glutamine synthetase isoforms in two differentially drought-tolerant rice (Oryza sativa L.) cultivars under water deficit conditions. Plant Cell Rep 32:183–193. https://doi.org/10.1007/s00299-012-1353-6

Socolow R (1999) Nitrogen management and the future of food: lessons from the management of energy and carbon. Proc Natl Acad Sci U S A 96:6001–6008

Strauch E, Wohlleben W, Pühler A et al (1988) Cloning of a phosphinothricin N-acetyltransferase gene from Streptomyces viridochromogenes Tü494 and its expression in Streptomyces lividans and Escherichia coli. Gene 63:65–74

Su J, Zhang X, Yan Q et al (1995) Construction of plant expression vectors carrying glnA gene encoding glutamine synthetase and regeneration of transgenic rice plants. Sci China B 38:963–970

Suárez R, Márquez J, Shishkova S et al (2003) Overexpression of alfalfa cytosolic glutamine synthetase in nodules and flowers of transgenic Lotus japonicus plants. Physiol Plant 117:326–336. https://doi.org/10.1034/j.1399-3054.2003.00053.x

Sun H, Huang Q-M, Su J (2005a) Highly effective expression of glutamine synthetase genes GS1 and GS2 in transgenic rice plants increases nitrogen-deficiency tolerance. Zhi Wu Sheng Li Yu Fen Zi Sheng Wu Xue Xue Bao 31:492–498. (in Chinese)

Sun H, Huang QM, Su J (2005b) Overexpression of glutamine synthetases confers transgenic rice herbicide resistance. High Technol Lett 11:75–79. (in Chinese)

Sun H, Qian Q, Wu K et al (2014) Heterotrimeric G proteins regulate nitrogen-use efficiency in rice. Nat Genet 46:652–656. https://doi.org/10.1038/ng.2958

Sun Y, Zhang X, Wu C et al (2016) Engineering herbicide-resistant Rice plants through CRISPR/Cas9-mediated homologous recombination of Acetolactate synthase. Mol Plant 9:628–631

Suzuki N, Rivero RM, Shulaev V et al (2014) Abiotic and biotic stress combinations. New Phytol 203:32–43. https://doi.org/10.1111/nph.12797

Swarbreck SM, Defoin-Platel M, Hindle M et al (2011) New perspectives on glutamine synthetase in grasses. J Exp Bot 62:1511–1522. https://doi.org/10.1093/jxb/erq356

Tabuchi M, Sugiyama K, Ishiyama K et al (2005) Severe reduction in growth rate and grain fill-ing of rice mutants lacking OsGS1;1, a cytosolic glutamine synthetase1;1. Plant J 42:641–651

Tabuchi M, Abiko T, Yamaya T (2007) Assimilation of ammonium ions and reutilization of nitro-gen in rice (Oryza sativa L.). J Exp Bot 58:2319–2327

Tachibana K, Watanabe T, Sekizawa Y, Takematsu T (1986) Accumulation of ammonia in plants treated with bialaphos. J Pestic Sci 11:33–37. https://doi.org/10.1584/jpestics.11.33

Taira M et al (2004) Arabidopsis thaliana GLN2-encoded glutamine Synthetase is dual targeted to leaf mitochondria and chloroplasts. Plant Cell 16:2048–2058. https://doi.org/10.1105/tpc.104.022046

Teixeira J, Fidalgo F (2009) Salt stress affects glutamine synthetase activity and mRNA accumulation on potato plants in an organ-dependent manner. Plant Physiol Biochem 47:807–813. https://doi.org/10.1016/j.plaphy.2009.05.002

Teixeira J, Pereira S (2007) High salinity and drought act on an organ-dependent manner on potato glutamine synthetase expression and accumulation. Environ Exp Bot 60:121–126. https://doi.org/10.1016/j.envexpbot.2006.09.003

Temple SJ, Knight TJ, Unkefer PJ et al (1993) Modulation of glutamine synthetase gene expression in tobacco by the introduction of an alfalfa glutamine synthetase gene in sense and antisense orientation: molecular and biochemical analysis. Mol Gen Genet 236:315–325

Temple SJ, Bagga S, Sengupta-Gopalan C (1994) Can glutamine synthetase activity be modulated in transgenic plants by the use of recombinant DNA technology? Biochem Soc Trans 22:915–920

Temple SJ, Bagga S, Sengupta-Gopalan C (1998) Down regulation of specific members of the glutamine synthetase gene family in alfalfa by antisense RNA technology. Plant Mol Biol 37:535–547

Thompson CJ, Moval NR, Tizard R et al (1987) Characterization of the herbicide-resistance gene bar from Streptomyces hygroscopicus. EMBO J 6:2519–2523. https://doi.org/10.1021/jf703567t

Thomsen HC, Eriksson D, Møller IS et al (2014) Cytosolic glutamine synthetase: a target for improvement of crop nitrogen use efficiency? Trends Plant Sci 19:656–663. https://doi.org/10.1016/j.tplants.2014.06.002

Tian Y-S, Xu J, Zhao W et al (2015) Identification of a phosphinothricin-resistant mutant of rice glutamine synthetase using DNA shuffling. Sci Rep 5:15495. https://doi.org/10.1038/srep15495

Tobin AK, Yamaya T (2001) Cellular compartmentation of ammonium assimilation in rice and barley. J Exp Bot 52:591–604. https://doi.org/10.1093/jexbot/52.356.591

Torreira E, Seabra AR, Marriott H et al (2014) The structures of cytosolic and plastid-located glutamine synthetases from *Medicago truncatula* reveal a common and dynamic architecture. Acta Crystallogr Sect D Biol Crystallogr 70:981–993. https://doi.org/10.1107/S1399004713034718

Uhrig RG, Ng KKS, Moorhead GBG (2009) PII in higher plants: a modern role for an ancient protein. Trends Plant Sci 14:505–511

Unno H, Uchida T, Sugawara H et al (2006) Atomic structure of plant glutamine Synthetase: a key enzyme for plant productivity. J Biol Chem 281:287–296. https://doi.org/10.1074/jbc.M601497200

Urriola J, Rathore KS (2015) Overexpression of a glutamine synthetase gene affects growth and development in sorghum. Transgenic Res 24:397–407. https://doi.org/10.1007/s11248-014-9852-6

Van Rooyen JM, Abratt VR, Belrhali H et al (2011) Crystal structure of type III glutamine Synthetase: surprising reversal of the inter-ring Interface. Structure 19:471–483. https://doi.org/10.1016/j.str.2011.02.001

Veeranagamallaiah G, Chandraobulreddy P, Jyothsnakumari G et al (2007) Glutamine synthetase expression and pyrroline-5-carboxylate reductase activity influence proline accumulation in two cultivars of foxtail millet (*Setaria italica* L.) with differential salt sensitivity. Environ Exp Bot 60:239–244. https://doi.org/10.1016/j.envexpbot.2006.10.012

Vincent R, Fraisier V, Chaillou S et al (1997) Overexpression of a soybean gene encoding glutamine synthetase in shoots of transgenic *Lotus corniculatus* L. plants triggers changes in ammonium assimilation and plant development. Planta 201:424–433

Vinocur B, Altman A (2005) Recent advances in engineering plant tolerance to abiotic stress: achievements and limitations. Curr Opin Biotechnol 16:123–132

Wallsgrove RM, Turner JC, Hall NP et al (1987) Barley mutants lacking chloroplast glutamine synthetase - biochemical and genetic analysis. Plant Physiol 83:155–158

Wang ZQ, Yuan YZ, Ou JQ et al (2007) Glutamine synthetase and glutamate dehydrogenase contribute differentially to proline accumulation in leaves of wheat (Triticum aestivum) seedlings exposed to different salinity. J Plant Physiol 164:695–701. https://doi.org/10.1016/j.jplph.2006.05.001

Wang H, Zhang M, Guo R et al (2012) Effects of salt stress on ion balance and nitrogen metabolism of old and young leaves in rice (Oryza sativa L.). BMC Plant Biol 12:194. https://doi.org/10.1186/1471-2229-12-194

Wang Y, Fu B, Pan L et al (2013) Overexpression of Arabidopsis Dof1, GS1 and GS2 enhanced nitrogen assimilation in transgenic tobacco grown under low-nitrogen conditions. Plant Mol Biol Report 31:886–900. https://doi.org/10.1007/s11105-013-0561-8

Wang X, Wei Y, Shi L et al (2015) New isoforms and assembly of glutamine synthetase in the leaf of wheat (Triticum aestivum L.). J Exp Bot 66:6827–6834. https://doi.org/10.1093/jxb/erv388

Watanabe A, Takagi N, Hayashi H et al (1997) Internal Gln/Glu ratio as a potential regulatory parameter for the expression of a cytosolic glutamine Synthetase gene of radish in cultured cells. Plant Cell Physiol 38:1000–1026. https://doi.org/10.1093/oxfordjournals.pcp.a029263

Wild A, Manderscheid R (1984) The effect of phosphinothricin on the assimilation of ammonia in plants. Z Naturforsch Sect C Biosci 39:500–504. https://doi.org/10.1515/znc-1984-0539

Willekens H, Chamnongpol S, Davey M et al (1997) Catalase is a sink for H2O2 and is indispensable for stress defence in C3 plants. EMBO J 16:4806–4816. https://doi.org/10.1093/emboj/16.16.4806

Wingler A, Quick WP, Bungard RA et al (1999) The role of photorespiration during drought stress: an analysis utilizing barley mutants with reduced activities of photorespiratory enzymes. Plant Cell Environ 22:361–373. https://doi.org/10.1046/j.1365-3040.1999.00410.x

Wingler A, Lea PJ, Quick WP et al (2000) Photorespiration: metabolic pathways and their role in stress protection. Philos Trans R Soc Lond Ser B Biol Sci 355:1517–1529. https://doi.org/10.1098/rstb.2000.0712

Woods DR, Reid SJ (1993) Recent developments on the regulation and structure of glutamine synthetase enzymes from selected bacterial groups. FEMS Microbiol Rev 11: 273–283

Wray LV, Fisher SH (2010) Functional roles of the conserved Glu304 loop of Bacillus subtilis glutamine synthetase. J Bacteriol 192:5018–5025. https://doi.org/10.1128/JB.00509-10

Yamaya T, Obara M, Nakajima H et al (2002) Genetic manipulation and quantitative-trait loci mapping for nitrogen recycling in rice. J Exp Bot 53:917–925. https://doi.org/10.1093/jexbot/53.370.917

Yan S, Tang Z, Su W et al (2005) Proteomic analysis of salt stress-responsive proteins in rice root. Proteomics 5:235–244. https://doi.org/10.1002/pmic.200400853

Yanagisawa S, Akiyama A, Kisaka H et al (2004) Metabolic engineering with Dof1 transcription factor in plants: improved nitrogen assimilation and growth under low-nitrogen conditions. Proc Natl Acad Sci U S A 101:7833–7838. https://doi.org/10.1073/pnas.0402267101

Yousfi S, Márquez AJ, Betti M et al (2015) Gene expression and physiological responses to salinity and water stress of contrasting durum wheat genotypes. J Integr Plant Biol 58:48–66

Yu LH, Wu J, Tang H et al (2016) Overexpression of Arabidopsis NLP7 improves plant growth under both nitrogen-limiting and -sufficient conditions by enhancing nitrogen and carbon assimilation. Sci Rep 6:27795. https://doi.org/10.1038/srep27795

Zhu C, Fan Q, Wang W et al (2014) Characterization of a glutamine synthetase gene DvGS2 from Dunaliella viridis and biochemical identification of DvGS2-transgenic Arabidopsis thaliana. Gene 536:407–415. https://doi.org/10.1016/j.gene.2013.11.009

Zhu C, Zhang G, Shen C et al (2015) Expression of bacterial glutamine synthetase gene in Arabidopsis thaliana increases the plant biomass and level of nitrogen utilization. Biologia 70:1586–1596. https://doi.org/10.1515/biolog-2015-0183

Zozaya-Hinchliffe M, Potenza C, Ortega JL et al (2005) Nitrogen and metabolic regulation of the expression of plastidic glutamine synthetase in alfalfa (Medicago sativa). Plant Sci 168:1041–1052. https://doi.org/10.1016/j.plantsci.2004.12.001

Chapter 15
Understanding the Phytohormones Biosynthetic Pathways for Developing Engineered Environmental Stress-Tolerant Crops

Sameh Soliman, Ali El-Keblawy, Kareem A. Mosa, Mohamed Helmy, and Shabir Hussain Wani

Abstract Plants are significantly subject of diverse environmental stresses. Abiotic stresses are mainly due to nonliving environmental factors such as drought, heat, cold, and salinity, whereas biotic stresses are mainly caused by other living organisms in the surrounding environment such as bacteria, fungi, viruses, nematodes, and insects. A long series of investigations has now developed beyond the doubt that major phytohormones such as auxins, cytokinins (CK), gibberellins (GAs), abscisic acid (ABA), ethylene (ET), brassinosteroids (BRs), salicylic acid (SA), jasmonates (JAs), and strigolactones and their biosynthetic and signaling pathways play central roles in integrating and coordinating the whole plant stress responses. Understanding the mechanisms and the biosynthetic pathways of different phytohormones that can

S. Soliman
Department of Medicinal Chemistry, College of Pharmacy, University of Sharjah, P.O. Box 27272, Sharjah, UAE

Department of Pharmacognosy, Faculty of Pharmacy, University of Zagazig, Zagazig, Egypt

A. El-Keblawy
Department of Applied Biology, College of Sciences, University of Sharjah, P.O. Box 27272, Sharjah, UAE

K. A. Mosa (✉)
Department of Applied Biology, College of Sciences, University of Sharjah, P.O. Box 27272, Sharjah, UAE

Department of Biotechnology, Faculty of Agriculture, Al-Azhar University, Cairo, Egypt
e-mail: kmosa@sharjah.ac.ae

M. Helmy
The Donnelly Centre for Cellular and Biomedical Research, University of Toronto, Toronto, ON, M5S 3E1, Canada

S. H. Wani
Division of Genetics and Plant Breeding, Sher-e-Kashmir University of Agricultural Sciences and Technology of Kashmir, J&K, India

© Springer International Publishing AG, part of Springer Nature 2018
S. S. Gosal, S. H. Wani (eds.), *Biotechnologies of Crop Improvement, Volume 2*,
https://doi.org/10.1007/978-3-319-90650-8_15

enhance plant stress tolerance could lead to developing an environmental stress-tolerant crop through engineering the target phytohormones biosynthetic pathways. This chapter provides an overview on the relationships between different types of phytohormones and plant response to environmental stresses. We emphasize the significant contribution of transgenerational effects (maternal and epigenetic) on phytohormones biosynthesis. Additionally, the molecular mechanisms and regulation of phytohormones biosynthetic pathways are discussed in details. Omics and metabolic engineering prospective for developing environmental stress-tolerant crops are also highlighted.

Keywords Phytohormones · Stress · Tolerance · Crops · Metabolic engineering · Biosynthetic pathways

15.1 Introduction

Plants are sessile organisms and are, therefore, constantly challenged by varied abiotic stresses, such as changes in temperature, light intensity, and nutrient and water availability, as well as by biotic stresses, such as pathogens and insects (Kinoshita and Seki 2014). In addition, soil salinization and drought are growing problems affecting crop productivity worldwide, especially in arid climate, where fresh water is limited. Many crop species are negatively affected by soil salinity and drought (Wang and Chen 2008; Golldack et al. 2014). Additionally, global warming would increase temperature and drought stresses, which exacerbate the future impact of these stresses on plant productivity (Zandalinas et al. 2017). Interestingly, both salinity and drought stresses are greatly affected by the increased temperatures. The combination of two or three of these stresses is different from that of the exposure of any of them individually (Rizhsky et al. 2004; Mittler 2006). Such adverse environmental stresses disrupt the growth, development, and productivity of plants. Consequently, abiotic stresses lead to continuous loss of arable land and decrease crop yields (Hawkesford et al. 2013).

Cellular homeostasis, which is required for acclimation of plants to changes in the surrounding environment, could be disrupted during water and salt stresses, especially when the cell or the entire plant is exposed to a rapid decrease in water potential (Mittler et al. 2006). In order to respond and acclimatize with environmental stresses, plants have developed elaborate sensing mechanisms mediated by signaling cascades and gene transcription networks (Shinozaki and Yamaguchi-Shinozaki 2007; Fu and Dong 2013). Signaling pathways involve plant hormones, such as abscisic acid (ABA), jasmonic acid (JA), salicylic acid (SA), ethylene (ET), gibberellin (GA), nitric oxide and auxin, and others that play a central role in integrating and coordinating the whole plant stress responses (Smith and Boyko 2007; Wu et al. 2007). However, some of these hormones, such as ABA, SA, and JA, are traditionally known to be involved in the responses to both abiotic and biotic stresses

(Berkowitz et al. 2016). Induction and modulation of plant hormones are among the very important mechanisms that enhance stress tolerance (Parida and Das 2005). Plant hormones play a protective signaling role in responses to environmental stresses by activating acclimation responses such as stomatal closure, hydraulic conductivity responses to drought and salinity, and regulation of developmental processes that affect stress tolerance such as senescence and abscission (Sakamoto et al. 2008; Miller et al. 2010). Consequently, there are needs for a deep understanding of the mechanisms that enhance stress tolerance through some of these phytohormones. This might help in developing stress-tolerant crops, which is a target for several kinds of researches.

15.2   Phytohormones and Plant Response to Environmental Stress

15.2.1   Role of Phytohormones in Water Stress

When water uptake and water loss cannot be balanced by primary adaptive responses, different mechanisms may be exploited to avoid and/or tolerate dehydration, which involve regulation of stress-responsive gene expression through ABA and other signaling pathways (Zhu 2003) (Fig. 15.1). In the early 1970s, it has been reported that ABA levels rise substantially after water deprivation (Walton 1980). The major role of ABA in water relations emerged in controlling the response of guard cell during water deficit (Sirichandra et al. 2009). For example, wetly tomato

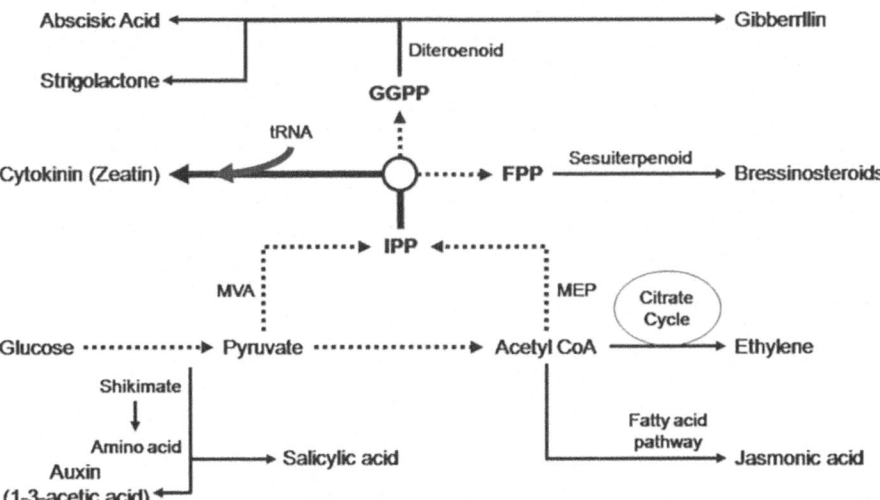

Fig. 15.1 Phytohormones and plant response to environmental stress. *C3 mode* C3 mode of photosynthesis, *CAM* crassulacean acid metabolism

flacca mutant was deficient in ABA; exogenous application of ABA rescues this mutant (Imber and Tal 1970). In addition, exogenous application of ABA caused stomatal closure in *Xanthium* (Jones and Mansfield 1970).

Plant responses to water stress are controlled by complex regulatory events mediated by ABA signaling, ion transport, and the activities of transcription factors (Hou et al. 2016). In addition, stomatal movements are controlled by complex signaling networks that respond to environmental and endogenous signals (Sierla et al. 2016). ABA is perceived by guard cells to minimize loss of water through transpiration. ABA functions to control turgor pressure changes in guard cells and stimulate stomatal closure (Osakabe et al. 2014). The regulation of stomatal responses involves both long-distance transport and modulation of ABA concentration at the guard cells (Wilkinson and Davies 2002). In addition, it has been reported that factors that are involved in ABA modulation usually increased under other stresses that induce high evaporative demand, such as high deficit in water vapor pressure of the air, high light intensity, and high leaf temperature (Chaves et al. 2003).

Stomatal closure and organ drop are among the common mechanisms for drought tolerance. These mechanisms are regulated by ABA which plays a key role in modulating the intensity of the physiological response with the stress pressure (Gómez-Cadenas et al. 1996). Stomatal closure is the first reaction to drought stress in most plants that prevents water loss from transpirational pathways. During water-deficit stress, decrease in cytokinin levels leads to an increase in shoot responses to ABA, leading to stomatal closure (Goicoechea et al. 1997). These stress-induced changes in cytokinin and ABA promote early leaf senescence leading to leaf abscission, thus decreasing the plant's canopy and reducing water loss (Pospisilová et al. 2000).

Closure of stoma is mainly controlled by chemical signals such as ABA production in dehydrating roots. ABA is accumulated in the vascular tissue of roots and then transported to guard cells via passive diffusion in response to pH changes and by specific transporters (Shabala et al. 2016). It is well known that xylem sap and leaf tissue pH are among the important factors involved in ABA modulation (Karuppanapandian et al. 2017). The ABA transport system from roots to leaves plays a significant role in water-deficit tolerance and growth adjustment. For example, *ABCG25*, which is an ABA export transporter, was induced by drought stress and exhibited vascular tissue specificity (Kuromori et al. 2010). In addition, the ABA import transporter A*BCG40* was expressed in guard cells (Kang et al. 2010), suggesting the possibility that the ABA is synthesized in the vasculature during drought stress can be imported into the guard cells.

15.2.2 *Role of Phytohormones in Salt Stress*

Phytohormones play a crucial role in regulating plant responses to salt stress, even at low concentrations. Changes in level of phytohormones like JA, GAs, ET, and ABA and enzymes related to their biosynthesis like allene oxide cyclase (AOC), lipoxygenase (LOX) (JA biosynthesis), DWARF3 (GA biosynthesis), SAMS (ET biosynthesis), and NCED (ABA biosynthesis) have been reported in many halophytic species in response to high salt concentration. The main phytohormone which shows abundance in expression is ABA (Kumari et al. 2015). It has been reported that ABA is responsible for the regulation of salt-stress-induced genes (de Bruxelles et al. 1996). ABA-inducible genes play an important role in the mechanism of salt tolerance in economic crops such as rice (Gupta et al. 1998). In addition, salt stress induces the increase of the ABA biosynthesis enzyme NCED level in *Thellungiella salsuginea* (Taji et al. 2004). The immediate ABA response directs a decrease in pH causing loss of turgor in stomatal guard cells leading to stomatal closure (Fig. 15.1). In addition, a delayed response for ABA includes induction of many ABA-responsive transcription factors that are bind to ABA-responsive promoter elements (ABRE) in the promoters of delayed response genes. The products of these genes accumulate in plant cells to high levels to confer high salt tolerance (Taji et al. 2004). However, some investigators have suggested that the salinity-induced increase in endogenous ABA could be due to water deficit rather than specific salt effect (Zhang et al. 2006).

Salt stress imposes a water deficit because of osmotic effects on a wide variety of metabolic activities (Parida and Das 2005; Parihar et al. 2015). Consequently, salt tolerance in some halophytes is aimed to increase water use efficiency under salinity (Shabala 2013). In order to conserve water, facultative halophytes such as *Mesembryanthemum crystallinum* shift their C3 mode of photosynthesis to crassulacean acid metabolism (CAM) (Cushman et al. 1989). This change allows the plant to reduce water loss by opening stomata at night, thus decreasing transpiratory water loss under prolonged salinity conditions (Fig. 15.1). It has been reported that ABA promotes such switch from C3 to CAM under salt stress in *M. crystallinum* (Thomas et al. 1992). A possible role for ABA has been also proposed for shifting the C3 to the water conservative C4 pathway in response to salinity in the halophyte *Atriplex lentiformis* (Meinzer and Zhu 1999).

It has been reported that NaCl can stabilize the growth-repressing DELLA protein repressor of GA in the roots (Conti et al. 2014). This effect is dependent upon ABA signaling (Achard et al. 2006). The increase in ABA biosynthesis under salt stress conditions is mainly to mediate growth suppression (Achard et al. 2006). In response to high salinity, some phytohormones are immediately synthesized in roots, especially in the root cortex cells, approximately 3 hours after application of salt stress (Geng et al. 2013). For example, the induction of ABA prevents lateral root elongation into surrounding media when salinity stress increases (Duan et al. 2013). In addition, halotropism, in which plant exposed to salinity stress directs their root growth to less saline areas, is salt-triggered auxin responses, not osmotic stress (Galvan-Ampudia et al. 2013). Further, ET developed in roots was recently shown to improve the Na^+/K^+ ratio in shoots, which enhance plant salt tolerance in soil-grown *Arabidopsis* plants (Jiang et al. 2013).

Changes in abundance of several enzymes involved in phytohormone metabolism such as JA, GA, ET, and ABA biosyntheses have been detected in salt-treated plants. For example, the increase in the relative abundance of ABA biosynthesis (9-cis-epoxycarotenoid dioxygenase) found in *Thellungiella salsuginea* corresponded with enhanced ABA levels observed in salt-treated plants and with an increased expression of several early and delayed ABA-responsive genes (Taji et al. 2004; Parihar et al. 2015). Salt stress can increase levels of ABA, aminocyclopropane-1-carboxylic acid, and ethylene production in *Citrus sinensis* (Gomez-Cadenas et al. 1998).

15.2.3 Reactive Oxygen Species as Stress Signals

When different pathways are uncoupled, electrons that have a high-energy state are transferred to molecular oxygen (O_2) to form reactive oxygen species (ROS, such as $^1O_2$, H_2O_2, $O_2^{\cdot-}$, and $HO\cdot$) (Takahashi and Asada 1988; Mittler 2002). ROS are toxic molecules capable of causing oxidative damage to proteins, DNA, and lipids (Apel and Hirt 2004). Under optimal growth conditions, ROS are mainly produced at a low level in organelles such as chloroplasts, mitochondria, and peroxisomes. However, water deficit and salinity stresses, especially under high light intensity, disrupt photosynthesis and increase photorespiration and cause an increased production of ROS. Despite ROS are toxic by-products of stress metabolism, they are important signals involved in the regulation of stomatal closure (Murata et al. 2015).

Recent research advances suggest and support a regulatory role of ROS in the cross talks of stress-triggered hormonal signaling such as the ABA pathway and endogenously induced redox and metabolite signals (Golldack et al. 2014). ABA-stimulated ROS accumulation induced stomatal closure via activation of plasma membrane calcium channels that lead to an increase in cytosolic Ca^{2+} (Pei et al. 2000). Experimental evidence showed that the increase of Ca^{2+} uptake is associated with the rise of ABA under salt stress, which contributes to the maintenance of membrane integrity that enables plants to regulate uptake and transport under high levels of external salinity in the longer term (Chen et al. 2001).

Several reports indicated that regulation of stomatal aperture requires coordinated activity of ROS-generating enzymes (Choi et al. 2007; Daudi et al. 2012). Accumulation of ROS in the apoplast and chloroplasts leads to an increase in cytoplasmic Ca^{2+} concentration and governs the activity of multiple kinases that regulate the activity of ROS-producing enzymes and ion channels (Sierla et al. 2016). For example, anion efflux, which is activated by two distinct types of anion channels (S- and R-types), results in a depolarization of the plasma membrane. This further results in a decrease in the inward K^+ channels (KAT1/KAT2) and H^+-ATPase, which are involved in stomatal opening, and the activation of outward K^+ channels that has a role in K^+ efflux (Osakabe et al. 2014). The efflux of anion and K^+ from guard cells reduces turgor of guard cell, which causes stomatal closure (Negi et al. 2008). Various *Arabidopsis* mutants have been used to dissect ABA and ROS signaling in guard cells. In the growth controlled by ABA (gca2) mutant, ABA increased ROS production, but H_2O_2-induced calcium channel activation and stomatal closure were absent in the mutant (Pei et al. 2000).

ABA was shown to increase the expression and the activity of ROS network genes and increase H_2O_2 levels in maize embryos, seedlings, and leaves (Guan and Scandalios 2000). In addition, ABA-deficient mutants of *Arabidopsis*, tomato, and tobacco plants lack the activities of cytosolic aldehyde oxidase and xanthine dehydrogenase that produce, respectively, H_2O_2 and O_2^- (Leydecker et al. 1995; Schwartz et al. 1997). It has been suggested that drought stress can increase ROS accumulation in plants through XDH and AO in an ABA-dependent manner (Yesbergenova et al. 2005).

15.3   Transgenerational Effects on Phytohormones Biosynthesis

Transgenerational induction has been defined as the change in offspring phenotype that is caused by an environmental signal in the parental generation and is expressed independently of changes in the offspring genotype (Holeski et al. 2012). Such effects can occur through maternal and/or epigenetic effects.

15.3.1   Maternal Effect

In maternal effects, progeny phenotypes are modified as a function of the environment caused by or experienced by the maternal parent (Donohue 2009). Adaptive maternal effects are expected to evolve as mechanisms to ameliorate factors that reduce plant fitness under certain stress factors (Galloway 2005). In addition, maternal effects often provide a mechanism for adaptive transgenerational phenotypic plasticity, in which the environment experienced by the mother is translated into phenotypic variation in the offspring (Mousseau and Fox 1998). The transgenerational maternal effect may probably be adaptive, in the sense that it could increase offspring reproductive success (Donohue and Schmitt 1998; Munir et al. 2001). Whereas environmental maternal effects are usually diminish in the first generation, epigenetic contribute to transmit heritable plastic responses to environmental cues (Jablonka and Raz 2009).

15.3.2   Epigenetic Effects and Stress Responses

In epigenetic effects, stable heritable phenotype modifications result from changes in chromosomes without alterations in DNA sequence (e.g., DNA methylation) (Herman and Sultan 2011). There are growing evidences that epigenetic effects contribute to stress responses and memory in plants. Epigenetic mechanisms include transcription, replication, DNA repair, gene transposition, and cell differentiation (Chinnusamy and Zhu 2009). It has recently been shown that plants can remember past environmental events and can use these memories to aid responses when these events recur (Kinoshita and Seki 2014). For example, Ding et al. (2012) reported that multiple exposures to drought stress conditions enable the progeny plants to respond to a new stress by more rapid adaptive changes to gene expression patterns compared with plants not previously exposed to a drought stress. Both RD29B and RAB18 transcripts accumulate in a progressive manner, i.e., the accumulated levels of transcripts are higher than in the previous stress treatment (Ding et al. 2012). Epigenetic effect is modulated by some hormones. For example, NCED3, which encodes a key enzyme in the ABA biosynthesis pathway under drought stress conditions (Ding et al. 2011), is activated with an increase in H3K4me3 marked by the histone methyltransferase ATX1 (Kim et al. 2015). As ABA is a central regulator of plant growth under stress, it has been suggested that its adjustments are likely a mechanism for transgenerational environmental effects on offspring growth and development (Herman and Sultan 2011).

Exposure of plants to various pathogens would also result in a longer-lived epigenetic "memory" mechanism. The initial exposure to the pathogen may activate transient defense systems involving some hormones, such as the defense hormone SA, and genes that interfere with growth of the pathogen (Vlot et al. 2009). However, if the plants are then exposed to a second pathogen, the genes are activated more rapidly than after the first inoculation. In *Mimulus*, for example, heritable alterations in expression of a gene controlling trichome production have been induced as a response to artificial herbivory. This has resulted in corresponding transgenerational changes in trichome density (Holeski et al. 2010). In addition, the possibility that pathogen-induced changes are epigenetic is further reinforced by the extensive and dynamic changes to the methylation of genomic DNA in *Arabidopsis* infected with bacteria or treated with the defense hormone salicylic acid (Dowen et al. 2012). A similar defense priming effect is induced by b-amino butyric acid (Baulcombe and Dean 2014).

Plants can also acclimatize and modulate the physiological processes such as flowering time and photosynthesis at the onset of stress through the epigenetic action (Yaish et al. 2011). For example, DNA methylation profiles of dandelion (*Taraxacum officinale*) genetically identical clones subjected to environmental stresses revealed that the fraction of altered loci in stressed groups was much higher than that of the control (Verhoeven et al. 2010). In addition, the genome-wide pattern of DNA methylation is modified if the parental plants are exposed to environmental stress and the progeny show modifications of root/shoot biomass ratio, P content, leaf morphology, and stress tolerance relative to the control. Furthermore, an increase in methylation was observed in nuclear genome of *Mesembryanthemum crystallinum* plants during their exposure to high salinity stress (Dyachenko et al. 2006), and this was associated with the switching of C3 photosynthesis to the CAM photosynthetic pathway, which is adapted to more stressful condition.

15.3.3 Seed Provisioning Control Maternal Effects

Seed is the first stage that connects mothers to the progeny generation. Maternal control of seed development is a complex developmental event influenced by both genetic and epigenetic processes (Chaudhury and Berger 2001). Environmental maternal effects can be transmitted to the next generation through the individual effect of seed provisioning and epigenetic mechanisms and their interaction

(Herman and Sultan 2011). Seed provisioning, which are the resources that mother plants allocate to seeds, are environmentally dependent (EL-Keblawy and Lovett-Doust 1998, 1999). Maternal plants control seed characteristics through their contribution to the organelles, seed coat, endosperm, and provisioning resources and phytohormones, which all play important roles in determining seed dormancy and seedling establishment (Roach and Wulff 1987). The amount and quality of the resources stored within seeds can greatly affect germination and early development of plants (Metz et al. 2010), indicating that seed provisioning is an important transmission vehicle of environmental maternal effects, but its effect is usually restricted to one generation and normally diminishes with seedling age (EL-Keblawy and Lovett-Doust 1998). Conversely, epigenetic mechanisms, which are a set of molecular processes that modulate the phenotype by modifying gene expression, contribute to transmit heritable plastic responses to environmental cues (Jablonka and Raz 2009). Epigenetic changes may persist throughout the life cycle and even across multiple generations (Thayer and Kuzawa 2011).

Seed germination and dormancy are also the consequence of complex interactions between maternal plants and the offspring (Bewley 1997). Consequently, these seed traits could be among the life history traits that are highly responsive to environmental stresses experienced during seed maturation on the maternal plant (Donohue and Schmitt 1998). For example, water deficiency during seed development reduced dormancy and improved germination of several wild species, such as *Amaranthus retroflexus* (Karimmojeni et al. 2014), *Sorghum halepense* (Benech-Arnold et al. 1992), *Biscutella didyma* and *Bromus fasciculatus* (Lampei Sr 2008), *Bromus tectorum* (Christensen et al. 1996), and *Sinapis arvensis* (Luzuriaga et al. 2005).

Endogenous hormonal levels that affect dormancy and germination are affected by many environmental stresses, such as salinity and drought (Kabar 1987). Debeaujon and Koornneef (2000) reported that dormancy and germination are the net result of a balance between many promoting and inhibiting factors, including GA and ABA, which have the embryo and the testa as targets (Debeaujon and Koornneef 2000). According to the growth regulator theory, the control of dormancy has been attributed to various growth inhibitors, such as ABA, and promoters, such as GAs, cytokinins, and ET. It has been reported that the interactions between ABA, GAs, ET, brassinosteroids, auxin, and cytokinins regulate the interconnected molecular processes that control dormancy alleviation and germination (Kucera et al. 2005).

ABA regulates seed germination and seedling growth, and it is required for plant resistance to biotic and abiotic stresses, such as drought, salinity, and pathogen infection (Kang et al. 2010). ABA is present in dormant seeds and plays an important role in maintaining seed dormancy (Finkelstein et al. 2008). Maruyama et al. (2016) studied the effect of drought on germination success of three populations of *Impatiens capensis* and found a negative relationship between final germination and ABA level, i.e., drought reduced germination and increased ABA level. They reported enforced dormancy due to maternal drought in two populations, which represent a desiccation avoidance/drought tolerance strategy for the heterogeneously dry sites. If maternal plants are experiencing drought stress, it is possible they may supply embryos with more endogenous ABA to help them tolerate harsh conditions during establishment (Maruyama et al. 2016). The accumulation of ABA has been also reported to occur during seed development of *Arabidopsis thaliana* (Karssen et al. 1983). The ABA accumulated during mid-maturation of *Arabidopsis* seeds is synthesized in both zygotic and maternal tissues. The ABA synthesized in maternal tissues, such endosperm, is most likely transported from the mother plant to the offspring through seeds (Karssen et al. 1983). The maternal ABA that is developed as a result of water deficiency during seed development of Arabidopsis seeds was involved in the germination inhibition (Karssen et al. 1983; Koornneef et al. 1989).

The increase in drought led to an increase in ABA levels in seeds of several plants, such as barley, wheat, and sorghum (Fenner 1991). In Sorghum bicolor, for example, drought resulted in lower levels of ABA at maturity and less dormant seeds, compared to control (Benech-Arnold et al. 1991). In addition, many studies reported that salinity stress usually increases ABA/GA ratio (Kabar 1987; Tuna et al. 2008). In addition, phytohormones could be carried over from the maternal plants to the seed progeny (Boyko and Kovalchuk 2011; Migicovsky et al. 2014). For example, Maruyama et al. (2016) showed that *Impatiens capensis* seeds from drought-stressed maternal plants had higher levels of ABA that resulted in a significant increase in progeny seed dormancy (Maruyama et al. 2016). In two indifferent habitat halophytes (*Suaeda aegyptiaca* and *Anabasis setifera*), which are able to grow equally well in both saline and nonsaline habitats, El-Keblawy et al. (2016, 2017) found that seeds produced under nonsaline habitats germinated greater than those produced under saline habitats. They attributed this to a possible higher ratio of GA:ABA in seeds developed under nonsaline habitats, but higher ABA:GA ratio in seeds developed in saline habitats (El-Keblawy et al. 2016, 2017). Furthermore, exogenous application of GA has resulted in greater salinity tolerance in many species such as *Atriplex griffithii* (Khan and Ungar 2000), *Salicornia rubra* (Khan et al. 2002), *Suaeda salsa* (Li et al. 2005), and *Prosopis juliflora* (El-Keblawy et al. 2005).

Any model which seeks to describe the control of seed dormancy by external factors needs to take into account the influence of the environment both on the ABA levels and on the altered sensitivity of the embryo. It also needs to be borne in mind that the measurement of ABA in whole seeds may be less useful than measurements in isolated embryos, since dormancy may be controlled by embryonic rather than maternal ABA, as in *Arabidopsis thaliana* (Karssen et al. 1983). For example, an important effect of drought during seed development in *Sorghum* is the reduction in the sensitivity of the embryo to ABA by about a factor of ten (Benech-Arnold et al. 1991).

15.4   Molecular Mechanisms and Regulation of Phytohormones Biosynthetic Pathways

Plants are adapting biotic and abiotic stresses by developing molecular mechanisms that are controlled by phytohormones in order to perceive the stress signals and hence optimize the plant growth and defense responses. Plants use their own chemistry for the biosynthesis of the secondary metabolic compounds, phytohormones, in order to turn on/off their multi-biological aspects via complex signaling networks in a key-lock manner. Major phytohormones include auxins, CK, GAs, ABA, ET, BRs, SA, JAs, and strigolactones. Genetic approaches such as collections of large knockout and activation-tagged mutants have significantly enhanced the understanding of the molecular basis of phytohormones biosynthesis and actions (Browse 2009a). The genetic approaches side to side with advances in combinatorial synthesis and chemical libraries have enabled access to highly diverse and wide range of phytohormones biological targets (Hicks and Raikhel 2012). Natural product chemistry and physicochemical methods such as high-performance liquid chromatography (HPLC) and GC-MS play major roles in the identification and measurement of phytohormones.

The mechanism by which phytohormones act can be summarized into four major steps: phytohormones biosynthesis, transportation, perception, and downstream response.

15.4.1 Phytohormones Biosynthesis Mechanisms

Plant phytohormones biosynthetic pathways are tightly regulated and integrated to control responses to diverse developmental and environmental stresses. The integration of multiple signaling pathways and hormonal effects determines how plants developed and grow; however, the range of responses requires activation of phytohormones biosynthesis. Essential primary metabolic pathways provide the plant with building blocks required for the biosynthesis of the secondary metabolic compounds including the plant growth regulator, phytohormones (Fig. 15.2). For instance, amino acid metabolism contributes to the synthesis of ethylene, auxin, and salicylic acid (Wang et al. 2002). Aromatic amino acids are precursors of auxin and salicylic acid biosynthesis (Woodward and Bartel 2005; Tanaka et al. 2006). Lipid and isoprenoid pathways provide the required precursors for the biosynthesis of JAs (Wasternack and Hause 2013), CK (Frébort et al. 2011), BRs (Piotrowska and Bajguz 2011), ABA (Nambara and Marion-Poll 2005), strigolactones (Ruyter-Spira et al. n.d.), and GAs (Hedden and Thomas 2012) (Fig. 15.2).

15.4.1.1 Auxin

The most abundant auxin, indole-3-acetic acid (IAA), is biosynthesized either through tryptophan (Trp) (Zhao 2014) by multiple enzymatic pathways or Trp-independent routes. For the Trp-dependent pathways, generally, tryptophan-derived IAA biosynthesis can occur through indole-3-pyruvic acid (IPyA) pathway, indole-3-acetamide (IAM) pathway, or indole-3-acetonitrile (IAN) pathway. In the case of indole-3-pyruvic acid (IPyA) pathway, IAA biosynthesis can be summarized in three prominent steps, the conversion of L-tryptophan to IPyA by tyrosine aminotransferase (AAT) (Oyama et al. 1997; Tao et al. 2008), followed by the conversion of IPyA to IAAld catalyzed by indole-3-pyruvate decarboxylase (IPDC) (Koga

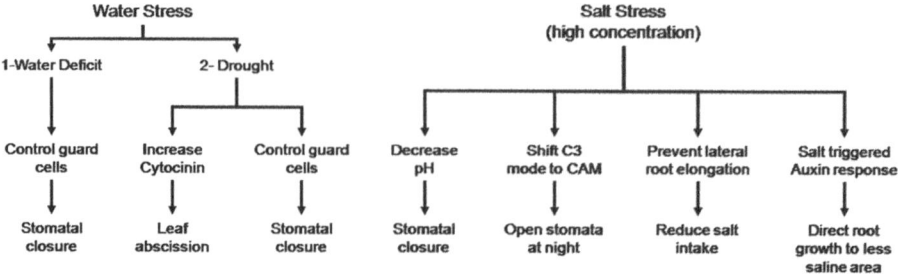

Phytohormones in response to environmental stress

Fig. 15.2 Biosynthetic pathways of plant phytohormones and their integrated network. Auxin and salicylic acid are biosynthesized from amino acids via shikimate pathway. Jasmonic acid is biosynthesized from fatty acid pathway via polyketides. Ethylene is biosynthesized through citrate cycle. Cytokinin is biosynthesized from both terpenoids and RNA. Brassinosteroids are biosynthesized from FPP via sesquiterpenoid pathway. Gibberellin, abscisic acid, and strigolactone are biosynthesized from GGPP via diterpenoid pathway. All pathways are primarily cytosolic except diterpenoids which primarily occur in plastids with structure modifications and oxidation in the cytosol. *MVA* mevalonate pathway, *MEP* methylerythritol phosphate pathway, *IPP* isopentenyl pyrophosphate, *FPP* farnesyl pyrophosphate, *GGPP* geranylgeranyl pyrophosphate

1995). Finally, IAA is biosynthesized by the oxidation of indole-3-acetaldehyde (IAAld) by aldehyde oxidase (Seo et al. 1998). The previously described pathway is the major biosynthetic pathway of IAA in higher plants (Normanly et al. 1995). However, indole-3-acetonitrile (IAN) pathway is rarely characterized in plants (Hull et al. 2000). On the other hand, the indole-3-acetamide (IAM) pathway does not occur in plants. In contrast, Trp-independent pathway is a unique pathway to plants. The primary precursor for this pathway is likely indole-3-glycerol phosphate (IPG) or indole.

15.4.1.2   Cytokinins (CK)

Cytokinin is biosynthesized by the addition of a prenyl moiety from dimethylallyl diphosphate to ATP/ADP to produce N6-isopentenyladenine (iP) ribotides by isopentenyltransferase (IPT) enzyme (Hitoshi 2006). The iP ribotides are subsequently converted to trans-zeatin (tZ)-type cytokinins by hydroxylation of the isoprenoid by cytochrome P450 enzymes (Takei et al. 2004). The active cytokinins base is then biosynthesized by cytokinin nucleoside 59 monophosphate phosphoribohydrolase enzyme (Kuroha et al. 2009).

15.4.1.3   Gibberellins (GAs)

Gibberellins biosynthesis primarily occurs in the plastids through methylerythritol phosphate (MEP) pathway (Kasahara et al. 2002). First geranylgeranyl diphosphate (GGPP) is biosynthesized from isoprene unite by GGPP synthase. Second, the produced GGPP is cyclized by terpene cyclase (ent-kaurene synthase) to ent-kaurene (Aach et al. 1997). The released ent-kaurene is then subjected to several oxidative and structural modifications in the endoplasmic reticulum and cytosol in order to produce the final products (Helliwell et al. 2001a, b; Mizutani and Ohta 2010).

15.4.1.4   Salicylic Acid (SA)

Salicylic acid is biosynthesized through shikimate pathway from chorismate through two reactions catalyzed by isochorismate synthase (ICS) and isochorismate pyruvate lyase (IPL) (Wildermuth et al. 2001). Chorismate is first converted to phenylalanine which in its turn converted to cinnamate by phenylalanine ammonia lyase (PAL) (Lee et al. 1995). Cinnamate in its turn converted to SA likely through benzoate step (Coquoz et al. 1998).

15.4.1.5   Abscisic Acid (ABA)

Abscisic acid biosynthesis can be summarized as follows: first the biosynthesis of uncyclized C40 carotenoid phytoene from MEP pathway (Mandel et al. 1996). Second, the carotenoid precursor is cleaved into the major C15 skeleton of ABA, xanthoxal (Sindhu et al. 1990). Third, xanthoxal is oxidized to xanthoxic acid by a molybdenum-containing aldehyde oxidase in two oxidation steps (Seo et al. 2000).

15.4.1.6   Ethylene (ET)

Ethylene is biosynthesized through three major steps, the conversion of methionine amino acid into S-adenosyl methionine (SAM) by SAM synthase (Yang and Hoffman 1984). The produced SAM is then converted to 1-aminocyclopropane-1-carboxylic acid (ACC) by ACC synthase (Sato and Theologis 1989) which is followed by oxidation to ethylene by ACC oxidase (Hamilton et al. 1991).

15.4.1.7   Brassinosteroids (BRs)

Brassinolide (BL), the most active BR, is a steroidal compound biosynthesized from mevalonate pathway through two parallel-branched oxidative pathways, the early and late C6-oxidation pathways (Fujioka et al. 2000). First, campesterol (CR) is converted to campestanol (CN) (Noguchi et al. 1999). CN is then oxidized to 6-oxocampestanol (6-OxoCN) followed by several steps toward cathasterone (CT) (Fujioka et al. 2000). CT is then followed by several other steps for the biosynthesis of castasterone (CS), and finally CS is converted to BL (Noguchi et al. 2000).

15.4.1.8 Strigolactones (SLs)

SLs are mainly biosynthesized from the key precursor carlactone (CL), a key intermediate for SL biosynthesis (Al-Babili and Bouwmeester 2015). CL is derived from all-trans β-carotene via the action of an isomerase and two carotenoid cleavage dioxygenases (CCD7 and CCD8). CL is biosynthesized in the plant plastids which is then transported to the cytosol for further structural modifications including oxidation and ring closures in order to produce the final SLs and SL-like compounds (Zhang et al. 2014).

15.4.2 Phytohormones Translocation Mechanisms

Phytohormones are mobile small molecules whose inter- or intracellular transport is required for function and control of physiological responses. Phytohormones are transported either through short or long distances.

15.4.2.1 Phytohormone Transported Across Short Distances

Phytohormones such as auxins are active in tissues where they are produced. Auxin is usually released at the plant-affected area and so it is transported across short distance. Auxin transportation is assessed by aux1/lax (auxin resistant 1/like auxin resistant) (Yang et al. 2006), abcb/mdr/pgp (atp-binding cassette subfamily b/multidrug resistance/p-glycoprotein) (Ma and Robert 2014), and pins (pin-formed) (Gälweiler et al. 1998). Auxin perception and transportation are difficult to be separated since perception of auxin allows regulation of endogenous intracellular auxin and hence modifying its transporters (Tsuda et al. 2011).

15.4.2.2 Phytohormones Transported Across Long Distances

Phytohormones biosynthesized in defined organs of the plant and then translocated to other sites, where it triggers specific biochemical, physiological, and morphological responses. These include ABA, cytokinin, strigolactones, jasmonates, and gibberellins. These phytohormones are detected in the plant vascular elements including phloem or xylem, suggesting their transportation to the affected tissues by diffusion from the vascular system (Kohlen et al. 2011). Plant wounding activates systemic responses which allow the de novo synthesis of JA and jasmonate-isoleucine (JA-Ile) in distal leaves (Wasternack and Hause 2013). However, Me-JA, JA and JA-Ile are all detected in plant vascular elements including phloem and xylem (Matsuura et al. 2012) indicating the migration within the plant vascular system (Liu and Sang 2013). Gibberellins are actively transported through energy-dependent mechanisms via accumulation within the endodermal cells (Shani et al. 2013).

15.4.3   Phytohormones Perception Mechanisms

Phytohormones deliver their actions by controlling the activity of individual proteins or protein families in spatial-temporal manner which is initiated by unique perception strategies (Kumari and van der Hoorn 2011). Perception is the first step for the activation of downstream signaling cascades. Phytohormones are active at very low concentrations due to their high-affinity recognition systems.

Recognition mechanisms can be divided into phytohormones that act as molecular glue by holding their receptor complexes together and phytohormones that bind directly within a cavity of specific receptor where it induces conformational changes and promote their interactions (Melcher et al. 2010).

15.4.3.1   Phytohormones Perception by Promoting the Formation of Receptor Complex Through a "Molecular Glue" Mechanism

1. Auxin perception requires not only its binding with the F-box tir1 (transport inhibitor response 1) and AFBs (auxin signaling f-box) receptors but also the interaction of the co-receptors Aux/IAA (auxin resistant/indole-3-acetic acid inducible) and the inositol hexakisphosphate (IP6) cofactor. The structural modifications produced by the formation of the tetramer stabilize the hormone perception (Mockaitis and Estelle 2008).
2. Jasmonate induces the formation of a receptor tetramer complex by JA-Ile, the F-box COI1 (coronatine insensitive1), the co-receptor JAZ (jasmonate ZIM-domain protein), and the inositol pentakisphosphate (IP5) cofactor (Sheard et al. 2010).
3. Gibberellins promote the establishment of glue complex by GID1 (gibberellin insensitive dwarf1) receptor and the F-box SLY1 (sleepy1) (Shimada et al. 2008).

15.4.3.2   Phytohormones Perception by Direct Binding with Specific Protein Receptor Complex

1. ABA binds directly to PYR1 (pyrabactin resistant 1) and PYL (pyrabactin resistant-like) receptors in cooperation with the co-receptors type 2C protein phosphatases, such as ABI1 (aba insensitive 1) and ABI2 (aba insensitive 2). The subsequent inactivation of the phosphatases induces the SNF1-type kinase activity, which in turn regulates ABA-dependent gene expression and downstream signaling cascades (Weiner et al. 2010).
2. CK binds directly to the membrane-located CRE1 (cytokinin response 1) (Arata et al. 2010) and histidine kinase (AHK) receptors. This initiates a phosphoryl cascade where a phosphoryl group is translocated via the histidine-containing phosphotransfer proteins (AHPs) and then to the response regulator (ARRs)

transcription factors. Type B ARRs regulate the transcription of cytokinin-responsive genes and type A ARRs acting as negative feedback regulators to desensitize plants to excess cytokinin (Kieber and Schaller 2014).

3. Brassinolide (BL) is a potent brassinosteroid (BR) that binds the BR receptor BRI1 directly and induces the interaction between BRI1 and SERK1 (somatic embryogenesis receptor-like kinase1) (Santiago et al. 2013). Perception of BR occurs at the plasma membrane by the receptor brassinosteroid insensitive (BRI1), followed by endocytosis of the receptor-ligand complex. Trafficking and endocytosis of the BRI1-AFCS complex is dependent on clathrin, ARF GTPases, and the Rab5 GTPase pathway. Retention of active BRI1 at the plasma membrane, rather than in endosomes, is an important factor in the activation of BR signaling. BR could bind directly and inhibits a subset of the GSK3 (glycogen synthase kinase 3) kinase family (De Rybel et al. 2009) and the auxin efflux transporter PGP19 (p-glycoprotein 19) (Rojas-Pierce et al. 2007).

15.4.4   *Phytohormones Specificity and Responses Mechanisms*

Phytohormones responses mechanisms can be divided into those responses related to plant developmental processes and those related to defensive processes.

15.4.4.1   Phytohormones and Plant Development Mechanisms

Auxin/Cytokinin Interactions on Plant Developmental Processes
Auxins and cytokinins play major roles in regulating plant development. Auxin and cytokinin act together to control root/shoot patterning, branching architecture, and vascular development (Hwang et al. 2012) as follows:

Auxin/Cytokinin Interactions on Root Development

1. Formation and maintenance of the root apical meristem (RAM): RAM is a group of undifferentiated cells that drives the growth of roots. Within this meristem there are self-renew cells named quiescent center (QC) that can produce daughter cells, which then differentiate with other specific cell to pattern the root (van den Berg et al. 1997). The growth of the meristem is a dynamic process, and the rates of cell division to differentiation change over time. Auxin and cytokinin control the balance between the rate of cell division and differentiation which is essential for the continuous root growth and to maintain an appropriately meristem size. Cytokinin reduces the meristems size and promotes cell differentiation (Dello Ioio et al. n.d.), while auxin increases meristem size and promotes cell division in the proximal meristem (Blilou et al. 2005). The mitotic inactivity and function of the QC appears to require a high auxin/low cytokinin environment.

2. Anatomical patterning of the root: The patterning of xylem in a single axis and phloem in two perpendicular poles is mediated by hormonal responses; hence xylem axis shows high auxin response (Bishopp et al. 2011), while cells within the phloem show high cytokinin response (Mähönen et al. 2006).
3. Architectural patterning of the root
   Root branching plays primary role in determining the overall size of the root and root opportunities for resources acquisition. Auxin has a stimulatory role in the formation of new lateral root primordia (Casimiro et al. 2003). However, cytokinin has an inhibitory role in lateral root organogenesis. Plants with reduced cytokinin display a greater number of lateral roots (Riefler et al. 2006).

Auxin/Cytokinin Interactions on Shoot Development

1. Formation and maintenance of the shoot apical meristem (SAM): Similar to RAM, SAM is a group of undifferentiated cells that drives the growth of shoots. Within this meristem there are self-renewed cells named organizing center (OC) that can produce daughter cells, which then differentiate with other specific cells to pattern the shoot (Zürcher et al. 2013). The OC cells that are directed toward the base of the central zone give rise to the cells that form the stem, and cells moving toward the peripheral zone give rise to the lateral organs. However, the function of auxin to cytokinin in the SAM is the reverse to that in RAM, where higher cytokinin levels are found in OC. Cytokinin promotes the proliferation of undifferentiated cells in the SAM, while auxin acts in the peripheral zone to induce cellular differentiation and organ outgrowth (Chickarmane et al. 2012).
2. Organogenesis and establishment of phyllotaxy: Auxin plays an important role in promoting organogenesis. Changes in meristem size, due to reduction in cytokinin, can indirectly alter leaf phyllotaxy (Bartrina et al. 2011).
3. Development of female gametophyte: Auxin plays a critical role in the apical-basal patterning of gynoecium, with high levels of auxin at the apical part (Hawkins and Liu 2014), while cytokinin acts by affecting polar auxin transport in the developing gynoecium. On the other hand, auxin defines the specifications of the cells in the female gametophyte (Pagnussat et al. 2009).

Gibberellins Interactions on Plant Developmental Processes

1. GA plays an important role in internode elongation (Ross et al. 1997). It stimulates cell division and expansion in response to light or dark (Gallego-Bartolomé et al. 2011).
2. GA regulates flower initiation and its development but not differentiation of floral organs. GAs are essential for male and female fertility (Griffiths et al. 2006).
3. GA is important constituent to regulate the temporal organization of embryo maturation phase (White et al. 2000).

Strigolactones Interactions on Plant Developmental Processes
SLs can modulate multiple aspects of plant growth and development, either independently or via interactions with other hormonal and environmental pathways,

including reduced secondary growth, delay in leaf senescence, or modified root architecture (Yamada et al. 2014; Ueda and Kusaba 2015). SLs are best known for their role in repressing shoot branching (Braun et al. 2012). Strigolactones stimulate the growth of the parasitic *Striga* and *Orobanche* on several crops. Structure-activity relationship analyses showed that different strigolactones derivatives are required to regulate different processes such as seed germination, hyphal branching of arbuscular mycorrhizal fungi, and shoot branching inhibition (Zwanenburg and Pospíšil 2013).

Brassinosteroids Interactions on Plant Developmental Processes
BRs have been shown to be involved in numerous plant processes including promotion of cell expansion and cell elongation (Clouse and Sasse 1998) and vascular differentiation (Caño-Delgado et al. 2004). Furthermore, BRs are involved in the acceleration of senescence in dying tissues (Clouse and Sasse 1998). Additionally, it has a role in cell division and cell wall regeneration (Clouse and Sasse 1998) and pollen elongation for pollen tube formation (Hewitt et al. 1985). Also, it can provide some protection to plants during chilling and drought stress (Clouse and Sasse 1998).

15.4.4.2 Phytohormones Defensive Responses Mechanisms

The mechanism of phytohormones production in response to both abiotic and biotic stresses can be summarized as follows: (i) Abiotic and biotic stresses initiate several complex signaling pathways in plants including alteration of intracellular Ca^{2+} concentration, production of secondary signaling molecules such as inositol phosphate and reactive oxygen species (ROS, "the oxidative burst"), as well as the activation of kinase cascades. (ii) The increase in intracellular Ca^{2+} levels is detected by calcium binding proteins sensors (Kudla et al. 2010). (iii) The activated Ca^{2+} sensors can either bind to cis-elements in the promoters of major stress-responsive genes, resulting in their activation or suppression, or activate calcium-dependent protein kinases (CDPKs), calcium/calmodulin-dependent protein kinases (CCaMKs), or phosphatases that in turn can phosphorylate/dephosphorylate specific transcription factors, regulating the expression levels of stress-responsive genes (Reddy et al. 2011). (iv) The produced phytohormones serve as the key endogenous factors that control a wide range of plant physiological processes and stress responses.

ABA, SA, JA, and ET are the major mediators in plant defense responses against both biotic (pathogens) and abiotic stresses (Bari and Jones 2009).

Phytohormones Defensive Responses Mechanisms Against Abiotic Stresses
Environmental osmotic conditions such as drought, salinity, cold, and heat stress trigger the increase in ABA levels (Zhang et al. 2006). ABA in its turn stimulates short-term responses like closure of stomata, resulting in maintenance of water balance (Zhang et al. 1987), and longer-term growth responses through regulation of stress-responsive genes.

Phytohormones Defensive Responses Mechanisms Against Pathogens
SA, JA, and ET play significant roles in regulating plant defense responses against various pathogens and pests (Bari and Jones 2009). SA is involved in the defense responses against biotrophic and hemi-biotrophic pathogens (Loake and Grant 2007), while JA and ET are involved in the defense responses against necrotrophic pathogens and herbivorous insects (Wasternack and Hause 2013) as follows:

1. SA biosynthesis is activated at the site of infection which in its turn triggers a defense response in the distal plant parts to protect undamaged tissues. This response is long-lasting and broad-spectrum induced resistance and named as systemic acquired resistance (SAR). On the other hand, SA levels in pathogen-infected tissues induce the expression of pathogenesis-related (PR) genes that encode proteins with antimicrobial activities against a wide range of pathogens (Loake and Grant 2007).
2. JA plays an important role in plant defense against herbivores including caterpillars, spider mites, beetles, thrips, and mirid bugs (Wasternack and Hause 2013). Expression of JA-defensive gene is controlled by a transcription factor, jasmonate insensitive 1/myc2 (JIN1/MYC2) (Eulgem and Somssich 2007), and a repressor protein, jasmonate ZIM-domain (JAZ) (Chini et al. 2007). Stimulation of JA production leads to binding of JA-Ile to its receptor COI1, which in its turn activates 26S proteasome and degradation of JAZ, resulting in stimulation of MYC2 and the increase in the expression level of JA-target genes (Chini et al. 2007).
3. The perception of ET by its receptor ethylene response 1 (ETR1) relieves the repression of ethylene insensitive2 (EIN2) and activates the ET signaling (Ju et al. 2012). ET can crosstalk with SA and JA pathways either agonistically or antagonistically.

The signaling pathways of ABA, SA, JA, and ET are known to interact with themselves and cross talk with the major growth-promoting hormones, auxins, GAs, and CKs in regulating plant defense response (Bari and Jones 2009; Nishiyama et al. 2013) as follows:

1. The cross talk of GA with ABA, mediated by DELLAs, regulates the balance between seed dormancy and germination.
2. SA and JA regulate biotic stress responses antagonistically (Bari and Jones 2009).
3. JA and ET pathways induce/stabilize EIN3 and thus exhibit synergy in root hair development and resistance to necrotrophs (Caballero et al. 2017).
4. ET also counteracts ABA action in seeds and thereby improves dormancy release and germination (Arc et al. 2013).
5. Auxins and ET regulate root development and architecture, a key aspect of drought and salinity tolerance (Kohli et al. 2013).
6. Auxin promotes disease susceptibility, and thus repression of auxin signaling is necessary to diseases resistance. SA represses the expression of the transport

inhibitor resistant 1 (TIR1)/auxin signaling f-box (AFB) genes, resulting in stabilization of auxin repressor protein AUX/IAA and thus repression of auxin responses (Wang et al. 2007).

7. Stabilized CK levels exhibited resistance against infection with hemi-biotrophic pathogens (Reusche et al. 2013).
8. SA and CK showed synergistic interaction to increase the resistance to the blast fungus *Magnaporthe oryzae* (Jiang et al. 2012).
9. DELLA mediates the cross talk between GA and JA pathways by inhibiting JAZ1 (Hou et al. 2010).
10. DELLAs integrate ET signaling in promoting salt tolerance (Achard et al. 2006).

15.4.5   Major Regulatory Mechanisms of Phytohormones Biosynthesis

15.4.5.1   Plant Phytohormones Methylation/Demethylation

The methylation and demethylation of plant phytohormones rapidly switch their biological activities. Phytohormones methylation increases the volatility and hence aids in long-distance translocation but dramatically reduces the activity and requires removal to retain activity (Browse 2009b). Furthermore, methylation can disturb hormonal homeostasis including IAA, gibberellins, and JA (Seo et al. 2001; Qin et al. 2005; Varbanova et al. 2007). In plants, the SABATH methyltransferases catalyze the addition of methyl groups to a range of phytohormones (Pott et al. 2004). On the other hand, demethylation of methylated phytohormones by methylesterases (MESs) leads to their activation and hence retains their specific biological functions (Yang et al. 2008).

15.4.5.2   Plant Phytohormones-Amino Acid Conjugation/Hydrolysis

Conjugation of amino acids to IAA and JA dramatically changes their biological activities (Ljung et al. 2002; Ludwig-Müller 2011). The biological activities of IAA depend on the type of amino acids conjugation. Conjugation of either alanine or leucine to IAA leads to an inactive but hydrolyzable storage form (LeClere et al. 2002). Conjugation of IAA with either aspartate or glutamate leads to degradation (LeClere et al. 2002). Conjugation of IAA with tryptophan (IAA-Trp) acts as an anti-auxin that inhibits plant growth effects of IAA but does not compete with auxin receptor (Staswick 2009). On the other hand, isoleucine conjugation to JA leads to the formation of the biologically active JA-Ile, which increases its effects (Staswick et al. 2002, Sheard et al. 2010). Moreover, JA-Trp acts as an anti-auxin and hence suggests possible cross talk (Staswick et al. 2002; Finkelstein and Lynch 2000). The enzymes that catalyze amino acid conjugation of plant phytohormones belong to the

GH3 (Gretchen Hagen 3) family of acyl acid-amido synthetases (Hagen and Guilfoyle 1985). Amino acid-conjugated phytohormones can be hydrolyzed back to the free hormone and amino acid by hydrolase enzyme (Bartel and Fink 1995).

15.5   Omics and Metabolic Engineering for Developing Environmental Stress-Tolerant Crops

Phytohormones are signaling molecules produced by the plant cells to regulate the plant growth and response to different stress. Therefore, research in phytohormones field harnesses the modern approaches in investigating cell signaling and gene regulation including omics approaches such as genomics, transcriptomics, proteomics, and metabolomics as well as computational methods (Rigal et al. 2014; Dejonghe and Russinova 2017). The omics approaches are characterized by their high-throughput nature and the ability to investigate/screen thousands (or even millions) of molecules in a single analysis. The employment of omics approaches in studying phytohormones biosynthesis and its related signaling pathways and genes resulted in outstanding advance in understanding how phytohormone regulates plant growth and response to stress (Pérez-Alfocea et al. 2011; Albacete et al. 2014; Yoshida et al. 2015). Subsequently, this knowledge could be employed for engineering the phytohormones biosynthetic pathways to develop stress-tolerant plants.

Genomic and transcriptomic analysis using next-generation sequencing (NGS) enabled revealing sequences and functions of several plant molecules including phytohormones (El-Metwally et al. 2014). Tanigaki and colleagues used RNA-seq to measure gene expression in tomato leaves every 2h for 48h and identified temporally expressed genes in the hormone synthesis pathways for salicylic acid, abscisic acid, ethylene, and jasmonic acid (Tanigaki et al. 2015). Kakei et al. employed a transcriptomic approach to investigate the hormone-induced gene expression in *Brachypodium distachyon* and compare its expression profiles with rice and *Arabidopsis* (Kakei et al. 2015). RNS-seq, coupled with bioinformatics, was also used to analyze the genome-wide expression analysis of jasmonate-treated plants and plant cultures (Pollier et al. 2013). Watanabe et al. identified the 533 new genes regulated by plant hormones through treating the rice aleurone cells with different phytohormones (ABA, GA, or both) and performing RNA-seq for the treated cells and then analyzing the RNA-seq data using new algorithm that they developed (Watanabe et al. 2014).

Mass spectrometry (MS)-based proteomics is the technology of choice for studying proteins and identifying their sequences and posttranslational modifications (PTMs) (Tohge et al. 2014; Misra et al. 2014); also, mass spectrometry is widely applicable in the field of metabolomics (Misra et al. 2014). Mass spectrometry was employed to study plant hormones through identifying the proteins involved in the signaling pathways that response to hormones. Zhang et al. presented rapid sample preparation method coupled with bioinformatics workflow to use MS-based proteomics in investigating the response to plant hormone and use it to compare the response to hormones from two different classes of plant growth regulators (Zhang et al. 2012). MS-based proteomics also is the best approach to address the PTMs (phosphorylation and ubiquitination) involved in the signaling transductions upon stimulations with plant hormones. An intensive review of MS-based plant hormone studies can be found at Wlaton et al. (2015).

After the identification of the genes that regulate the response to the plant hormones and then the proteins involved in the response signaling pathways, it is important to identify the interacting partners of those proteins. There are several methods to investigate protein-protein interactions involving large-scale screening techniques. Two main techniques are widely employed in identifying interacting partners with proteins involved in response to plant hormones: (1) yeast two-hybrid (Y2H) and (2) affinity purification-MS (AP-MS) approaches (Zhang et al. 2010; Van Leene et al. 2010). The Y2H techniques are a heterologous expression system that allows screening of cDNA libraries of other organisms in side yeast to find potential interactors, while AP-MS identifies members of a protein complex purified from plant material using MS-based identification methods (Walton et al. 2015). Several methods for utilizing these two approaches in plant hormones response study are available (Fu et al. 2011; Chini 2014).

15.6   Conclusion and Future Prospective

Plant hormones are involved in different aspects of plant development as well as in the response of plants to environmental stresses. The superposition in hormone-regulated pathways and interactions indicates a compound network of extensive cross talk among the various hormone signaling pathways such as ABA, JA, SA, ET, and GA, in response to environmental stresses. Understanding the molecular mechanisms and regulation of phytohormones biosynthetic pathways is crucial for engineering the candidate phytohormones biosynthetic pathways to develop environmental stress-tolerant crops. With the aid of the recently developed omics technologies including genomics, transcriptomics, and metabolomics, several phytohormones biosynthesis and signaling pathways and genes have been elucidated. However, more efforts should be done to discover more phytohormones associated with environmental stress responses and to unravel more of the key biosynthetic pathways and genes. Developing stable phytohormone-engineered crops for the important cash crops such as wheat, rice, soybean, and corn is still a big challenge to be confronted (Wani et al. 2016). In the future efforts should be made to employ genome editing strategies using ZFNs (zinc finger nucleases), TALENs (transcription activator-like effector nucleases), or the powerful CRISPR-Cas9 (clustered regularly interspaced short palindromic repeats) system to produce environmental stress-tolerant crops via engineering the phytohormones pathways.

References

Aach H, Bode H, Robinson DG, Graebe JE (1997) ent-Kaurene synthase is located in proplastids of meristematic shoot tissues. Planta 202:211–219

Achard P, Cheng H, De Grauwe L et al (2006) Integration of plant responses to environmentally activated phytohormonal signals. Science 311(80):91–94. https://doi.org/10.1126/science.1118642

Al-Babili S, Bouwmeester HJ (2015) Strigolactones, a novel carotenoid-derived plant hormone. Annu Rev Plant Biol 66:161–186

Albacete AA, Martínez-Andújar C, Pérez-Alfocea F (2014) Hormonal and metabolic regulation of source–sink relations under salinity and drought: from plant survival to crop yield stability. Biotechnol Adv 32:12–30. https://doi.org/10.1016/j.biotechadv.2013.10.005

Apel K, Hirt H (2004) Reactive oxygen species: metabolism, oxidative stress, and signal transduction. Annu Rev Plant Biol 55:373–399. https://doi.org/10.1146/annurev.arplant.55.031903.141701

Arata Y, Nagasawa-Iida A, Uneme H et al (2010) The phenylquinazoline compound S-4893 is a non-competitive cytokinin antagonist that targets Arabidopsis cytokinin receptor CRE1 and promotes root growth in Arabidopsis and rice. Plant Cell Physiol 51:2047–2059

Arc E, Sechet J, Corbineau F et al (2013) ABA crosstalk with ethylene and nitric oxide in seed dormancy and germination. Front Plant Sci. https://doi.org/10.3389/fpls.2013.00063

Bari R, Jones JDG (2009) Role of plant hormones in plant defence responses. Plant Mol Biol 69:473–488. https://doi.org/10.1007/s11103-008-9435-0

Bartel B, Fink GR (1995) ILR1, an amidohydrolase that releases active indole-3-acetic acid from conjugates. Science 268(80):1745–1748. https://doi.org/10.1126/science.7792599

Bartrina I, Otto E, Strnad M et al (2011) Cytokinin regulates the activity of reproductive meristems, flower organ size, ovule formation, and thus seed yield in Arabidopsis thaliana. Plant Cell 23:69–80. https://doi.org/10.1105/tpc.110.079079

Baulcombe DC, Dean C (2014) Epigenetic regulation in plant responses to the environment. Cold Spring Harb Perspect Biol. https://doi.org/10.1101/cshperspect.a019471

Benech-Arnold RL, Fenner M, Edwards PJ (1991) Changes in germinability, ABA content and ABA embryonic sensitivity in developing seeds of Sorghum bicolor (L.) Moench. induced by water stress during grain filling. New Phytol 118:339–347. https://doi.org/10.1111/j.1469-8137.1991.tb00986.x

Benech-Arnold RL, Fenner M, Edwards PJ (1992) Changes in dormancy level in Sorghum halepense seeds induced by water stress during seed development. Funct Ecol 6:596–605. https://doi.org/10.2307/2390058

Berkowitz O, De Clercq I, Van Breusegem F, Whelan J (2016) Interaction between hormonal and mitochondrial signalling during growth, development and in plant defence responses. Plant Cell Environ 39:1127–1139

Bewley JD (1997) Seed germination and dormancy. Plant Cell 9:1055–1066. https://doi.org/10.1105/tpc.9.7.1055

Bishopp A, Help H, El-Showk S et al (2011) A mutually inhibitory interaction between auxin and cytokinin specifies vascular pattern in roots. Curr Biol 21:917–926. https://doi.org/10.1016/j.cub.2011.04.017

Blilou I, Xu J, Wildwater M et al (2005) The PIN auxin efflux facilitator network controls growth and patterning in Arabidopsis roots. Nature 433:39–44. doi: http://www.nature.com/nature/journal/v433/n7021/suppinfo/nature03184_S1.html

Boyko A, Kovalchuk I (2011) Genome instability and epigenetic modification-heritable responses to environmental stress? Curr Opin Plant Biol 14:260–266

Braun N, de Saint Germain A, Pillot J-P et al (2012) The pea TCP transcription factor PsBRC1 acts downstream of strigolactones to control shoot branching. Plant Physiol 158:225–238

Browse J (2009a) The power of mutants for investigating jasmonate biosynthesis and signaling. Phytochemistry 70:1539–1546. https://doi.org/10.1016/j.phytochem.2009.08.004

Browse J (2009b) Jasmonate passes muster: a receptor and targets for the defense hormone. Annu Rev Plant Biol 60:183–205

Caballero R, Utrilla RG, Amorós I et al (2017) Tbx20 controls the expression of the KCNH2 gene and of hERG channels. Proc Natl Acad Sci 114:E416–E425. https://doi.org/10.1073/pnas.1612383114

Caño-Delgado A, Yin Y, Yu C et al (2004) BRL1 and BRL3 are novel brassinosteroid receptors that function in vascular differentiation in Arabidopsis. Development 131:5341–5351. https://doi.org/10.1242/dev.01403

Casimiro I, Beeckman T, Graham N et al (2003) Dissecting Arabidopsis lateral root development. Trends Plant Sci 8:165–171. https://doi.org/10.1016/S1360-1385(03)00051-7

Chaudhury AM, Berger F (2001) Maternal control of seed development. Semin Cell Dev Biol 12:381–386. https://doi.org/10.1006/scdb.2001.0267

Chaves MM, Maroco JP, Pereira JS (2003) Understanding plant responses to drought – from genes to the whole plant. Funct Plant Biol 30:239–264

Chen S, Li J, Wang S et al (2001) Salt, nutrient uptake and transport, and ABA of Populus euphratica, a hybrid in response to increasing soil NaCl. Trees – Struct Funct 15:186–194. https://doi.org/10.1007/s004680100091

Chickarmane VS, Gordon SP, Tarr PT et al (2012) Cytokinin signaling as a positional cue for patterning the apical–basal axis of the growing Arabidopsis shoot meristem. Proc Natl Acad Sci 109:4002–4007. https://doi.org/10.1073/pnas.1200636109

Chini A (2014) Application of yeast-two hybrid assay to chemical genomic screens: a high-throughput system to identify novel molecules modulating plant hormone receptor complexes. Methods Mol Biol (Clifton, NJ) 1056:35–43

Chini A, Fonseca S, Fernandez G et al (2007) The JAZ family of repressors is the missing link in jasmonate signalling. Nature 448:666–671. doi: http://www.nature.com/nature/journal/v448/n7154/suppinfo/nature06006_S1.html

Chinnusamy V, Zhu J-K (2009) Epigenetic regulation of stress responses in plants. Curr Opin Plant Biol 12:133–139. https://doi.org/10.1016/j.pbi.2008.12.006

Choi HW, Kim YJ, Lee SC et al (2007) Hydrogen peroxide generation by the pepper extracellular peroxidase CaPO2 activates local and systemic cell death and defense response to bacterial pathogens. Plant Physiol 145:890–904. https://doi.org/10.1104/pp.107.103325

Christensen M, Meyer SE, Allen PS (1996) A hydrothermal time model of seed after-ripening in Bromus tectorum L. Seed Sci Res 6:155–164. https://doi.org/10.1017/S0960258500003214

Clouse SD, Sasse JM (1998) Brassinosteroids: essential regulators of plant growth and development. Annu Rev Plant Physiol Plant Mol Biol 49:427–451

Conti L, Nelis S, Zhang C, et al (2014) Small ubiquitin-like modifier protein SUMO enables plants to control growth independently of the phytohormone gibberellin. Dev Cell 28:102–110. doi: https://doi.org/10.1016/j.devcel.2013.12.004

Coquoz J-L, Buchala A, Métraux J-P (1998) The biosynthesis of salicylic acid in potato plants. Plant Physiol 117:1095–1101

Cushman JC, Meyer G, Michalowski CB et al (1989) Salt stress leads to differential expression of two isogenes of phosphoenolpyruvate carboxylase during Crassulacean acid metabolism induction in the common ice plant. Plant Cell 1:715–725. https://doi.org/10.1105/tpc.1.7.715

Daudi A, Cheng Z, O'Brien JA et al (2012) The apoplastic oxidative burst peroxidase in arabidopsis is a major component of pattern-triggered immunity. Plant Cell 24:275–287. https://doi.org/10.1105/tpc.111.093039

de Bruxelles GL, Peacock WJ, Dennis ES, Dolferus R (1996) Abscisic acid induces the alcohol dehydrogenase gene in Arabidopsis. Plant Physiol 111:381–391. https://doi.org/10.1104/pp.111.2.381

De Rybel B, Audenaert D, Vert G et al (2009) Chemical inhibition of a subset of Arabidopsis thaliana GSK3-like kinases activates brassinosteroid signaling. Chem Biol 16:594–604

Debeaujon I, Koornneef M (2000) Gibberellin requirement for Arabidopsis seed germination is determined both by testa characteristics and embryonic abscisic acid. Plant Physiol 122:415–424

Dejonghe W, Russinova E (2017) Plant chemical genetics: from phenotype-based screens to synthetic biology. Plant Physiol 174:5–20. https://doi.org/10.1104/pp.16.01805

Dello Ioio R, Linhares FS, Scacchi E et al Cytokinins determine Arabidopsis root-meristem size by controlling cell differentiation. Curr Biol 17:678–682. https://doi.org/10.1016/j.cub.2007.02.047

Ding Y, Avramova Z, Fromm M (2011) The Arabidopsis trithorax-like factor ATX1 functions in dehydration stress responses via ABA-dependent and ABA-independent pathways. Plant J 66:735–744. https://doi.org/10.1111/j.1365-313X.2011.04534.x

Ding L, Ley T, Larson D et al (2012) Clonal evolution in relapsed acute myeloid leukaemia revealed by whole-genome sequencing. Nature 481:506–510. https://doi.org/10.1038/nature10738

Donohue K (2009) Completing the cycle: maternal effects as the missing link in plant life histories. Philos Trans R Soc B Biol Sci 364:1059–1074. https://doi.org/10.1098/rstb.2008.0291

Donohue K, Schmitt J (1998) Maternal environmental effects in plants: adaptive plasticity. Matern Eff as Adapt:137–158

Dowen RH, Pelizzola M, Schmitz RJ et al (2012) Widespread dynamic DNA methylation in response to biotic stress. Proc Natl Acad Sci 109:E2183–E2191. https://doi.org/10.1073/pnas.1209329109

Duan L, Dietrich D, Ng CH et al (2013) Endodermal ABA signaling promotes lateral root quiescence during salt stress in *Arabidopsis* seedlings. Plant Cell 25:324–341. https://doi.org/10.1105/tpc.112.107227

Dyachenko OV, Zakharchenko NS, Shevchuk TV et al (2006) Effect of hypermethylation of CCWGG sequences in DNA of Mesembryanthemum crystallinum plants on their adaptation to salt stress. Biochemistry (Moscow) 71:461–465. https://doi.org/10.1134/S000629790604016X

EL-Keblawy ALI, Lovett-Doust JON (1998) Persistent, non-seed-size maternal effects on life-history traits in the progeny generation in squash, Cucurbita pepo. New Phytol 140:655–665. https://doi.org/10.1046/j.1469-8137.1998.00305.x

EL-Keblawy ALI, Lovett-Doust JON (1999) Maternal effects in the progeny generation in zucchini, Cucurbita pepo (Cucurbitaceae). Int J Plant Sci 160:331–339. https://doi.org/10.1086/314136

El-Keblawy A, Al-Ansari F, Al-Rawai A (2005) Effects of dormancy regulating chemicals on innate and salinity induced dormancy in the invasive Prosopis juliflora (Sw.) DC. shrub. Plant Growth Regul 46:161–168. https://doi.org/10.1007/s10725-005-7356-3

El-Keblawy A, Gairola S, Bhatt A (2016) Maternal salinity environment affects salt tolerance during germination in Anabasis setifera: a facultative desert halophyte. J Arid Land 8:254–263. https://doi.org/10.1007/s40333-015-0023-2

El-Keblawy A, Gairola S, Bhatt A, Mahmoud T (2017) Effects of maternal salinity on salt tolerance during germination of Suaeda aegyptiaca, a facultative halophyte in the Arab Gulf desert. Plant Species Biol 32:45–53. https://doi.org/10.1111/1442-1984.12127

El-Metwally S, Ouda OM, Helmy M (2014) Next generation sequencing technologies and challenges in sequence assembly, 1st edn. Springer New York, New York

Eulgem T, Somssich IE (2007) Networks of WRKY transcription factors in defense signaling. Curr Opin Plant Biol 10:366–371. https://doi.org/10.1016/j.pbi.2007.04.020

Fenner M (1991) The effects of the parent environment on seed germinability. Seed Sci Res. https://doi.org/10.1017/S0960258500000696

Finkelstein RR, Lynch TJ (2000) The Arabidopsis abscisic acid response gene ABI5 encodes a basic leucine zipper transcription factor. Plant Cell 12:599–609. https://doi.org/10.1105/tpc.12.4.599

Finkelstein R, Reeves W, Ariizumi T, Steber C (2008) Molecular aspects of seed dormancy. Annu Rev Plant Biol 59:387–415. https://doi.org/10.1146/annurev.arplant.59.032607.092740

Frébort I, Kowalska M, Hluska T et al (2011) Evolution of cytokinin biosynthesis and degradation. J Exp Bot 62:2431–2452. https://doi.org/10.1093/jxb/err004

Fu ZQ, Dong X (2013) Systemic acquired resistance: turning local infection into global defense. Annu Rev Plant Biol 64:839–863. https://doi.org/10.1146/annurev-arplant-042811-105606

Fu J, Sun X, Wang J et al (2011) Progress in quantitative analysis of plant hormones. Chin Sci Bull 56:355–366

Fujioka S, Noguchi T, Watanabe T et al (2000) Biosynthesis of brassinosteroids in cultured cells of Catharanthus roseus. Phytochemistry 53:549–553. https://doi.org/10.1016/S0031-9422(99)00582-8

Gallego-Bartolomé J, Alabadí D, Blázquez MA (2011) DELLA-induced early transcriptional changes during etiolated development in Arabidopsis thaliana. PLoS One 6:e23918. https://doi.org/10.1371/journal.pone.0023918

Galloway LF (2005) Maternal effects provide phenotypic adaptation to local environmental conditions. doi.org 166:93–100. doi: https://doi.org/10.1111/j.1469-8137.2004.01314.x

Galvan-Ampudia CS, Julkowska MM, Darwish E et al (2013) Halotropism is a response of plant roots to avoid a saline environment. Curr Biol 23:2044–2050. https://doi.org/10.1016/j.cub.2013.08.042

Gälweiler L, Guan C, Müller A et al (1998) Regulation of polar auxin transport by AtPIN1 in Arabidopsis vascular tissue. Science 282(80):2226–2230

Geng Y, Wu R, Wee CW et al (2013) A spatio-temporal understanding of growth regulation during the salt stress response in Arabidopsis. Plant Cell 25:2132–2154. https://doi.org/10.1105/tpc.113.112896

Goicoechea N, Antolin MC, Sánchez-D'iaz M (1997) Gas exchange is related to the hormone balance in mycorrhizal or nitrogen-fixing alfalfa subjected to drought. Physiol Plant 100:989–997

Golldack D, Li C, Mohan H, Probst N (2014) Tolerance to drought and salt stress in plants: unraveling the signaling networks. Front Plant Sci. https://doi.org/10.3389/fpls.2014.00151

Gómez-Cadenas A, Tadeo FR, Talon M, Primo-Millo E (1996) Leaf abscission induced by ethylene in water-stressed intact seedlings of Cleopatra mandarin requires previous abscisic acid accumulation in roots. Plant Physiol 112:401–408

Gomez-Cadenas A, Tadeo FR, Primo-Millo E, Talon M (1998) Involvement of abscisic acid and ethylene in the responses of citrus seedlings to salt shock. Physiol Plant 103:475–484

Griffiths J, Murase K, Rieu I et al (2006) Genetic characterization and functional analysis of the GID1 gibberellin receptors in Arabidopsis. Plant Cell 18:3399–3414. https://doi.org/10.1105/tpc.106.047415

Guan LM, Scandalios JG (2000) Catalase transcript accumulation in response to dehydration and osmotic stress in leaves of maize viviparous mutants. Redox Rep 5:377–383. https://doi.org/10.1179/135100000101535951

Gupta S, Chattopadhyay MK, Chatterjee P et al (1998) Expression of abscisic acid-responsive element-binding protein in salt-tolerant indica rice (Oryza sativa L. cv. Pokkali). Plant Mol Biol 37:629–637. https://doi.org/10.1023/A:1005934200545

Hagen G, Guilfoyle TJ (1985) Rapid induction of selective transcription by auxins. Mol Cell Biol 5:1197–1203. https://doi.org/10.1128/mcb.5.6.1197

Hamilton AJ, Bouzayen M, Grierson D (1991) Identification of a tomato gene for the ethylene-forming enzyme by expression in yeast. Proc Natl Acad Sci 88:7434–7437. https://doi.org/10.1073/pnas.88.16.7434

Hawkesford MJ, Araus J-L, Park R et al (2013) Prospects of doubling global wheat yields. Food Energy Secur 2:34–48. https://doi.org/10.1002/fes3.15

Hawkins C, Liu Z (2014) A model for an early role of auxin in Arabidopsis gynoecium morphogenesis. Front Plant Sci 5:327. https://doi.org/10.3389/fpls.2014.00327

Hedden P, Thomas SG (2012) Gibberellin biosynthesis and its regulation. Biochem J 444:11–25. https://doi.org/10.1042/bj20120245

Helliwell CA, Chandler PM, Poole A et al (2001a) The CYP88A cytochrome P450, ent-kaurenoic acid oxidase, catalyzes three steps of the gibberellin biosynthesis pathway. Proc Natl Acad Sci 98:2065–2070

Helliwell CA, Sullivan JA, Mould RM et al (2001b) A plastid envelope location of Arabidopsis ent-kaurene oxidase links the plastid and endoplasmic reticulum steps of the gibberellin biosynthesis pathway. Plant J 28:201–208

Herman JJ, Sultan SE (2011) Adaptive transgenerational plasticity in plants: case studies, mechanisms, and implications for natural populations. Front Plant Sci. https://doi.org/10.3389/fpls.2011.00102

Hewitt FR, Hough T, O'Neill P et al (1985) Effect of brassinolide and other growth regulators on the germination and growth of pollen tubes of Prunus avium using a multiple hanging-drop assay. Funct Plant Biol 12:201–211. https://doi.org/10.1071/PP9850201

Hicks GR, Raikhel NV (2012) Small molecules present large opportunities in plant biology. Annu Rev Plant Biol 63:261–282

Hitoshi S (2006) Cytokinins: activity, biosynthesis, and translocation. Annu Rev Plant Biol 57:431–449. https://doi.org/10.1146/annurev.arplant.57.032905.105231

Holeski LM, Chase-Alone R, Kelly JK (2010) The genetics of phenotypic plasticity in plant defense: trichome production in *Mimulus guttatus*. Am Nat 175:391–400. https://doi.org/10.1086/651300

Holeski LM, Jander G, Agrawal AA (2012) Transgenerational defense induction and epigenetic inheritance in plants. Trends Ecol Evol 27:618–626

Hou X, Lee LYC, Xia K et al (2010) DELLAs modulate jasmonate signaling via competitive binding to JAZs. Dev Cell 19:884–894. https://doi.org/10.1016/j.devcel.2010.10.024

Hou Q, Ufer G, Bartels D (2016) Lipid signalling in plant responses to abiotic stress. Plant Cell Environ 39:1029–1048

Hull AK, Vij R, Celenza JL (2000) Arabidopsis cytochrome P450s that catalyze the first step of tryptophan-dependent indole-3-acetic acid biosynthesis. Proc Natl Acad Sci 97:2379–2384. https://doi.org/10.1073/pnas.040569997

Hwang I, Sheen J, Müller B (2012) Cytokinin signaling networks. Annu Rev Plant Biol 63:353–380

Imber D, Tal M (1970) Phenotypic reversion of flacca, a wilty mutant of tomato, by abscisic acid. Science 169(80):592–593

Jablonka E, Raz G (2009) Transgenerational epigenetic inheritance: prevalence, mechanisms, and implications for the study of heredity and evolution. Q Rev Biol 84:131–176. https://doi.org/10.1086/598822

Jiang C-J, Shimono M, Sugano S et al (2012) Cytokinins act synergistically with salicylic acid to activate defense gene expression in rice. Mol Plant Microbe Interact 26:287–296. https://doi.org/10.1094/MPMI-06-12-0152-R

Jiang WK, Liu Y, Xia EH, Gao LZ (2013) Prevalent role of gene features in determining evolutionary fates of whole-genome duplication duplicated genes in flowering plants. Plant Physiol 161:1844–1861. https://doi.org/10.1104/pp.112.200147

Jones RJ, Mansfield TA (1970) Suppression of stomatal opening in leaves treated with abscisic acid. J Exp Bot 21:714–719. https://doi.org/10.1093/jxb/21.3.714

Ju C, Yoon GM, Shemansky JM et al (2012) CTR1 phosphorylates the central regulator EIN2 to control ethylene hormone signaling from the ER membrane to the nucleus in Arabidopsis. Proc Natl Acad Sci U S A 109:19486–19491. https://doi.org/10.1073/pnas.1214848109

Kabar K (1987) Alleviation of salinity stress by plant growth regulators on seed germination. J Plant Physiol 128:179–183. https://doi.org/10.1016/S0176-1617(87)80193-1

Kakei Y, Mochida K, Sakurai T et al (2015) Transcriptome analysis of hormone-induced gene expression in Brachypodium distachyon. Sci Rep 5:14476. https://doi.org/10.1038/srep14476

Kang J, Hwang J-U, Lee M et al (2010) PDR-type ABC transporter mediates cellular uptake of the phytohormone abscisic acid. Proc Natl Acad Sci 107:2355–2360. https://doi.org/10.1073/pnas.0909222107

Karimmojeni H, Bazrafshan AH, Majidi MM et al (2014) Effect of maternal nitrogen and drought stress on seed dormancy and germinability of *Amaranthus retroflexus*. Plant Species Biol 29:E1–E8. https://doi.org/10.1111/1442-1984.12022

Karssen CM, Brinkhorst-van der Swan DLC, Breekland AE, Koornneef M (1983) Induction of dormancy during seed development by endogenous abscisic acid: studies on abscisic acid deficient genotypes of Arabidopsis thaliana (L.) Heynh. Planta 157:158–165. https://doi.org/10.1007/BF00393650

Karuppanapandian T, Geilfus CM, Mühling KH et al (2017) Early changes of the pH of the apoplast are different in leaves, stem and roots of Vicia faba L. under declining water availability. Plant Sci 255:51–58. https://doi.org/10.1016/j.plantsci.2016.11.010

Kasahara H, Hanada A, Kuzuyama T et al (2002) Contribution of the mevalonate and methylerythritol phosphate pathways to the biosynthesis of gibberellins in Arabidopsis. J Biol Chem 277:45188–45194

Khan MA, Ungar IA (2000) Alleviation of innate and salinity-induced dormancy in Atriplex griffithii Moq. var. stocksii Boiss. Seed Sci Technol 28:29–37

Khan MA, Gul B, Weber DJ (2002) Improving seed germination of Salicornia rubra (Chenopodiaceae) under saline conditions using germination-regulating chemicals. West North Am Nat 62:101–105

Kieber JJ, Schaller GE (2014) Cytokinins. Arab B 12:e0168

Kim S, Song M-H, Wei W, Yun Y-S (2015) Selective biosorption behavior of Escherichia coli biomass toward Pd(II) in Pt(IV)-Pd(II) binary solution. J Hazard Mater 283:657–662. https://doi.org/10.1016/j.jhazmat.2014.10.008

Kinoshita T, Seki M (2014) Epigenetic memory for stress response and adaptation in plants. Plant Cell Physiol 55:1859–1863

Koga J (1995) Structure and function of indolepyruvate decarboxylase, a key enzyme in indole-3-acetic acid biosynthesis. Biochim Biophys Acta Protein Struct Mol Enzymol 1249:1–13. https://doi.org/10.1016/0167-4838(95)00011-I

Kohlen W, Charnikhova T, Liu Q et al (2011) Strigolactones are transported through the xylem and play a key role in shoot architectural response to phosphate deficiency in nonarbuscular mycorrhizal host Arabidopsis. Plant Physiol 155:974–987

Kohli A, Sreenivasulu N, Lakshmanan P, Kumar PP (2013) The phytohormone crosstalk paradigm takes center stage in understanding how plants respond to abiotic stresses. Plant Cell Rep 32:945–957. https://doi.org/10.1007/s00299-013-1461-y

Koornneef M, Hanhart CJ, Hilhorst HW, Karssen CM (1989) In vivo inhibition of seed development and reserve protein accumulation in recombinants of abscisic acid biosynthesis and responsiveness mutants in Arabidopsis thaliana. Plant Physiol 90:463–469. https://doi.org/10.1104/pp.90.2.463

Kucera B, Cohn MA, Leubner-Metzger G (2005) Plant hormone interactions during seed dormancy release and germination. Seed Sci Res 15:281–307. https://doi.org/10.1079/SSR2005218

Kudla J, Batistič O, Hashimoto K (2010) Calcium signals: the lead currency of plant information processing. Plant Cell 22:541–563. https://doi.org/10.1105/tpc.109.072686

Kumari S, van der Hoorn RAL (2011) A structural biology perspective on bioactive small molecules and their plant targets. Curr Opin Plant Biol 14:480–488

Kumari A, Das P, Parida AK, Agarwal PK (2015) Proteomics, metabolomics, and ionomics perspectives of salinity tolerance in halophytes. Front Plant Sci. https://doi.org/10.3389/fpls.2015.00537

Kuroha T, Tokunaga H, Kojima M et al (2009) Functional analyses of LONELY GUY cytokinin-activating enzymes reveal the importance of the direct activation pathway in Arabidopsis. Plant Cell 21:3152–3169. https://doi.org/10.1105/tpc.109.068676

Kuromori T, Miyaji T, Yabuuchi H et al (2010) ABC transporter AtABCG25 is involved in abscisic acid transport and responses. Proc Natl Acad Sci 107:2361–2366. https://doi.org/10.1073/pnas.0912516107

Lampei Sr C (2008) COS 86-3: the effect of maternal environment on seed dormancy and its support for an evolutionary stable strategy. In: The 93rd ESA Annual Meeting

LeClere S, Tellez R, Rampey RA et al (2002) Characterization of a family of IAA-amino acid conjugate hydrolases from Arabidopsis. J Biol Chem 277:20446–20452

Lee HI, León J, Raskin I (1995) Biosynthesis and metabolism of salicylic acid. Proc Natl Acad Sci U S A 92:4076–4079

Leydecker MT, Moureaux T, Kraepiel Y et al (1995) Molybdenum cofactor mutants, specifically impaired in xanthine dehydrogenase activity and abscisic acid biosynthesis, simultaneously overexpress nitrate reductase. Plant Physiol 107:1427–1431. doi: 107/4/1427 [pii]

Li W, Liu X, Khan MA, Yamaguchi S (2005) The effect of plant growth regulators, nitric oxide, nitrate, nitrite and light on the germination of dimorphic seeds of Suaeda salsa under saline conditions. J Plant Res 118:207–214. https://doi.org/10.1007/s10265-005-0212-8

Liu S, Sang R (2013) Bioactive fluorescent jasmonate designed by molecular modeling and its migration in tomato visualized by fluorescent molecular imaging. Tetrahedron 69:844–848

Ljung K, Hull AK, Kowalczyk M et al (2002) Biosynthesis, conjugation, catabolism and homeostasis of indole-3-acetic acid in Arabidopsis thaliana. Plant Mol Biol 50:309–332

Loake G, Grant M (2007) Salicylic acid in plant defence—the players and protagonists. Curr Opin Plant Biol 10:466–472. https://doi.org/10.1016/j.pbi.2007.08.008

Ludwig-Müller J (2011) Auxin conjugates: their role for plant development and in the evolution of land plants. J Exp Bot 62:1757–1773. https://doi.org/10.1093/jxb/erq412

Luzuriaga AL, Escudero A, Olano JM, Loidi J (2005) Regenerative role of seed banks following an intense soil disturbance. Acta Oecologica 27:57–66. https://doi.org/10.1016/j.actao.2004.09.003

Ma Q, Robert S (2014) Auxin biology revealed by small molecules. Physiol Plant 151:25–42

Mähönen AP, Bishopp A, Higuchi M et al (2006) Cytokinin signaling and its inhibitor AHP6 regulate cell fate during vascular development. Science 311(80):94–98. https://doi.org/10.1126/science.1118875

Mandel MA, Feldmann KA, Herrera-Estrella L et al (1996) CLA1, a novel gene required for chloroplast development, is highly conserved in evolution. Plant J 9:649–658. https://doi.org/10.1046/j.1365-313X.1996.9050649.x

Maruyama C, Goepfert Z, Squires K et al (2016) Effects of population site and maternal drought on establishment physiology in impatiens capensis meerb.(Balsaminaceae). Rhodora 118:32–45

Matsuura H, Takeishi S, Kiatoka N et al (2012) Transportation of de novo synthesized jasmonoyl isoleucine in tomato. Phytochemistry 83:25–33

Meinzer FC, Zhu J (1999) Efficiency of C4 photosynthesis in Atriplex lentiformis under salinity stress. Funct Plant Biol 26:79–86

Melcher K, Zhou XE, Xu HE (2010) Thirsty plants and beyond: structural mechanisms of abscisic acid perception and signaling. Curr Opin Struct Biol 20:722–729

Metz J, Liancourt P, Kigel J et al (2010) Plant survival in relation to seed size along environmental gradients: a long-term study from semi-arid and Mediterranean annual plant communities. J Ecol 98:697–704. https://doi.org/10.1111/j.1365-2745.2010.01652.x

Migicovsky Z, Yao Y, Kovalchuk I (2014) Transgenerational phenotypic and epigenetic changes in response to heat stress in Arabidopsis thaliana. Plant Signal Behav 9:e27971. https://doi.org/10.4161/psb.27971

Miller G, Suzuki N, Ciftci-Yilmaz S, Mittler R (2010) Reactive oxygen species homeostasis and signalling during drought and salinity stresses. Plant Cell Environ 33:453–467. https://doi.org/10.1111/j.1365-3040.2009.02041.x

Misra BB, Assmann SM, Chen S (2014) Plant single-cell and single-cell-type metabolomics. Trends Plant Sci 19:637–646. https://doi.org/10.1016/j.tplants.2014.05.005

Mittler R (2002) Oxidative stress, antioxidants and stress tolerance. Trends Plant Sci 7:405–410. https://doi.org/10.1016/S1360-1385(02)02312-9

Mittler R (2006) Abiotic stress, the field environment and stress combination. Trends Plant Sci 11:15–19

Mittler R, Kim Y, Song L et al (2006) Gain- and loss-of-function mutations in Zat10 enhance the tolerance of plants to abiotic stress. FEBS Lett 580:6537–6542. https://doi.org/10.1016/j.febslet.2006.11.002

Mizutani M, Ohta D (2010) Diversification of P450 genes during land plant evolution. Annu Rev Plant Biol 61:291–315

Mockaitis K, Estelle M (2008) Auxin receptors and plant development: a new signaling paradigm. Annu Rev Cell Dev Biol 24:55–80

Mousseau T, Fox C (1998) The adaptive significance of maternal effects. Trends Ecol Evol 13:403–407. https://doi.org/10.1016/S0169-5347(98)01472-4

Munir J, Dorn LA, Donohue K, Schmitt J (2001) The effect of maternal photoperiod on seasonal dormancy in Arabidopsis thaliana (Brassicaceae). Am J Bot 88:1240–1249. https://doi.org/10.2307/3558335

Murata Y, Mori IC, Munemasa S (2015) Diverse stomatal signaling and the signal integration mechanism. Annu Rev Plant Biol 66:369–392. https://doi.org/10.1146/annurev-arplant-043014-114707

Nambara E, Marion-Poll A (2005) Abscisic acid biosynthesis and catabolism. Annu Rev Plant Biol 56:165–185

Negi J, Matsuda O, Nagasawa T et al (2008) CO2 regulator SLAC1 and its homologues are essential for anion homeostasis in plant cells. Nature 452:483–486. https://doi.org/10.1038/nature06720

Nishiyama R, Watanabe Y, Leyva-Gonzalez MA et al (2013) Arabidopsis AHP2, AHP3, and AHP5 histidine phosphotransfer proteins function as redundant negative regulators of drought stress response. Proc Natl Acad Sci 110:4840–4845. https://doi.org/10.1073/pnas.1302265110

Noguchi T, Fujioka S, Takatsuto S et al (1999) Arabidopsis det2 is defective in the conversion of (24R)-24-methylcholest-4-en-3-one to (24R)-24-methyl-5α-cholestan-3-one in brassinosteroid biosynthesis. Plant Physiol 120:833–840. https://doi.org/10.1104/pp.120.3.833

Noguchi T, Fujioka S, Choe S et al (2000) Biosynthetic pathways of brassinolide in Arabidopsis. Plant Physiol 124:201–210

Normanly J, Slovin JP, Cohen JD (1995) Rethinking auxin biosynthesis and metabolism. Plant Physiol 107:323–329. https://doi.org/10.1104/pp.107.2.323

Osakabe Y, Osakabe K, Shinozaki K, Tran L-SP (2014) Response of plants to water stress. Front Plant Sci. https://doi.org/10.3389/fpls.2014.00086

Oyama T, Shimura Y, Okada K (1997) The Arabidopsis HY5 gene encodes a bZIP protein that regulates stimulus-induced development of root and hypocotyl. Genes Dev 11:2983–2995

Pagnussat GC, Alandete-Saez M, Bowman JL, Sundaresan V (2009) Auxin-dependent patterning and gamete specification in the Arabidopsis female gametophyte. Science 324(80):1684–1689. https://doi.org/10.1126/science.1167324

Parida AK, Das AB (2005) Salt tolerance and salinity effects on plants: a review. Ecotoxicol Environ Saf 60:324–349

Parihar P, Singh S, Singh R et al (2015) Effect of salinity stress on plants and its tolerance strategies: a review. Environ Sci Pollut Res 22:4056–4075. https://doi.org/10.1007/s11356-014-3739-1

Pei Z-M, Murata Y, Benning G et al (2000) Calcium channels activated by hydrogen peroxide mediate abscisic acid signalling in guard cells. Nature 406:731–734

Pérez-Alfocea F, Ghanem ME, Gómez-Cadenas A, Dodd IC (2011) Omics of root-to-shoot signaling under salt stress and water deficit. Omi A J Integr Biol 15:893–901. https://doi.org/10.1089/omi.2011.0092

Piotrowska A, Bajguz A (2011) Conjugates of abscisic acid, brassinosteroids, ethylene, gibberellins, and jasmonates. Phytochemistry 72:2097–2112. https://doi.org/10.1016/j.phytochem.2011.08.012

Pollier J, Rombauts S, Goossens A (2013) Analysis of RNA-Seq data with TopHat and cufflinks for genome-wide expression analysis of jasmonate-treated plants and plant cultures. Methods Mol Biol (Clifton, NJ) 1011:305–315

Pospisilová J, Synková H, Rulcová J (2000) Cytokinins and water stress. MINIREVIEW. Biol Plant 43:321–328. https://doi.org/10.1023/A:1026754404857

Pott MB, Hippauf F, Saschenbrecker S et al (2004) Biochemical and structural characterization of benzenoid carboxyl methyltransferases involved in floral scent production in Stephanotis floribunda and Nicotiana suaveolens. Plant Physiol 135:1946–1955. https://doi.org/10.1104/pp.104.041806

Qin G, Gu H, Zhao Y et al (2005) An indole-3-acetic acid carboxyl methyltransferase regulates Arabidopsis leaf development. Plant Cell 17:2693–2704. https://doi.org/10.1105/tpc.105.034959

Reddy ASN, Ali GS, Celesnik H, Day IS (2011) Coping with stresses: roles of calcium- and calcium/calmodulin-regulated gene expression. Plant Cell 23:2010–2032. https://doi.org/10.1105/tpc.111.084988

Reusche M, Klásková J, Thole K et al (2013) Stabilization of cytokinin levels enhances Arabidopsis resistance against Verticillium longisporum. Mol Plant Microbe Interact 26:850–860. https://doi.org/10.1094/MPMI-12-12-0287-R

Riefler M, Novak O, Strnad M, Schmülling T (2006) Arabidopsis cytokinin receptor mutants reveal functions in shoot growth, leaf senescence, seed size, germination, root development, and cytokinin metabolism. Plant Cell 18:40–54. https://doi.org/10.1105/tpc.105.037796

Rigal A, Ma Q, Robert S (2014) Unraveling plant hormone signaling through the use of small molecules. Front Plant Sci 5:373. https://doi.org/10.3389/fpls.2014.00373

Rizhsky L, Davletova S, Liang H, Mittler R (2004) The zinc finger protein Zat12 is required for cytosolic ascorbate peroxidase 1 expression during oxidative stress in Arabidopsis. J Biol Chem 279:11736–11743. https://doi.org/10.1074/jbc.M313350200

Roach DA, Wulff RD (1987) Maternal effects in plants. Annu Rev Ecol Syst 18:209–235. https://doi.org/10.1146/annurev.es.18.110187.001233

Rojas-Pierce M, Titapiwatanakun B, Sohn EJ et al (2007) Arabidopsis P-glycoprotein19 participates in the inhibition of gravitropism by gravacin. Chem Biol 14:1366–1376

Ross JJ, Murfet IC, Reid JB (1997) Gibberellin mutants. Physiol Plant 100:550–560. https://doi.org/10.1111/j.1399-3054.1997.tb03060.x

Ruyter-Spira C, Al-Babili S, van der Krol S, Bouwmeester H   The biology of strigolactones. Trends Plant Sci 18:72–83. https://doi.org/10.1016/j.tplants.2012.10.003

Sakamoto M, Munemura I, Tomita R, Kobayashi K (2008) Involvement of hydrogen peroxide in leaf abscission signaling, revealed by analysis with an in vitro abscission system in Capsicum plants. Plant J 56:13–27. https://doi.org/10.1111/j.1365-313X.2008.03577.x

Santiago J, Henzler C, Hothorn M (2013) Molecular mechanism for plant steroid receptor activation by somatic embryogenesis co-receptor kinases. Science 341(80):889–892

Sato T, Theologis A (1989) Cloning the mRNA encoding 1-aminocyclopropane-1-carboxylate synthase, the key enzyme for ethylene biosynthesis in plants. Proc Natl Acad Sci 86:6621–6625

Schwartz SH, Leon-Kloosterziel KM, Koornneef M, Zeevaart JAD (1997) Biochemical characterization of the aba2 and aba3 mutants in Arabidopsis thaliana. Plant Physiol 114:161–166

Seo M, Akaba S, Oritani T et al (1998) Higher activity of an aldehyde oxidase in the auxin-overproducing superroot1 mutant of Arabidopsis thaliana. Plant Physiol 116:687–693

Seo M, Koiwai H, Akaba S et al (2000) Abscisic aldehyde oxidase in leaves of Arabidopsis thaliana. Plant J 23:481–488. https://doi.org/10.1046/j.1365-313x.2000.00812.x

Seo HS, Song JT, Cheong J-J et al (2001) Jasmonic acid carboxyl methyltransferase: a key enzyme for jasmonate-regulated plant responses. Proc Natl Acad Sci 98:4788–4793. https://doi.org/10.1073/pnas.081557298

Shabala S (2013) Learning from halophytes: Physiological basis and strategies to improve abiotic stress tolerance in crops. Ann Bot 112:1209–1221

Shabala S, White RG, Djordjevic MA et al (2016) Root-to-shoot signalling: integration of diverse molecules, pathways and functions. Funct Plant Biol 43:87–104

Shani E, Weinstain R, Zhang Y et al (2013) Gibberellins accumulate in the elongating endodermal cells of Arabidopsis root. Proc Natl Acad Sci 110:4834–4839

Sheard LB, Tan X, Mao H et al (2010) Jasmonate perception by inositol-phosphate-potentiated COI1-JAZ co-receptor. Nature 468:400–405

Shimada A, Ueguchi-Tanaka M, Nakatsu T et al (2008) Structural basis for gibberellin recognition by its receptor GID1. Nature 456:520–523

Shinozaki K, Yamaguchi-Shinozaki K (2007) Gene networks involved in drought stress response and tolerance. J Exp Bot 58(2):221–227

Sierla M, Waszczak C, Vahisalu T, Kangasjärvi J (2016) Reactive oxygen species in the regulation of stomatal movements. Plant Physiol 171:1569–1580. https://doi.org/10.1104/pp.16.00328

Sindhu RK, Griffin DH, Walton DC (1990) Abscisic aldehyde is an intermediate in the enzymatic conversion of xanthoxin to abscisic acid in Phaseolus vulgaris L. leaves. Plant Physiol 93:689–694

Sirichandra C, Wasilewska A, Vlad F et al (2009) The guard cell as a single-cell model towards understanding drought tolerance and abscisic acid action. J Exp Bot 60:1439–1463. https://doi.org/10.1093/jxb/ern340

Smith CM, Boyko EV (2007) The molecular bases of plant resistance and defense responses to aphid feeding: current status. Entomol Exp Appl 122:1–16

Staswick PE (2009) The tryptophan conjugates of jasmonic and indole-3-acetic acids are endogenous auxin inhibitors. Plant Physiol 150:1310–1321

Staswick PE, Tiryaki I, Rowe ML (2002) Jasmonate response locus JAR1 and several related Arabidopsis genes encode enzymes of the firefly luciferase superfamily that show activity on jasmonic, salicylic, and indole-3-acetic acids in an assay for adenylation. Plant Cell 14:1405–1415. https://doi.org/10.1105/tpc.000885

Taji T, Seki M, Satou M et al (2004) Comparative genomics in salt tolerance between Arabidopsis and aRabidopsis-related halophyte salt cress using Arabidopsis microarray. Plant Physiol 135:1697–1709. https://doi.org/10.1104/pp.104.039909

Takahashi M, Asada K (1988) Superoxide production in aprotic interior of chloroplast thylakoids. Arch Biochem Biophys 267:714–722. https://doi.org/10.1016/0003-9861(88)90080-X

Takei K, Yamaya T, Sakakibara H (2004) Arabidopsis CYP735A1 and CYP735A2 encode cytokinin hydroxylases that catalyze the biosynthesis of trans-zeatin. J Biol Chem 279:41866–41872. https://doi.org/10.1074/jbc.M406337200

Tanaka M, Takei K, Kojima M et al (2006) Auxin controls local cytokinin biosynthesis in the nodal stem in apical dominance. Plant J 45:1028–1036. https://doi.org/10.1111/j.1365-313X.2006.02656.x

Tanigaki Y, Higashi T, Takayama K et al (2015) Transcriptome analysis of plant hormone-related tomato (Solanum lycopersicum) genes in a sunlight-type plant factory. PLoS One 10:e0143412. https://doi.org/10.1371/journal.pone.0143412

Tao Y, Ferrer J-L, Ljung K et al (2008) Rapid synthesis of auxin via a new tryptophan-dependent pathway is required for shade avoidance in plants. Cell 133:164–176. https://doi.org/10.1016/j.cell.2008.01.049

Thayer ZM, Kuzawa CW (2011) Biological memories of past environments: epigenetic pathways to health disparities. Epigenetics 6:798–803. https://doi.org/10.4161/epi.6.7.16222

Thomas JC, McElwain EF, Bohnert HJ (1992) Convergent induction of osmotic stress-responses : abscisic acid, cytokinin, and the effects of NaCl. Plant Physiol 100:416–423. doi: 0032-0889/92/100/041

Tohge T, de Souza LP, Fernie AR (2014) Genome-enabled plant metabolomics. J Chromatogr B 966:7–20. https://doi.org/10.1016/j.jchromb.2014.04.003

Tsuda E, Yang H, Nishimura T et al (2011) Alkoxy-auxins are selective inhibitors of auxin transport mediated by PIN, ABCB, and AUX1 transporters. J Biol Chem 286:2354–2364

Tuna AL, Kaya C, Dikilitas M, Higgs D (2008) The combined effects of gibberellic acid and salinity on some antioxidant enzyme activities, plant growth parameters and nutritional status in maize plants. Environ Exp Bot 62:1–9. https://doi.org/10.1016/j.envexpbot.2007.06.007

Ueda H, Kusaba M (2015) Strigolactone regulates leaf senescence in concert with ethylene in Arabidopsis. Plant Physiol 00325:2015

van den Berg C, Willemsen V, Hendriks G et al (1997) Short-range control of cell differentiation in the Arabidopsis root meristem. Nature 390:287–289

Van Leene J, Hollunder J, Eeckhout D et al (2010) Targeted interactomics reveals a complex core cell cycle machinery in Arabidopsis thaliana. Mol Syst Biol 6:397. https://doi.org/10.1038/msb.2010.53

Varbanova M, Yamaguchi S, Yang Y et al (2007) Methylation of gibberellins by Arabidopsis GAMT1 and GAMT2. Plant Cell 19:32–45. https://doi.org/10.1105/tpc.106.044602

Verhoeven KJF, Jansen JJ, van Dijk PJ, Biere A (2010) Stress-induced DNA methylation changes and their heritability in asexual dandelions. New Phytol 185:1108–1118. https://doi.org/10.1111/j.1469-8137.2009.03121.x

Vlot AC, Dempsey DA, Klessig DF (2009) Salicylic acid, a multifaceted hormone to combat disease. Annu Rev Phytopathol 47:177–206. https://doi.org/10.1146/annurev.phyto.050908. 135202

Walton DC (1980) Biochemistry and physiology of abscisic acid. Annu Rev Plant Physiol 31:453–489

Walton A, Stes E, De Smet I et al (2015) Plant hormone signalling through the eye of the mass spectrometer. Proteomics 15:1113–1126. https://doi.org/10.1002/pmic.201400403

Wang J, Chen C (2008) Biosorbents for heavy metals removal and their future. Biotechnol Adv 27:195–226. https://doi.org/10.1016/j.biotechadv.2008.11.002

Wang KL-C, Li H, Ecker JR (2002) Ethylene biosynthesis and signaling networks. Plant Cell 14:S131–S151. https://doi.org/10.1105/tpc.001768

Wang D, Pajerowska-Mukhtar K, Culler AH, Dong X (2007) Salicylic acid inhibits pathogen growth in plants through repression of the auxin signaling pathway. Curr Biol 17:1784–1790. https://doi.org/10.1016/j.cub.2007.09.025

Wani SH, Kumar V, Shriram V, Sah SK (2016) Phytohormones and their metabolic engineering for abiotic stress tolerance in crop plants. Crop J 4:162–176

Wasternack C, Hause B (2013) Jasmonates: biosynthesis, perception, signal transduction and action in plant stress response, growth and development. An update to the 2007 review in Annals of Botany. Ann Bot 111:1021–1058. https://doi.org/10.1093/aob/mct067

Watanabe KA, Ringler P, Gu L, Shen QJ (2014) RNA-sequencing reveals previously unannotated protein- and microRNA-coding genes expressed in aleurone cells of rice seeds. Genomics 103:122–134. https://doi.org/10.1016/j.ygeno.2013.10.007

Weiner JJ, Peterson FC, Volkman BF, Cutler SR (2010) Structural and functional insights into core ABA signaling. Curr Opin Plant Biol 13:495–502

White CN, Proebsting WM, Hedden P, Rivin CJ (2000) Gibberellins and seed development in maize. I. evidence that gibberellin/abscisic acid balance governs germination versus maturation pathways. Plant Physiol 122:1081–1088. https://doi.org/10.1104/pp.122.4.1081

Wildermuth MC, Dewdney J, Wu G, Ausubel FM (2001) Isochorismate synthase is required to synthesize salicylic acid for plant defence. Nature 414:562–565. http://www.nature.com/nature/journal/v414/n6863/suppinfo/414562a_S1.html

Wilkinson S, Davies WJ (2002) ABA-based chemical signalling: the co-ordination of responses to stress in plants. Plant Cell Environ 25:195–210

Woodward AW, Bartel B (2005) Auxin: regulation, action, and interaction. Ann Bot 95:707–735. https://doi.org/10.1093/aob/mci083

Wu G, Shao H-B, Chu L-Y, Cai J-W (2007) Insights into molecular mechanisms of mutual effect between plants and the environment. A review. Agron Sustain Dev 27:69–78. https://doi.org/10.1051/agro:2006031

Yaish MW, Colasanti J, Rothstein SJ (2011) The role of epigenetic processes in controlling flowering time in plants exposed to stress. J Exp Bot 62:3727–3735

Yamada Y, Furusawa S, Nagasaka S et al (2014) Strigolactone signaling regulates rice leaf senescence in response to a phosphate deficiency. Planta 240:399–408

Yang SF, Hoffman NE (1984) Ethylene biosynthesis and its regulation in higher plants. Annu Rev Plant Physiol 35:155–189

Yang Y, Hammes UZ, Taylor CG et al (2006) High-affinity auxin transport by the AUX1 influx carrier protein. Curr Biol 16:1123–1127

Yang Y, Xu R, Ma C et al (2008) Inactive methyl indole-3-acetic acid ester can be hydrolyzed and activated by several esterases belonging to the AtMES esterase family of Arabidopsis. Plant Physiol 147:1034–1045. https://doi.org/10.1104/pp.108.118224

Yesbergenova Z, Yang G, Oron E et al (2005) The plant Mo-hydroxylases aldehyde oxidase and xanthine dehydrogenase have distinct reactive oxygen species signatures and are induced by drought and abscisic acid. Plant J 42:862–876. https://doi.org/10.1111/j.1365-313X.2005.02422.x

Yoshida T, Mogami J, Yamaguchi-Shinozaki K (2015) Omics approaches toward defining the comprehensive abscisic acid signaling network in plants. Plant Cell Physiol 56:1043–1052. https://doi.org/10.1093/pcp/pcv060

Zandalinas SI, Mittler R, Balfagón D et al (2017) Plant adaptations to the combination of drought and high temperatures. Physiologia Plantarum 162: 2–12. 2018

Zhang J, Schurr U, Davies WJ (1987) Control of stomatal behaviour by abscisic acid which apparently originates in the roots. J Exp Bot 38:1174–1181. https://doi.org/10.1093/jxb/38.7.1174

Zhang J, Jia W, Yang J, Ismail AM (2006) Role of ABA in integrating plant responses to drought and salt stresses. F Crop Res 97:111–119. https://doi.org/10.1016/j.fcr.2005.08.018

Zhang Y, Gao P, Yuan JS (2010) Plant protein-protein interaction network and interactome. Curr Genomics 11:40–46. https://doi.org/10.2174/138920210790218016

Zhang Y, Liu S, Dai SY, Yuan JS (2012) Integration of shot-gun proteomics and bioinformatics analysis to explore plant hormone responses. BMC Bioinformatics 13:S8. https://doi.org/10.1186/1471-2105-13-S15-S8

Zhang Y, Van Dijk ADJ, Scaffidi A et al (2014) Rice cytochrome P450 MAX1 homologs catalyze distinct steps in strigolactone biosynthesis. Nat Chem Biol 10:1028–1033

Zhao Y (2014) Auxin biosynthesis

Zhu J-K (2003) Salt and drought stress signal transduction in plants. Annu Rev Plant Biol 53:247–273

Zürcher E, Tavor-Deslex D, Lituiev D et al (2013) A robust and sensitive synthetic sensor to monitor the transcriptional output of the cytokinin signaling network in planta. Plant Physiol 161:1066–1075. https://doi.org/10.1104/pp.112.211763

Zwanenburg B, Pospíšil T (2013) Structure and activity of strigolactones: new plant hormones with a rich future. Mol Plant 6:38–62

Chapter 16
Transgenic Crops: Status, Potential, and Challenges

Tejinder Mall, Lei Han, Laura Tagliani, and Cory Christensen

Abstract Since the commercialization of the first GM crop in mid-1990s, agricultural biotechnology has enjoyed remarkable growth in product development, commercialization, and global adoption. Areas planted with GM crops in the last 20 years have increased more than 100-fold, making crop biotechnology one of the fastest adopted agricultural technologies. World population is 7.3 billion today and is expected to reach 9.5 billion in 2050. To sustain this ever-growing population, we will be required to produce 70% more food than what we produce today (Headrick Res Technol Manag 59:3, 2016). Agricultural biotechnology has been and will continue to play an important role in meeting the challenge. This chapter covers a brief overview of agricultural biotechnology, starting with the development of *Agrobacterium* and gene gun-mediated transformation technologies. Input, output, and agronomic biotechnology traits are discussed with emphasis on the major crops being cultivated around the world. A brief overview of the next generation of precision transformation technologies is given with emphasis on site-specific nucleases, i.e., meganucleases, ZFNs (zinc finger nucleases), TALENs (transcription activator-like effector nucleases), and CRISPR/Cas (clustered regulatory interspaced short palindromic repeats/CRISPR-associated). Specific examples of the use of these technologies resulting in commercially important traits are discussed. Lastly, challenges associated with further adoption of GM crops are discussed with an emphasis on risk assessment of GM crops and food, perception of risk and benefits, regulation of GM products and policy development, international trade concerns and policy decisions, and social concerns.

Keywords Transgenic crops · Transgenic traits · *Agrobacterium* · Gene gun · Risk assessment · GM regulation

T. Mall (✉) · C. Christensen
Dow AgroSciences LLC, 1281 Win Hentschel Blvd., West Lafayette, 47906 IN, USA
e-mail: TKMall@dow.com

L. Han · L. Tagliani
Dow AgroSciences LLC, 9330 Zionsville Road, Indianapolis, 46268 IN, USA

© Springer International Publishing AG, part of Springer Nature 2018
S. S. Gosal, S. H. Wani (eds.), *Biotechnologies of Crop Improvement, Volume 2*,
https://doi.org/10.1007/978-3-319-90650-8_16

16.1   Genetic Transformation - A Brief Historic Overview

Genetic transformation is a well-established and widely used technology today with applications ranging from functional genomics to the introduction of desired traits in the plants. The journey of the development of this technology started with researchers trying to understand the mechanism of crown gall disease in plants. In their quest of knowledge, researchers discovered something very surprising which was inter-kingdom transfer of genes by the *Agrobacterium* into woody plants. This discovery became the stepping stone into the field of genome manipulation of plants that has changed the face of agriculture today.

Crown gall is a disease that forms tumors at the crown of woody plants. The disease affects the vascular system and hence interferes with normal transport of water and nutrients. Severe infections lead to death of the plant and can result in economic losses. The disease has been known for a long time; however it was in 1907 when Smith and coworkers while working on the crown gall of marguerite established *Agrobacterium* to be the causative agent of the disease (Smith et al. 1907). Initially it was thought that the irritation caused or the chemicals released by the bacterium led to the formation and growth of tumors. However a study reported by White and Braun (1942) contradicted these hypotheses by showing that although the bacteria led to the initiation of the gall, the gall has the potential to grow further even if the *Agrobacterium* is no longer present. Therefore, a new hypothesis was formed that the bacterium transforms "something" into the plant cells that continues functioning independent of the bacterium. Braun (1958) called this "something" as tumor-inducing principle (TIP); however the nature of TIP was still a mystery. Another important milestone in the field was the report of Menage and Morel (1964) which showed that plants infected with *Agrobacterium* produce opines which are used by *Agrobacterium* as a source of nitrogen and carbon. This indicated that *Agrobacterium* transferred TIP into the plant for its own advantage. Later it was established that the *Agrobacterium* has an extra chromosomal plasmid, named tumor-inducing (Ti) plasmid that provides the bacterium its tumor-forming ability (Zaenen et al. 1974). Mary-Dell Chilton et al. (1977) were the first to demonstrate that a small portion of Ti plasmid called T-DNA was transferred into the host plant genome and was responsible for producing the disease. The discovery revolutionized this field, and many research laboratories around the world further characterized the nature of this T-DNA and established it as a tool for genetic transformation.

Though *Agrobacterium*-mediated transformation has achieved widespread use, the technology has its own limitations. *Agrobacterium* does not have a wide host range to facilitate gene transfer to every genotype of crops. It was further complicated by the lack of tissue culture and regeneration protocols for a wide range of crop species. Hence there was a need to find alternate methods of transformation. John Sanford, a plant breeder at Cornell University, wanted to develop an easier method of gene transfer since it takes years to do so using cross pollination. Along with his colleagues, he developed a crude BB gun-based particle acceleration technique that was used to bombard onion cells (Sanford et al. 1991). This early version of a gene gun used 0.22 caliber bullets for acceleration of tiny, DNA-coated particles

(Klein et al. 1987). The plastic bullets used in this process were propelled toward the target tissue and were stopped by a stopping screen which had a hole in the middle. The plastic bullet stopped, and the DNA-coated particles passed through the hole, hit the target tissue, and released the DNA inside the cells. Initial experiments were crude, resulting in splashing and dying of onion cells. With some optimizations they were able to show transient gene expression in the cells. However, the use of this version of the gene gun was very cumbersome. It required frequent cleaning and used gunpowder to propel the particles. Therefore efforts were initiated to improve the technology. BioRad developed a new version of the gun called PDS-1000/He that replaced the gunpowder discharge with the blast of the inert gas helium. Further, either to reduce the cost of this device or to develop proprietary technologies, other forms of particle acceleration devices were invented, e.g., Accel gene gun (McCabe and Christou 1993), particle inflow gun (PIG) (Finer et al. 1992), etc. However despite all these innovations for the development of different types of gene guns, the earlier version of the gene gun developed by BioRad that used compressed helium to generate a blast remained most popular.

In the meantime, some other gene delivery methods were also established. However due to inherent constraints of gene transfer or due to limiting plant regeneration potential of the explants leading to poor transformation frequencies, those methods did not become very popular. Overall, *Agrobacterium*-mediated and gene gun-mediated gene transformation methods remained most commonly used.

The enablement of these gene delivery technologies coincided well with the advent of other related technologies. In vitro culture and regeneration protocols were being established for a large number of crops. Paul Berg produced the first recombinant DNA when he combined SV40 monkey virus and λ virus called λ bacteriophage (Jackson et al. 1972). This set the stage for recombinant DNA technology to join different pieces of DNA together to get desired plant gene expression cassettes. (Mullis et al. 1986) invented polymerase chain reaction (PCR) that enabled routine and easy amplification of DNA fragments in the lab. All these technologies coming together led to the dawn of the field of genetic transformation. With this continued effort, it was in 1983 when the first transgenic plant was successfully regenerated in tobacco followed by many other crops. The first transgenic plant commercially grown was virus-resistant tobacco in 1992 in China followed by FLAVR SAVR™ tomato approved for commercial cultivation in the USA in 1994 (James 1997). Since then, global GM crop acreage has made a phenomenal increase of more than 100-fold in a span of 20 years, making crop biotechnology one of the fastest adopted agricultural technologies (ISAAA 2016).

16.2 Commercial Biotechnology-Based Traits/Crops

Since the commercialization of the first GM crop in mid-1990s, GM traits have been produced in 15 crop plants (maize, soybeans, cotton, canola, alfalfa, sugar beets, eggplant, papaya, potato, pineapple, squash, apple, plum, eucalyptus, and poplar)

(James 2015). Many more transgenic crops and traits have been tested but have not yet been commercialized (Fernandez-Cornejo et al. 2014). In 2015, 18 million farmers from 28 countries planted almost 440 million acres of GM crops (James 2015).

The growth and development of commercial GM crops over the past 20 years has been based on a number of contributing factors such as potential market size of the traited crop, discovery of genes expressing the desired traits, and public and farmer acceptability of the product. Due to the high costs of developing transgenic crops, the majority of commercial products today are targeted at the largest agricultural markets: maize, soybeans, cotton, and canola in the Americas (Phillips McDougall 2011). There are several significant crops which are notably absent from GM commercialization today, in particular wheat and rice. GM traits have been developed and tested in both wheat and rice (e.g., glyphosate-tolerant wheat (Zhou et al. 2003) and Golden Rice (Ye et al. 2000)), and although they have substantial market sizes, no commercial products have made it to market yet due in part to public wariness about GM traits in crops for direct human consumption.

In addition to the commercialized GM crops, many more GM plants have been developed and field tested in the USA and elsewhere. For the years 1985 through 2013, USDA issued 17,000 release permits for testing the GM crops in field in USA (Fernandez-Cornejo et al. 2014). The four key commercial crops, maize, soybeans, cotton, and canola comprised two-thirds of the 17,000 total releases. In looking at types of traits being developed and tested, slightly more than two-thirds were for testing input traits, primarily insect-resistant(IR) and/or herbicide-tolerant (HT) traits. A snapshot of 2014 finds that there were 49 GM events being cultivated commercially and 53 more events that were in the late stages of development (Parisi et al. 2016). Of those 102 GM events, more than 75% of the traits were IR and HT.

The earliest commercial GM crop products were primarily single-trait events, meaning the product contained a single trait of interest, for example, glyphosate herbicide tolerance or lepidopteran insect resistance. Many single-trait events carry more than one gene, such as selectable markers that are used during transformation but do not confer an end-user trait in the final product, and thus are still classified as single-trait events. Increasingly, single-trait products have been combined into stacked trait products conferring two or more value-added traits to bring greater value to farmers in managing their crops. Stacked trait products are developed either by transforming more than one gene linked together in a single construct (referred to as molecular stacks) or by combining traits carried by two or more independent transformation events via breeding (referred to as breeding stacks) (Que et al. 2010). There are examples of both types of stacks among commercial products today, and nearly all GM maize, soybean, and cotton products on the market now carry more than one trait. Overall, stacked GM trait products across all crops were planted on 145 million acres in 2015, amounting to 33% of all the GM acres globally (James 2015).

For the simplicity of discussion, we have separated the types of traits into three categories: input traits, agronomic traits, and output traits. Input traits address the need for farmer inputs into the cropping system (e.g., an insect-resistant trait eliminates the need to apply insecticides in the field). Agronomic traits improve crop productivity by modifying intrinsic physiological properties of the plant (e.g., abiotic stress tolerance and improved yield). Output traits provide consumer-oriented benefits (e.g., enhanced nutritional quality). Some authors include herbicide, disease, and insect tolerance under the category of agronomic traits. Here we consider these traits as input traits since even though these traits improve yield performance; they do so without significant modifications to plant physiology.

16.2.1   Input Traits

The first herbicide-tolerant traits, glyphosate-tolerant Roundup Ready® from Monsanto and glufosinate-tolerant Liberty Link® from Bayer, were commercialized in the USA in the mid-late 1990s, enabling farmers to control weeds in maize, soybeans, cotton, and canola with glyphosate or glufosinate herbicides without injury to the crop plants. HT traits experienced broad farmer adoption such that by 2016, 94% of soybeans and 89% of both cotton and maize grown in the USA were genetically modified to be tolerant to one or more herbicides (USDA-ERS 2016).

In 2015, HT crops were planted on nearly 240 million acres globally with an increased value to farmers of $8 billion and a cumulative value of $63 billion for the 19 years since commercialization (1996–2014) (ISAAA 2016). However, the broad adoption of glyphosate-tolerant traits has more recently led to some weed species developing resistance to the herbicide (Heap 2014). This has created a new challenge for researchers, driving the development of a number of new herbicide-tolerant traits which began entering the market in 2016, including tolerance to dicamba (soybeans and cotton from Monsanto), 2,4-D (maize, soybeans, cotton from Dow AgroSciences), imidazolinone (soybeans from BASF and Embrapa), and HPPDs (soybeans from Bayer, MS Tech, and Syngenta). For most of these new HT traits, commercial products are stacks of multiple herbicide-tolerant traits to provide farmers with more options for managing hard-to-control weeds on their farms.

Insect-resistant traits in maize and cotton were also first commercialized in the mid-late 1990s, with the first products carrying single lepidopteran insect resistance genes derived from *Bacillus thuringiensis* (Bt). Bt is a soil bacteria that produces proteins, often referred to as Bt toxins that are toxic to specific classes of insects. A key advantage to farmers from the use of IR traits is the season long protection they provide against the target pests. With the rapid adoption of HT traits, single gene IR products quickly progressed to stacks of IR + HT traits. In the mid-2000s, Bt genes providing control of corn rootworm, a coleopteran insect pest of maize, were introduced. Following the same stacking trends, the coleopteran IR traits were rapidly converted to stacked products with lepidopteran IR and HT traits.

Depending on the geography and specific regulatory requirements, farmers planting IR crops are often required to plant a refuge, which is a portion of the crop without IR traits to serve as a refuge for insects to reproduce without selection for resistance to the IR trait (Huang et al. 2011). Refuge requirements can range from 20% to 50% of the crop area for single IR traits, with specific requirements for how the structured refuge areas are to be laid out relative to the IR field. A breakthrough in streamlining the management of refuge fields came in the late 2000's with multiple IR mode of action stacks in maize and cotton. These stacks, often called pyramids, have two or more IR genes with different modes of action against the same key pest with the advantage of providing greater durability against the development of resistance by the insect pests (Storer et al. 2012). They also enabled advantages for farmers because with the greater durability, less refuge was required, and in many cases, it could be planted intermixed with the IR seed. For this reason, "refuge in a bag" products were swiftly adopted by farmers due to the ease in managing refuge compliance.

Commercial IR (including IR + HT) trait products were found to be very effective, even under heavy insect pressure, and thus were broadly adopted across the Americas. The percentage of maize acres in the USA that are planted to IR traits has grown from 19% in 1997 to 79% of the 90 million acres grown in 2016, while IR cotton acres have increased from 15% in 1997 to 84% of the 10 million acres grown in 2016 (USDA-ERS 2016). In Brazil, adoption has grown even faster, with IR maize introduced commercially in 2008, and by 2015 85% of the 36 million maize acres were planted to IR traits. Similarly, IR cotton in Brazil has grown to 73% of the total cotton acres since it was first introduced in 2006 (ISAAA Brazil 2017). Due to the need for some amount of refuge for IR traits, adoption rates can never reach 100%. Commercially available IR traits had been limited to maize and cotton until 2013 when IR soybean and IR eggplant (brinjal) were commercialized for farmers in Brazil and Bangladesh, respectively. Adoption of IR soybeans in Brazil has grown dramatically since launch in 2013, with nearly 40% of the 80 million soybean acres planted in 2015 devoted to IR soy. Overall, in 2014 all insect-resistant crops globally provided an increased value of $9.8 billion, with a cumulative value of $86.9 billion over the 19 years of commercialization from 1996 to 2014 (James 2015).

16.2.1.1 Major Crops with Biotechnology-Based Input Traits

Maize

The earliest commercially grown GM maize traits were single gene traits expressing either HT or IR genes separately. In 1996, Bayer launched Liberty Link® maize with tolerance to glufosinate herbicide and Mycogen Seeds introduced event 176, the first maize providing resistance to lepidopteran insect pests. One year later in 1998, Monsanto launched Roundup Ready® maize with tolerance to glyphosate as well as the first stacked IR + HT maize product, YieldGard® + Roundup Ready®. Similar IR + HT products soon followed from other GM trait developers, namely, Dow AgroSciences and DuPont Pioneer (Herculex®) and Syngenta (Agrisure®). Each of the maize trait products has slightly different expression characteristics or features depending on the expressed genes.

GM maize can be divided into three groups of products for specific markets: HT only, lepidopteran IR + HT (often referred to as aboveground IR), and lepidopteran + coleopteran IR + HT (above- and belowground IR). HT only maize is primarily used as a refuge for IR products and in some niche markets where insect pressure is low. Lepidopteran IR + HT products target aboveground insects such as European corn borer (*Ostrinia nubilalis*), corn earworm (*Helicoverpa zea*), and fall armyworm (*Spodoptera frugiperda*), which are pests of maize globally. Lepidopteran and coleopteran IR + HT products target the aboveground pests as well as the belowground pest, corn rootworm (*Diabrotica virgifera*), which is primarily a pest in North America. Today, most of the IR products on the market are, in addition to being stacks with HT, trait pyramids with multiple modes of action targeting the key pests. Most of the current pyramid products have been achieved by creating breeding stacks that combine traits from two or more trait developers, for example, SmartStax®, developed by Monsanto and Dow AgroSciences, is a stack of four events, two from Monsanto and two from Dow AgroSciences, resulting in a product with three lepidopteran IR traits, two coleopteran IR traits, and two HT traits. Multiple IR + HT products have been developed by DuPont Pioneer (AcreMax®) and Syngenta (Agrisure®) as well, using similar stacking approaches.

GM maize was planted on 130 million or 29% of the 450 million acres of maize grown globally in 2015. The top 3 GM maize-producing countries in acres and percent adoption were the USA (82 million, 92%), Brazil (32 million acres, 89%), and Argentina (7.2 million, 70%), with 14 additional countries each producing 5 million acres or less of GM maize in 2015. Cumulative income benefits to farmers for the years 1996 to 2014 total $50.6 billion (ISAAA Crop 2017).

Soybean

The glyphosate-tolerant Roundup Ready® trait by Monsanto has been the predominant soybean trait since it was commercialized in 1996. By 2000, more than 50% of the US soybean acres had the trait, and by 2007 adoption was above 90%, where it remains today (Fernandez-Cornejo et al. 2014). This widespread, rapid adoption has been seen in nearly all of the geographies where the HT soybean has been introduced. Additional soybean traits did not arrive on the market until 2009, with a second-generation glyphosate-tolerant trait delivering improved yield over the original trait (Monsanto) and glufosinate-tolerant Liberty Link® soybeans (Bayer). In 2013, Monsanto launched the first IR soybean, Intacta™, a single Bt gene conferring resistance to lepidopteran pests and stacked with the Roundup Ready® HT trait. Key lepidopteran pests of soybean, in particular soybean looper (*Pseudoplusia includens*) and *velvetbean caterpillar* (*Anticarsia gemmatalis*), are a significant problem in South America but are not of widespread concern in the USA, and thus IR soybean has not been commercialized in the USA but has predominantly been commercialized in South America. A number of new GM traits in soybeans have been recently launched or are expected to be on the market soon, which include several new HT traits: dicamba (Monsanto), 2,4-D (Dow AgroSciences), imidazolinone (BASF and Embrapa), and HPPDs (Bayer, MS Tech, and Syngenta).

Driven by the extensive adoption of Roundup Ready® soybeans globally, in 2015 GM soybeans accounted for just over half of all the GM crop acres in the world.

The top GM soybean-growing countries with over 90% adoption in 2015 were the USA with 80 million acres, Brazil with 75 million acres, and Argentina with 52 million acres. These three countries, along with eight additional countries growing GM soybeans in 2015, amounted to 227 million acres, or 83% of the 274 million total soybean acres. The income benefit for farmers growing GM soybeans from 1996 to 2014 has been calculated to be $47.8 billion dollars (ISAAA Crop 2017).

Cotton

In cotton, IR + HT traits have become a mainstay in commercial production globally with 75% of the world's cotton acreage (59 out of 79 million acres) planted with GM traits in 2015. GM cotton is grown in 15 countries: the top three being India with 29 million acres, China with 9 million, and USA with 8 million acres in 2015. The cumulative value to farmers in the 19 years from 1996 to 2014 was $46.5 billion (ISAAA Crop 2017).

The first GM IR cotton (Bollgard®, Monsanto) was launched in the USA in 1996 with HT (Roundup Ready®, Monsanto) and the IR + HT stacked product (Bollgard + Roundup Ready®) launching a year later in 1997. Rapid adoption of GM traits in cotton led to more than 90% of US cotton acres planted to GM cotton by 2010 (Fernandez-Cornejo et al. 2014). After glyphosate tolerance, the next HT trait introduced was Bayer's glufosinate-tolerant Liberty Link® cotton in 2004, followed by two new HT traits, Xtend® dicamba-tolerant cotton from Monsanto and Enlist™ 2,4-D-tolerant cotton from Dow AgroSciences, commercialized in 2015 and 2016, respectively. As with other crops, the newer traits are being commercialized as stacked products with IR and multiple HT traits to provide farmers greater flexibility in controlling weeds and pests in their fields. The IR traits in cotton are targeted at lepidopteran pests, such as tobacco budworm (*Heliothis virescens*) and cotton bollworm (*Helicoverpa armigera*). Similar to stacking in maize, additional cotton IR traits with multiple modes of action against the key pests were developed by mid-2000 with Bollgard II® (Monsanto) and WideStrike® (Dow AgroSciences). Those products were also stacked with HT traits. Cotton products being commercialized today are breeding stacks of traits from different companies to combine even more IR traits in stacked combinations with multiple HT traits.

Canola

GM canola was planted on 21 million acres globally in 2015, comprising 24% of the world's canola acres. Canada, USA, and Australia are the primary areas of production, and the cumulative value in terms of the farmer's income benefits from GM canola was $4.9 billion (1996 through 2014) (ISAAA Crop 2017). GM traits in canola have been limited to HT and male sterility to date. Glufosinate-tolerant InVigor® canola was launched by Bayer in Canada in 1996, and glyphosate-tolerant Roundup Ready® canola was introduced by Monsanto in 1997. Bayer also incorporated a GM male sterility system with their HT canola trait to assist in production of hybrid canola seed. New glyphosate tolerance traits and glyphosate + glufosinate stacked products are nearing launch by several trait developers (Monsanto, DuPont Pioneer, Bayer). Dicamba-tolerant canola to provide a new herbicide mode of action HT trait is in earlier stages of development in Monsanto's trait development pipeline. The important insect pests of canola do not include the lepidopteran pests that Bt traits control today, and thus currently there are no IR canola traits.

Other Crops

GM input traits have been commercialized in several smaller crops as well. Traits that confer resistance to specific diseases have been commercialized in papaya, potato, plum, and squash with mixed success. Some disease resistance traits in these smaller crops have enjoyed wide market penetration and are credited with saving particular cultivation industries (e.g., papaya ring spot virus resistance trait by University of Hawaii and Cornell Univ.); others failed to achieve significant market share (e.g., potato virus Y and potato leaf roll virus in NewLeaf® potato). Herbicide-tolerant traits in smaller crops have also been met with challenges. The USDA deregulations of Roundup Ready® sugar beets (Monsanto) and Roundup Ready® alfalfa (Monsanto) were both challenged in the courts after the products were initially launched, forcing a hold on commercial sales until those challenges were resolved. Today both traits are grown commercially and enjoy wide adoption by farmers. IR eggplant (Mahyco) was first commercialized in Bangladesh in 2014, but other key markets, specifically India and the Philippines, have been met with challenges by critics of the technology and thus have not yet approved the product for sale.

Male Sterility

Hybrid crops, such as maize and canola, take advantage of heterosis to increase crop yields but also require additional inputs to produce the hybrid seed. Production of hybrid seed requires cross-fertilization of two parental lines using approaches ranging from hand detasseling to exploiting native male sterility systems. However, more recently GM male sterility systems have been developed for these crops, the first of which was the barstar/barnase system in canola (Bayer). In this system, the barnase gene confers male sterility by preventing pollen production, and the barstar gene inhibits barnase to restore fertility. DuPont Pioneer has developed a GM male sterility system in maize, termed Seed Production Technology (SPT), which combines male sterility with a seed color marker enabling segregation of the transgenic male sterile maize from the desired non-GM fertile hybrid seed (Wu et al. 2016). Monsanto is also working on a GM male sterility system named RHS in which a transgenic plant produces non-transgenic pollen that is killed by the application of glyphosate (Feng et al. 2014).

16.2.2 Agronomic Traits

For a variety of reasons, agronomic traits have not enjoyed the same level of market penetration in major crops as input traits to date. Nevertheless, there are specific examples of successful products on the market and some compelling traits in late stages of development that are approaching launch. As technology and knowledge of plant biology overcome current challenges, it is expected that an increasing number of these types of traits make it to the market.

Starting in the late 1990s with the advent of the first complete plant genome sequence (*Arabidopsis thaliana*), significant investments were made by multiple biotechnology start-ups (e.g., Paradigm Genetics, Ceres, Inc., Mendel Biotechnology, Cereon, Crop Design, etc.) and large multinational agricultural companies (e.g., Monsanto, Bayer, BASF, DuPont Pioneer, Syngenta, etc.) in the field of functional genomics. Thousands of genes were identified and then systematically mis-expressed (e.g., overexpression or antisense expression) as transgenes in *Arabidopsis thaliana* and other model and crop species. These transgenic events were then tested for their ability to confer tolerance to abiotic stresses, improved performance under nutrient-limiting conditions, or improved growth characteristics under non-limiting conditions using a variety of approaches.

As a result of these efforts coupled with the ongoing work of academic scientists, a large number of candidate genes were identified and evaluated in crops of interest such as maize and soybean for their commercial product potential. Of these hundreds of candidate genes, several advanced far enough in company product development pipelines to become publicly known through investor presentations and scientific publications (e.g., cspB and Nfy-B drought tolerance leads developed by Monsanto and AlaT, a nitrogen use efficiency lead developed by Arcadia Biosciences). However, as shown in Table 16.1, only two have been successfully commercialized to date. One is a cold shock protein from *B. subtilis* (CspB) that is marketed as Genuity® DroughtGard® (MON87460) (Castiglioni et al. 2008). This trait was planted on 810,000 hectares in 2015 and was donated by Monsanto to the public-private partnership, Water Efficient Maize for Africa (WEMA). It is expected to be available for African farmers in select countries in 2017 (James 2015). This trait is being stacked with current IR + HT products to provide farmers with additional yield protection. The second is an endo-1,4-β-glucanase from *A. thaliana* (*cel1*) that is expressed in *Eucalyptus* spp. to increase woody biomass (Shani et al. 2003). This trait has been brought to market by FuturaGene Group and was approved for cultivation in Brazil in 2015.

As discussed in the previous section, commercialization of input traits has enjoyed a great success. However, those traits act independently without interfering in plant endogenous cellular processes (e.g., CP4 EPSPS confers tolerance to glyphosate due to decreased binding affinity for the herbicide and Bt toxins act through the formation of a pore in insect midgut epithelial cells). On the other hand, agronomic traits exert their effects through interactions with endogenous cellular processes such as nutrient utilization or stress response pathways. A beneficial

Table 16.1 Commercialized GM crops

Trait type	Crop	Trait description	Developer	Availability
Input traits disease resistance	Papaya	Virus resistance	Cornell University, South China agricultural university	Commercial[a]
	Plum	Virus resistance	USDA ARS	Not launched[b]
	Potato	Virus resistance	Simplot, Monsanto	Commercial
	Squash	Virus resistance	Monsanto	Commercial
Input traits herbicide tolerance	Alfalfa	Glyphosate tolerance	Monsanto	Commercial
	Canola	Glufosinate tolerance	Bayer	Commercial
		Glyphosate tolerance	Monsanto	Commercial
	Cotton	2,4-D tolerance	Dow AgroSciences	Commercial
		Dicamba tolerance	Monsanto	Commercial
		Glufosinate tolerance	Bayer	Commercial
		Glyphosate tolerance	Monsanto, Bayer	Commercial
	Maize	2,4-D, 'fop tolerance	Dow AgroSciences	Commercial
		Glufosinate tolerance	Bayer	Commercial
		Glyphosate tolerance	Monsanto	Commercial
	Rice	Glufosinate tolerance	Bayer	Not launched
	Soybean	2,4-D tolerance	Dow AgroSciences	Not launched
		Dicamba tolerance	Monsanto	Commercial
		Glufosinate tolerance	Bayer	Commercial
		Glyphosate tolerance	Monsanto	Commercial
		Isoxaflutole tolerance	Syngenta	Not launched
		Mesotrione tolerance	Syngenta and Bayer	Not launched
		Sulfonylurea tolerance	BASF	Commercial
	Sugar beet	Glyphosate tolerance	Monsanto	Commercial
Inputs traits insect resistance	Cotton	Lepidopteran resistance	Bayer, Dow AgroSciences, Monsanto, Syngenta	Commercial
	Eggplant	Lepidopteran resistance	MAHYCO	Commercial
	Maize	Lepidopteran resistance	Dow AgroSciences, DuPont, Monsanto, Syngenta	Commercial
		Coleopteran resistance	Dow AgroSciences, DuPont, Monsanto, Syngenta	Commercial
	Potato	Lepidopteran resistance	Monsanto	Sales ended[c]
		Coleopteran resistance	Monsanto	Sales ended
	Soybean	Lepidopteran resistance	Dow AgroSciences, Monsanto	Not launched, commercial

(continued)

Table 16.1 (continued)

Trait type	Crop	Trait description	Developer	Availability
Male sterility	Canola	Male sterility system	Bayer	In use[d]
	Maize	Male sterility system	DuPont, Monsanto	In use, not launched[e]
Agronomic traits	Maize	Drought tolerance	Monsanto	Commercial
	Eucalyptus	Volumetric wood increase	FuturaGene group	Commercial
Output traits	Alfalfa	Altered lignin	Monsanto	Commercial
	Apple	Non-browning	Okanagan	Commercial
	Maize	Modified alpha-amylase	Syngenta	Commercial
		Increased lysine	Renessen	Sales ended
	Pineapple	High lycopene	Del Monte	Commercial
	Potato	Altered starch	BASF	Sales ended
		Reduced acrylamide	Simplot	Commercial
	Soybean	Modified oil	Monsanto	Commercial
		Modified oil/fatty acid	DuPont	Commercial
	Canola	Modified oil/fatty acid	Monsanto	Sales ended
		Phytase production	BASF	Sales ended
	Tomato	Delayed fruit softening	Monsanto	Sales ended

Data compiled from http://www.isaaa.org/gmapprovaldatabase/default.asp and http://cera-gmc.org/GMCropDatabase

[a]Commercial indicates trait is commercially available at the time of this writing

[b]Not launched indicates most of the regulatory approvals have been obtained, but product has not yet been made commercially available

[c]Sales ended indicates the trait was previously available commercially but has been removed from the market

[d]In use indicates that the male sterility system is currently in use but is not a commercial product for farmers to purchase

[e]In use, not launched indicates that the DuPont male sterility system in maize is currently in use but is not a commercial product for farmers to purchase. The Monsanto male sterility system is still awaiting final regulatory approvals prior to use

effect on crop performance in the case of agronomic traits depends on effective modification of a complex system. A primary challenge in the commercialization of agronomic traits is identifying target genes capable of consistently delivering a significant performance improvement across diverse genetic backgrounds (in elite commercial germplasm) and across diverse environmental conditions.

Much is known on the molecular, biochemical, and physiological level about plant responses to stress and about the source-sink relationships that impact yield. However, the precise system perturbations that are required to fine-tune a plant response or redirect metabolic flux onto preferred pathways, for example, without resulting in undesired changes or no change at all is not always well understood. For this reason, many of the candidate genes that show promise in model systems or under controlled laboratory conditions do not provide consistent results when tested in different genetic backgrounds under field conditions.

From a logical point of view, improving plant performance (agronomic traits) relies on the assumption that either native plant responses to environmental conditions are not optimized to maximize economic yield or that native plants lack certain characteristics that would be beneficial to yield. That the biotechnology industry has not been more successful in delivering agronomic traits to market is not an indication that the solution is intractable. It merely indicates that technical capabilities to make molecular modifications to plants have temporarily exceeded the understanding of the biological system.

The continuous development of new methods for measuring the influence of the genome on the phenome will eventually enable more sophisticated approaches that more precisely control transgene expression or combine the effects of multiple transgenes, for example, to deliver traits with sufficient impact to be economically viable. A recent paper by Sun et al. (2017) where a potential trait gene only delivers beneficial effects when its expression is spatially restricted is indicative of the increasing levels of sophistication that will be required. In this study, the maize *PLASTOCHRON1* gene, which is involved in the regulation of cell division, was driven by a *GA2-oxidase* gene promoter, which is preferentially expressed in the growth zone where there is a transition from cell division to cell expansion of the leaf. The resulting transgenic events demonstrated increased plant height and leaf area with positive impacts on overall plant biomass and yield. However, when the PLA1 gene was expressed with a strong constitutive promoter (*UBIL*), severe developmental abnormalities ensued including failure to flower (Sun et al. 2017). Whether traits such as these are delivered using what may now be considered as traditional transgene technology or using newer gene editing tools such as zinc finger nucleases (ExZACT™) or CRISPR-Cas will depend at least in part on whether the target genes are present in the crop species. These genome-editing tools and their use will be discussed later in this chapter.

Given the diversity of environments and germplasm backgrounds that an agronomic biotechnology trait will encounter, perhaps it is unrealistic to expect the same kind of cross-crop and broad geographic penetration of particular traits that have been seen for input traits. If this limitation on agronomic traits is fundamental, their development will have to be tailored to germplasm and environment niches which will decrease the potential market size thus negatively impacting the trait valuation. Compensatory decreases in other product development costs would be needed in order to warrant investment by the agricultural biotechnology industry.

16.2.3 Output Traits

In agronomic traits multiple target genes have been identified that are involved in key physiological processes, but the precise perturbations required to deliver a quantum change in economic yield across germplasm and environments remain for the most part elusive. In contrast, output traits in most cases target metabolic endpoints or key effector proteins in accessible and well-defined pathways. Examples include oil, starch, amino acid, and antioxidant biosynthesis as well as antigens and ripening signals. Modifications to these pathways are designed to deliver characteristics beneficial to consumers that can be grouped into several categories: enhanced nutritional content, food/feed safety, and forage quality.

While adoption of agronomic traits remains largely a technical challenge in generating products with desired effects, the delivery of output traits is primarily a market challenge. In the first instance, there is the problem of public acceptance of new GM products with the Flavr Savr™ tomato (developed by Monsanto) being a well-known example. Amflora® potatoes (developed by BASF) with a modified starch content favorable for industrial starch production also experienced a short commercial lifespan due to public concerns in the European Union. New attempts that will test public acceptance of GM produce have recently been launched including the Arctic® family of apple products featuring a non-browning trait (developed by Okanagan) that will debut in the Midwest US market in 2017 and the Innate® family of potatoes (developed by J.R. Simplot Co.) launched in 2015 that feature non-browning, black spot bruise resistance and reduced acrylamide formation potential as consumer benefits.

In commodity crops, there are a few examples of output traits that have been commercialized in "closed loop cultivation" including Plenish® and Vistive® Gold and high oleic soybean varieties from DuPont Pioneer and Monsanto, respectively. Another example is the Enogen® maize trait from Syngenta that uses an alpha-amylase enzyme to improve starch breakdown for bioethanol production (Urbanchuk et al. 2009). Many others have been developed, but not yet commercialized (e.g., high omega-3 canola (Walsh et al. 2016) and vitamin A-enriched "golden rice" (Stone et al. 2017), and some have been discontinued (Laurical™ canola, which is enriched for the fatty acid laurate and Phytaseed™ canola, which expresses an enzyme to degrade phytate developed by Monsanto and BASF, respectively). Of all the non-input traits that have reached product launch, 11 out of 13 are output traits (Table 16.1).

The National Academies report (National Academies of Sciences 2016) speculates that, "Many potential future genetically engineered traits are predicted to be output traits, engineered specifically to change the quality of a crop. Most output traits developed soon will probably not require the use of chemical agents and should not require substantial changes in agricultural practices other than the requirement for identity protection and control of gene flow."

A more fundamental problem to the industry is the potential return on investment for products that address specialty or niche markets. The return has to be weighed against the significant investments in product development and deregulation associated with bringing a biotechnology trait to market. In most cases, consumer-oriented output trait products will exist in the market place alongside traditional products and must be kept in separate distribution channels to preserve identity and value. The added effort associated with this means the market must be of sufficient size to warrant the investment. For output traits that deliver broadly recognized consumer value, it may be possible to convert the distribution channels such that the biotechnology product predominates and identity preservation is no longer necessary. However, this is unlikely to be a common occurrence.

16.2.4 Predicting Traits in the Near Future

Projecting the traits of the future is inherently challenging due to limited availability of public information relating to industry R&D pipelines. A recent paper by Parisi and coworkers (Parisi et al. 2016) outlines an exhaustive approach that relies on public databases that collect information about GM crops, databases from government regulatory agencies, information available on company websites, and an international workshop convened in 2014 with representation from key constituencies to validate and correct the gathered information. Given the limitations of the digital resources available, the vetting of compiled information by a body of industry and government representatives is the only reliable way to ensure the quality of the data. However, this approach is not easily replicated. Since it was published in 2016, it is not anticipated that significant changes in the forecast have accrued by the time of this publication. By gathering information on the biotechnology events at several stages of product development (commercial cultivation, pre-commercial, regulatory, advanced R&D, and early R&D), the authors were able to generate a prediction of the biotechnology products that may become commercialized within the next several years. Remarking on their findings, the authors state, "The number of GM events at the commercial cultivation, pre-commercial or regulatory stages has more than doubled between 2008 and 2014. Although current GM commercial varieties and the outlook for 2020 are still dominated by a few arable crops (usually for feed or industrial use) and certain [input] traits, there is a nascent growth in quality traits, with a focus on bio-fortified food and industrial applications. Also, more specialty crops are being introduced into the pipeline and bean, rice, potatoes, and sugarcane may be cultivated by 2020" (Parisi et al. 2016).

A report compiled by the National Academies of Science, Engineering, and Medicine (National Academies of Sciences 2016) offers a more circumspect tone and an important disclaimer on predicting the biotechnology crops of the future stating, "It is not possible to predict with certainty the traits that will and will not make it to market or be diffused through nonmarket mechanisms in the future. The outcome will depend on environmental challenges that need to be addressed (for example, climate change), political-economic drivers, the regulatory landscape, and the rate of scientific advances, which is in part a function of the availability of public and private science funding."

Further investment and progress in at least one category, output traits, may depend heavily on how well the new produce (apple and potato) and commodity (high oleic acid soybean) crops are received by the public. Agronomic traits that improve farm productivity may need to wait on further advancements in the understanding of how the genome exerts its influence on the phenome before we see many new products on the market. Targeted opportunities to use transgenic biotechnology to deliver disease resistance are likely to be pursued so long as the business valuation of the trait exceeds the development costs. Overall, these differences in outcome merely underscore an important point of emphasis in any discussion of biotechnology traits – the market will pick the winners regardless of how clever, sophisticated, or well-adapted a particular technical solution may be. The costs associated with the discovery, development, deregulation, launch, and maintenance of biotechnology traits simply demand a market share or product premium that is commensurate. Traits that cannot meet those hurdles will not be commercially viable.

16.3   From Random Gene Insertions Toward Designer Crops

As we have discussed, significant progress has been made in the development and commercialization of transgenic plants in a large number of crops. However, random gene insertion has been the main approach for expressing foreign genes in plants. Multigenic products have been generated as breeding stacks where multiple transgenic events are crossed to bring all the genes together in one plant. However, since all the transgenes being combined are present in multiple events and are located at random locations in the genome, it has been challenging for the plant breeders to introgress these genes into elite varieties for product development. Moreover, random gene insertions have been used to produce desired traits in crop plants, but it is not effective in modifying any existing gene. Therefore methods for precise gene addition or modification at predetermined locations of the genome were required. Significant progress has been made in this field of study as well and a wide array of precision genome modification technologies are available today. These technologies have been demonstrated to design the genomes effectively in both plant and animal systems.

Cre/loxP is one of the earliest systems that was discovered for modifications at a single locus in the genome (Sternberg 1978). The working principle of this system is simple since it requires only Cre recombinase to initiate recombination at the pre-engineered loxP site and does not need any other cofactor for the reaction (Nagy 2000). This system has been shown to work effectively for targeted gene insertion or deletion in plant as well as animal systems (Vergunst et al. 1998; Schaart et al. 2004; Jia et al. 2006). However, mainly the system has been used for removal of selectable marker cassette from transgenic plants, such as by Monsanto and Rennessen for removal of the *nptII* gene from the high lysine maize event, LY038 (Lucas et al. 2004). Later an analogous system called FLP/FRT was discovered. It is analogous to Cre/loxP system and has similar applications (Luo et al. 2002; Li et al. 2010). Though these systems can be used for continued gene additions and removals at a pre-designed locus, these have not been able to modify the existing sequence in the genome. In addition to the above two, some other recombinase systems were also discovered which did not become very popular. A brief overview of those is nicely described by Wang et al. (2011).

A revolution in the field of precise gene modification, insertion, gene stacking, or removal came with the adaptation of site-directed nucleases, i.e., meganucleases, ZFNs (zinc finger nucleases), TALENs (transcription activator-like effector nucleases), and CRISPR/Cas (clustered regulatory interspaced short palindromic repeats/CRISPR-associated). Though the mechanism of all these nucleases is different, they all introduce double-strand breaks in the genome in specific targeted sequences. In response to this break, the host cell initiates its double-strand break repair mechanism. Scientists exploit this mechanism by inserting donor DNA into the cell which either gets integrated at the double-strand break site or can be used as a template for precise modification of a single or few base pairs.

The first nucleases discovered in this category were meganucleases. Meganucleases, also known as homing endonucleases, can be divided into multiple families. However, the LAGLIDADG family of meganucleases are the most studied and have been used extensively for gene targeting (Silva et al. 2011). These nucleases recognize DNA sequences ranging from 12 to 40 bp long and then insert a double-strand break. Their high degree of specificity provides higher accuracy and lower cellular toxicity. However, since a long sequence of DNA is recognized by meganucleases, it leaves very few sites in the genomes that can be modified (Rinaldo et al. 2015). The redesigning of these nucleases to read new target sequences has also been a challenge since the DNA recognition and cleavage functions of these enzymes are present in the same domain. Any changes to the DNA-binding domain may affect the cleavage activity of the enzyme (Chandrasegaran et al. 2016). Therefore, although reengineering of some meganucleases has been done to recognize new sites, largely the method has been very cumbersome and complex.

Another class of nucleases with greater flexibility are the ZFNs. These were based on the discovery of the zinc finger DNA-binding domain in a large number of transcription factors providing them the DNA-binding specificity (Diakun et al. 1986). Each finger has its own unique recognition sequence which is provided by amino acids at position -1, $+2$, $+3$, and $+6$ relative to the start of the alpha-helix in the zinc fingers (Osakabe et al. 2015). Amino acids at these positions can be modified to alter its DNA recognition specificity. Therefore developing and joining multiple fingers in order to derive an array of fingers to recognize a desired target sequence became the basis of developing a DNA-recognizing protein. Further, these DNA-binding domains were combined with a non-specific cleavage domain from Fok1 restriction enzyme to generate sequence-specific cuts in the DNA. Though ZFN is an efficient method of introducing a double-strand break at the target sequence, it requires rigorous development and screening of ZFN arrays to find the efficient ones. Since their discovery, these ZFNs have been used in a large number of organisms for targeted genome modifications.

TALENs are another designed nuclease for generating double-strand breaks at target sequence. TALENS were discovered in the bacteria of genus *Xanthomonas*. *Xanthomonas* is a pathogen of crops like rice, pepper, cotton, and tomato. It secretes effector proteins (TALEs) into the cytoplasm of plant cells that binds the specific DNA sequences to modify the plant processes in order to make the plant more susceptible to infection (Nemudryi et al. 2014). The DNA-binding property of TALENs has been exploited for the development of site-specific restriction enzyme by attaching a non-specific restriction enzyme Fok1 to it. TALENS are similar to ZFNs in that they have a DNA-binding domain attached to the Fok1 domain, and both work in dimers. However the DNA-binding domain of TALENs is different from ZFNs. TALENs consist of a series of repeated domains, each of which is about 33–35 amino acids long. Most of these amino acids are highly conserved except at position 12 and 13. Amino acids at these two positions are highly variable and are responsible for target nucleotide specificity and can be modified to change the target recognition. TALENs are comparatively easier to build as compared to ZFNs due to its straightforward DNA interaction code and the modular nature of the array. However the challenge with TALENs is their large size (about 950 amino acids for each protein) and repetitive nature of the DNA sequence due to conserved sequence of the multiple domains joined together (Baltes et al. 2014). Therefore delivery of these proteins into plants becomes a challenge. Despite the challenges, TALENs have been used for targeted genome modifications of a large number of organisms.

Recently a new RNA-guided nuclease system called CRISPR/Cas (clustered regulatory interspaced short palindromic repeats/CRISPR-associated) was developed, and it has gained widespread attention in a short period of time. The system has been adopted from bacteria where it provides acquired immunity against invading nucleic acids such as bacteriophage and plasmids (Rinaldo et al. 2015). Bacteria acquire small fragments (called spacers) from invading DNA and incorporate them into the CRISPR loci. These CRISPR repeats along with the spacers are then transcribed into pre-CRISPR RNA (pre-crRNA) which is further processed to create a restriction enzyme that consists of a spacer-based guide RNA and a Cas enzyme. The guide RNA pairs with the invading DNAs and destroys it by generating double-strand cuts with the action of Cas enzyme. This mechanism was used to create an engineered CRISPR/Cas enzyme where guide RNA is designed to recognize the desired target sequence. Target recognition by this complex requires the presence of a protospacer adjacent motif (PAM) followed by crRNA recognition sequence on the target DNA (Gaj et al. 2013; Khatodia et al. 2016). Therefore, the system's only limitation is the required presence of a PAM sequence at the target site. Since its discovery, the system has been used extensively to obtain double-strand breaks in a wide range of organisms, and the literature is replete with reports mentioning the use of this technology.

These site-specific nucleases are transformative tools and are revolutionizing the entire field of biology. These nucleases can not only be used for targeted gene insertions or stacking but also for making small changes in the genome to generate desired traits. Though traits generated with targeted foreign gene insertions would be called transgenic, traits generated by inserting small changes in the endogenous genes have the potential to be considered non-transgenic. Plants regenerated with the use of these nucleases may contain the DNA from these nucleases in addition to the intended change. However these can be easily segregated out in subsequent generations, and plants homozygous for the intended change and free of any other unintended gene integration in the genome can be obtained in the progeny plants. Though these technologies have been extensively used and a large number of reports have been published, some selected examples that have commercial and economic importance are discussed below.

Examples of developing input traits using gene editing include ALS (acetolactate synthase) herbicide resistance. Chlorsulfuron and bispyribac are some of the herbicides that are used to control weeds in crops. These herbicides kill plants by inhibiting the activity of acetolactate synthase, an enzyme involved in amino acid biosynthesis. Transgenic lines containing mutated ALS enzyme have been generated that are resistant to these herbicides. Sun et al. (2016) took a different approach to regenerate resistant plants by editing the endogenous ALS gene instead of inserting a resistant ALS transgene into the crop plant. The researchers used the CRISPR/Cas system to edit this gene in rice callus. A donor fragment that had desired mutations in the sequence and had homology arms for homology-based repair of the endogenous gene was also transformed into the callus along with a CRISPR/Cas cassette designed to cut in the endogenous ALS gene. They successfully regenerated rice plants that showed tolerance to the application of herbicide. Leaves of the herbicide sprayed wild-type plants withered and died, while gene-edited plants showed complete resistance to the herbicide. Researchers showed the regeneration of homozygous herbicide-resistant rice plants in the T0 generation itself due to biallelic modifications created by CRISPR/Cas showing the specificity and effectiveness in recognizing the target sequence.

In a similar effort, Li and coworkers edited the rice genome to impart bacterial blight resistance. Bacterial blight is an economically important disease of rice since outbreak of this disease may lead up to 50% of the crop yield loss and may even go up to 70% in case of severe infections (Cernadas et al. 2014). OsSWEET14 is a bacterial blight susceptibility gene in rice. The effectors AvrXa7 and PthXo3 produced by *X. oryzae pv oryzae* bind to the effector binding element in the promoter of OsSWEET14 gene. This upregulates the gene which favors infection by this pathogen. Li et al. (2012) used TALENs and mutated the effector binding sites in the promoter of the OsSWEET14 gene. Inability of effectors to bind the target sequence resulted in bacterial blight resistant in the crop. Another desired trait in crop plants is the development of male sterility which is extensively used for the development of hybrids. However this trait is not easily developed in all the genotype backgrounds using traditional breeding programs. Djukanovic et al. (2013) designed a homing endonuclease to target a 22 bp sequence in the fifth exon of MS26 gene (a maize fertility gene) in corn. The enzyme led to targeted mutagenesis resulting in small deletions and insertions leading to the disruption of coding sequence. The mutation is recessive and the resulting homozygous plants for the mutation were male sterile.

Some output traits have also been generated using gene editing. Haun et al. (2014) used TALENs to disrupt the FAD2-1A and FAD2-1B genes, thus reducing the polyunsaturated fatty acid content in soybean oil. The resulting plants showed increased oleic acid from 20% to 80% and decreased linoleic acid from 50% to 4%. Reduced content of polyunsaturated fatty acids improves the shelf life of soybean oil. This eliminates the need of partial hydrogenation which is an industrial process and results in production of trans-fatty acids which are known for certain health risks. Similarly Shan and coworkers (2015) used TALENs to improve the aroma in rice grains. 2AP (2-acetyl-1-pyrroline) is the compound responsible for fragrance in rice, and BADH2 (betaine aldehyde dehydrogenase) inhibits the synthesis of 2AP. Researchers designed TALENs and disrupted the BADH2 DNA sequence. Homozygous lines in T1 and T2 generation showed increase in the levels of 2AP from 0.35 to 0.75 mg/kg. In another similar effort, TALENs were used to improve the cold storage of potato tubers. Low-temperature warehouses are used for potato storage to extend shelf life. However cold storage induces the accumulation of reducing sugars in potato tubers. When these tubers are processed at high temperature, these sugars react with amino acids and lead to brown, bitter tasting products and even increased levels of acrylamide which is a potential carcinogen. It is known that vacuolar invertase gene (Vinv) is responsible for accumulation of reducing sugars in potato. As discussed above, transgenes have been introduced to downregulate the expression of this gene. However, Clasen et al. (2016) used TALEN technology to knock out Vinv gene. Researchers showed that the chips made from modified potato contain reduced levels of acrylamide and were light in color.

Deciphering the gene function in polyploid crops has been challenging due to the presence of multiple homoeo alleles. In order to determine the function of the gene by a typical reverse genetics approach, all of them would need to be silenced. RNAi has been used to simultaneously knock down (mRNA degradation) multiple alleles; however this technique gives variable results, and gene silencing is not always complete. Therefore a gene knockout (coding sequence mutation) strategy would be more effective as compared to knockdown of the genes. Loss of function alleles (*mlo*) of MLO locus are known to provide broad-spectrum resistance against powdery mildew to barley (Piffanelli P et al. 2004). Wang et al. (2014) used TALENs to knock out the three homoeo alleles encoding MILDEW RESISTANCE LOCUS (MLO) proteins in hexaploid bread wheat. TALENs were designed to target a conserved region on exon 2 to create a simultaneous mutation in all the three alleles. The mutations were successfully created in all three MLO genes which conferred powdery mildew resistance to the plants. Recently the same strategy was used in tomato using CRISPR/Cas technology. The MLO alleles were mutated resulting in the regeneration of a powdery mildew-resistant plant which was named as Tomelo (Nekrasov et al. 2017). This shows the efficiency of these gene modification technologies in reading a specific DNA sequence in the genome and inserting a double-strand break.

Recently a gene-edited mushroom received widespread attention. The white button mushroom is prone to browning shortly after picking which reduces its market value. Waltz and coworkers from Penn State University manipulated the genome of this fungus using CRISPR/Cas technology (Waltz 2016). They mutated the polyphenol oxidase (PPO) gene that resulted in a delay in browning. They demonstrated that the target gene had been mutated, and no other gene fragment related to CRISPR/Cas has been integrated into the genome. When they enquired about the regulatory assessment needs of the modified mushroom, the USDA stated that since it does not contain any foreign sequence, no plant-pest sequence was used to create the intended change, and no foreign sequence was present in the resulting product in addition to the change that was induced; the agency does not consider that the product needs to be regulated (USDA 2016). This is one major step that would substantially ease out the commercialization of the gene-edited crops by reducing the timeline between the discovery of traits and releasing the crops for commercial cultivation in the field. Earlier, the USDA had given a similar determination to some other gene-edited crops as well, e.g., disease-resistant rice (Iowa State University), potato with better processing attributes (Cellectis), and reduced phytate corn (Dow AgroSciences) (Wolt et al. 2016). However, the USDA stated that when a template is inserted into the cell to repair a gene, the template may likely get integrated in the genome as well. Therefore, the regulatory requirements for any such product will be determined on a case-by-case basis. Though the USDA has opted out of the regulation of some such products, it is not known how the other agencies in the USA and rest of the world will treat the situation and what regulatory guidelines will be established to oversee such products.

Another similar development is the CRISPR-induced waxy maize developed by DuPont Pioneer. Normal field maize contains two types of starches, ~78% amylopectin and ~22% amylose, while waxy maize contains 100% amylopectin (Eriksson 1969). To develop waxy maize, scientists at DuPont Pioneer knocked out the Wx1 gene that encodes granule-bound starch synthase responsible for synthesis of amylose. As the USDA stated that CRISPR-/Cas-edited mushroom did not need regulatory approval, waxy maize edited by a similar methodology may also receive a similar finding of nonregulated status and may reach the market earlier.

16.4 Challenges Associated with Further Adoption of GM Crops

In the Americas, the USA, Brazil, Argentina, and Canada are the major producers of GM crops, including soybean, maize, cotton, and canola, representing 80% of the total global GM crop production. Elsewhere in the Americas, countries such as Paraguay, Uruguay, Bolivia, Mexico, Colombia, Honduras, and Costa Rica have each planted one or more of the major GM crops (James 2010). In Asia, Bt cotton is grown in India and China, accounting for the largest GM crop hectare plantings in the region, and GM papaya is widely adopted in southern China (James 2010; USDA FAS 2016a). GM cotton and canola have been adopted in Australia since 2008 (USDA FAS 2016b). Bt maize with stacked traits is grown in the Philippines for commercial use, and Bt brinjal/eggplant, the first locally developed GM crop which was developed through the USAID-ABSP support project, is poised for future commercialization (USDA FAS 2016c). GM maize with stacked insect resistance and herbicide-tolerant traits was planted for the first time in Vietnam in 2015 (ISAAA 2016). In Africa, Egypt, and Burkina, Faso joined South Africa in the adoption of GM crops by planting GM maize and Bt cotton, respectively, in 2008 (James 2010; USDA FAS 2015). In Europe, Bt maize, the only GM crop approved for cultivation, enjoyed low, but nevertheless, stable level of adoption, primarily in Spain (USDA FAS 2016d). The four major countries cultivating GM crops are also the leading exporters of soybean, maize, cotton, and canola. These nations trade internationally with the major destinations including China, EU, Japan, Mexico, and Southeast Asia depending on the products (USDA 2017).

Despite the economic, social, and environmental benefits of GM crops to global society (Qaim 2009; Anderson 2010; Carpenter 2013; Qaim and Kouser 2013;

Barfoot and Brookes 2014; Brookes and Barfoot 2015), adoption of GM crops in large parts of the world, such as Africa and Europe, remains compromised. This opposition is derived from a multitude of complex and intermingled concerns which have persisted ever since the adoption of GM foods and has as much to do with social and political values as with concerns about health and safety (WHO 2005). Some of the underlying concerns to general acceptance have themselves become a driving force for GM crop regulations and policy development. This section highlights risk and benefit perception of GM crops and food, regulatory and political development, international trade protection, and social concerns.

16.4.1 Risk Assessment of GM Crops and Food

Regardless of the method used (traditional breeding or recombinant DNA techniques) or the traits developed (herbicide tolerance, insect resistance, yield improvement, and/or improved nutritional value), for any crop with new traits, the potential exists for safety risks. Risks associated with GM crops, when the introduced genes and traits are safe, are no greater than conventionally bred crops. The process used to introduce genes into crops, with a history of safe use, is unrelated to risk (OECD 1986; White House OSTP 1986; US NAS 1987). While new varieties of conventionally bred crops are not usually subject to regulatory scrutiny for potential safety concerns prior to marketing, GM crops undergo risk assessment with extensive toxicological and nutritional evaluation.

Risk assessment, risk management, and risk communication are three components of risk analysis (Codex 2003a). Two international regulatory instruments, Cartagena Protocol on Biosafety (CBD 2000) and Codex principles and guidelines on foods derived from modern biotechnology (Codex 2003a, b), cover environmental safety of living modified organisms and GM food safety, respectively. The concepts and principles outlined in the Cartagena Protocol on Biosafety and in Codex (2003a, b) are intended to provide international consistency in the assessment of environment and food safety of GM crops. One of the risk assessment principles laid down in the Cartagena Protocol on Biosafety is that risks should be considered in the context of the risks posed by the nonmodified recipients or parental organisms in the environment. The concept of "familiarity" incorporated in environmental assessment of GM plants facilitates risk/safety assessments. The term "familiar" in this context expressly identifies the means to having enough information to be able to make a safety or risk judgment (CBD 2000).

The assessment approach to GM food safety is based on the principle, referred to as "substantial equivalence," that the safety of foods derived from new plant varieties, including GM plants, is assessed relative to the conventional counterpart having a history of safe use (Codex 2003a, b). Risk assessment of GM food is designed to identify whether a hazard (nutritional or other safety concern) is present, and if present, to determine its nature and severity. The safety assessment includes a comparison between the food derived from GM crop and the conventional counterpart, taking into account both intended and unintended effects. If a new or altered hazard is identified by the safety assessment, the risk associated with it is characterized to determine its relevance to human health.

GM crops, released for commercial use and traded on the international markets, have demonstrated that they are as safe and nutritious as conventional counterparts. To date, there has not been a single confirmed case of an adverse health issue for humans or animals due to consumption of approved GM products. To create awareness and to emphasize the safety of GM crops, in June 2016, 123 Nobel Laureates signed an open letter to the leaders of Greenpeace, the United Nations, and governments around the world supporting the efficacy and safety of GMO food products. In the letter, they reiterated conclusions made by scientists and regulatory agencies around the world that assert that crops and foods improved through biotechnology are safe.

16.4.2 Perception of Risk and Benefits

Over the years, various surveys across geographies have been conducted in an attempt to gain insights into public perception and acceptance of GM food. It was widely interpreted and generally accepted that public opposition to GM crops and food is due to a general misperception of potential risks (Gaskell et al. 2004; DeFrancesco 2013). Attempts to address the disconnect between real and perceived risk have led to strategies for broader communication and public education of the technology to clarify the true risk through communication from trusted independent sources.

Interestingly, survey results from regions where acceptance of GM food is low indicate lack of "perceived" benefits as an important factor leading to their mistrust (Gaskell et al. 2010; USDA FAS 2016d; WHO 2005). A study commissioned by WHO (2005) indicated that "people do not react so much to genetic modification as a specific technology, but rather to the context in which GMOs are developed and the purported benefits they are to produce." The survey conducted by the European Commission in 2010 appears to echo the same sentiment. The survey indicated that objections to GM food are related to concerns regarding safety in the context of a lack of perceived product benefit (Gaskell et al. 2010; USDA FAS 2016d). Modern medicines made from GMOs (bacteria and plants) are generally well received (WHO 2005), while GM foods continue to meet strong opposition in many parts of the world. Patients needing medical care place greater emphasis on the benefits of medicines. Modern medicines made from GMOs, such as insulin, growth hormones, and vaccines, come with added, but nevertheless important, benefits of affordability and availability.

Farmers who are dependent on abundant harvests view crop yield, efficacy, reduction in pesticide use, and overall input cost as primary benefits. First-generation GM crops with herbicide tolerance and/or insect resistance traits provided farmers economic benefits because of increased crop yield as a result of improved weed and pest management as well as reduced input costs. Farmers worldwide in both developed and developing nations where they are free to choose often embrace GM crops. The findings from these early studies showed farmers benefited from the traits in the first-generation GM crops as evidenced by their rapid adoption (ISAAA 2016). Studies comparing yields of adopters and non-adopters showed that smallholder farmers in developing countries have benefited the most, especially in terms of yield, averaging 16% increase in yield for insect-resistant maize, 21% for herbicide-tolerant soybean, and 30% for insect-resistant cotton (Carpenter 2010, 2013). Yield improvement plus reduced input cost drove the value-added profitability which in turn brings social welfare gains.

However, first-generation GM crops did not project the potential benefits easily to consumers. Crop yield impacts commodity price which eventually impacts the food price consumers pay at stores. It has been estimated that consumers realize a significant portion of the total economic benefit of the first-generation GM crops (Carpenter 2013). GM crops containing IR traits provide consumers potential health benefits in addition to cost benefits. IR crops require lower insecticide usage than conventional crops which are impacted by insect pests (Shelton et al. 2002; Qaim et al. 2008). This, in turn, results in lower pesticide residues in food and water. In some circumstances, GM traits can also directly benefit human health. For example, traits such as IR maize controls mycotoxin contamination caused by insect damage to plant tissues (Wu 2006; Qaim et al. 2008). Insect damage predisposes maize tissue to mycotoxin contamination as insect pests create pores through which fungal spores enter maize kernels. Field studies have demonstrated that IR maize contains significantly lower levels of certain mycotoxins, which can cause adverse health effects in humans and livestock (Wu 2006 and references therein). If lack of perceived benefits is an important contributor to low-level acceptance, second-generation GM crops with nutrient quality traits might improve the image, and therefore acceptability, of GM food.

16.4.3   Regulatory and Policy Development

Progression in public acceptance of GM crops and food has been generally less well-received than the developers of these products had anticipated. During the period of 1996–2010, there was an overall downward trend in the percentage of GM food supporters in Europe (Gaskell et al. 2010). In China, public attitude has turned from largely neutral to negative (Jayaraman and Jia 2012; Li et al. 2016). Several food scares and crises in Europe and in China unrelated to GM crops negatively impacted public confidence in food safety and trust in the regulatory bodies charged to protect consumers. Facing increasing consumer skepticism, low level of public trust, pressure from organizations opposing crop biotechnology, and political requirements, authorities in EU and China in particular resorted to legislative changes and regulation enhancements aiming to bolster public confidence in the regulatory processes evaluating GM crops.

However, these new regulatory measures resulted in delayed regulatory approvals of otherwise safe products which result in far-reaching consequences including affecting new product development and innovation, limiting farmer access to useful technologies, and stymieing international trade of these products. In Europe, several major private European plant biotechnology companies relocated R&D operations to the USA because of the more favorable regulatory climate in the USA (USDA FAS 2016d). In 2013, BASF withdrew the application for authorization of phytophthora-resistant potato for food and feed uses, processing, and cultivation. In his assessment of regulatory triggers for products of biotechnology, McHughen (2016) asserted "Science must form the foundation for effective regulation but it is not and should not be the sole determinant of public regulatory policy. Other considerations, such as social policy, ethics, economics, etc., maybe constructed upon the scientific foundation, but they should not drive public policy in the absence of scientifically sound foundation, any more than science alone should direct policy in the absence of these other aspects." The following sections highlight the evolution of the regulatory and policy developments in EU and China and potential impact on the adoption of GM crops.

16.4.3.1   European Union

European environmental policies in 1970s established a regulatory policy that was based on the precautionary principle. This principle emphasizes an awareness of scientific uncertainty about potential negative effects resulting from a phenomenon, product, or process (Freestone and Hey 1996). The concept was later adopted in Directive 2001/18/EC concerning authorization for cultivation, the first major change to EU biotechnology legislation since 1990. Also included in the directive is mandatory post-market monitoring. In the subsequent years, the European Parliament and the Council of the European Union released several regulations, including (1) Regulation (EC) No 1829/2003 concerning authorization for import, distribution, or processing, (2) Regulation (EC) No 641/2004 on implementation of Regulation (EC) No 1829/2003, (3) Regulatory (EC) No 1946/2003 concerning transboundary movements of GMOs, and (4) Directive (EU) 2015/412 allowing member states to restrict or ban the cultivation of EU-authorized GM plants in their territories for nonscientific reasons. These regulations are complemented by 11 guidance documents released between 2005 and 2015. The complicated regulatory procedures and voluminous data requirements delay regulatory submission, risk assessment, and product approval. Delay in bringing products to market and high regulatory cost have particularly large ramifications on continuous innovation and participation by public institutions and small private companies in product development.

Since 1997, EC regulation on labeling requires that products intentionally containing GM ingredients must be labeled, whatever the level of GM content. In 2003, the European Parliament and the Council of the European Union released Regulation (EC) No 1830/2003 concerning traceability and labeling. The traceability and process-based labeling requirements for all food and feed derived from GM plants are among the most demanding in the world. The labeling requirements are also applicable to highly refined processed oil and sugar in which no trace of introduced genetic material or protein can be detected. GM labeling was intended to give consumers the right to choose. However, amid the region's preference for "naturalness" and negative publicity from those who oppose GM food products, retailers started to avoid labeled food products to protect their market. Generally, requirements that demand traceability and identification of GM food products have a negative stigma attached that discourages their acceptance by retailers which in turn discourages farmer adoption. Romania was one of a handful of countries that adopted GM maize, the only approved GM crop, but farmers have chosen to grow a conventional variety in 2016, amid complex traceability rules (USDA FAS 2016d).

16.4.3.2 China

Facing a huge population and potential food shortages, China positioned agricultural biotechnology as one of the important strategic tools for food security. The government of China invested heavily in biotechnology research and seed development. In 2001, the State Council of China decreed a general policy for regulation of GMO biosafety titled "Regulations on Safety Administration of Agricultural Genetically Modified Organisms" which replaced the first biosafety regulation for agricultural biotechnology issued in 1993 (USDA FAS 2016a). Following the State Council Regulations, the Ministry of Agriculture (MOA) announced a series of implementing regulations. In 2009, MOA issued biosafety certificates for two Bt rice lines and GM maize expressing phytase, paving the way for production trials of GM products in China prior to commercialization. China was on the brink of commercializing a first genetically modified staple crop (rice) and a feed crop (maize) when public sentiment toward GM crops turned negative. Issuance of safety certificates for Bt rice and GM maize by MOA prompted outcries from professionals in humanities and social sciences who signed a public petition asking MOA to withdraw the biosafety certificates (Jia 2010). The petition states "the approval for the commercialization of GM rice and maize enables China to become the world's first country to plant a GM staple food, threatening the national safety." The petition represented one of the most high-profile challenges to China's policy toward the adoption of GM crops. In the following year, the news of golden rice tests in children provoked public outrage amid negative portrayal of the study intent by GMO opponents (Jayaraman and Jia 2012). The test, after a successful trial involving US adults, was designed to assess whether beta-carotene, a precursor of Vitamin A, would be converted efficiently to Vitamin A in children eating golden rice (Tang et al. 2012).

In 2016, MOA revealed a road map for commercialization of GM crops in China. The determined order of priority was as follows: cash crops not for food use, crops with input traits for feed and industrial use, food crops, and finally staple food crops including rice, wheat, and soybean (USDA 2016a). This order of priority indicated Bt rice commercialization would likely remain undetermined in the near future.

Also in 2016, MOA released a revised "Regulations on Safety Administration of Agricultural Genetically Modified Organisms" and guidelines pertaining to the conduct of risk assessment. China's regulatory procedures for GM crops for either cultivation or for importation for food and feed are complicated and lengthy.

This level of complexity has created a challenging environment toward successfully approving products in China. These requirements include approval from the country of origin prior to submission, in-country environment safety and food and feed safety studies, and multiple submissions and assessments of the same product. Presumably, the complicated process and additional requirements were intended to demonstrate the rigor of risk assessment and bolster public confidence in the regulatory system. However, by requiring data in excess of what is typically required for product approval in other nations, this bureaucratically complicated process may inadvertently reaffirm public perception that GM crops are inherently risky.

16.4.4 International Trade Concerns and Policy Decisions

Fear of trade-related impact and loss of access to export markets is another concern for adoption of GM crops (WHO 2005). In parts of Africa and Asia, there is a perception that avoiding cultivation of GM crops might give the region a marketing edge by guaranteeing that agricultural exports are "GM-free." This is especially true of European markets where consumer skepticism toward GM food is relatively high and regulatory climate is particularly challenging. A de facto moratorium imposed by the EU in 1998 on the importation of food products that might contain GMO followed by EU traceability and labeling rules implemented in 2003 did nothing but substantiate the concern that adoption of GM crops would result in a loss of European market access.

In the sub-Sahara region of Africa, the export risks are related to cash crops, including tea, coffee, sugar, bananas, and a wide range of horticultural products (Wafula and Gruére 2013). To date, no GM varieties for these African cash crops are available, nor will they likely become the main driver for commercial interest in the near future given the situation in Africa. Moreover, their fear is that the genetic elements in the GM crops might enter these indigenous crops. However, these fears are unfounded because these crops are not biologically compatible with these cash crops. Export risk was even cited as a reason for rejection of food aid during a famine situation in southern Africa in 2002 (WHO 2005; Gruére and Sengupta 2009). Several countries, including Zambia and Zimbabwe, were concerned that accepting food aid potentially containing GM maize could risk exports of organic vegetables and horticultural products to European markets. However, the perceived export risk is not fully supported by actual trade flows which show only a small trade volume of select products with countries outside Africa (Wafula and Gruére 2013).

In Asia, export concerns for fruits, in particular papaya, were heightened in Thailand when reports of possible gene escape from GM papaya field trials began circulating in 2005. Under pressure from exporters, the Thai Department of

Agriculture instituted a temporary moratorium on all GM field trials (Gruére and Sengupta 2009). In 2006, both the Vietnam food association and Thai rice exporters announced their decisions to ban the use of any GM rice coinciding with the widespread international rejection of US rice out of fear of GM contamination (Gruére and Sengupta 2009). These decisions were largely driven by the concern of rice exports to Europe and Japan. The Thai government responded to the decision by adopting a GM-free clause in the Thailand 2007–2011 rice strategic plan. Elsewhere in Asia, rice exporters in India supported a ban on GM rice for fear of losing market access to Europe and denounced GM rice field trials when the US rice situation was unfolding (Gruére and Sengupta 2009).

Anderson (2010) examined potential economic impacts of GM crop adoption in sub-Saharan Africa and Asia. The study considered several adoption scenarios of GM coarse grains, oilseeds, rice, wheat, and cotton by key countries with or without EU policy response as well as global full adoption. The analyses revealed economic welfare gains by countries willing to adopt GM crops and multiplication of economic gains if next-generation GM crops with traits alleviating nutritional deficiency were to be adopted. More importantly, economic benefits from GM crop adoption by countries in sub-Saharan Africa and Asia would not be greatly impacted by developed countries banning imports of agricultural products from the adopting countries.

16.4.5   Social Concerns

Social concerns involving agricultural biotechnology are complex. Among the concerns expressed by some is that GM crop adoption would put some consumers who assert their wish to maintain a GM-free diet in a position where they are unable to apply their values (Thompson 2000). This expressed concern leads to "right to know" so consumer preference can be considered when it comes to addressing a non-GM food choice. Currently, European regulation requires mandatory labeling of GM food products. The US law permits voluntary labeling of food products containing ingredients from GM crops, but labeling of GM foods is required only if the food has a nutritional or food safety property that is significantly different from what consumers would expect of that food (FDA 1992, 2015). There are important differences between the two labeling systems at technical and practical levels. Mandatory labeling is often used to warn consumers of specific health risks, while voluntary labeling is commonly used to differentiate products for marketing purposes (Qaim 2009). At a practical level, mandatory labeling requires food products containing any GM ingredients above a certain threshold for trace amounts to indicate their presence. There are no specific requirements for voluntary labeling. Labeling requires a system of market segregation and identity preservation which comes at a significant cost to the product. This layering of complex identification has implications on international trade of such products. In the EU, the high degree of complexity, uncertainty, and direct incurred cost because of the labeling and traceability rules provide no incentives to farmers to plant GM crops who are willing to adopt the technology (USDA FAS 2016d).

Despite nearly 20 years of cultivation of a variety of GM crops and intense research on the safety of those crops, there has yet to be any identified significant hazard directly linked to GM crops (Nicolia et al. 2013). Regardless, some remain concerned as to their safety (Verma 2013; Zilberman et al. 2013). Several of these concerns regard the linking of GM crops to potential adverse environmental impacts. The first of these concerns is the "weediness potential" in which engineered plants become agriculture weeds or invasive in natural habitats, displacing other crops or native plants. A second concern focuses on horizontal gene transfer, in which plant genes move into other organisms, genetically altering the compromised plant. Another identified concern is the potential for outcrossing between plants in which traits are transferred from GM crops to wild relatives. Finally, the impact on nontarget organisms, such as beneficial insects, which become exposed to insecticidal traits expressed in GM crops, has also been identified as an expressed concern (Shelton et al. 2002; WHO 2005; Verma 2013; Zilberman et al. 2013).

These concerns may have stemmed from non-GM related events and early preliminary studies involving GM plants. There are cases of non-GM human-released organisms, including plants and animals intended to be used as ornamentals or biological controls, that have become widely established and threaten indigenous organisms in many habitats worldwide (Stemke 2004). Preliminary environmental impact studies on nontarget organisms (Carpenter 2011) and the gene flow study concerning the wild relatives of maize in Mexico (Quist and Chapela 2001) heightened the awareness of potential negative impacts although the findings were unsupported upon further investigations (Carpenter 2011 and references therein). For example, although outcrossing with wild relatives is not unique to GM crops, concern for potential outcrossing of trait(s) from GM maize to traditional landraces and wild relatives of maize in Mexico evokes strong emotions (USDA FAS 2016e). A systematic review of the scientific literature spanning the years between 2002 and 2012 on GM crop safety has failed to detect any significant hazards through the use of GM crops (Nicolia et al. 2013). From a scientific point of view, potential outcrossing can be managed by spatial isolation as demonstrated by Baltazar et al. (2015). However, using the example of maize in Mexico, it is possible to illuminate the complexity involved. In Mexico, maize is a symbol of national heritage and holds culture and tradition values. These factors drive the reluctance to adopt GM maize for cultivation in Mexico, the center of origin for maize. This may explain why bringing GM crops into Mexico may not be easily overcome by local skepticism even if sound scientific persuasion is employed.

16.5   Conclusions

Ever since the development of transformation technology, scientists have made rapid progress. Transgenic traits have been generated in a large number of crops and have been adopted by farmers around the world. The main commercial crops have been soybean, maize, cotton, and canola with some acreage devoted to papaya and sugar beet cultivation. Most of the commercialized traits have been input traits that mainly include herbicide tolerance and insect resistance. Development of agronomic and output traits have been challenging either due to complex gene interactions that need to be managed to confer such traits or due to poor public acceptance. So far, transgenic crops have been generated with random integration of the genes in the genome. With the advent of designed nucleases that can introduce double-strand breaks in the genome, a new generation of gene-edited products is being developed. With this technology, site-specific gene integrations, single locus gene stacking and genome editing have become a reality. A large number of traits have already been generated using these nucleases in lab-based experiments but have not been evaluated in the field yet. Since the technology can be used to edit endogenous genes to confer the desired traits, such products have the potential to be called non-transgenic and may not be as tightly regulated as the traditional transgenic crops. Though transgenic crops have enjoyed wide commercial success around the world, public perception will continue to affect the demand for such crops into the foreseeable future. Negative public perception in some countries has resulted in tougher policies resulting in prolonged product development timelines. Nevertheless, the science behind the development of genetically modified traits is strong. There is a need for academia and industry to do a better job in educating people so that they can better understand the technology and can make more informed decisions about their food choices.

Disclaimer The views and opinions expressed in this article are drawn from scientific literature and the author's professional experience. These are views of the authors and do not reflect the official views or policy of Dow AgroSciences or any other organization.

References

Anderson K (2010) Economic impacts of policies affecting crop biotechnology and trade. New Biotechnol 27:558–564

Baltazar BM et al (2015) Pollen-mediated gene flow in maize: implications for isolation requirements and coexistence in Mexico, the center of origin of Maize. PLoS One 10(7):e0131549

Baltes NJ, Voytas DF (2014) Enabling plant synthetic biology through genome engineering. Trends Biotechnol 33:1–12

Barfoot P, Brookes G (2014) Key global environmental impacts of genetically modified (GM) crops use 1996-2012. GM Crops Food 5:149–160

Braun AC (1958) A physiological basis for autonomous growth of the crown gall tumor cell. PNAS 44:344–349

Brookes G, Barfoot P (2015) Global income and production impacts of using GM crop technology 1996-2013. GM Crops Food 6:13–46

Carpenter JE (2010) Peer-reviewed surveys indicate positive impact of commercialized GM crops. Nat Biotechnol 28:319–321

Carpenter JE (2011) Impact of GM crops on biodiversity. GM Crops 2:7–23

Carpenter JE (2013) The socio-economic impacts of currently commercialised genetically engineered crops. Int J Biotechnol 12:249–268

Castiglioni P, Warner D, Bensen RJ, Anstrom DC, Harrison J, Stoecker M, Abad M, Kumar G, Salvador S, D'Ordine R, Navarro S, Back S, Fernandes M, Targolli J, Dasgupta S, Bonin C, Luethy MH, Heard JE (2008) Bacterial RNA chaperones confer abiotic stress tolerance in plants and improved grain yield in maize under water-limited conditions. Plant Physiol 147:446–455

CBD (Convention on Biological Diversity) (2000) Cartagena protocol on biosafety. Montreal. https://www.cbd.int/doc/legal/cartagena-protocol-en.pdf

Cernadas RA, Doyle EL, Nino-Liu DO, Wilkins KE, Bancroft T, Wang L, Schmidt CL, Caldo R, Yang B, White FF, Nettleton D, Wise RP, Bogdanove AJ (2014) Code assisted discovery of TAL effector targets in bacterial leaf streak of rice reveals contrast with bacterial blight and a novel susceptibility gene. PLoS Pathog 10(2):1–24

Chandrasegaran S, Carroll D (2016) Origins of programmable nucleases for genome engineering. J Mol Biol 428(5):963–989

Chilton MD, Drummond MH, Merlo DJ, Sciaky D, Montoya AL, Gordon MP, Nester EW (1977) Stable incorporation of plasmid DNA into higher plant cells: the molecular basis of crown gall tumorigenesis. Cell 11:263–271

Clasen BM, Stoddard TJ, Luo S, Demorest ZL, Li J, Cedrone F, Tibeby R, Davison S, Ray EE, Daulhac A, Coffman A, Yabandith A, Retterath A, Haun W, Baltes NJ, Mathis L, Voytas DF, Zhang F (2016) Improved cold storage and processing traits in potato through targeted gene knockout. Plant Biotechnol J 14:169–176

Codex Alimentarius Commission (Codex) (2003a) Principles for the risk analysis of food derived from modern biotechnology. Food and Agriculture Organization of the United Nations/World Health Organization

Codex Alimentarius Commission (Codex) (2003b) Guideline for the conduct of food safety assessment of foods produced using recombinant-DNA plants. Food and Agriculture Organization of the United Nations/World Health Organization

DeFrancesco L (2013) How safe does transgenic food need to be? Nat Biotechnol 31:794–802

Diakun GP, Fairall L, Klug A (1986) EXAFS study of the zinc binding sites in the protein transcription factor IIIA. Nature 324:698–699

Djukanovic V, Smith J, Lowe K, Yang M, Gao H, Jones S, Nicholson MG, West A, Lape J, Bidney D, Falco SC, Jantz D, Lyznik LA (2013) Male-sterile maize plants produced by targeted mutagenesis of the cytochrome P450-like gene (MS26) using a re-designed I- I homing endonuclease. The Plant Journal 76 (5):888–899

Eriksson GA (1969) The waxy character. Hereditas 63(1–2):180–204

FDA (1992) Statement of policy: foods derived from new plant varieties. Federal Register 57:22–984

FDA (2015) Guidance for industry: voluntary labeling indicating whether foods have or have not been derived from genetically engineered plants. https://www.fda.gov/Food/GuidanceRegulation/GuidanceDocumentsRegulatoryInformation/LabelingNutrition/ucm059098.htm

Feng PCC, Qi Y, Chiu T, Stoecker MA, Schuster CL, Johnson SC, Fonseca AE, Huang J (2014) Improving hybrid seed production in corn with glyphosate-mediated male sterility. Pest Manag Sci 70:212–218

Fernandez-Cornejo J, Wechsler SJ, Livingston M, Mitchell L (2014) Genetically engineered crops in the United States. United States Department of Agriculture-Economic Research Service, Washington, DC

Finer JJ, Vain P, Jones MW, McMullen MD (1992) Development of the particle inflow gun for DNA delivery to plant cells. Plant Cell Rep 11:232–238

Freestone D, Hey E (1996) Origins and development of the precautionary principle. In: Freestone D, Hey E (eds) The precautionary principle and international law. Kluwer Law International, The Hague, pp 3–15

Gaj T, Gersbach CA, Barbas CF (2013) ZFN, TALEN and CRISPR/Cas based methods for genome engineering. Trends Biotechnol 31(7):397–405

Gaskell G et al (2004) GM foods and the misperception of risk perception. Risk Anal 24:185–194

Gaskell G et al (2010) Europeans and biotechnology in 2010: winds of change? A report to the European Commission's Directorate-General for Research

Gruére G, Sengupta D (2009) Biosafety decisions and perceived commercial risks. IFPRI Discussion Paper 00847

Haun W, Coffman A, Clasen BM et al (2014) Improved soybean oil quality by targeted mutagenesis of the fatty acid desaturase 2 gene family. Plant Biotechnol J 12:934–940

Headrick D (2016) Rethinking mealtime for 9 billion. Res Technol Manag 59(2):3

Heap I (2014) Global perspective of herbicide-resistant weeds. Pest Manag Sci 70(9):1306–1315

Huang F, Andow DA, Buschman LL (2011) Success of the high-dose/refuge resistance management strategy after 15 years of Bt crop use in North America. Entomol Exp Appl 140:1–16

ISAAA (2016) Global status of commercialized biotech/GM crops 2016. http://www.isaaa.org/resources/publications/briefs/52/download/isaaa-brief-52-2016.pdf

ISAAA Brazil (International Service for the Acquisition of Agri-Biotech Applications) (2017) Biotech country facts & trends, Brazil. http://www.isaaa.org/. Accessed March 2017

ISAAA Crop (International Service for the Acquisition of Agri-Biotech Applications) (2017) Biotech crop annual updates. http://www.isaaa.org/. Accessed March 2017

Jackson DA, Symons RH, Berg P (1972) Biochemical method for inserting new genetic information into DNA of simian virus 40: circular SV40 DNA molecule containing lambda phage genes and the galactose operon of Escherichia coli. PNAS 69(10):2904–2909

James C (1997) Global status of transgenic crops in 1997. ISAAA Briefs No. 5:13

James C (2010) A global overview of biotech (GM) crops: adoption, impact and future prospects. GM Crops 1:8–12

James C (2015) 20th anniversary (1996 to 2015) of the global commercialization of biotech crops and biotech crop highlights in 2015. ISAAA brief no. 51. ISAAA, Ithaca

Jayaraman K, Jia H (2012) GM phobia spreads in South Asia. Nat Biotechnol 30:1017–1019

Jia H (2010) Chinese green light for GM rice and maize prompts outcry. Nat Biotechnol 28:390–391

Jia H, Pang Y, Chen X, Fang R (2006) Removal of the selectable marker gene from transgenic tobacco plants by expression of cre recombinase from a tobacco mosaic virus vector through agroinfection. Transgenic Res 15:375–384

Khatodia S, Bhatotia K, Passricha N, Khurana SMP, Tuteja N (2016) The CRISPR/Cas genome-editing tool: application in improvement of crops. Front Plant Sci 7:1–13

Klein TM, Wolf ED, Wu R, Sanford JC (1987) High velocity microprojectiles for delivering nucleic acids into living cells. Nature 327:70–73

Li B, Li N, Duan X, Wei A, Yang A, Zhang J (2010) Generation of marker free transgenic maize with improved salt tolerance using the FLP/FRT recombination system. J Biotechnol 145(2):206–213

Li T, Liu B, Spalding MH, Weeks DP, Yang B (2012) High efficiency TALEN based gene editing produces disease resistant rice. Nat Biotechnol 30:390–392

Li Y et al (2016) The development and status of Bt rice in China. Plant Biotechnol J 14:839–848

Lucas D, Glenn K, Bu J-Y (2004) Petition for determination of nonregulated status for Lysine Maize LY038. USDA https://www.aphis.usda.gov/brs/aphisdocs/04_22901p.pdf

Luo H, Kausch AP (2002) Application of FLP/FRT site specific DNA recombination system in plants. Genet Eng 24:1–16

McCabe D, Christou P (1993) Direct DNA transfer using electrical discharge particle acceleration (Accell technology) plant cell tissue organ. Culture 33:227–236

McDougall P (2011) The cost and time involved in the discovery, development and authorization of a new plant biotechnology derived trait: a consultancy study for CropLife International. Phillips McDougall, Midlothian

McHughen A (2016) A critical assessment of regulatory triggers for products of biotechnology: product vs. process. GM Crops Food 7:125–158

Menage A, Morel G (1964) Sur la presence doctopine dans les tissue de crown gall. C R Acad Sci Paris 259:4795–4796

Mullis K, Faloona F, Scharf S, Saiki R, Horn G, Erlich H, (1986) Specific Enzymatic Amplification of DNA In Vitro: The Polymerase Chain Reaction. Cold Spring Harbor Symposia on Quantitative Biology 51 (0):263–273

Nagy A (2000) Cre recombinase: the universal reagent for genome tailoring. Gebesus 26:99–109

NAS (National Academies of Sciences, Engineering, and Medicine) (2016) Genetically engineered crops: experiences and prospects. The National Academies Press, Washington, DC. https://doi.org/10.17226/23395

National Academies of Sciences, E., and Medicine (2016) Genetically engineered crops: experiences and prospects. The National Academies Press, Washington, DC

Nekrasov V, Wang C, Win J, Lanz C, Weigel D, Kamoun S (2017) Rapid generation of a transgene free powdery mildew resistant tomato by genome deletion. Sci Rep 7:1–6

Nemudryi AA, Valetdinova KR, Medvedev SP, Zakian SM (2014) TALEN and CRISPR/Cas genome editing systems: tools of discovery. Acta Nat 3(22):19–40

Nicolia A et al (2013) An overview of the last 10 years of genetically engineered crop safety research. Crit Rev Biotechnol 34:77–88

Organization for Economic Cooperation and Development (OECD) (1986) Recombinant DNA safety consideration. OECD Publications and Information Centre, Washington, DC

Osakabe Y, Osakabe K (2015) Genome editing with engineered nucleases in plants. Plant Cell Physiol 56(3):389–400

Parisi C, Tillie P, Rodriguez-Cerezo E (2016) The global pipeline of GM crops out to 2020. Nat Biotechnol 34:31–36

Piffanelli P, Ransay L, Waugh R, Benabdelmouna A, D'Hont A, Hollricher K, Jorgensen JH, Lefert P, Panstruga R (2004) A barley cultivation-associated polymorphism conveys resistance to powdery mildew. Nature 430:887–891

Qaim M (2009) The economics of genetically modified crops. Ann Rev Resour Econ 1:665–693

Qaim M, Kouser S (2013) Genetically modified crops and food security. PLoS One 8:e64879

Qaim M et al (2008) Economic and social considerations in the adoption of Bt crops. In: Romeis J, Shelton AS, Kennedy GG (eds) Integration of insect-resistant genetically modified crops within IPM programs. Springer, New York

Que Q, Chilton MDM, Fontes CM, He C, Nuccio M, Zhu T, Wu Y, Chen JS, Shi L (2010) Trait stacking in transgenic crops, challenges and opportunities. GM Crops 1(4):220–229

Quist D, Chapela IH (2001) Transgenic DNA introgressed into traditional maize landraces in Oaxaca, Mexico. Nature 414:541–543

Rinaldo AR, Ayliffe M (2015) Gene targeting and editing in crop plants: a new era of precision opportunities. Mol Breed 35:1–15

Sanford JC, Wolf ED, Allen NK (1991) Method for transporting substances into living cells and tissues and apparatus therefor. United States Patent number 5036006

Schaart JG, Krens FA, Pelgrom KT, Mendes O, Rouwendal GJ (2004) Effective production of marker free transgenic strawberry plants using inducible site specific recombination and a bifunctional selectable marker gene. Plant Biotechnol J 2:233–240

Shan Q, Zhang Y, Chen K, Zhang K, Gao C (2015) Creation of fragrant rice by targeted knockout of the gene using TALEN technology. Plant Biotechnology Journal 13(6):791–800

Shani Z, Dekel M, Cohen B, Barimboim N, Kolosovski N, Safranuvitch A, Cohen O, Shoseyov O (2003) Cell wall modification for the enhancement of commercial eucalyptus species. In: Sundberg B (ed) IUFRO tree biotechnology. Umea Plant Science Center, Umea, pp S10–S26

Shelton AM et al (2002) Economic, ecological, food safety, and social consequences of the deployment of Bt transgenic plants. Annu Rev Entomol 47:845–881

Silva G, Poirot L, Galetto R, Smith J, Montoya G, Duchateau P, Paques F (2011) Meganucleases and other tools for targeted genome engineering perspectives and challenges for gene therapy. Curr Gene Ther 11:11–27

Smith EF, Townsend CO (1907) A plant tumor of bacterial origin. Science 25:671

Stemke D (2004) Genetically modified organisms: biosafety and bioethical issues. In: The GMO handbook: genetically modified animals, microbes, and plants in biotechnology. Humana Press, Totowa

Sternberg N (1978) Demonstration and analysis of P1 site specific recombination using ʎ-P1 hybrid phages constructed in vitro. Cold Spring Harb Symp Quant Biol 43:1143–1146

Stone GD, Glover D (2017) Disembedding grain: golden Rice, the green revolution, and heirloom seeds in the Philippines. Agric Hum Values 34(1):87–102

Storer NP, Thompson GD, Head GP (2012) Application of pyramided traits against Lepidoptera in insect resistance management for Bt crops. GM Crops Food 3(3):154–162

Sun Y, Zhang X, Wu C, He Y, Ma Y, Hour H et al (2016) Engineering herbicide resistant rice plants through CRISPR/Cas9-mediated homologous recombination of acetolactate synthase. Mol Plant 9:628–631

Sun XH et al (2017) Altered expression of maize PLASTOCHRON1 enhances biomass and seed yield by extending cell division duration. Nat Commun 8:14752

Tang G et al (2012) β-Carotene in Golden Rice is as good as b-carotene in oil at providing vitamin A to children. Am J Clin Nutr 96:658–664

Thompson PB (2000) Bioethics issues in a biobased economy. In: Eaglesham A, Brown WF, Hardy RW (eds) The biobased economy of the twenty-first century: agriculture expanding into health, energy, chemicals, and materials. National Agricultural Biotechnology Council, Ithaca

Urbanchuk JM, Kowalski DJ, Dale B, Kim S (2009) Corn amylase: improving the efficiency and environmental footprint of corn to ethanol through plant biotechnology. AgBioforum 12(2):149–154

US National Academy of Science (1987) Introduction of recombinant DNA-engineered organisms into the environment. The National Academies Press, Washington, DC

USDA (2016) Re-Request for confirmatin that transgene-free, CRISPR-edited mushroomis not a regulated article. https://www.aphis.usda.gov/biotechnology/downloads/reg_loi/15-321-01_air_response_signed.pdf

USDA (2017) World agricultural supply and demand estimates. USDA. http://usda.mannlib.cornell.edu/usda/waob/wasde//2010s/2017/wasde-03-09-2017.pdf

USDA FAS (2015) Francophone West Africa Biotechnology Report

USDA FAS (2016a) China moving towards commercialization of its own biotechnology crops. GAIN Report Number: CH16065

USDA FAS (2016b) Australia agricultural biotechnology annual. GAIN Report Number: AS1619

USDA FAS (2016c) Philippine agricultural biotechnology situation and outlook. GAIN Report Number: RP1617

USDA FAS (2016d) EU-28 agricultural biotechnology annual. GAIN Report Number: FR1624

USDA FAS (2016e) Mexico agriculture biotechnology annual. GAIN Report Number: MX6044

USDA-ERS (United States Department of Agriculture – Economic Research Service) 2016. Adoption of genetically engineered crops in the U.S. recent trends in GE adoption. https://www.ers.usda.gov/data-products/adoption-of-genetically-engineered-crops-in-the-us/recent-trends-in-ge-adoption/ Last updated November 2016

Vergunst AC, Hooykaas PJ (1998) Cre/lox-mediated site specific integration of Agrobacterium T-DNA in Arabidopsis thaliana by transient expression of cre. Plant Mol Biol 38:393–406

Verma SR (2013) Genetically modified plants: public and scientific perceptions. ISRN Biotechnol 2013:1–11

Wafula D, Gruére G (2013) Genetically modified organisms, exports, and regional integration in Africa. In: IFPRI book chapters,in: Falck-Zepeda, Benjamin J, Gruare, Guillaume P, Sithole-Niang I (eds) Chap. 5: Genetically modified crops in Africa: Economic and policy lessons from countries south of the Sahara. International Food Policy Research Institute (IFPRI), pp 143–157

Walsh TA et al (2016) Canola engineered with a microalgal polyketide synthase-like system produces oil enriched in docosahexaenoic acid. Nat Biotechnol 34(8):881

Waltz E (2016) Gene edited CRISPR mushroom escapes US regulation. Nature 532:293

Wang Y, Yau YY, Balding DP, Thomson JG (2011) Recombinase technology: applications and possibilities. Plant Cell Rep 30:267–285

Wang Y, Cheng X, Shan Q, Zhang Y, Liu J, Gao C, Qiu JL (2014) Simultaneous editing of three homoeoalleles in hexaploid bread wheat confers heritable resistance to powdery mildew. Nat Biotechnol 32:947–951

White PR, Braun AC (1942) A cancerous neoplasm of plants. Autonomous bacteria free crown gall tissue. Proc Am Phil Soc 86:467–469

White House, Office of Science and Technology Policy (1986) Coordinated framework for regulation of biotechnology. https://www.aphis.usda.gov/brs/fedregister/coordinated_framework.pdf

Wolt JD, Wang K, Yang B (2016) The regulatory status of genome edited crops. Plant Biotechnol J 14:510–518

World Health Organization (2005) Modern food biotechnology, human health, and development: an evidence-based study. World Health Organization, Geneva

Wu F (2006) Mycotoxin reduction in Bt Corn: potential economic, health, and regulatory impacts. Transgenic Res 15(3):277–289

Wu Y, Fox TW, Trimnell MR, Wang L, Xu R-J, Cigan AM, Huffman GA, Garnaat CW, Hershey H, Albertsen MC (2016) Development of a novel recessive genetic male sterility system for hybrid seed production in maize and other cross-pollinating crops. Plant Biotechnol J 14:1046–1054

Ye X, Al-Bbili S, Kloti A, Zhang J, Lucca P, Beyer P, Potrykus I (2000) Engineering the Provitamin A (β-Carotene) biosynthetic pathway into (carotenoid-free) Rice endosperm. Science. 14 Jan 2000 287:303–305

Zaenen I, Van N, Teuchy H, Van M, Schell J (1974) Supercoiled circular DNA in crown gall inducing *Agrobacterium* strains. J Mol Biol 86:109–127

Zhou H, Berg JD, Blank SE, Chay CA, Chen G, Eskelsen SR, Fry JE, Hoi S, Hu T, Isakson PJ, Lawton MB, Metz SG, Rempel CB, Ryerson DK, Sansone AP, Shook AL, Starke RJ, Tichota JM, Valenti SA (2003) Field efficacy assessment of transgenic roundup ready wheat roundup and roundup ready are trademarks of Monsanto company. Crop Sci 43:1072–1075

Zilberman D et al (2013) Continents divided: understanding differences between Europe and North America in acceptance of GM crops. GM Crops Food 4:202–208

Printed by Printforce, the Netherlands